The History of Modern Physics

1800 – 1950

Volume 8

The History of Modern Physics, 1800–1950

The History of Modern Physics, 1800–1950

TITLES IN SERIES

VOLUME 1
Alsos
by Samuel A. Goudsmit

VOLUME 2
Project Y: The Los Alamos Story
Part I: Toward Trinity
by David Hawkins
Part II: Beyond Trinity
by Edith C. Truslow and Ralph Carlisle Smith

VOLUME 3
American Physics in Transition: A History of Conceptual
Change in the Late Nineteenth Century
by Albert E. Moyer

VOLUME 4
The Question of the Atom: From the Karlsruhe
Congress to the First Solvay Conference, 1860–1911
by Mary Jo Nye

VOLUME 5
Physics For a New Century:
Papers Presented at the 1904 St. Louis Congress

VOLUME 6
Basic Bethe:
Seminal Articles on Nuclear Physics, 1936–1937
by Hans A. Bethe, Robert F. Bacher, and M. Stanley Livingston

VOLUME 7
History of the Theories of Aether and Electricity:
Volumes I and II
by Sir Edmund Whittaker

VOLUME 8
Radar in World War II
by Henry Guerlac

INTRODUCTORY NOTE

The Tomash/American Institute of Physics series in the History of Modern Physics offers the opportunity to follow the evolution of physics from its classical period in the nineteenth century when it emerged as a distinct discipline, through the early decades of the twentieth century when its modern roots were established, into the middle years of this century when physicists continued to develop extraordinary theories and techniques. The one hundred and fifty years covered by the series, 1800 to 1950, were crucial to all mankind not only because profound evolutionary advances occurred but also because some of these led to such applications as the release of nuclear energy. Our primary intent has been to choose a collection of historically important literature which would make this most significant period readily accessible.

We believe that the history of physics is more than just the narrative of the development of theoretical concepts and experimental results: it is also about the physicists individually and as a group—how they pursued their separate tasks, their means of support and avenues of communication, and how they interacted with other elements of their contemporary society. To express these interwoven themes we have identified and selected four types of works: reprints of "classics" no longer readily available; original monographs and works of primary scholarship, some previously only privately circulated, which warrant wider distribution; anthologies of important articles here collected in one place; and dissertations, recently written, revised, and enhanced. Each book is prefaced by an introductory essay written by an acknowledged scholar, which, by placing the material in its historical context, makes the volume more valuable as a reference work.

The books in the series are all noteworthy additions to the literature of the history of physics. They have been selected for their merit, distinction, and uniqueness. We believe that they will be of interest not only to the advanced scholar in the history of physics, but to a much broader, less specialized group of readers who may wish to understand a science that has become a central force in society and an integral part of our twentieth-century culture. Taken in its entirety, the series will bring to the reader a comprehensive picture of this major discipline not readily achieved in any one work. Taken individually, the works selected will surely be enjoyed and valued in themselves.

The History of Modern Physics
1800 - 1950

Volume **8**

RADAR in World War II

HENRY E. GUERLAC

Sections A – C

Tomash Publishers

American Institute of Physics

Library of Congress Cataloging in Publication Data

Guerlac, Henry.
    Radar in World War II.

(The History of modern physics, 1800–1950; v. 8)
    Includes bibliographies and index.
    1. World War, 1939–1945—Radar. 2. United States. Office of Scientific
Research and Development. National Defense Research Committee. Division 14,
Radar—History. 3. Massachusetts Institute of Technology. Radiation Laborato-
ry—History. I. Title. II. Series.
D810.R33G84 1985   940.54′41   85-28752
ISBN 0-88318-486-9

CONTENTS

Preface.. xi
Foreword (1946) xiv
Foreword (1987) xv
Introduction.. xix

SECTION A
Early History of Radar
in the United States and Great Britain

Chapter 1
The Case History of a Secret Weapon 3
Chapter 2
How Radar Works.. 6
Chapter 3
The Radio Background of Radar..27
Chapter 4
Early Radar Research in the U.S. Navy 59
Chapter 5
Radar Research by the Signal Corps 93
Chapter 6
Radio Detection in Great Britain 122
Chapter 7
British Disclosures to Her Friends
    and Allies—French Radar...175
Chapter 8
The Early History of Microwaves..185

SECTION B
The Establishment of the NDRC Radar Program

Chapter 9
Inauguration of the NDRC Program 243
Chapter 10
The Foundation of the Radiation Laboratory 253

SECTION C
Technical Developments at the Radiation Laboratory

Part 1—Radar Systems
Chapter 11
Technical Developments in Division 14 309
Chapter 12
Airborne Development on Ten Centimeters...........................318
Chapter 13
Airborne Development on Three Centimeters........................341
Chapter 14
The Development of Radar Beacons and IFF 353
Chapter 15
Blind Bombing 375
Chapter 16
Ship Search Radar Systems...396
Chapter 17
Ground Systems and Their Shipboard Counterparts 427
Chapter 18
Fire–Control Radar 473
Chapter 19
Blind Landing 497
Chapter 20
K-Band Radar at One Centimeter.......................................507
Chapter 21
Loran...525
Chapter 22
Project Cadillac...537
Chapter 23
Trainers...551

Part 2—Radar Components
Chapter 24
Magnetrons and Modulators 561
Chapter 25
Antennas and rf Components 578
Chapter 26
Receivers, Indicators, and Precision Range Circuits..............591
Chapter 27
Power Equipment and the Test Laboratory 606

Part 3—Fundamental Research
Chapter 28
Experimental Systems Groups..615

Chapter 29
History of the Theory Group at the
    Radiation Laboratory *by Albert E. Heins* 625
Chapter 30
Research on the Propagation of Microwaves 633

SECTION D
Administrative History of Division 14

Chapter 31
The Organization and Functions of Division 14
    of the National Defense Research Committee......................647
Chapter 32
Growth of the Radiation Laboratory: Plant and Personnel 661
Chapter 33
Education and Information 675
Chapter 34
The Radiation Laboratory: A Big Business............................681

SECTION E
Operational History—Radar as a Combat Weapon

Chapter 35
Division 14 Overseas...695
Chapter 36
The War Against the U-Boat..708
Chapter 37
Overseas Cooperation with the British—The Founding of
    BBRL 731
Chapter 38
Radar Strategic Bombing of Germany 757
Chapter 39
Radar in the Invasion...816
Chapter 40
Radar on the Continent 863
Chapter 41
Carrier Warfare in the Pacific...909
Chapter 42
Army Radar in the Pacific 994

APPENDICES

Appendix A
German Radar 1073
Appendix B
Range–Limited Bombing Methods 1076
Appendix C
15th Air Force Notes 1084
Appendix D
The Organization of Civilian Scientists
  in the Pacific 1105
Appendix E
Project Dolphin 1118
Appendix F
Joint Army and Navy Equipment Designations 1121

Glossary 1123
Abbreviations and Code Names 1137
Name Index 1155
Subject Index 1163

Preface

This is first and foremost a report on the radar program of the National Defense Research Committee for the period between June 1940 and August 1945. Despite the inclusion of much additional material, it does not pretend to be a definitive account of the development of radar, still less a complete history of its operational use. It seemed desirable to include chapters on the origins of radar and on those microwave developments which led up to the creation of the Radiation Laboratory at the Massachusetts Institute of Technology. And it very soon became clear, since the activities of Division 14 personnel were by no means confined to the laboratory in Cambridge, but led them to all the theaters of war, that at least a provisional and incomplete account of the use of radar in military operations must be included.

It is with reluctance that I have entered the preserve of the Army's and Navy's historians, far better qualified, better situated, and better staffed to handle these operational questions. It is obvious that any real attempt at evaluating the overall influence of radar as a weapon must take into account many other factors, technical and military, that I have not been in a position to consider. This evaluation future historians of this war will some day be obliged to undertake. When that time comes perhaps this account, based so extensively upon the testimony of the scientists themselves and their own special channels of information, may be of use in the final appraisal.

This study was prepared in the first instance as a fully documented, classified report, of which copies are on file in the War and Navy Departments and with the records of OSRD. Much of the material was assembled at the Radiation Laboratory between March 1943 and August 1945 with the help of one assistant. The overseas installations of NDRC in England and France were visited in the spring of 1945. The final report was prepared after the cessation of hostilities with the help of an enlarged staff. The half-dozen additions to the staff—and others who contributed special reports or memoranda for our use—were all persons who had dealt with radar during the war as scientists or technical writers or in other capacities. None laid claim to historical training. Their diligence and patience in striving for a professional standard of documentation, and their good humor in the face of apparently trivial criticism, this author-editor will not soon forget. Our principal sources for the work of Division 14 were the voluminous official records of the Radiation Laboratory, supplemented at all possible points by interviews with scientific and administrative personnel. To this must be added much first-hand information, and the generous loan of several private diaries. Technical and operational information from the active theaters came partly from Army and Navy sources, but in great measure was supplied by reports, official or informal, of scientific field personnel. Considerable information on the use of Laboratory equipment in the field was made available to the Laboratory during the war by the U.S. Navy, the U.S. Army Air Forces, and the British.

It should be emphasized that no restriction of any sort was imposed during the preparation of the classified report; on the contrary every reasonable request for travel, clearance, assistance, interviews, etc., was granted without hesitation.

Nevertheless the decision to publish this study has meant extensive condensation and made it necessary to repair the ravages of a rather crippling bout with military and naval censorship, and to smooth over other deletions made at the insistence of the Government's patent advisors. Inexcusable gaps and startling omissions will leap to the eye of anyone familiar with the work of the Radiation Laboratory at home and in the field. Some at least can be laid to the apparent necessity of self-protection against the Government's zealous allies, foreign and domestic.

Dr. F. Wheeler Loomis, Associate Director of the Radiation Laboratory, was the first to express the desire that this history might become the case history of a secret weapon, by being broadened to include the technical background of radar, the work of other agencies, and as far as possible the operational use of the devices. It is to him in the first instance that this study is due. President James P. Baxter III, Historian of the OSRD, and Rear Admiral Julius A. Furer were instrumental in obtaining the authorization from the Secretary of the Navy, the late Mr. Frank Knox, and from Mr. Stimson, the Secretary of War, which made possible an inter-agency approach to the problem of the origins of radar. Thus the chapter dealing with the Navy's early work in radar could be based upon official records of the Naval Research Laboratory and on interviews with its engineers and scientists, while the chapter on early Signal Corps work on radar could draw very heavily upon a History of Signal Corps Radar prepared by my friend Harry M. Davis. This was kindly placed at my disposal through the good offices of Lt. Col. C. J. McIntyre, Chief, Special Activities Branch of the War Department. Additional evidence not available to him led me to depart somewhat from Mr. Davis' conclusions.

To Professor Bruce Hopper, Historian of the U. S. Strategic Tactical Air Force, and to Dr. John Trump and Mr. David Griggs of the Advisory Specialist Group, I am indebted for the invitation to carry out, with the assistance of Mr. Hugh Odishaw, a study of the use of radar for blind bombing by the 8th and 9th Air Forces. Professor Philip M. Morse and

Dr. Jacinto Steinhardt, of the Operations Evaluation Group attached to the Office of Chief of Naval Operations, generously made available, with the approval of the Navy Department, documents on the anti-submarine war in the Atlantic, and were kind enough to read the chapter that resulted. To the officers in charge of the Historical Division, Army Air Forces, I am grateful for the privilege of using their records in our attempt to summarize the Air Forces' use of radar in the Pacific, and for material to supplement that which we consulted in London and on the Continent. The chapter on the use of radar in the Fleet could not have been prepared without access to material in the Bureau of Ships, the Bureau of Ordnance, and other agencies of the Navy Department.

In the account of the British Technical Mission of 1940, I made extensive use of the papers of the Mission, put at my disposal by the British Central Scientific Office, and of details supplied by Dr. E. G. Bowen and Brigadier F. C. Wallace, the radar representatives on the Mission. The story of early British radar, to a greater extent than I might have wished, was pieced together a number of years after the events from the oral testimony of some of the chief participants, several of them members of the British Air Commission in the United States. Nevertheless for the account of the earliest period of radar development in England I was able to use the Minutes of the Committee on the Scientific Survey of Air Defense, as well as some documents kindly put at my disposal in London by Sir Robert Watson-Watt, and others transmitted to me from Mr. P. A. Rowe of TRE. A balanced account, which this chapter undoubtedly is not, can only be prepared after a perusal of British sources not accessible to me. For information on French radar developments I am almost entirely indebted to E. M. Deloraine and Gerard Lehmann of the IT&T.

The reader may wonder at the almost complete absence of any discussion of enemy radar or of radar countermeasures. This material, however, belongs more properly in the history of Division 15, NDRC, which specialized in those matters. Besides this, the greatest single omission is without doubt a separate treatment of the radar research and development carried on during the war by the principal electronic research laboratories, chief among them the Bell Telephone Laboratories. Understandably, some of the larger concerns were distinctly cool toward the possibility of being included in a detailed survey of this sort, so this aspect of the program had to be abandoned, though in every case they were generous when it came to answering specific requests.

Department of History   Henry E. Guerlac
Cornell University
May 1947

Foreword
1946

This volume records the history of Division 14, Radar, of the National Defense Research Committee from its origin in July 1940 until June 1946.

The MIT Radiation Laboratory was the center for the Division's microwave radar and Loran navigation development program and its work is quite naturally emphasized. In fact, this volume is also the history of the Radiation Laboratory. A large number of other colleges and universities, research institutions, and the principal radio and electronic concerns throughout the country actively participated on specific radar research and systems projects under OSRD sponsorship. Particular credit, except in a few instances, has been impossible. The Division, therefore, takes this opportunity of congratulating all of the organizations and the individuals who contributed to the program.

OSRD did not invent radar. It is only proper, therefore, that this volume should contain a tribute to the prior developments by the U.S. Navy, the Army Signal Corps, and the British. Radar equipment in use during the war could only be the result of the combined intense efforts of the Army and Navy in procurement planning, personnel training and supply and maintenance, and industry in engineering and production as well as the laboratories in research and development. Since OSRD was only one of the contributors we can say that the result was a remarkable national achievement.

As its first obligation, OSRD has supplied complete government reports on its radar program to the Army and Navy. The principal technical developments have, however, been made generally available in the *Radiation Laboratory Series*, a twenty-eight volume set of technical monographs prepared by members of the staff of the Radiation Laboratory and published by the McGraw-Hill Book Company.

I should like to express my gratitude to the members of the Division 14 Committee for the services which they generously contributed in the planning and direction of the Division's activities. Finally, I should like to thank Dr. Guerlac and his staff for their efforts in the preparation of this volume.

New York, N.Y.   Alfred L. Loomis
August 1946   Chief, Division 14, NDRC

Foreword
1987

Henry Guerlac graduated from Cornell University in 1932 with an under-graduate premedical major and a BS in chemistry, and he continued there for an MS in biochemistry. After work for two summers at the Marine Biology Laboratory in Woods Hole he moved to Harvard in the fall of 1933 and began work for a PhD in biochemistry. In 1935 he was elected to a Junior Fellowship, originally intended by Harvard as a substitute for the conventional doctorate and ranking above it—a credential hardly recog-nized west of the Hudson, Guerlac once wrote. He continued work to-ward the PhD in biochemistry but he began to have doubts about a career in a laboratory and thoughts about a radical change in career plan began to tempt him. On a trip to Sweden he visited Uppsala and the Linnaeus garden and he visited the pharmacy in the town where the chemist Carl Wilhelm Scheele worked and where he documented his independent dis-covery of oxygen. These experiences tipped the scale and Guerlac decided to become a historian.

His Junior Fellowship gave him the opportunity to work across disci-plines and in ways not available to most students. After what he described as "my delightfully prolonged graduate study," in the rich Harvard atmo-sphere of library resources and graduate courses in far-ranging fields and in association with stimulating scholars, Henry Guerlac received his PhD in European history in 1941.

During his graduate study he was already attracted to the history of scientific ideas and by the time he received his degree he was committed to the history of science as a field. It was natural that his training as a chemist and his French heritage would eventually lead him to Lavoisier, and the chemical revolution deriving from Lavoisier, as a focus of his career. Through his interest and influence, Cornell, where he spent most of his professional career, has one of the world's preeminent Lavoisier library collections.

In 1941 Dr. Guerlac moved to the University of Wisconsin as Chair-man of a new Department of the History of Science and in early 1943 he was recruited by Wheeler Loomis, the Assistant Director of the MIT Radiation Laboratory, to prepare an official history of the Laboratory, intended to justify, should there be a congressional investigation, the large amount of money spent by the country's preeminent radar development operation. And so he moved back to Cambridge, to the other end of Mas-sachusetts Avenue.

Although the Laboratory employed thousands of people by that time, more than two years after its founding, and although he was a junior, nontechnical staff member, he had the support of Wheeler Loomis and that of Lee DuBridge, the Lab Director, and he had access to all aspects of the operation. He set about the task of documenting everything that was going on—and turning himself into a professional contemporary historian.

I was aware of him and his operation but I did not know him then. I joined the Laboratory in its infancy in January 1941, but at the end of 1942 I moved to Washington as a Laboratory representative with the U.S. Army Air Force. I returned to Cambridge for a day or two every week or two until mid-1945 and remained in the midst of laboratory work, but with a focus on the operational end.

Henry Guerlac grew up in Ithaca in a Cornell family—his father was a professor of French literature there. When an opportunity came for him to return to Cornell in the fall of 1946, as a full Professor of History and with an opportunity to become Cornell's primary historian of science, he was quick to accept the offer.

I also went to Cornell in the fall of 1946 as an Assistant Professor of Physics. I am not sure when I made, or remade, his acquaintance but it was early in our careers there. He showed me his Radiation Lab manuscript sometime in those first years and I read parts of it with complete absorption, especially the parts relating to my own work.

The manuscript has never been published until now, 40 years later, and I do not fully understand the reasons. The work of the Laboratory was brilliant and it played a critical and probably decisive role in the war. How the Lab came about; where it fit in the total defense effort; identification of those who made the major contributions; how the staff worked; all these matters are discussed in the manuscript. It would surely be important for those organizing any future operation of a similar nature in a similar time of national crisis to have easy access to such an account for the lessons it might hold. The manuscript is especially useful because it documents radar development in this country and in England prior to 1941.

Professor Guerlac himself was somewhat hesitant about his work. He was not a technical person and he was only beginning his career as a historian. He was not at the laboratory for the first two years and he had to reconstruct what had happened from technical reports and from discussions with busy people then engaged in other activities. He was comfortable with the study of 18th-century science, with the objectivity lent by the distance of time, but the construction of living history made him insecure. Over-arching judgments were hard to make. It was not a history, he said, but the basis of a history. I think he need not have had such reservations.

Otto Neugebauer once wrote, "The common belief that we gain 'historical perspective' with increasing distance seems to me utterly to misrepresent the actual situation. What we gain is merely confidence in generalizations which we would never dare to make if we had access to the real wealth of contemporary evidence."

After Tomash Publishers and the American Institute of Physics made the decision to publish the manuscript, Professor Guerlac went to New York a short time before his death for a discussion of the publication plans. Among other things, he wanted to write an Introduction that would state his hesitation about the manuscript. He died before he had a chance to do it.

In September 1984 I spoke about the Radiation Laboratory at the University of Rochester in an American Physical Society Symposium honoring Lee DuBridge. It was the 50th anniversary of DuBridge's move to Rochester (from Washington University, St. Louis) to become Chairman of the Physics Department there. From Rochester he had gone to MIT for an outstanding career as Director of the Radiation Laboratory. In preparing my paper I read the Guerlac manuscript in the Cornell Library, where one of the few extant copies resided, and I had a long talk with Professor Guerlac about the Laboratory and about the manuscript. It was the last conversation I ever had with him.

The Radiation Laboratory was staffed largely by physicists—the senior staff consisted almost exclusively of physicists in the early days of the Lab. It was a 1940 who's who of physics: Luis Alvarez, Ed Purcell, Ed McMillan, Hans Bethe, Robert Bacher, Curry Street, Ken Bainbridge, Ernest Pollard, Ray Herb, Julian Schwinger, Robert Dicke, Jerrold Zacharias, and I. I. Rabi.

Why should the physicists have dominated the staff to such a degree? I do not know, other than for the hubris of those organizing the Laboratory and the war research effort generally: Ernest Lawrence, Karl Compton, Alfred Loomis, James Conant, and Vannevar Bush. Conant was a chemist and Bush was an electrical engineer, but they apparently supported the concentration on physicists.

Henry Guerlac writes:

No full understanding of the administrative eccentricities of the Radiation Laboratory is possible without recognizing that one of its outstanding merits, in the eyes of its own management, was that it was a physicist's world, run for, and as completely as possible by, physicists. Everything was subordinated to producing an environment for research as free and as untrammeled as in a university and to preventing research from becoming entangled or impeded by the growing responsibilities thrust upon the organization. One consequence of this basic article of policy was that physicists assumed a large number of administrative duties, which might perhaps have been assigned to the lay brethren, and carried them out in some instances with extraordinary success, and in other cases not quite so effectively. A second consequence was that when service departments were set up, they operated at something of a disadvantage, and not without justice felt that insufficient confidence was reposed in them. On the other hand, no policy could have been better designed to rid the research man of unwise interference, and to give him unsurpassed opportunities for creative work. The Radiation Laboratory came close to realizing a scientist's dream of a scientific republic, whose only limitation was the supply of scientists.

So here, at long last, is Henry Guerlac's book on the history of "Radar in World War II," documenting a brilliant case of American scientific and technological genius, discussing the operational mode that made the Radiation Laboratory so successful, illuminating the personalities of those responsible, and reporting the decisive role that the Lab played. I only regret that he did not live long enough to see his work published—and to write the Introduction for it.

Dale R. Corson
Ithaca, New York

Introduction

by

John S. Rigden and I. I. Rabi

For the British, the summer of 1940 was grim. France fell on 17 June and the victorious German armies were looking further to the west—to Great Britain. In mid-August the Battle of Britain began with a vengeance: on 14 August alone, the German *Luftwaffe* flew 489 sorties against England. Winston Churchill rallied his nation with the somber words, "...if we fail, then the world, including the United States, including all that we have known and cared for, will sink into the abyss of a new Dark Age..."

At various locations along the east coast of England stood tall, spindly towers. Atop each tower, some 200–300 feet from the ground, was a *radio-detection-and-ranging* system with its antenna directed towards the eastern horizon. Each radar unit was capable of emitting radio waves and, if the emitted waves struck an object, the small portion of the wave reflected back to the radar unit could be detected. This radar network, called the Chain Home, monitored the skies over the Channel and the North Sea. It was a primitive system; however, with the warning it provided, the Royal Air Force met the German *Luftwaffe* head on.

In the summer of 1940, physicists in the United States watched the expanding conflict in Europe with growing concern. During most of the 1930s, American physicists had had an insider's view of the events that followed Hitler's rise to power in 1933. Physicists throughout the United States had received letters from their European counterparts, physicists who were desperate to leave their homelands; the Americans had received refugee physicists onto their campuses and into their homes; they had listened to their European colleagues give first-hand accounts of the horrors that had forced their departure. In August 1938, physicist P. P. Ewald had just finished a visiting professorship at Columbia University. On the day before he sailed for his European home, he wrote a warm letter to his Columbia host. "Things are getting worse and worse in Germany," wrote Ewald, "I received word yesterday that Lise Meitner had been 'eloped'—because otherwise she would not have been able to leave. A warning was added for me not to attempt to return."

American physicists were alarmed. Jerrold Zacharias remembers a conversation that took place on the corner of 116th Street and Broadway in New York City. France had just fallen and Zacharias was talking to a colleague about the war. "My Lord in Heaven, what will we do?," Zacharias remembers asking. "I don't know... I don't know," came the answer.

Many physicists felt compelled to do something, but what? They wanted to take action against Hitler, but how? These questions were answered in the autumn of 1940.

"The most valuable cargo ever brought to our shores" arrived from England in September 1940. The cargo was an odd-looking little thing: three small disks sandwiched together with wires and glass tubes sticking out of it. The object could easily be held in one's hand. This small device was a cavity magnetron.

The cavity magnetron was invented by two British physicists, Henry A. Boot and John T. Randall, in the early weeks of 1940. The magnetron was a source of microwaves. In addition to being an ingeniously designed device, the magnetron represented a technical breakthrough on two important counts. First, it could produce electromagnetic waves with a wavelength of only 10 centimeters (such waves are called microwaves). Second, these microwaves had large intensities of 10,000 watts.

Why are these figures—10,000 watts and 10 centimeters—significant? In radar detection, a beam of radio waves is sent out towards a target. The waves spread out as they move towards the target so that only a small fraction of the initial energy is reflected. The reflected wave, moving back towards the radar system, also spreads out so that only a small fraction of the reflected energy is available for detection. Obviously, the more intense the starting signal, the more intense will be the returning echo, and the easier it is to detect it.

Reflectors are used to reduce the extent to which the waves spread out as they travel from source to target. Thus, reflectors concentrate the energy emanating from the radar system and increase its working range. The size of the reflectors, however, depends on the wavelengths being used: the longer the wavelength, the larger the reflector. The Chain Home, with its 10-meter radio waves, required large reflectors. The 10-cm microwaves produced by the magnetron were one hundred times shorter than the radio waves used by Chain Home and, as a result, compact radar systems could be imagined. In fact, implicit in the cavity magnetron, with its powerful, short waves, was a full range of military applications: airborne radar systems small enough to fit inside a night fighter; gunlaying radar to fix on a target and aim artillery to hit it; shipboard radar to guide an armada in the thick of fog or the dark of night.

The British delegation demonstrated their magnetron for the Americans at the Bell Telephone Laboratories on 6 October. This demonstration was dramatic and its effect on both events and people was equally dramatic.

Ten days after the fall of France, on 27 June 1940, President Franklin D. Roosevelt formed the National Defense Research Committee

(NDRC). The purpose of the NDRC was to coordinate the scientific research activities of the armed services and to mobilize civilian science for wartime needs. With Vannevar Bush as its chairman, the NDRC became active immediately.

The first request that came to the NDRC came from the Air Corps who wanted to be able to carry out bombing raids even when targets were obscured by clouds. In response to their request, the NDRC formed the Microwave Committee. The committee members, headed by Alfred L. Loomis, worked through the summer of 1940 and, at summer's end, they were ready to write a report—"A sure sign," said one of the members, "that we didn't know what to do next."

Then came the magnetron.

At the time the magnetron was demonstrated, the Mark I radar system was under development in the United States. The Mark I system was based on vacuum-tube sources that generated only 2000 watts of power and did so at the longer wavelength of 40 centimeters. The cavity magnetron, with its 10,000 watts and 10-cm wavelengths, was such a dramatic breakthrough that it stimulated immediate action.

One week after the 6 October demonstration, the Microwave Committee decided to establish a central laboratory to develop microwave radar. Furthermore, the committee set priorities and adopted three projects for the new, as-yet-unnamed laboratory:

Project I—Highest priority—An airborne 10-cm radar system that a pilot could use to detect other aircraft;

Project II—A precision gunlaying radar system to track on a target and aim artillery;

Project III—A long-range navigational radar system.

In quick succession, Lee A. DuBridge was named director of the new Laboratory (16 October), a plan was formed (16 October) to gather key physicists together at MIT for the period 28–31 October, and, two days later (18 October), the Massachusetts Institute of Technology was selected as the site for the new laboratory.

As it happened, a conference on Applied Nuclear Physics was already scheduled for 28–31 October at, of all places, MIT. DuBridge and Ernest Lawrence called select physicists and "invited" them to come to the conference. They came. The conference was ordinary in the sense that papers were presented and physicists talked about their work. On the other hand, a perceptive conferee would have noticed that the subject of microwaves was the common denominator of those sessions added to the program at the last minute. Further, their curiosity might have been piqued about certain luncheon meetings.

The British scientists who had brought the magnetron to the United States were housed in the Algonquin Club of Boston. (The whole afffair was so hush-hush that a secret side entrance had been arranged for their use.) A luncheon meeting was held at the Algonquin Club where select American physicists met with the British delegation. At this luncheon meeting, the plans for a microwave laboratory were announced and an overall view of the radar problem was given.

The MIT conference ended on 31 October. Six days later, one day after election day and only one month after the first demonstration of the magnetron, physicists began arriving at the MIT Radiation Laboratory. In less than one week, these physicists had arranged to leave their homes and their university positions in order to begin the development of microwave radar systems. The promptness of their response was a measure of their concern. The United States did not formally enter the conflict until December 1941, but for the "Rad Lab" physicists, the war began in November 1940.

This book, *Radar in World War II*, is, as the title accurately suggests, first about radar and the war. Second, however, this book is about the MIT Radiation Laboratory and the remarkable group of people who assembled there during the years 1940–1945.

It started small—a handful of physicists. They were predominantly nuclear physicists. In order to exploit the cavity magnetron for radar purposes they had to create, almost from scratch, the field of microwave electronics. Entirely new methods had to be developed in order to transport electrical power at microwave frequencies. Yet, on 4 January 1941, less than two months after the first physicists arrived at the Rad Lab, a 10-cm radar beam was transmitted from the roof of MIT Building 6 and the echoes from buildings across the Charles River in Boston were detected.

By the end of the war, the Rad Lab was big: it employed approximately 4000 people and spent about $125,000 per day. By 1945, the Rad Lab physicists and engineers had developed more than 100 radar systems along with the necessary ancillary equipment; they had designed almost half of all the radar used during the war. The radar systems designed at the Rad Lab had made vital contributions to the Allied effort: from the beachhead at Anzio to the Battle of the Philippine Sea, from the D-Day landings in June 1944 to the Pacific-island campaigns of Guam, Iwo Jima, and Okinawa. Because of radar, the war was won before it ended. The Rad Lab alumni say it differently. They say, "The bomb ended the war, but radar won the war."

The MIT Radiation Laboratory had a direct and decisive influence on the outcome of World War II. It also had a direct and telling influence on the peace that followed.

As soon as security restrictions were lifted, civilian use of radar increased. From that time until the present, the principles and methods found and developed at the Rad Lab have been further developed and applied in manifold ways. From the modern kitchen to the police cruiser, from the ocean liner to jet aircraft, from military defense warning systems to meteorological warning systems, the legacy of the Rad Lab can be observed.

Science itself, however, was perhaps the biggest beneficiary of the wartime radar work.

In a microwave radar system, there are three basic elements: a source of microwaves, a detector of microwaves, and the means of getting microwaves from the source, to the target, and back to the detector. In each of these three major areas—source, detection, and propagation—the Rad Lab physicists confronted new questions of basic physics and developed new techniques to improve radar operation. They created new sources of radiation. They started with the 10-cm cavity magnetron, but soon they had 3-cm magnetrons and later 1-cm sources. They invented the circuit elements that were necessary to shuttle microwave power from one location in a circuit to another. They designed solid-state devices to detect feeble microwave signals. A vast knowledge had accumulated in taking radar from the 1000-cm radio-frequency systems used at the beginning of the war to the 1-cm microwave systems that were operating at the end.

By late in the summer of 1944 it was apparent that the Axis powers were on the defensive and that it was only a matter of time before their surrender. With the end anticipated, a few of the physicists left their laboratories and they began a new activity—writing. The published material that resulted from this writing effort started with the following paragraph written by Lee DuBridge:

> The tremendous research and development effort that went into the development of radar and related techniques during World War II resulted not only in hundreds of radar sets for military (and some for possible peacetime) use but also in a great body of information and new techniques in the electronics and high-frequency fields. Because this basic material may be of great value to science and engineering, it seemed most important to publish it as soon as security permitted.

So extensive was the information compiled at the Rad Lab, and so numerous were the techniques developed by the Rad Lab scientists, that 28 thick volumes were required to provide an adequate description.

The 28 volumes of the Massachusetts Institute of Technology Radiation Laboratory Series established an entire field of science and engineering. They were the authoritative references in the field of microwave electronics. In this series there are books such as *Radar System Engineering* that are general in scope; there are volumes like *Principles of Microwave Circuits* that provide the basic theory used in the field; and there are books such as *Microwave Mixers* that focus on specific components that are a vital part of microwave electronics. Some of these books, even to the present day, remain classics in the field.

The Rad Lab Series established the field of microwave electronics; but more, the methods described in its volumes were broadly applicable to established areas of physical research and, even more, they opened up entirely new fields of research. As a result, these volumes became standard references in scores of American physics laboratories. Many research physicists as well as generations of graduate students studied from their pages.

The microwave methods were particularly useful for probing properties of molecules, atoms, and nuclei. The post-war research field of microwave spectroscopy, a research field that provided incredibly precise and highly detailed information about molecular structure, resulted directly from the MIT Radiation Laboratory. The entire field of resonance phenomena, a field initiated just before the war at Columbia University, blossomed after the war because of the new techniques made available by the Rad Lab. Nuclear magnetic resonance, for example, was not only valuable to physicists, it quickly became an indispensable tool for chemists and, more recently, the magnetic imaging method promises to become a powerful diagnostic tool for physicians. The microwave methods also opened new windows to the universe: the whole field of radio astronomy has its roots in the wartime radar work.

While many specific research fields were born in the post-war years because of the microwave techniques developed for radar, scientific research in a much more general sense benefitted from the Rad Lab experience. The general advances in electronics made at the Rad Lab were applied in many laboratories to a great variety of experiments. Further, the knowledge acquired about such things as signal-to-noise characteristics in electronic circuits enables post-war scientists to design experiments in which they detected the feeblest of signals. This knowledge carried the frontiers of science into new areas.

Finally, and perhaps the most important of all, the Rad Lab experience had a profound effect on the Rad Lab physicists themselves and, subsequently, on physics. American physics came into its own shortly before World War II and, at that time, it was a profession dominated by youth. (At 42, I. I. Rabi was a "senior citizen" at both the Rad Lab and Los Alamos.) With the future of the nation at risk, many of the country's most able young physicists joined their more senior colleagues and they spent five years thinking together, learning together, and working together. At the Rad Lab, these young physicists received a most unusual apprentice-ship. When the war ended, many of them became professors at prestigious universities and directors of industrial laboratories. The Rad Lab class of '45 is peerless.

Radar altered the course of a war and, in so doing, it changed the course of the world. For this, the alumni of the MIT Radiation Laboratory can be proud. But the MIT Rad Lab has served the peace as effectively as it had served the war. For this, they can be even prouder.

St. Louis, Missouri   New York, New York

Publishers' Note

The original manuscript for this book, *Radar in World War II*, exists as both a "long" and a "short" version. The long version was offered to the McGraw-Hill Publishing Company as part of the Radiation Laboratory Series of monographs; it was declined by McGraw-Hill chiefly because it was a historical rather than a technical treatise. It was then condensed and offered to Little Brown & Company, but the changes required for security and patent reasons delayed submission until after the deadline date. The present monograph, with the approval of the author, is a merger of both versions.

We are grateful to Dr. Henry E. Guerlac for the encouragement and assistance that he gave us towards the publishing of his manuscript in the final months before his death in May 1985. Dr. Guerlac wished to give credit for help in preparing the original manuscript to Marie Boas Hall, Mildred Brown Hyde, Thomas F. Farrell, Lucy Marshall, Arthur Musgrave, and Harry Remde.

The publishers also acknowledge with gratitude the help of Helen Samuels of the M.I.T. Archives and the staff of the Niels Bohr Library of the American Institute of Physics for the loan of the manuscripts, Heidi Saraceno of the M.I.T. Museum and Dr. Marjorie Ciarlante of the National Archives for their help in obtaining copies of photographs, and Dr. David Corson of Cornell University and Dr. Helen Thomas, Assistant Historian of the Radiation Laboratory, for their advice and support.

Sterilis materia...narratur, et haec sordissima sui parte, ac plurimarum rerum aut rusticis vocabulis aut externis, immo barbaris, etiam cum honoris praefatione ponendis. Praeterea iter est non trita auctoribus via nec qua peregrinari animus expetat: nemo apud nos qui idem temptaverit invenitur, nemo apud Graecos qui omnia ea tractaverit.

Plinius Secundus, *Naturalis Historia* (Prefatio)

Section A

Early History of Radar in the United States and Great Britain

CHAPTER 1

The Case History of a
Secret Weapon

It is now common knowledge that the name "radar" designates special radio equipment operating on the higher radio frequencies and used to detect objects hidden from sight by mist, fog, or darkness.[1]

The main purpose of this volume is to record, as thoroughly as space permits, the program of research and development in the field of radar conducted between 1940 and 1945 under the National Defense Research Committee. In line with the policy set forth by Vannevar Bush and his advisors in June 1940, the work was conducted under contracts placed with educational institutions and commercial firms, principally the former. It was directed and carried out almost exclusively by men and women from the faculties and graduate schools of American universities, working in close cooperation with the procurement agencies of the Army and Navy, with the development and production departments of American industry, and—on a basis of notable candor, intimacy, and mutual respect—with the scientists of Great Britain and the Dominions.

It will be evident from the early chapters of this volume that radar had reached a fairly advanced stage of development in America and in England before the creation of NDRC. The radar division of NDRC, Division 14, had been originally charged, and was throughout the war almost wholly concerned, with developing radically improved radar equipment for the armed services. Before taking up the detailed historical narrative it may be helpful to the readers to give them some notion of the scope of the program and the varied activities that were involved.

Although the scientists of NDRC entered the field only a year before the attack on Pearl Harbor, they participated in the development of very nearly half the $3,000,000,000 worth of radar and associated equipment delivered to the armed services by July 1945. Their activity ranged from fundamental research on the behavior of superhigh-frequency waves (microwaves), through the development of new vacuum tubes, new circuits, and new radar components, to the design of complete radar systems serving widely differing military purposes. In cases where a small number of units was urgently required by the Services these were manufactured with the utmost speed by NDRC facilities. Finally, the work of the division included, in many instances, an important share in introducing this new equipment into operational use in the field.

The radar research and development of Division 14 was mainly concentrated in a single, large, secret laboratory, the Radiation Laboratory in Cambridge, Massachusetts, created by virtue of a contract with the Massachusetts Institute of Technology, housed in buildings on the Institute grounds. Several smaller contracts were placed with other educational institutions, notably Columbia University, and with industrial concerns. During the war the number of Division 14 contracts outstanding averaged about 50, though many of these were in fact, though not in name, subcontracts of the Radiation Laboratory. The manufacturing facilities of industry were relied upon by the Services for all full-scale production under Army and Navy contract; but Division 14 had its own model shop, the Research Construction Corporation (RCC) in Cambridge, Massachusetts, which worked in close cooperation with the Radiation Laboratory and shared with it the burden of manufacture under "crash" programs, the name given to rush orders for a small number of specially built sets. By the end of August 1945, approximately $25,000,000 worth of radar equipment had been directly supplied to the Services by RCC and the Radiation Laboratory, slightly less than half of which had been produced by RCC.

Approximately 150 distinct radar systems were developed as a result of this research program for installation on land, on shipboard, and in aircraft and for purposes ranging from early warning against enemy planes to blind bombing and anti-aircraft fire control. The only section of the Radiation Laboratory not devoted to radar was responsible for the development of Loran, a pulsed long-range navigational aid widely used by the armed forces of the United States and Great Britain. A total of $71,000,000 worth of Loran equipment—all but one item of which had been developed in whole or in part by the Radiation Laboratory—had been purchased by the Army and Navy by the end of July 1945.

The field activities of Radiation Laboratory personnel took them to all principal theaters of war: to the European and Mediterranean fighting fronts, the China–Burma–India theater, the Pacific. A British branch of the Radiation Laboratory, a small group in Australia, one at the Mediterranean Allied Air Forces Headquarters at Caserta, Italy, and an Advanced Service Base in Paris backed up the efforts of the field representatives and drew in turn upon the resources of the Radiation Laboratory. When the war ended, some twenty-five Radiation Laboratory men were in the Pacific or en route, while many others were standing by to man the radar laboratory planned to be established in Manila.

At the end of the war, nearly 5000 persons were engaged in radar development under Division 14 contracts; of these over 3900 were employees of the Radiation Laboratory. The nucleus of nearly a thousand scientists at this laboratory was drawn from universities and colleges in all parts of the country.

During the fiscal year 1944 – 1945 the NDRC was allocating over $4 million each month to Division 14. At the end of the war, the Division had received allocations amounting to $141 million for the development of radar and Loran equipment since November 1940. When the amount spent on research and development is weighed against the dollar value of equipment actually delivered by the end of the war, the sums seem relatively modest. Every dollar spent for research and development produced a little over ten dollars worth of military equipment.

This statistical picture may serve to convey some idea of the scale on which the enterprise was conducted, but can give only an imperfect idea of its military importance. The value of the NDRC program to the Allied war effort, should military historians some day turn their hands to assessing it, will be found to consist even more in the character of the new developments than in their magnitude. It was the successful development and exploitation of radar using microwaves, which means, as we shall see later, radio waves only 10 cm in length or shorter, that distinguishes most sharply the NDRC development program from that of the Army and Navy laboratories. Although developed and perfected after 1940, microwave radar nevertheless had an extensive operational history, and was largely responsible for making modern radar equipment as versatile and flexible a weapon as it became in the course of World War II. This is the story with which we shall be mainly concerned. Nevertheless to place the NDRC development in proper perspective, as well as to give proper credit to the scattered pioneers in the field of radio who paved the way unknown to themselves for the emergence of this extraordinary weapon, we have thought it appropriate to include chapters describing, all too briefly, some of the early work leading to the first successful radar. We make no claim that we have covered the field exhaustively, and only hope that the main conclusions will not be too far removed from what later historians may find to be the truth. Imperfect though we are conscious of its being, the whole may be taken to constitute the case history or the biography of a secret weapon.

NOTES

1. The name RADAR, the United States Navy code word standing for Radio Detection And Ranging, was officially adopted by the Navy in November 1940 as the designation for what had been previously called, among other things, "radio echo equipment." The code name was suggested by Commander (later Captain) S. M. Tucker, USN. The adoption of this term was recommended by Captain F. R. Furth, USN. The name was adopted by the U.S. Army in 1942. In July 1943 it was substituted by the British for their own term R.D.F., presumably standing for radio direction finding. In France the device was called D.E.M. (*détection électromagnétique*), and in Germany it was called *Funkmessgerät*.

CHAPTER 2
How Radar Works

In the preceding chapter it was emphasized that radar is a device which not only detects invisible or distant objects by means of reflected radio waves, but is capable of locating them accurately in space. Although, in the course of its steady improvement, radar has been built to work on a succession of higher and hitherto inaccessible radio frequencies, the basic method of its operation is independent of the frequencies that are employed or the special functions for which the particular radar set is designed. This method depends upon certain familiar principles of elementary physics which it may be well to review at this point.

Wave Motion and the Electromagnetic Spectrum

Physics is often preoccupied with phenomena that are periodic, i.e., in which certain parameters vary cyclically as a function of time. These are often referred to as wave motions, whether or not there is an actual material displacement. As early as the 17th century sound traveling in air was clearly identified with compressional waves in the air produced by the vibrations of sound-producing bodies. By the last third of the 19th century the study of sound had culminated in a highly rigorous and complete treatment of wave motion which reduced the description of such periodic phenomena to a mathematical expression known as the *wave equation.*

In the course of the 19th century it came to be recognized that light was also a periodic disturbance, and that certain of the characteristics of light, particularly the phenomenon described as interference, could best be understood, by analogy with certain similar occurrences in the field of sound, on the assumption that light consisted of transverse waves, each color of the spectrum representing waves of a different length. With the demonstration of the existence of radio waves, and a little later, the discovery of x-rays, it became evident that the visible spectrum was only a small region of a much more extensive spectrum of electromagnetic waves, the greater part of which man is unable to perceive directly by his senses. This spectrum is shown schematically in Fig. 2-1. The magnified portion shows the bands of ultrahigh-frequency and superhigh-frequency waves used for radar. By a band, the radio engineer understands a delimited region of adjacent frequencies, such as are reserved by law for special purposes.

These electromagnetic vibrations occur transversely, more closely resembling water waves of small amplitude than sound waves, which are

6

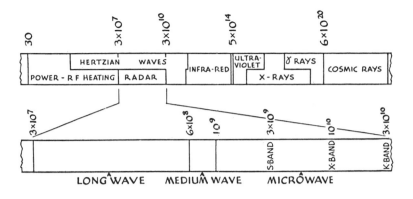

FIG. 2-1. Electromagnetic and radio spectrum.

longitudinal waves of compression and rarefaction. Electromagnetic waves vibrate in a plane perpendicular to the direction of the ray (which is only a geometrical representation of the axis of propagation of a train of waves). These waves consist of a periodic electrical disturbance and a periodic magnetic disturbance at right angles to one another, which are often called the electric and magnetic vectors of the electromagnetic waves. See Fig. 2-2.

If we do not push the analogy too far, we need not worry about asking what physical material it is that vibrates. The 19th century physicists postulated a purely gratuitous luminiferous or light-carrying ether which, for lack of experimental confirmation, soon dropped out of physics. To-day when we refer to wave motion, we mean only that certain measurable quantities vary with time in a periodic and wave-like manner when we plot them graphically. To demonstrate the existence of these waves the radio engineer has only to set up standing waves—which we shall shortly de-scribe—by the interference effects of a direct and reflected wave. When this is done it can be shown that these standing waves consist of periodic variations in the electrical and the magnetic intensity as measured along the line of propagation. In the case of the electrical vector, a succession of nodes, or high values of electrical potential, are separated by a series of minima or antinodes. Twice the distance between each pair of nodes or antinodes is spoken of as the *wavelength.* An identical value for the wave-length is obtained if the value of the magnetic field strength is similarly measured by means of standing waves.

The velocity of a wave moving out from a source can be measured by determining the time required for a wave of length λ to pass a given point, or $V = \lambda /T$. The number of complete oscillations or cycles passing per second is called the *frequency*, defined by the relation $f = 1/T$, where T is

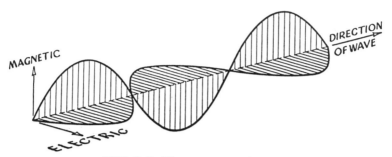

FIG. 2-2. Electromagnetic wave.

the time for a single cycle. Substituting the second result in the first equation, and remembering that the velocity of all electromagnetic waves is given by the universal constant c, the velocity of light, we get the important relation between wavelength and frequency:

$$c = \lambda f.$$

If λ is in meters and f is in megahertz, this becomes

$$\lambda = 300/f.$$

The wavelength is always given in metric units, and frequency in hertz (cycles per second), abbreviated Hz. The wavelength of the longest low frequency waves of the spectrum is usually measured in kilometers (thousands of meters) and the frequency in hertz. In the broadcast and shortwave bands the wavelengths are expressed in kilometers or meters and the frequencies in kilohertz (thousands of hertz). In the ultrahigh-frequency and superhigh-frequency radar bands, wavelengths are given in meters or centimeters and frequencies are expressed in megahertz (millions of hertz). The following table gives the principal radio bands.

For the purposes of this book, the electromagnetic spectrum may be divided as follows:[1]

	Frequency			Wavelength		
Audio frequencies:	30 Hz	–	10 000 Hz	10 000 000 m	–	30 000 m
Low frequencies:	10 000 Hz	–	500 kHz	30 000 m	–	600 m
Broadcast radio:	500 kHz	–	1 500 kHz	600 m	–	200 m
Shortwave radio:	1 500 kHz	–	30 MHz	200 m	–	10 m
Long-wave radar:	30 MHz	–	600 MHz	10 m	–	50 cm
Medium-wave radar:	600 MHz	–	1 000 MHz	50 cm	–	30 cm
Microwave radar:	1 000 MHz	–	30 000 MHz	30 cm	–	1 cm

Figure 2-3 compares schematically the wavelengths commonly used for radar.

Radar has usually been built to operate on wavelengths below 5 m, although experimental equipment was first tried out by the British on 50

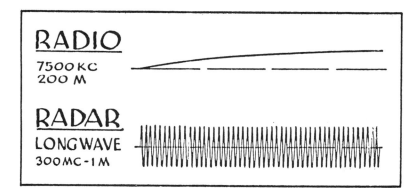

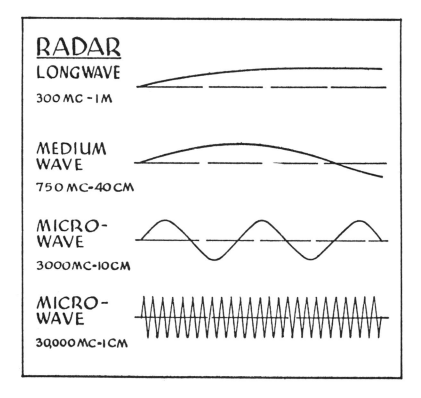

FIG. 2-3. Top: Comparative wavelengths, radio and radar. Bottom: Comparative wavelengths, radar (long-wave, medium-wave, microwave).

m. Useful radar designed before 1940 operated on wavelengths between 10 m and 50 cm (corresponding to 30 and 600 MHz). This ultrahigh-frequency band is sharply distinguishable from the rest of the radio spectrum, for 30 MHz is the frequency above which waves are no longer reflected from the ionosphere but usually pass on through; while 500 MHz is about the highest frequency for which conventional transmitting triode vacuum tubes can be persuaded to operate successfully. Above this frequency are the superhigh frequencies (centimeter waves or microwaves) for which special techniques had to be evolved to produce and handle the waves.

The waves we have described were assumed to be unmixed waves from a single oscillatory source. All such waves are plane polarized, i.e., the electrical vector is confined to a single plane. The nature and orientation of the oscillating source alone determines whether the electric vector is perpendicular to the ground (vertical polarization) or parallel to the ground (horizontal polarization). When two plane-polarized wave systems of the same frequency, but slightly out of phase, are propagated in the same direction they combine to form elliptically polarized waves, as shown in Fig. 2-4. Circular polarization is a special case of elliptical polarization, where the amplitudes are the same. The choice of proper polarization is an important consideration in the design of radar systems, as will become evident.

Interference phenomena provide perhaps the most striking and conclusive evidence of the periodic nature of sound and light. When two sets of transverse waves traveling in the same direction encounter each other, the effect is additive and the amplitude of the resulting wave is the sum of the amplitudes of the separate waves, as shown in Fig. 2-5. Interference effects, properly speaking, are observed when the paths of waves from two sources intersect: the effect of one wave at some points reinforces, and at others destroys, the effect of the second. This results in the production of a new periodic motion which is the algebraic sum of the other two, and which, if the two sources are of equal intensity and of slightly different frequency, may be stronger than either source alone. The frequency of the new oscillations is equal to the difference in the frequencies of the two sources. These "beat notes" can readily be detected in the throbbing sound of an out-of-tune piano, and may be experimentally demonstrated by sounding simultaneously two tuning forks differing only slightly in frequency.

A similar phenomenon can be observed when waves from the same source reach a distant point after traveling by two different paths. Only if the paths are the same length or differ by an integral number of whole wavelengths will the waves arrive with their maxima and minima, their humps and depressions, precisely coinciding, i.e., *in phase*. If they arrive *out of phase* the same interference effects occur as have just been described. The effect is like the intersecting of wave trains from two separate sources. These interference effects due to difference in path length can often be experienced when a sound reaches a listener nearly simultaneously by a direct route and by means of an echo from some nearby surface. In concert and lecture halls the effect of such interference is to blur and distort the sounds of the human voice or of musical instruments. We shall find these effects of great importance in the fields of radio and radar. In both fields the shaping of transmitted radio frequency (rf) energy into a beam is performed by taking advantage of this property. For example, a

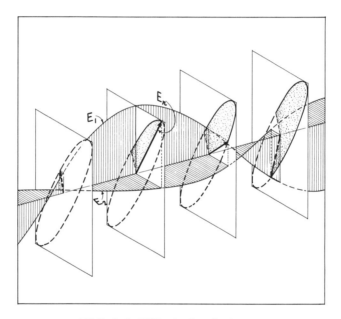

FIG. 2-4. Elliptical polarization.

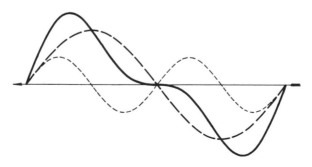

FIG. 2-5. Addition of two transverse waves.

row of dipoles such as make up a *linear array* can be used to concentrate the beam in the horizontal plane when the dipoles are so spaced that the transmitted waves reinforce one another along a desired axis of propagation and destructively interfere in other directions.

A final and important case of interference remains to be described. When two trains of traveling waves move through one another, in opposite directions, *standing waves* are set up. For radio, as for acoustics, a particularly important and useful case arises when standing waves are set up by the interference of a wave and its own reflection, where, of course, the interfering waves have the same wavelength and nearly the same amplitude. This is equivalent to arresting a wave for purposes of study, for the standing wave has the same form and frequency as the waves from which it is created.

The Pulse Method[2]

With this review behind us, we can proceed to describe the mode of operation of a radar set. A beam of ultrahigh- or superhigh-frequency radio waves is sent out from a transmitting station. This is not unlike a small radio station, except that instead of sending the energy indiscriminately in all directions (broadcasting) by means of a nondirectional antenna, the transmitter concentrates the energy into a beam which can be pointed steadily in a chosen direction, i.e., "search-lighted," and can be made to sweep back and forth in an arc, rotate a full 360°, or otherwise "scan" a given region in which a target such as an aircraft or a ship is likely to be found. The target intercepts and reflects only a small fraction of the energy of the beam; and since most of this is scattered over a wide angle, a still smaller portion finds its way back to the neighborhood of the transmitter where it can be picked up and amplified by a suitable receiver located nearby. See Fig. 2-6. The radiation from the transmitter is not sent out as a continuous train of waves (continuous wave or cw radiation) but in short bursts or pulses lasting a few *millionths* of a second (microseconds). These bursts of energy are separated by relatively long intervals (two or three *thousandths* of a second in length); and during this period of silence the faint echo signal is received. Then the cycle is repeated. See Fig. 2-7. If we remember that power measures the *rate* at which energy is generated or delivered, Fig. 2-7 should help to make it clear that the power developed by a given transmitter in pulse operation *(peak power)* is many times greater than if energy were continuously radiated by the transmitter *(average power)*, the same amount of energy having been spread over the time of the full cycle. *Peak power* is the value generally used in comparing the performance of radar transmitters.

After the returning signal has been detected and amplified by the receiver, it is made to appear on the screen of a cathode-ray oscilloscope.[3]

This important instrument uses the motion of an almost inertialess pencil beam of electrons (cathode rays) to record or measure small and rapidly changing voltages. As shown in Fig. 2-8 the beam of electrons is projected from an electron gun which serves both as the source of electrons and as a means of focusing them electrically and magnetically into a sharp beam. This beam passes through two pairs of deflection plates and impinges on a *fluorescent screen* where it forms a spot of light. This screen is composed of special substances, called phosphors, deposited on the inside wall of the

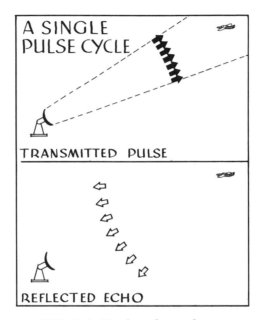

FIG. 2-6. Single pulse cycle.

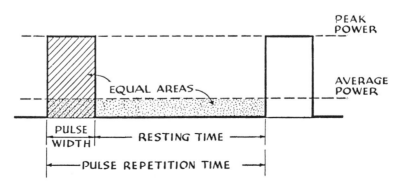

FIG. 2-7. Pulse length and resting time.

end of the tube. The phosphors have the property of transforming the energy of the electrons striking them into visible light.

When an alternating potential is put on the pair of *horizontal deflection plates,*[4] the spot is swept back and forth across the face of the tube at great speed, say 500 times a second. Because of the slow response of the eye, a steady bright line appears across the face of the tube. If a signal voltage, such as a returning echo from a radar receiver, is put on the *vertical deflection* plates of the tube, the beam is deflected sharply upwards or downwards to produce a "pip" or "blip" as shown in Fig. 2-9. This constitutes the simplest type of radar indicator, generally referred to as an A scope. In the A scope the sweep is synchronized with the pulsing; that is, the spot starts its journey across the face of the tube while the pulse is traveling out into space. The main pulse is recorded at the beginning of the trace and the echo will be picked up by the receiver after the spot has traveled some distance across the tube. This produces a second weaker blip at some point to the right of the one from the outgoing pulse. This process recurs many times a second and the succession of nearly identical transient images are added together by the eye to form a static image, just as a still picture can be projected by successive identical frames of a movie film.

The cathode-ray oscilloscope is not just an indicator, however. In radar, it serves as a precision timing device enabling us to determine (with great accuracy) the range to a target by measuring the length of time it takes the high-speed pulse to complete its round trip to the target and return. If the changing voltage that is put on the horizontal deflection plates to produce the sweep line increases during the course of each sweep as a precise function of time (usually a linear function) then the distance along the trace becomes an extremely accurate measure of an incredibly short time interval. Since the distance traversed at constant velocity is (by the definition of velocity itself) proportional to the time elapsed, the distance can be read off the sweep line in miles or yards if the sweep is

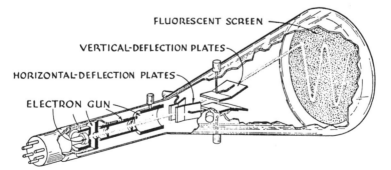

FIG. 2-8. Cathode-ray tube.

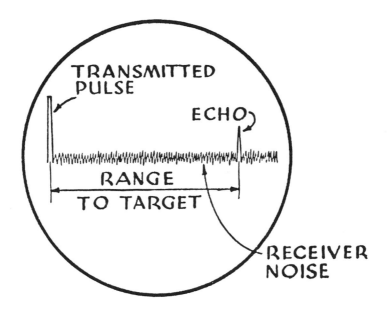

FIG. 2-9. The A scope.

properly calibrated by a superimposed scale or by little marker pips introduced electrically at known intervals. Since the pulses are traveling with
the speed of light, 186 000 miles per second, in round numbers, then the
distance covered in one microsecond is 0.186 mile. But since the range of
the target from the station is only half the to and fro path traversed by the
pulse, each microsecond recorded on the cathode-ray tube corresponds to
a range of about 0.093 mile or 164 yards.

Except for being limited by the horizon, the range at which a radar set
can detect distant objects depends upon the power it can transmit to the
neighborhood of the target; upon the nature of the target itself (its size,
shape, electrical properties); and upon the sensitivity of the receiver, that
is, its ability to pick up a faint echo. The power impinging on the target
depends not only on the number of watts of power generated by the transmitter but also upon the success with which the radio energy is concentrated into a narrow beam. This last depends upon the optical properties
of the antenna system. Under the ideal conditions of transmission in "free
space," and assuming the use of a parabolic reflector, these results are
summed up in the following expression for the power returned from a
target and detectable at the receiver:

$$P_r = \frac{P_t A^2 \rho}{4\pi r^4 \lambda^2}.$$

This indicates that the power at the receiver, P_r, is proportional to the transmitted power P_t; to the square of the area A of the parabolic reflector; and to the scattering cross section ρ of the target. But it falls off inversely as the fourth power of the range r to the target and inversely as the square of the wavelength λ of the radiation. If we solve this equation for the range, and replace P_r by P_{min}, the minimum signal strength to which the receiver will respond, then r becomes the maximum range of the system and we have

$$r_{max} = \left(\frac{P_t A^2 \rho}{4\pi P_{min} \lambda^2} \right)^{1/4}.$$

This equation indicates that range can be improved by raising the transmitter power, by using larger antenna reflectors (both of which tend to increase the weight of the equipment), and by increasing the receiver sensitivity, which though largely a matter of improved circuit design, involves an extremely complicated set of problems and severe theoretical limitations.

Power considerations apart, the maximum and minimum ranges at which a radar system can be used will depend upon the length of the pulses and upon the interval between the pulses. If the pulse lasts too long, it will obscure targets close by, for the transmitter will still be operating when the echo returns to the receiver. Thus a system with a one-microsecond pulse has a useful minimum range of 164 yards, while one with a four-microsecond pulse is not useful for targets nearer than about 650 yards away. In similar fashion, the interval of silence between the pulses determines the distance from which a target can send back an echo without having it obscured by the next outgoing pulse. The pulse recurrence rate is therefore chosen with care to make use of the maximum capabilities of the system. Since the range of the highest-powered radar systems is limited only by the horizon, and under certain atmospheric conditions can extend some distance beyond, such systems use a resting time of about 2000 microseconds, which enables objects to be seen out to 200 miles.

So much for finding range. The determination of bearing and elevation depends on the properties of the antenna system. Now the simplest type of radiating elements used on higher radio frequencies is a so-called Hertzian dipole cut to half the wavelength it is supposed to transmit. This has very low directivity, and in free space produces a doughnut-shaped pattern shown in cross section in Fig. 2-10. Antennas of this sort, broadcasting pulses in several directions at once, are perfectly capable of detecting objects and determining their range. Indeed, the very earliest experimental radar systems in America and in England were of this sort. But with such antennas it is extremely difficult and sometimes impossible to determine the direction or the height from which the echo is coming.

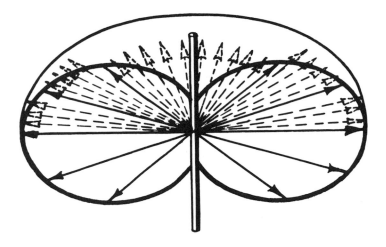

FIG. 2-10. Hertzian dipole.

If, instead, a sharply directed beam of energy is used, and the antenna is mounted so that it can rotate, it is only necessary to determine the angular position which gives the strongest signal on the indicator in order to determine the bearing of the target in approximate fashion. Special methods, to be discussed later, have been devised where high precision is required, but apart from these the sensitivity of the single beam method depends only upon the angular width of the beam. The sharper the beam, the more accurate the determination of bearing.

It would appear that similar conditions should govern the determination of elevation. Here, however, the situation is greatly complicated by the effect of ground reflections of the antenna beam. If the beam fails to intercept the ground at any point, either because the beam is very narrow or because it has been elevated sufficiently, the pattern is unaffected. But if the antenna beam is lowered, some of the transmitted energy hits the ground and is reflected back into the direction of the beam. The energy that strikes a target is made up of two components: waves that came directly from the antenna to the target, and those that took the longer path from the antenna to the ground and thence to the target. Waves that arrive simultaneously at the target after having followed two slightly different paths will arrive either in phase or to some extent out of phase. If the path difference is a half wavelength the fields cancel; if the path difference is a full wavelength the fields add. The effect of ground reflection and the resulting interference effects is to break up the single lobe into multiple lobes with gaps between. This distortion of the pattern as shown in Fig. 2-11 is so complicated that it is clearly impossible to determine altitude

FIG. 2-11. Long-wave beam pattern.

merely by changing the angle of the antenna. It is nevertheless possible to use the multiple lobes produced by these ground reflections for height finding, but this is a very complicated business even under ideal conditions. Yet the conditions are never ideal, for the ground is rarely perfectly flat and its conductivity—upon which the reflection largely depends—is variable. The ideal solution for height-finding is a beam that is narrow enough, or so high above the ground, that the reflection problem is eliminated.

The foregoing should help to make it clear that an important phase of the radar art must be concerned with building directive antenna systems to focus the radiation into beams as sharp as possible. It has already been pointed out that beams of radio energy are produced by making use of interference effects between radio waves from nearby sources. The patterns that result closely resemble those produced by optical diffraction gratings. The extent of the focusing that is obtained depends on two factors: the wavelength and the dimensions of the antenna system. For a given wavelength, the greater the aperture of the antenna system, the narrower and sharper the beam that can be produced. Conversely, using a given aperture we may obtain a narrower beam by using a shorter wavelength. These two statements can be recombined in the rule that *it is the number of wavelengths in a given linear dimension that determines the sharpness of the beam in that direction.* This is given quantitatively by the expression $\theta = k/N$, where θ is the beam width in radians, k is a proportionality constant, and N represents the number of wavelengths in the given aperture. In Fig. 2-12 the reader should be able to observe how a principal beam and sidelobes are produced by interference effects. The beam width θ is measured between the half-power points on a power diagram as shown in Fig. 2-13.

A quantity called *antenna gain* is used to express numerically the energy-directing properties of an antenna. This quantity compares the power radiated (per unit solid angle along the beam) with the power (per unit solid angle) for an ideal point source radiating isotropically, i.e., uniformly in all directions. Antenna gain is measured in decibels (dB), a unit which expresses a ratio between two power measurements.[5] A dipole without a reflector has a power gain of 3/2 when compared to the ideal point source. A dipole with a parabolic reflector has a gain that depends upon the aperture, i.e., the area A of the reflector, and upon the wavelength employed, λ, according to the following relation:

$$G = \frac{4\pi A}{\lambda^2}.$$

Since any gain or loss of power can be expressed in decibels, these units are used whenever it is necessary, by a power comparison, to comprehend the performance of a radar system. It is used to express *receiver gain*, i.e., the amplification factor of a receiver or one of its amplification stages; and it is used to express the progressive loss of power (attentuation) as rf energy passes along a transmission line. It is also used in studies of radio propagation, to measure the attenuation of radar signals by the atmosphere and by clouds and rain.

There are two principal ways by which the radiation can be concentrated in a desired direction, i.e., by which the antenna "gain" can be increased. The simplest is to use a parabolic metal reflector to focus the energy of a single dipole, much as similar reflectors are used to beam the light of an automobile headlamp. See Fig. 2-14. This is only practical, however, for waves considerably less than a meter in length, and it is really convenient for most purposes only in the microwave region. But directional effects can be produced with the longer waves of the ultrahigh-frequency region by using two or more dipoles so spaced and so phased that the radiation from the dipoles is intensified in a chosen direction and canceled out in others. One system of this sort, the linear array, has already been mentioned; another, sufficiently light to have proved useful in early forms of airborne radar, is the so-called Yagi array. See Fig. 2-15. Here the radiation from a half-wave dipole is picked up by an adjacent "parasitic" tuned reflector and by several neighboring "parasitic director" elements and reradiated with such phase relations that the field is built up in the forward direction. Another type of dipole assemblage used

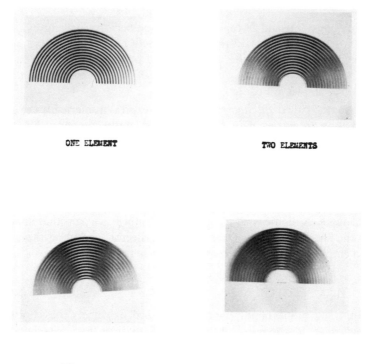

ONE ELEMENT   TWO ELEMENTS

FOUR ELEMENTS   SEVEN ELEMENTS

FIG. 2-12. Production of beams by interference effects. The decreasing width of the primary beam, together with the increasing number of side lobes, for 1, 2, 4, and 7 radiating elements, can be observed from the above photographs. These radiating elements may be considered to be either dipoles in a linear array of imaginary elements spaced across the aperture of a parabolic reflector, illuminated by a single dipole.

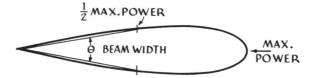

FIG. 2-13. Beam width.

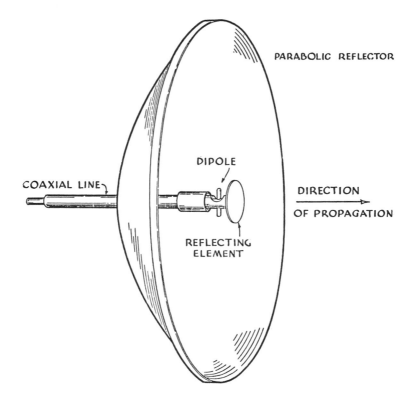

FIG. 2-14. Parabolic reflector.

to produce narrow beams is the stacked-dipole or broadside array, shown in Fig. 2-16. In this type, employed in several of the earliest Navy and Army radar search systems, several rows of dipoles are placed one above the other. This would normally yield a beam in both directions at right angles to the plane of the dipoles, but the array is made unidirectional by putting a second array of parasitic reflectors behind the antenna or better still by using an untuned screen reflector. A broadside array of reasonable size with screen reflector can give a well-defined beam about 20° wide.

The Operation of a Radar System[6]

A radar system is made up of distinct units, usually called component parts or components, each of which has a specialized function in the cycle of events by which a pulse of radio energy is transmitted and the returning echo received. We may follow the successive events during a single cycle and learn something of the duties of each component as they are illustrated in the block diagram given in Fig. 2-17.

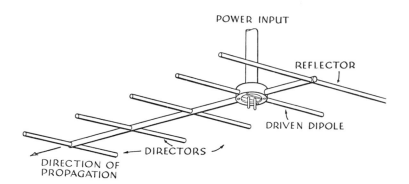

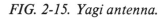

FIG. 2-15. Yagi antenna.

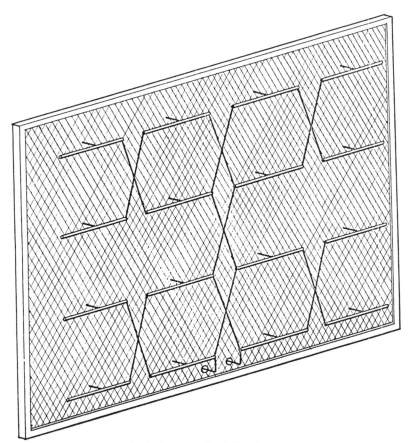

FIG. 2-16. Stacked dipole array.

When the radar system is turned on, the cycle is initiated by a unit (called the "control central" in Fig. 2–17) which sends a triggering impulse simultaneously to the modulator and to the sweep-generating circuits of the indicator. In the modulator, the low-voltage trigger impulse is transformed into a high-voltage, carefully shaped square pulse. This operates the transmitter, which consists of one or more vacuum-tube oscillators capable of generating ultrahigh-frequency or superhigh-frequency waves when properly excited. The pulse from the modulator excites the transmitter which proceeds to radiate rf energy in pulses of the same shape, and at the same repetition rate, as the pulses from the modulator. The rf pulses are then radiated into space from the antenna. While this is taking place the beam of the cathode-ray tube is tracing out a sweep line or time-base. While the transmitter is operating, the special switch called the T-R or transmit–receive box prevents all but a small portion of the outgoing rf energy from leaking into the receiver where it could do considerable damage. This small portion, however, passes through the receiver to the

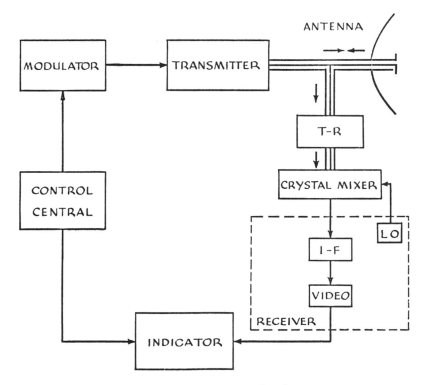

FIG. 2-17. Block diagram of radar system.

indicator where it appears as an initial deflection, colloquially called the main bang, marking the beginning of the sweep.

Meanwhile we shall assume that the radiated pulse has encountered a target and been reflected back to the antenna where it is picked up. The T-R box passes the weak rf signal to the receiver where it is amplified and converted into a *video signal* (i.e., a signal on video or audio frequencies), which appears as a pip on the sweep line of the indicator.

The receiver employed in radar is the *superheterodyne*, which is widely used for ultrahigh-frequency communication receivers. Such a receiver consists of the crystal detector and mixer, a local oscillator, and several amplification stages. The local oscillator, ordinarily a velocity-modulated type of vacuum tube, is a separate source of energy of a frequency that is different from the transmitted frequency by some chosen amount. In microwave radar this difference is 30 MHz. This energy is fed to the mixer where it is combined with the received signal. The output of the mixer is the beat frequency resulting from the combination of the two frequencies, namely 30 MHz. This intermediate frequency (IF) is then amplified. In the later stages of the receiver the signal is converted to a video frequency (that of the envelope of the IF) and further amplified. The output of the receiver is then fed to one or more cathode-ray tube indicators.

By variation of the scanning motion of the beam and the means of impressing the signal, a wide variety of types of indicators have been devised to serve particular purposes. Some of the more specialized types will be described when they are encountered in the course of the book. Here the discussion will be limited to the five most widely used types. Four of these are shown in Fig. 2-18, and one in Fig. 2-9.

The *Type A* indicator, or A scope (see Fig. 2-9), is only a simple cathode-ray oscilloscope with a horizontal time base, usually linear. The video signals are put on the vertical deflection plates.

The *Type J* indicator is a modification of the A scope in which the beam sweeps in a circle with uniform angular velocity. The video signal is applied to electrodes which produce a radial deflection of the beam. Because this permits a longer sweep than the Type A indicator it is used when more accurate range measurements are required.

In the *Type B* indicator the field being scanned is plotted in rectangular coordinates, range being estimated vertically and azimuth along the horizontal coordinate. This type of indicator is spoken of as being *intensity modulated,* for the video signal is applied to an intensity grid in the cathode-ray tube.

The *Type C* indicator is a similar intensity-modulated indicator in which azimuth is plotted horizontally and elevation is given vertically.

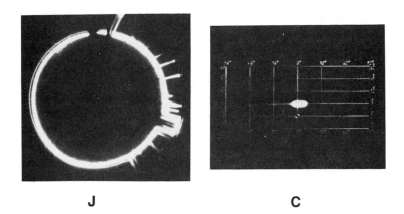

J   C

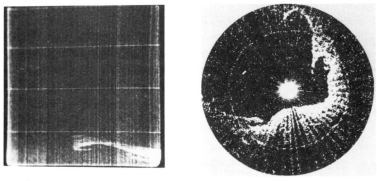

B   PPI

FIG. 2-18. The J scope indicates range only. The C scope shows bearing and elevation relative to plane. The B scope, presenting range and bearing in rectangular coordinates, resembles a map in Mercator's projection. The horizontal lines are range markers indicating 5-mile intervals. The tip of Cape Cod is visible at about 3 miles. The PPI (Plan Position Indicator) gives map-like indication. Compare the above PPI picture of Cape Cod with the B scope at left. The scale is of course different. Range circle is at 10 miles.

The most important of all indicators is probably the PPI or Plan Position Indicator. In this indicator the sweep line runs from the center of the tube, which represents the location of the observer, to the periphery. This line rotates about the face of the tube in synchronism with the angular rotation of the antenna. Signals appear as intensity modulations of this sweep line. Because of the long-persistence phosphors chosen for the fluorescent screen of the PPI, the rotating line sweeps out a map in polar coordinates of the region being scanned.

NOTES

1. This classification of the electromagnetic spectrum does not agree with the Federal Communications Commission's official classification. For example, the FCC has defined the region from 30 to 30 000 MHz as comprising three bands: very high frequencies, 30 to 300 MHz; ultrahigh frequencies, 300 to 3000 MHz; superhigh frequencies, 3 000 to 30 000 MHz.

2. In this book we shall limit the name radar to radio detection systems using reflected pulses, though other kinds of systems have at various times been devised. This accords with general usage. The derivation of the name emphasizes the importance of the ranging feature most readily obtained by the use of pulses.

3. The name *oscilloscope*, which has replaced the term *oscillograph*, is applied to a cathode-ray tube with its associated sweep circuits.

4. In special types of indicators coils are substituted for plates.

5. Actually the logarithm of the ratio: $dB = 10 \ \log P_1/P_2$.

6. Any operable assembly of radar components, no matter how experimental, will be referred to as a radar "system." A standardized type of system procured by the Army or Navy will be referred to as a "set."

CHAPTER 3

The Radio Background of Radar

Radio detection devices using the pulse-echo principle were developed independently and almost simultaneously during the 1930s by a majority of the great powers. In 1939 closely guarded secret programs were in various stages of advancement in Great Britain, France, Germany, Canada, and the United States. Russia, China, Japan, and Italy were at that time without the equipment, and seem to have acquired it after the outbreak of war, by capture and by disclosures from their allies. In France and Germany the equipment was developed in commercial laboratories under military supervision, whereas in Great Britain and the United States the pioneer development was in laboratories of the armed services.

Such a duplication of effort will surprise only those who cling to a hero theory of scientific progress and demand for each discovery or development a single putative inventor; or those who are unaware of the frequency—one is tempted to write, the regularity—with which such parallelisms are encountered in modern scientific work. The duplications are by no means rare in pure science; but they are much more common in industrial or applied research, where a significant discovery in pure science, simultaneously taken up by many hands, can spawn a numerous progeny of practical applications and useful embodiments. The chances of duplication are better in the present century than in the last, for in the final analysis they depend only upon the scope and intensity of the research effort. It has been well said that the greatest invention of the 19th century was the art of inventing; and this is an art which the 20th century has greatly extended and perfected, by widespread technical education, well-financed team research, generous budgets, and well-equipped laboratories. The element of chance in scientific progress (which includes also the chance of discoveries of isolated individuals being lost to society) is steadily diminishing, and the discontinuities in the upward curve of technical progress, though they are still present, are less pronounced.

Clearly such a striking instance of parallel and independent discovery raises a number of fundamental historical questions. When a real burgeoning takes place we are led to ask what were the conditions that favored it. The main lines of inquiry are obvious. First of all, when a serious effort is put behind a given development, as it was in the case of radar, it is evident that it satisfied a clear and urgent need. Second, since the principle of *nihil ex nihilo* applies quite as surely to the history of science as to biology, it is obvious that some key principles or central ideas in pure science must have been the point of departure. Lastly, when success is attained, it is obvious that the state of the art, that is, the perfection of engineering skills in this or neighboring fields, must have reached a point where success was fairly well assured.

It is proposed in this chapter first to analyze the military need for radio detection equipment, and then to explore the background of scientific ideas and the state of technical progress that together made possible or even inevitable the development of radar before World War II. It will become apparent that conditions were definitely propitious in the period 1930–1939; so well suited, in fact, that although there were important adumbrations long before, it is hard to imagine a successful development much before the period in question.

The Reply to Air Power

In the first chapter we referred to the many uses to which radar was put during the war. But it is essential not to lose sight of the fact that radar was first thought of, indeed was for some time solely envisaged, as a defensive weapon against hostile aircraft. In all the pioneer developments of which we have adequate information, the first types of equipment to be designed were long-range, higher-power sets to give early warning against attacks by enemy planes, or sets with shorter range to provide accurate information for the direction of searchlights or the pointing of anti-aircraft guns. Viewed in historical perspective, radar will probably appear as the reply, the countermeasure, to the threat of air power.

Perhaps we are still too close in time to realize fully that the ten years before World War II were the years that really introduced the age of air power. It requires a conscious effort of memory to recall that in 1930 we were only on the threshold of the time when, for peaceful purposes, flight became a commonplace necessity. In that year we were still marveling at the audacity of Lindbergh's solo flight to Le Bourget. By 1940 the situation had changed radically. A series of important technological changes had taken place which were largely responsible for the swift progress of civil aviation during that decade.

The introduction of multiengined planes; the steady improvement of aviation gasoline and the adoption of the octane rating; improved design of cowlings for wing engines; the steady improvement in aircraft instruments and aircraft engines; the total abandonment of wooden construction; and the revival and adoption of such earlier proposals as wing flaps, retractable landing gear, and the controllable pitch propeller: these technical, economic, and safety factors conspired to revolutionize air transport during the 1930s. Most of these features were incorporated in the Boeing 247 and the Douglas DC-2, which went into service in 1933–1934. Routine, dependable commercial air transportation became commonplace almost overnight. The cruising speeds of planes rose steadily. In 1931, Post and Gattie had flown around the world in eight days; in July 1938 Howard Hughes did somewhat better than cut their time in half, and maintained an average flying speed of 206 mph. On the eve of the war, the principal countries of the world all could point to two- or four-motored aircraft in routine service having cruising speeds in excess of 200 mph. In Germany the Junkers Ju-86 had a rated cruising speed of 224 mph.

These were unmistakable signs that the warnings of airpower prophets like Douhet and Mitchell must be heeded; they foreshadowed the importance of high-speed, long-range military aircraft in the future conflict that was unmistakably looming. Long before the 200-mph figure was reached as a norm of performance, there was no doubt that existing aircraft warning devices were hopelessly archaic and inadequate. By the midpoint of the decade efforts were being seriously directed toward finding a completely novel type of device.

Aircraft Detection in the First World War

Even the circumscribed role of aircraft in the First World War gave rise to methods of detection, not all of which were actually used, that depended in all cases upon telltale information involuntarily supplied by the attacking plane itself. Aside from the use of visual spotters, this was limited in practice to detecting the sound of the approaching plane and determining its direction; but methods were also proposed for detecting and amplifying electromagnetic radiation (on radio and infrared frequencies) emanating from the plane.

Sound detectors, designed to pick up and amplify the drone of a distant engine and to give a rough indication of its direction, seem to have begun, on the Allied side at least, with the "orthophone," a simple binaural device used in the French Army in 1917.[1] At roughly the same time an acoustical detector of the reflector type was produced in England by the Anti-Aircraft Section (under A. V. Hill) of the Munitions Invention Department.

Active development of sound detectors went forward between the wars in commercial concerns of various countries. Among the most highly perfected were those made by the Sperry Gyroscope Company in this country (using exponential horns); by Metropolitan-Vickers in England and Goertz in Vienna (using parabolic reflectors); and by various French firms which favored what they called *nids d 'abeilles*, reflectors made up of a cluster of smaller horns spaced about a central collector.

By the time they were supplanted, these acoustical detectors had been brought to a high pitch of perfection. In 1936 an error of only 0.25° in azimuth was claimed on sound from a fixed source; and for an airplane flying at moderate heights, all sound locator manufacturers quoted a 2° accuracy. In certain cases the sound was converted into visual information on the screen of a cathode-ray oscilloscope.

The deficiencies of these devices were numerous and fatal: they gave no range information; their performance depended upon the wind; and they were quite unreliable on gusty days. Lastly, and most important, their range was so short that with the high speed of modern planes the observers scarcely had time to alert the defenders before the attacking planes were overhead; certainly they did not give sufficient warning in most cases for intercepting fighters to be sent up to meet the attackers. Their main function was to direct searchlights. This range deficiency was enhanced by the low velocity at which sound travels in air, or rather by the fact that this velocity is not sufficiently large compared with that of the plane. When the sound reaches the detector the aircraft has already come much closer than when it produced the first sound picked up by the listening system.

While with the AEF in France during the First World War, the late Edwin H. Armstrong, then a Major in the Signal Corps, conceived the idea that perhaps the extremely short radio waves produced by the spark gaps of a gasoline aircraft engine could be picked up and used to detect the presence of enemy aircraft. Careful shielding of this disturbance was known to be necessary to prevent interference with radio equipment in the plane. The difficulty was to detect and amplify those waves, which were shorter than any hitherto familiar in radio communication. To solve this difficulty, Major Armstrong developed the first superheterodyne receiver, whose operation has been described in general terms in the preceding chapter. The superheterodyne receiver was never used in World War I for the purpose for which it was devised.[2]

Almost simultaneously a second method of attack—the attempt to detect the infrared radiation from airplane engines and exhaust gases—was adopted early in 1918 by Master Signal Electrician Samuel O. Hoffman and a group of enlisted men from the Signal Corps and Air Service working in the laboratory of Professor George B. Pegram at Columbia University.[3] Attempts were made to detect men and other objects at higher temperature than their background by means of a Hilger thermopile (a combination of thermocouples) mounted in the focus of a parabolic mirror. In the spring of 1918 men were detected at 600 feet; shortly after, in August 1918, an apparatus was sent overseas, but tests with this equipment in France were unsuccessful. Thermal equipment specially modified to detect aircraft was tested with moderate success in January 1919 at Langley Field, Virginia. Ranges of from 4000 ft to over a mile were reported, but there were complications from false indications caused by clouds. No further work on heat detection was carried on by the United States Army until 1926 when the Ordnance Department set up at the Frankford Arsenal a project for the "Investigation of Detection Devices using the Infra-Red Ray." This was under the direction of Captain William Sackville, C.A.C., and was a continuation of research he had been conducting with Lieutenant Jose E. Olivares of the Philippine Scouts, U.S. Army, at the Massachusetts Institute of Technology. The problem of how aircraft might be detected and located at night had been suggested to the two officers by Professors Vannevar Bush and J. W. Barker as a thesis subject for the Master of Science degree in Electrical Engineering. After reviewing the principal avenues of approach to the detection problem—all, that is, except the use of *reflected* electromagnetic waves from a ground source—they rejected sound because of the time lag, shortwave radio waves from the engine ignition system because of the ease of shielding, and for similar reasons visual observation of the exhaust flare, and focused their attention on detection of the infrared energy from the engine exhaust. These experiments at MIT and at the Frankford Arsenal were not much more successful than the closely similar Hoffman experiments.[4]

Detection with Reflected Energy

A moment's thought will reveal that the radar principle explained in the first chapter is in fact divisible for purposes of analysis into two elements: detection by means of reflected radio energy (the echo principle) and the determination of range by means of carefully timed pulses (the pulse principle). In what follows, we shall find it convenient to refer to any device which detects objects by means of reflected energy as an echo detection device, being careful to specify the sort of energy used in each case. We shall limit the term *radar* to echo detection devices using pulses of radio energy. It will be of some interest to trace in outline the histories of both ideas, which will sometimes be found separated and sometimes together. What follows is a mere sketch. A book could be written on the subject of this chapter.

The earliest echo detection device was of course the human voice. A stentorian halloo from the bridge of a ship has often given the skipper of a fog-bound ship some notion of his proximity to land. More sophisticated detection systems based on sound reflection were proposed and tried in the last century. John Tyndall, the British physicist, undertook at the request of the Elder Brethren of Trinity House to find a method of preventing ships from running ashore at night. He experimented with lights and with crude electrical contrivances and finally suggested the use of steam whistles.[5] In 1912 Sir Hiram S. Maxim was inspired by the sinking of the *Titanic* to propose a method of detecting such obstacles as the fatal iceberg by sending out low-frequency (below the audible range) sound waves, and receiving the returning echo on a diaphragm, the vibrations of which could either ring a series of bells or make a kymograph recording.[6] Maxim compared his method with the manner in which Abbé Spallanzani's blinded bats were able to avoid obstacles. He interpreted these 18th century experiments as indicating that bats depend upon sounds of low frequency produced by their wings, which he thought were reflected from obstacles and picked up by special sense organs situated in the center of the face. It is established today[7] that bats avoid obstacles by emitting supersonic notes which their ears are able to detect when they are reflected back to them from obstacles in their path. A highly developed binaural sense enables them to interpret these echoes correctly and so to locate obstacles in space with considerable accuracy. The use made by human beings of supersonic echo methods will be discussed below in connection with the pulse method.

The Reflection of Radio Waves

It is possible to detect invisible or distant objects by means of radio and infrared energy only because such radiation has, like light, the property common to the whole electromagnetic spectrum of being reflected from natural or man-made obstacles. It is only because such waves can be focused into more or less sharply directed beams of radiation, as described above, that efficient utilization of energy is possible, and that detecting devices can act as direction finders or locators. We shall refer briefly in another place to some early attempts to use reflected infrared rays for detection; here we shall confine ourselves to discussing this property of radio waves.

In the case of radio waves, this important property of the electromagnetic spectrum is easier to demonstrate the shorter the wavelengths of the radiation. The phenomenon is only readily observable for wavelengths a few meters in length or shorter. It is for this reason that attention was not drawn to reflection phenomena during the period when interest was focused on the broadcast frequencies. Although well understood in the time of Hertz, this property was largely overlooked, until work was resumed on the shorter wavelengths.

In his great *Treatise on Electricity and Magnetism* (1873) James Clerk Maxwell brought under a unified theoretical dominion the existing facts concerning light, electricity, and magnetism. The most striking consequence of Maxwell's electromagnetic theory of light was that there should exist other as yet undiscovered waves which should be propagated through space with a velocity equal to that of light; that these should be produced wherever oscillatory electric currents are set up, as, for example, in a discharging Leyden jar; and that they should be reflected from conducting surfaces and refracted by dielectrics according to the classical laws of geometrical optics. The existence of these radio waves was first experimentally verified in 1887–1888, nearly ten years after Maxwell's death, by Heinrich Hertz. Hertz's experiments, once he had succeeded in generating and detecting the earliest known radio waves, were designed to prove that these waves did in truth have the resemblance to light, or as we say today, the quasioptical properties, which Maxwell had predicted.

Hertz used wavelengths which made the optical properties easy to observe.[8] In his first experiments his spark gap gave him waves 10 m in length; subsequent experiments were made with a much smaller oscillator that yielded a wavelength of 66 cm. His transmitter was placed in the focal line of a cylindrical metal reflector; his receiver consisted of a dipole in the focal line of a receiving reflector. A spark gap served as a receiver. With this equipment Hertz was able to detect radiation when the cylinders were separated by several meters, and with it he was able to duplicate many of the phenomena encountered in the study of light. He showed that shadows were cast when objects opaque to the radiation were interposed; he observed the phenomenon of diffraction; and he showed by rotating the receiving cylinder through 90° that the waves emitted by his oscillator were polarized. This was confirmed by beautiful experiments in which he employed a grating of wires interposed between the transmitter and receiver: when the wires of the grating were at right angles to the linear radiator and to the receiving dipole the energy passed through, but when the screen was rotated until the wires were parallel the screen was opaque to the radiation. Hertz likewise showed that it is possible to refract electromagnetic waves. With a large prism of pitch cast in a wooden box he was able to observe a refraction of as much as 22°; and from this result he was able to calculate the index of refraction of the pitch for his waves to be 1.69, as compared with values of 1.5 and 1.6 for closely similar substances when transmitting light.[9]

In the decade following the publication of Hertz's results a number of different workers—Righi, Klemencic, A. D. Cole, Lebedev, and others—extended Hertz's work by using what we would now call microwaves.[10] With extremely short waves a few centimeters in length, generated at very low power, they were able to perform quasioptical experiments which had not been practical with Hertz's somewhat longer waves. For example, Peter Lebedev in Moscow succeeded in demonstrating, by means of 6-mm waves, that these waves could be doubly refracted by crystals.[11]

While the attention of physicists was quite generally focused on microwaves and their optical properties, interest was soon drawn to the practical application of these discoveries and to the possibilities of wireless telegraphy by means of the new Hertzian waves. With this shift of interest came an all but universal preoccupation with the very long radio waves. It will be recalled that Marconi's first successful transmissions on Salisbury Plain in 1896 were made with very short waves. Using parabolic reflectors behind both his transmitter and receiver, Marconi established communication at a distance of $1\frac{3}{4}$ miles using waves a meter in length.

Almost at once, however, Marconi turned to the use of much longer waves, principally because it was much easier to get high transmitter power at these wavelengths. It was soon discovered that they were much better adapted to long-range communication, shorter waves being strongly attenuated over land or water. Still later in 1901 Marconi's extraordinary success in bridging the Atlantic with radio waves revealed that certain of the lower frequencies, at least, were not limited by the earth's curvature.

The Return to Shorter Waves

At the time of the First World War there was a sudden revival of interest in directed transmission and hence in shorter wavelengths because of the military possibilities of secret point-to-point beam communication. These possibilities were explored, for example, in 1917 by the French,[12] by the Signal Corps Research Laboratory of the AEF in Paris,[13] and a few years later by the United States Navy.[14] The only work published in detail was that of Marconi and C. S. Franklin in 1916–1917.[15] These men worked in England and in Italy, in collaboration with the Italian government, and got successful transmission up to 20 miles using waves of 2–5 m. At the end of the war interest again shifted to longer waves, although the Marconi Company continued to explore the possibilities of directed radiation and experimental work continued for a time in military laboratories.

These early experiments all served to confirm the opinion that short waves—in practice any wavelength much less than 200 m—were worthless for long-range communication. In consequence, the amateurs, who continued to be severely restricted under the Radio Act of 1912, were relegated to the supposedly unprofitable region of 200 m and below. The credit for the exploring of the region in the spectrum below 200 m and discovering its remarkable properties belongs to the amateurs who made up the membership of the American Radio Relay League,[16] the French *Société des Amis de la T.S.F.*,[17] and other associations. In November 1923 the tight little world of radio enthusiasts was startled to hear that a French amateur from Nice, Léon Deloy, had established a two-way connection across the Atlantic with two American amateurs, F. H. Schnell and J. L. Reinartz.[18] This extraordinary and unexpected achievement served to advertise the lower wavelengths and especially to point to the peculiar advantages of "short waves" for long-distance communication. The achievement was soon repeated. In that year and the next, forcing their equipment to the limit, operating vacuum tubes often at ten times their rating, the amateurs of America, Europe, Asia, and Australia linked up the world. The communications industry followed closely on the heels of these pioneering amateurs. The Radio Corporation of America and the Telefunken Company's station at Nauen soon established commercial short wave transoceanic links; in 1923 the Westinghouse station KDKA began broadcasting on 100 m to Europe and America; in July 1924, the Marconi Company signed a contract with the British government and the Dominions to build four shortwave stations to connect the metropolis with India and the Dominions.

The amateur successes were also responsible for redoubled activity on the part of the U.S. Navy, whose radio engineers had just taken possession of the recently completed Naval Research Laboratory at Bellevue, D.C. Under the leadership of A. Hoyt Taylor, Superintendent of the Radio Division at the Naval Research Laboratory (NRL), a program was launched, in close cooperation with the members of the American Radio Relay League, with whom Taylor had been intimately associated, to explore the behavior of short waves in long-distance communication.[19]

Two consequences of this pioneer work on 100 m were of far-reaching importance. The immediate effect was to provoke research into the question of why these short waves gave such unexpectedly long range. This work, which served to open up the field of ionospheric research, will be mentioned below. The second consequence was to inspire everyone, especially the amateurs, to explore still shorter wavelengths.

Many persons were curious to see whether better transmission could be obtained below 100 m; whether that is, the favorable properties found in the short waves might not be accentuated as one went further down the spectrum. The exploration of shorter waves, touched off in 1923, continued unabated during the next decade. Work went forward in England and on the Continent as well as in America; it was carried out both by amateurs and by investigators in commercial or government laboratories. By the end of 1925 it had been demonstrated that long-distance daytime communication was possible on 20 m.

Despite the increasingly evident commercial importance of the short waves (and presumably of the very short waves as well) it was some time before the industrial concerns took the lead away from the amateurs. In July 1924 the amateurs were authorized to use four new bands: 75–80, 40–43, 20–22, and 4–5 m.[20] Very little work was done by them at the lowest wavelength. Those who did tackle it discovered that it did not enjoy the remarkable long-distance virtuosity characteristic of 20 m. Moreover, the techniques of producing and receiving 5-m waves required departures from customary methods, whereas conventional tubes served adequately for 20 m. In 1927 the amateurs were allotted a new band at 10–10.7 m. Soon after its allocation a few dozen amateurs tried it out, hoping it might duplicate the performance of 20-m waves. The results were disappointing—mainly, as it turned out later, because of the sunspot cycle which in 1929 caused serious magnetic disturbances and quite generally upset shortwave broadcasting—and the new frequency was pronounced quite unsatisfactory.

After 1930 the industrial laboratories put an increasing emphasis upon the study of the properties of short and ultrashort waves. Important papers appeared from industrial laboratories on directive antennas[21] and on propagation characteristics[22] of ultrahigh-frequency waves.

At about this time serious attempts were made rather widely to study transmission on waves 5 m or shorter. In 1930 the Germans reported the results of successful communication experiments on 3 m, obtaining a range of 180 miles with a 1.5-kW transmitter.[23] These workers demonstrated that line of sight limited the range on these frequencies. In France, tests showed the feasibility of ultrahigh-frequency (UHF) telephonic communication on 5 m between Nice and the Col de Teghine in Corsica. An English experimenter reported similar work on 2 m. Between 1930 and 1935 the commercial concerns made careful propagation measurements on these wavelengths—with application to television of these frequencies clearly in mind—and confirmed the German findings as to the limited range. RCA made plans in cooperation with the Mutual Telephone Company of Honolulu to explore the possibilities of UHF waves for the establishment of inter-island telephonic links, and decided it could be done on 7 m. In 1931, in the summer issues of QST,[24] a campaign was launched by Ross Hull to interest amateurs in tackling once again the problem of communication on 5 m. In 1934 Hull established an experimental 56-MHz link between Boston and West Hartford, publishing the results in the following year as an important pioneer study of line-of-sight propagation.[25]

Tubes and Techniques for Ultrahigh Frequencies

Thus, by the early 1930s, just as the need for radio detection equipment was becoming acute, the course of radio experimentation had reawakened an interest in those ultrahigh-frequency radio waves which, as we have seen, were the first to be studied. The developments in electronic television, which appeared on the scene at about this same time, tremendously stimulated progress in the ultrahigh-frequency field, and provided an important incentive for the development of new techniques and devices applicable to its problems, and the perfection of others which had lain neglected. In the latter category, the perfection of the cathode-ray tube deserves, perhaps, first mention.

It is hard to imagine either radar or television without the cathode-ray oscilloscope. These indispensable instruments first appeared about 1930, after a slow and uneventful development dating from the last century, as compact, lightweight, and, above all, reliable instruments. Thus improved, they promptly multiplied in scientific and engineering laboratories and were at once adopted for television, ionospheric research, and the earliest radar.

The cathode-ray oscilloscope was invented by Ferdinand Braun in 1897.[26] His was a very simple device with a flat disc cathode, a wire anode fed in from the side, an annular diaphragm to limit the beam, and a fluorescent screen of zinc sulfide. It contained air at low pressure, and closely resembled the tube used by J. J. Thomson in determining the value e/m for electrons. The Braun tube in its early form was used by Zenneck and his school in studying radio circuits and the transmission of radio waves. In 1905 an important improvement, not immediately effective, however, was made by Wehnelt when he suggested the use of a hot, lime-coated filament as the source of electrons. This ultimately made it possible to operate tubes at reasonable voltages. The use of early tubes was distinctly limited because the equipment was large and cumbersome, because it required such high voltages, and because of the difficulty of maintaining a constant vacuum. By about 1930, industrial research laboratories in the principal countries had produced compact hot-cathode tubes, operating on low voltages, that were sensitive, reliable, and moderate in price. In the next few years the cathode-ray oscilloscope supplanted the mechanical oscillograph as a laboratory instrument, and found applications in a new and important field where even the best mechanical oscillographs were inapplicable: the study of wave shapes in the higher radio frequencies.

The influence of television upon the development of radar was not as direct as the superficial similarity of the two types of equipment would lead one to expect at first glance. Television and radar were parallel developments, not successive ones; and because of the secrecy which surrounded radar they did not often interpenetrate, especially while work on the latter was confined almost entirely to military laboratories. Both, however, were products of the 1930s.

The earliest experimental television was accomplished in 1925–26 with equipment quite different from that of present-day television.[27] Both for the television "camera" or pickup device and for receiving the image, mechanical scanning was the rule, using devices such as the Nipkow disk or the later perforated drum. The early radio transmission of television images was accomplished on 200–300 m.

Two important developments in the history of television took place in the early 1930s. The first was the abandonment of the mechanical systems and the recognition that the future of television lay in using the cathode-ray tube both for pickup and for viewing, as A. A. Campbell-Swinton had suggested as early as 1908. The second was the decision to adopt ultrahigh frequencies for television transmissions, despite the propagation difficulties. Since television transmitters must radiate an extremely broad bandwidth of frequencies in order to transmit pictures of high definition, it was clearly advantageous to use ultrahigh frequencies. The region between 40 and 100 MHz was selected for television purposes and in this country seven bands in that region were assigned by the Federal Communications Commission for experimental television. The overall progress during the decade is indicated by the fact that 30-line television images were being produced in 1928 (60 lines shortly after), whereas pictures of between 400 and 500 lines were being broadcast in 1940.

After 1930, television provided the chief commercial incentive for opening up the ultrahigh-frequency region and learning its properties. Tube and circuit development, propagation studies, and above all systematic and continued development of the cathode-ray oscilloscope were the chief dividends reaped by radar from television research during the first years. In the decade before the war, important work went forward in improving the fluorescent screen and studying the properties of phosphors, and in improving the mechanical structure of cathode-ray tubes. To this should of course be added accumulated experience in the design of scanning and synchronizing circuits. Much of this influence was only felt on the eve of the war and during the war when television engineers, like ionospheric investigators, were recruited for radar work.

A large literature very soon grew up describing the special techniques required for using the ultrahigh-frequency waves, especially relating to the design of vacuum tubes to serve as sources of UHF radiation and for use in receivers. Beginning about 1930 a series of important studies began to appear which systematically explored the required characteristics that should be sought in transmitter or receiver tubes for ultrahigh frequency. It was evident from these studies that the upper limit was rapidly being approached, beyond which conventional tubes could not be dragooned into operating successfully.[28]

There are inherent limitations in the conventional vacuum tubes that become increasingly critical as the frequency is raised. With ordinary triodes, oscillations from a few cycles per second up to some 20 or 30 MHz can be generated with little or no decrease in output or efficiency. Somewhere in the 10–60 MHz range the efficiency of tubes begins to show a sharp decline as the frequency is raised. This is illustrated in Fig. 3-1. Heating effects become prominent, and decrease in output power and plate efficiency, attributable to altered circuit characteristics and to the influence of the electron transit time, make their appearance. It is possible to compensate for decreased output by using more than one tube in a push–pull circuit of some sort. One can use the ordinary tubes to go down to 10 m (30 MHz); and with relatively small improvements they can be made to work in the neighborhood of 3 m. Rather extensive redesign, however, is necessary if triodes are to function below 3 m.

In the 1930s the chief electronic concerns were putting on the market tubes specially designed to function on the ultrahigh frequencies. The Eitel–McCullough company brought out for the use of amateurs rugged transmitter tubes capable of being used in the UHF region.[29] In December 1933, Thompson and Rose of RCA announced the experimental production of vacuum tubes of very small dimension, the so-called acorn tubes.[30] Within a year or so the RCA acorn tubes and the Western Electric "doorknob" tubes were on the market. They had a low transit time, short leads

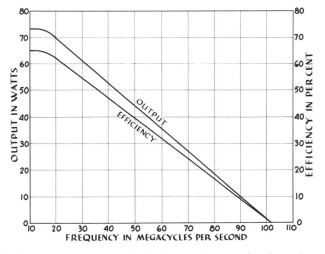

FIG. 3-1. Power output and efficiency of a standard triode as a function of frequency. (Kelly and Samuel, Electrical Engineering, 1934.)

with low inductance and capacitance, and very small interelectrode capacity.

These improvements disposed of the major obstacles that stood in the way of pushing down into the region between 3 and 1 m. Radio communication at these wavelengths, subject of course to certain limitations prescribed by the characteristics of the waves themselves, was thus clearly within reach. To go much below 1 m into the realm of the microwaves or centimeter waves was another story and clearly required a series of new departures or, one might fairly say, a whole new art. This need not concern us here, since the earliest radar did not presuppose proficiency in microwaves. We are consequently at liberty to postpone to a later chapter the story of early microwave research and to turn our attention to the elaboration of radio detection devices in the range between 10 m and $1\frac{1}{2}$ m since these were strongly influenced by the developments we have discussed so lengthily.

Observations on Reflection

With the shift of emphasis to the very long radio waves the phenomenon of reflection was no longer readily observed. While it would be absurd to describe such a fundamental fact of physics as having been forgotten, it was a fact not regularly encountered in everyday experience. There were, nevertheless, certain situations in which even the long waves showed these optical effects. Sir Henry Jackson, in the course of his pioneer experiments on the use of Hertzian waves for naval signaling, observed, and was the first to describe, the shielding effect that mountains and other large obstacles exerted on long radio waves.[31]

The reflection phenomenon came once more into prominence when work was resumed on ultrahigh frequencies during and just after the First World War. As we shall see, it was remarked on by Marconi and by A. Hoyt Taylor and his co-workers at the Naval Air Station. Marconi spoke of it in his New York address in 1922 (see below). Taylor referred to it as follows in an article that was published in QST early in 1924:

> Several years ago we made experiments with waves as short as 5 m with most interesting results. There, the production of standing waves and the influence of reflections, obstacles, etc., is very marked.[32]

Publication in such a journal as QST indicated that the phenomenon must henceforth be regarded as common knowledge among radio enthusiasts.[33] There are nevertheless numerous references in print coinciding with the revival of interest in ultrashort waves after about 1930. In that year W.J. Brown described it in connection with his experiments with waves of 2 m; it was reported by L. F. Jones when he described the RCA propagation experiments; Wenstrom commented upon it in his survey of the status of the ultrahigh-frequency problem[34]; and even earlier references exist. The phenomenon was also reported by engineers of the Bell Telephone Laboratories in a paper which has considerable historic significance as the first published account of the detection of radio energy reflected or reradiated from an aircraft. They spoke of the general phenomenon as being "well known," but emphasized their discovery that the effect "extends to unsuspected distances at times," notably in the case of reflections from aircraft, which they studied with care.[35] Identical but unpublished observations, as we shall see, had an important influence upon and led to the radio detection programs at the U.S. Naval Research Laboratory and in Great Britain.

It was only a step from these observations to the idea that this reflection could be used to detect the presence of objects in fog, darkness, or limited visibility. The earliest proposal of this sort that has come to light is the patent granted in several countries to an engineer of Düsseldorf, Christian Hulsmeyer, in 1904 for a collision prevention device. (See Fig. 3-2.) This consisted of a spark-gap transmitter sending out a directional beam of very short radio waves focused by means of a parabolic reflector located on the ship's mast. The echo is picked up by a second, receiving antenna also provided with a parabolic reflector. The signal is detected and amplified by a shielded receiver. The device was not pulsed.

No other proposal seems to have been made until after the First World War. On 20 June 1922 the celebrated father of radio, Guglielmo Marconi, was guest of honor at a joint meeting in New York of the Institute of Electrical Engineers and the Institute of Radio Engineers. The occasion was the presentation to Marconi of the IRE Medal of Honor in recognition of his work in wireless telegraphy. Marconi delivered a long address, published soon after, in which he described the communication experiments in which he had been engaged, using waves a few meters in length

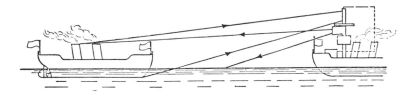

FIG. 3-2. Collision prevention device. (Hulsmeyer patent, 1904) British patent No. 13,170.

and directional transmitters. In an interesting passage at the close of his talk Marconi called attention to the phenomenon of reflected radio waves and suggested their use for radio detection:

> As was first shown by Hertz, electric waves can be completely reflected by conducting bodies. In some of my tests I have noticed the effects of reflection and deflection of these waves by metallic objects miles away.
>
> It seems to me that it should be possible to design apparatus by means of which a ship could radiate or project a divergent beam of these rays in any desired direction, which rays, if coming across a metallic object, such as another steamer or ship, would be reflected back to a receiver screened from the local transmitter on the sending ship, and thereby immediately reveal the presence and bearing of the other ship in fog or thick weather.[36]

The earliest experimental confirmation of these speculations took place in the early autumn of 1922 at the Naval Aircraft Radio Laboratory—later to become the Naval Research Laboratory—at Anacostia, D.C. The observers were Albert Hoyt Taylor and Leo Clifford Young. About the middle of September 1922 they began an informal investigation to explore the possibilities of 5-m waves for communication, and to discover the properties and propagation characteristics of these waves.

The equipment included a vacuum-tube transmitter with a tube which had never been used below 100 m, equipped with special circuits that enabled it to go below 5 m. This was modulated by a 500-Hz current applied to the plate. The antenna was nondirectional—which made the interference effects so much the more noticeable—and consisted of a 46-in. straight wire connected to the plate coil. The wavelength was determined by a Lecher wire method. A simple receiver with audio output picked up and amplified the signal from a half-wavelength wire receiving antenna.

The receiver was installed in an automobile and the first experiments were made on the grounds of the Naval Air Station, the transmitter being set up near the door of the laboratory, only a few feet above the ground. Almost at once interference effects were noticeable. As the car drove away from the transmitter and passed certain steel buildings from which some of the radiation was reflected, a very sharp series of maxima and minima were observed. The sound was now intensified and now weakened, so that a fluctuation was heard as the car moved away from the transmitter. This effect was strongly pronounced, and clearly resulted from the simultaneous reception of energy coming directly from the transmitter and energy reflected from the buildings. Since the path lengths were different the waves alternately tended to reinforce and to destroy one another. The frequency of the fluctuation varied with the speed of the car.

The shielding effect of other objects was also clearly evident. Such things as a passing automobile, a screen door, or a network of wires, such as the backstop of a tennis court, caused marked effects. Sometimes if signals were not received, moving the receiver a few feet would bring them in strongly. When the transmitter, in order to clear as many natural obstructions as possible, was installed on the compass house at the Air Station, an increased range was obtained, but reflections and interference patterns continued to be observed.

Experiments on the transmission of the energy over water were tried by leaving the transmitter on the compass house and driving the car with the receiver to Haines' Point across the Potomac from the station. The same interference effects were also noticed, and it was discovered that they came from clumps of willow trees near the receiver.

While these experiments were in progress and unobstructed signals were being received from the other side, the steamer *Dorchester*, a wooden vessel of no great size, passed down the channel. Fifty feet before the bow of the steamer crossed the line of vision between transmitter and receiver, the signals jumped to nearly twice the previous intensity. When the steamer actually passed across this line they dropped to half the normal value. Again when the stern of the vessel had passed 50 ft further downstream, the signals rose to normal intensity, then up to about twice normal, and then dropped back down again.

On September 27, a memorandum drawn up by Taylor was transmitted from the Commanding Officer of the Station to the Bureau of Engineering. The experiments described above were carefully recorded, and the memorandum spoke encouragingly about the possibility of directive communication on this wavelength, and of plans to repeat the experiments with directive antennas, which they thought might greatly increase the range. They also referred to the possibility of using such highly directive beams for landing aircraft at night or through overcast. More important, however, was their suggestion that the phenomena of reflection and interference could be used on board ship to detect other ships. The report said, "Possibly an arrangement could be worked out whereby destroyers located on a line a number of miles apart could be immediately aware of the passage of an enemy vessel between any two destroyers of the line, irrespective of fog, darkness, or smoke screen."[37] The writer observed that it was impossible to say whether the idea was a practical one at the present stage of the work, but that it seemed worthy of investigation. The memorandum requested that the Bureau approve the continuance of the 5-m work as an officially sanctioned project with a "problem number," the principal aim being to continue the initial experiments using directed radiation. No approval or encouragement, however, was forthcoming from the Navy Department and the 5-meter work was dropped. This can be considered the first proposal for the use of radio detection as a military device.

It was not until 1930 that the subject of radio detection raised its head once more, this time in print. In the July issue of the *Proceedings of the Institute of Radio Engineers* an English contributor wrote:

> It has already been suggested that icebergs, etc. might be detected by short-wave radio and it is possible, for instance, that this would be done by projecting a short-wave beam and observing whether any energy were reflected back to the source. A similar method might be employed for estimating the height of aircraft above ground.[38]

Pulse Ranging and the Pulse-Echo Principle

The earliest efforts to measure distance by the careful timing of pulses of energy were associated with the use of sound for depth measurement.[39] The first suggestion appears to have been made by Arago in 1807. In 1855 M.F. Maury, U.S.N., Superintendent of the National Observatory and pioneer oceanographer, attempted unsuccessfully to measure the ocean depths by detonating from the surface a charge of gunpowder on the bottom of the sea and determining the time it took the sound to be heard. This method seems to have been successfully used by Behn in measuring the depths of Lake Plön, Germany, in 1912. A more sophisticated application of this idea was the British Admiralty depth-sounding machine, where a steel hammer striking a steel plate in the bottom of a ship sent out a highly damped compressional wave; this was picked up by a hydrophone after reflection . In a similar device, the "fathometer" of Fessenden, the transmitter was an oscillator that sent out a short pulse of sound on a few thousand hertz. In the years following the First World War the principal maritime countries all announced successful trials of echo-sounding equipment, sonic or supersonic.

The *Titanic* disaster had led to Lewis Richardson's proposal that high-frequency supersonic beams could be used for depth measurement and obstacle detection at sea. Such devices became practical after the Langevin–Chilowsky system had been devised which produced supersonic beams at 30,000–40,000 Hz. Their use of the piezoelectric crystal found an important application in a supersonic pulse device for the detection of submarines and other submerged objects. In World War I development work on these devices went forward in America in the New York and San

Pedro groups under the National Research Council. Although it was developed too late to be of use in the war, a supersonic device that could detect submarines half a mile or more away had been built by the end of the war. The Navy continued work on this device between the wars. With many improvements, it became the modern sonar equipment of World War II. By a simple extension of the basic principle, supersonic altimeters for aircraft were soon thought of.

A large number of patent applications were filed between 1920 and 1940 and many articles were published on possible methods of using radio for distance measurement. This work had three separate ends in view: (1) point-to-point surveying and geodetic measurements by radio means: (2) altimeters for aircraft; and (3) exploration of the Kennelly–Heaviside layer or ionosphere. In each of these the use of pulses was proposed at one time or another.

About 1930 a program of research into the possibility of finding the distance between two fixed points by means of radio was undertaken in the Soviet Union.[40] Several different methods were considered by the Russians, including the use of pulses, but the method finally adopted by the Russians was as follows. Radio waves on a frequency f_1 in the neighborhood of 200–300 m in length are sent out from Station I. These waves are received by a Station II a number of miles away. This station replies on a frequency f_2 which bears some precise numerical relationship to f_1. At the first station the two frequencies are compared on a cathode-ray oscilloscope. The resulting Lissajous figure is determined by the transformation coefficient of the two frequencies and by the phase displacement undergone by the waves during the transit to and from the second station. Analysis of the figure permits the determination of this phase difference, though not without ambiguity. The ambiguity is eliminated by changing the frequency f_1 between known limits, keeping the frequency transformation coefficient constant at Station II and counting the number of phase reversals that take place. Using this method the Russians in 1934 made a series of 900 measurements in the North Caucasus, shooting between mountain peaks and from peaks to valleys. Measurements over salt water were made the next year on the Black Sea near Odessa and in 1936 between islands 44 km apart in the North Sea. Observations over fresh water were made across Lake Ilmen in 1934, a distance of 27 km.

Many proposals were made during the decade after World War I that reflected radio waves might be used to design aircraft altimeters to give, instead of the height above sea level as in the case of barometric altimeters, the height above the ground immediately below the plane. The patent literature before 1930 yields a wide variety of suggestions for utilizing the reflection of waves directed downward from the planes[41]: these included triangulation methods[42]; methods based on the variation of the frequency and output of a transmitter due to the effect of ground reflections[43]; methods based on the variation in radiation characteristics of an antenna as a function of the height above the earth[44]; and a wide variety of methods based on interference effects between the transmitted wave and the reflected wave, such as the use of standing waves, and the determination of a phase difference between the two waves.[45] Of considerable greater interest, and in general of greater simplicity, were devices depending on simple measurement of the time required for radiation to travel from a transmitter to a reflecting surface and back. It is here that we find a number of paper patents which closely resemble or actually anticipate the pulse method.

The earliest disclosure of this sort appears to be that of an Austrian, Heinrich Löwy, in a patent application filed in July 1923.[46] The patentee describes his proposal as the electric counterpart of Fizeau's method[47] for determining the velocity of light. In Löwy's scheme an electronic switch is used to key a transmitter alternately on and off and synchronously to block and unblock a receiver, so that while the transmitter is sending out a short pulse, the receiver is off; and when the transmitter is shut off, the receiver is unblocked and ready to receive the echo signal. The blocking and unblocking is produced by an imposed sinusoidal modulation, the frequency of which can be varied, but which is on the order of broadcast frequencies. In radar parlance, this device would have a pulse width roughly equal to the resting time. When the modulation frequency is chosen so that the half-wavelength corresponds to twice the distance to the reflecting object, then the situation corresponds to Fig. 3-3(a). The reception of the reflected energy begins immediately at the end of the transmission. When the distance is less than a quarter of the modulation wavelength, the situation corresponds to Fig. 3-3(b). When it is greater it corresponds to Fig. 3-3(c), i.e., the returning echo reaches the receiver while the transmitter is still operating. Conversely, when the distance is greater, the received echo returns some time after the cessation of the transmitter, and continues to arrive after the receiver has been shut off. The output of the receiver is put on a meter or a gas discharge tube as indicator. Since the maximum signal indicated above is received only when the modulation frequency is properly chosen, varying this frequency within convenient limits makes it possible to determine the altitude. The essential difference between this pulse device and radar should be evident.

In May 1930 two interesting patent applications were filed by Ezekiel Wolf and by Robert W. Hart and assigned to the Submarine Signal Company of Boston, Massachusetts. Both of these applications were for schemes of distance measurement by reflected radio waves, principally for altimeters for aircraft, and both provided for the use of pulses of energy.[48]

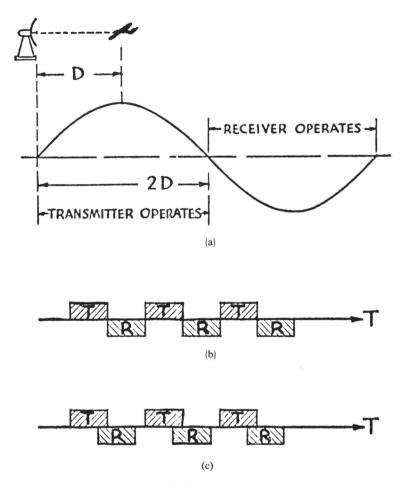

FIG. 3-3. Löwy's pulse method. Transmitted pulses are represented by the time axis, with the received below.

Wolf's patent applied to continuous waves as well as pulses. In his pulse scheme of distance measurement high-frequency waves, 5 m or less in length, would be sent out as discrete impulses at regular intervals. The interval is determined by a keying or modulating circuit, operating on a frequency corresponding to a wavelength equal to twice the greatest distance at which the device would be used. The chief feature of Wolf's patent is the method of timing the reflected wave. By means of a delay network one of the waves, usually the direct wave, is retarded and is brought into coincidence with the other. A gas tube indicator responds only when the direct and reflected impulses are brought into phase by the retardation.

The variable elements of the network can be driven through a cycle by a motor, and the neon tube is then operated periodically.

Hart's patent is for a pulse device corresponding in all essentials to that devised a few years before by Gregory Breit and Merle Tuve, which we shall discuss below. Though the language tends to obscure the fact—for example, he speaks of "trains of vibration" instead of pulses, and speaks of these trains as being produced by "overmodulation"—his patent is clearly for a pulse-echo altimeter. An oscillator for the production of high-frequency waves is fed to an amplifier biased at cutoff which is keyed or pulsed by a modulator feeding the plate circuit of the amplifier. The modulating frequency is a radio broadcast frequency. The pulses from the transmitter are reflected from the ground or other object and the echo is picked up by the receiver and is put on a cathode-ray tube with a circular trace. The cathode-ray trace makes one complete rotation for each cycle of the modulation frequency. The trace can be easily calibrated and the position of the "momentary serration" read as the distance to the ground or other target. It differs in no important respect from a nondirectional pulse altimeter.

It was the third purpose for which radio distance measurement was applied which made the pulse-echo method well known and widely employed throughout the scientific world. This was its use for studying the properties of the *ionosphere*, the ionized region of the upper atmosphere of which the lower layer is called the Kennelly–Heaviside layer, after the two workers who independently inferred its existence.

Marconi's dramatic success, on December 12, 1901, in receiving signals transmitted across the Atlantic from Cornwall to Newfoundland, led to speculations as to the mechanism which made this possible. Arthur Edwin Kennelly in America and Oliver Heaviside in England independently suggested within a short space of time that these waves were propagated around the curvature of the earth with the assistance of a conducting surface in the upper atmosphere; the waves moved, in Kennelly's words, "horizontally outwards in a 50-mile layer between the electrically reflecting surface of the ocean beneath, and an electrically reflecting surface, or successive series of surfaces, in the rarified air above."[49] This is illustrated in Fig. 3-4. The proposals of Kennelly and Heaviside were casual and descriptive. A quantitative theory was worked out by Eccles in 1912 based on the assumption that the waves were refracted by the ions in the layer. Although indirect evidence of various kinds began to accumulate in support of the theory of the ionized layer, experimental verification was delayed until interest was once more kindled by the extraordinary long-distance performance of 100-m waves.[50]

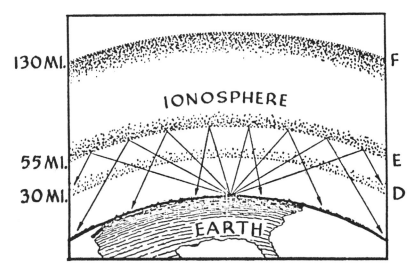

FIG. 3-4. Layers of the ionosphere, and reflection from the E layer.

The outstanding achievements of the amateurs in 1923–24 in spanning the world with shortwave links led various people to revive the earlier proposal of a conducting layer.[51] The interest was further heightened by the discovery during 1924 of what has been called the "skip" distance. Early in the year John L. Reinartz, one of the best-known radio amatures—he had shared with Schnell the distinction of receiving Léon Deloy on that historic night in November 1923, when communication had first been established on 100 m—began a series of long-range tests using wavelengths between 20 and 60 m. In his analysis of these results he made the observation that, with these very short waves, certain signals became weaker after sunset (instead of stronger, as experience with waves above 100 m had led him to expect) and that surprisingly enough these same signals were audible at great distances. Thus there seemed to be what he called a "dead belt" lying between two regions of good reception. This Reinartz described as lying between the region covered by the ground-wave and that reached by a skywave returned from the ionosphere.[52] These results were confirmed, from an analysis of the transmissions from the Naval Research Laboratory, by A. Hoyt Taylor,[53] who gave the dead belt the name of "skip" or "miss" region. With E. O. Hulburt, Superintendent of the Heat and Light Division of the Naval Research Laboratory, Taylor published a modified Eccles–Larmor theory to account for the new phenomenon.[54] It was clear that these very short waves were propagated for short distances in the conventional manner of the longer waves; but in addition to this ground ray, which quite rapidly became attenuated, skywaves reached great distances by being sharply refracted by the ionosphere. (See Fig. 3-4.) In regions where both the groundwave and a skywave are received, violent fading is encountered, due to the interference effects between the two waves.

The stage was now set for the experimental proof of the existence of the layer and for a systematic study of its properties. The first papers on this subject were published in 1925 by Smith-Rose and Barfield,[55] who sought to apply direction-finding methods on the assumption that there should be a detectable difference in angle between the electric (and of course between the magnetic) vectors of the groundwave and the skywaves. The results were inconclusive, though by the use of shorter waves and other improvements, they were able the following year[56] to confirm the results reported in the interim by two other English workers, E. V. Appleton and M. A. F. Barnett.

Appleton and Barnett were the first persons to use reflected radio waves for the radio-location of a distant surface and the determination of its range.[57] Their method was an adaptation of techniques used in Appleton's previous work on radio signal fading and involved intensity measurements instead of direction-finding methods. They believed that for wavelengths from 300 to 500 m there was a point about 100 miles from a given transmitter where the strength of the groundwave and the skywave would be sufficiently comparable for strong interference effects to be produced. By varying uniformly the frequency of the transmitter by a known small amount in a brief time—what we now refer to as frequency modulation—it was thought possible to observe directly the interference effects at that point and to determine the distance of the reflecting layer.[58] From two sets of experiments performed on the nights of December 11, 1924 and February 17, 1925, using the British Broadcasting Company's transmitter at Bournemouth, Appleton and Barnett obtained interference effects which were the first direct evidence of the existence of the reflecting layer. From these observations they estimated the layer to be situated at a height of 80–90 km.

An unambiguous, and in many respects simpler, way of exploring the reflecting layer by radio involves sending out the radio energy in discontinuous bursts or pulses and measuring the time for the returning echo to reach a receiver situated near the transmitter. Two American investigators, Merle A. Tuve and Gregory Breit, who undertook an ionospheric investigation without being aware of the similar work going forward in England,[59] were the first to suggest this method and the first to put it to use.

Preliminary notes describing the proposed method were published in the course of 1925.[60] In their first paper the authors acknowledge that their proposed method grew out of a method which had been thought of and to some extent tried out at the University of Minnesota by W. F. G. Swann and J. G. Frayne. This method was similar to the one in Löwy's altimeter patent discussed earlier, and consisted of an electrical shuttering scheme to block the receiver when the transmitter was sending and unblock it for the reception of the echo. Breit and Tuve offered a simplification. It was proposed to send out pulses of such a length and repetition rate that at a receiving station some distance away the pulses from the groundwave and the skywave could be distinguished. From the delay separating the two received pulses, and knowing the distance separating the two stations, the height of the ionosphere could be estimated. As proposed in the first paper the arrangement would be as shown in Fig. 3-5, where A_1 is the transmitting station, A_2 the receiving station, l the distance between the stations, and h the height of the receiving layer.

It was decided to rely upon existing shortwave transmitters, and arrangements were made with the Naval Research Laboratory for the use of their station NKF, with Westinghouse for the use of their pioneer station KDKA, and also with RCA and the Bureau of Standards.[61] The most definite results were obtained with the Navy's transmitter, partly because of the advantageous location of the two laboratories, and partly because the Navy transmitter was crystal controlled and had high stability. The workers received assistance from Gebhard, Schrenk, and Young at NRL in modifying the transmitter for pulse work and in building a receiver to

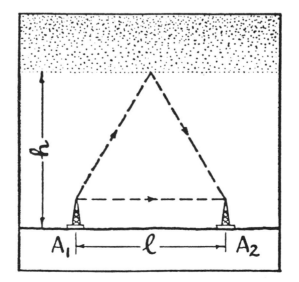

FIG. 3-5. Measuring the height of the ionosphere.

pick up the pulses. Experiments were begun on July 28, 1925, with the receiver located at the laboratory of the Department of Terrestrial Magnetism, 8 miles southeast of the Naval Research Laboratory. The transmitter sent out pulses of energy on 71.3 m, the pulses lasting about .001 s. The pulsing or modulation was produced by applying an ac frequency of about 500 Hz to the plate circuit of the transmitter. The received pulses, recorded by an oil-immersed General Electric oscillograph, were observed visually by means of a rotating mirror and also photographed. By measurement of the time delay from their photographs, Breit and Tuve estimated the height of the ionosphere as between 50 and 130 miles. Additional observations showed that reflections could be obtained with waves of about 40 m in length, but that no reflections were obtained with waves 20 m or less in length. The reflections apparently varied with the time of day and with the seasons. The fact that the reflections varied was of great importance, for it proved that the echoes could not come from ground reflections. For some days Breit and Tuve were troubled by the possibility that their nice echoes were attributable, not to the Kennelly–Heaviside layer, but to the Blue Ridge Mountains, which were at about the proper distance.[62]

This method of Breit and Tuve was widely adopted as the simplest and most direct approach to the problem. The British workers shortly abandoned the frequency-modulation method as too cumbersome. A worldwide network of investigators was soon busy exploring the upper atmosphere by radio methods and building up increasingly detailed knowledge of the fine structure of the ionosphere. An extensive literature has appeared in this extremely important field.[63] The ionosphere was discovered to consist, not of a simple region of ionization as was thought at first, but of two or three layers, depending upon whether it is day or night.

A succession of technical improvements in the pulse-echo method was introduced without altering the fundamental method. Two of these in particular deserve mention. The first was an improvement in the method of pulsing the transmitter. In 1928 Tuve and Dahl described an improvement suggested somewhat earlier by Breit for pulsing the transmitter by a better method than by means of the 500-Hz modulating emf, which had produced pulses very close together, separated by an interval scarcely greater than the pulse length.[64] This had led to difficulty in distinguishing multiple reflections, and poor resolution for small distances. The new method involved using the multivibrator circuit of Abraham and Block, later to play an important part in both television and radar, as a basic pulse-forming circuit.[65] With this important innovation it was possible to get shorter pulses (one or two ten-thousandths of a second) separated by relatively long intervals (a few hundredths of a second).

Of almost equal importance was the abandonment previously mentioned of the mechanical oscillograph and the adoption of the cathode-ray tube. Georg Goubau, in 1930, reported the use of a cathode-ray tube with a circular or elliptical trace, synchronized with the pulse cycle.[66] In 1931 Appleton reported simultaneously the adoption of the cathode-ray tube with linear trace as indicator for ionospheric experiments and the use of a self-pulsing or "squegging" transmitter.[67] Except for the frequencies employed, the chosen pulse lengths and intervals between the pulses, and the absence of directive antennas, the apparatus used for ionospheric work by the early 1930s bore considerable resemblance to the radar systems. It was the ionospheric investigations, moreover, which spread abroad most effectively knowledge of the pulse-echo principle, and reduced the art of radio locations by pulses to the level of daily experience.

Radar was developed by men who were familiar with the ionospheric work and the methods involved. It was a relatively straightforward adaptation for military purposes of a widely known scientific technique. This is not said to detract from the honor that goes to those who developed this new device. It goes a long way to explain the fact that this adaptation—the development of radar—took place simultaneously in several different countries.

NOTES

1. W. S. Tucker, "Direction finding by sound," Nature (London) Suppl. **138**, 111–118 1936); Vigneron, "Problèmes de la défense antiaerienne, Nature (Paris) **15**, 540–547 (1936); E. T. Paris, "Binaural sound-locators," Sci. Prog. **27**, 457ff (1932). For the Sperry locator see the Coast Artillery J. **74**, November–December (1931), and P. R. Bassett, Coast Artillery J. **75**, 200 (1932).

2. E. H. Armstrong, "Vagaries and elusiveness of invention," Electr. Eng. **62**, (4), 149 (1943).

3. S. O. Hoffman, "The detection of invisible objects by heat radiation", Phys. Rev. **14**, 163–166 (1919).

4. William Sackville and Jose E. Olivares, "An aircraft ranging device," M.S. thesis, Department of Electrical Engineering, Massachusetts Institute of Technology, 1926 (unpublished). An unclassified version is available at the University, and a complete version is in the Department of Defense.

5. "Systems of echo sounding," Nature (London) **135**, 896–897 (1935).

6. Sir Hiram Stevens Maxim, *A New System for Preventing Collisions at Sea* (Cassell, London, 1912).

7. Donald R. Griffin and Robert Galambos, "The sensory basis of the obstacle avoidance by flying bats," J. Exp. Zool. **86** (3), (1941).

8. Heinrich Hertz, *Untersuchungen über die Ausbreitung der elektrischen Kraft* (Barth, Leipzig, 1892).

9. Probably the best summary of early developments in radio is G. W. Pierce's old *Principles of Wireless Telegraphy* (McGraw-Hill, New York, 1910). See also William H. Wenstrom, "Historical review of ultra-short-wave progress," Proc. IRE **20**, 95–112 (1932),

especially pp. 96–98 for spark oscillators; E. Pierret: "Les ondes electriques ultra-courtes (production, application)," Onde Electr. **8**, 373–410 (1910), which has a brief historical summary and an excellent bibliography of the early work.

10. Augusto Righi, *L 'Ottica delle Oscillazioni Elettriche* (Zanichelli, Bologna, Italy, 1897). Ignaz Klemencic studied the reflection of microwaves from plates made of metal and of sulfur ["Über die Reflexion von Strahlen electrischer Kraft an Schwefel und Metalplatten," Ann. Phys. Chem. **45**, 62–79 (1892)]; A. S. Cole, working at Berlin, successfully applied Fresnel's equations for optical refraction to measurements of the refraction of 5-cm waves by water and alcohol ["Über den Brechungsexponenten und das Reflexionsvermögen von Wasser und Alcohol fuer elektrischen Wellen," Ann. Phys. Chem. **57**, 290–310 (1896)].

11. P. Lebedev, "Über die Doppelbrechung der Strahlen elektrischen Kraft," Ann. Phys. Chem. **56**, 1–17 (1895).

12. W. B. Eccles, *Wireless* (Butterworth, London, 1933), p. 216.

13. Report of the Chief Signal Officer, Fiscal Year Ended June 30, 1919, pp. 272, 288–289.

14. See below, Chap. 4.

15. These results are described by C. S. Franklin, "Short-wave directional wireless," J. Inst. Electr. Eng. **60**, 930ff (1922), and "Short-wave directional wireless telegraphy," Wireless World Rad. Rev. **10**, 219-225 (1922); by G. Marconi, "Radio telegraphy," Proc. IRE **10**, 215–238 (1922).

16. The story of the accomplishments of the radio "hams" is given in Clinton B. DeSoto, *Two Hundred Meters and Down* (American Radio Relay League, West Hartford, CN, 1936), the official history of the American Radio Relay League.

17. The French amateur organization, *La société des Amis de la T.S.F.*, was founded immediately after the First World War (1922) with the personal encouragement of General Ferrie, one of France's great radio pioneers; cf. R. Jouast: "General Ferrie," Onde Electr. **11**, 45–52 (1932), and C. Gutton, "Dix années de T.S.F., 1922–1932, les dix premieres années de la Societe des amis de la T.S.F., et de la Revue 'L'Onde Electrique'," Onde Electr. **11**, 397–404 (1932).

18. QST **8** (1) 9–12 (1924). For the French account see Léon Deloy, "Premiere communication transatlantique bilaterale entre postes d'amateurs," Onde Electr. **2**, 678–683; see also his "Communications transatlantiques sur ondes de 100 metres," Onde Electr. **3**, 38–42 (1924).

19. A. Hoyt Taylor, "The Navy's work on short waves," QST, **8** (5), 9–14 (1924).

20. QST, **8** (9), 24 (1924) gives the new regulations as summarized in the letter of 24 July from the Department of Commerce.

21. G. W. Southworth: "Certain factors affecting the gain of directive antennas," Proc. IRE. **18**, 1502–1536 (1930). This paper has a useful bibliography. E. J. Sterba, "Theoretical and practical aspects of directional transmitting systems," Proc. IRE **19**, 1184–1215 (1931). E. Bruce, "Developments in short-wave directive antennas," Proc. IRE **19**, 1406–1433 (1931). See Also A. E. Harper, *Rhombic Antenna Design* (Van Nostrand, New York, 1941); E. Bruce, A. C. Beck, and L. R. Lowry, "Horizontal rhombic antennas," Bell Syst. Tech. J. **14**, 135–158 (1935).

22. H. H. Beverage, H. P. Peterson, and C. W. Hansell, "Application of frequencies above 30,000 kilocycles to communication problems," Proc. IRE **19**, 484–485 (1931). L. F. Jones, "A study of the propagation of wavelength between three and eight meters," Proc. IRE **21**, 349–386 (1933). The airplane observations were made by Bertram Trevor and are described by Bertram Trevor and P. S. Carter, "Notes on propagation of waves below ten meters in length," Proc. IRE **21**, 387–426 (1933). It is interesting that ultrashort-wave broadcasting

(ca. 7 m) was experimentally begun in Germany. Cf. Wenstrom, *op. cit.* Note 9, p. 103. J. C. Schelling, C. R. Burrows, and E. B. Ferrell, "Ultra-short-wave propagation," Proc. IRE **21**, 427–463 (1933). Carl R. Englund, Arthur B. Crawford, and William W. Mumford, "Some results of a study of ultra-short wave transmission phenomena," Proc. IRE **21**, 464–492 (1933).

23. Abraham Esau and Walter H. Hahnemann, "Report on experiments with electric waves of about 3 meters: Their propagation and use," Proc. IRE **18**, 471–489 (1930). In 1927 Kruse and Phelps built various oscillators from 2.36 to 0.41 m. Successful beam transmission was reported on 474 m by Yagi in 1928. In the same year M. Ritz in France conducted 3-m transmission experiments with a small portable radiotelephone. Cf. Wenstrom, *op. cit.* Note 9, pp. 101–102.

24. The organ of the American Radio Relay League.

25. Ross A. Hull, "Air-mass conditions and the bending of ultra-high-frequency waves," QST **19**, (6), 13–18, 74, 76 (1935).

26. J. B. Johnson, "The cathode ray oscillograph," J. Franklin Inst. **212**, 687–717 (1931).

27. Sidney A. Moseley and H. J. Barton-Chapple: *Television Today and Tomorrow*, 5th ed. (Pitman, New York, 1940); V. K. Zworykin and G. A. Morton, *Television: the Electronics of Image Transmission* (Wiley, New York, 1940).

28. E. Marplus, "Communication with quasi-optical waves," Proc. IRE **19**, 1715–1730 (1931); Westrom, *op. cit.* Note 9, and "An experimental study of regenerative ultra-short-wave oscillators," Proc. IRE **20**, 113–130 (1932); E. D. McArthur and E. E. Spitzer, "Vacuum tubes as high-frequency oscillators," Proc. IRE **19**, 1971–1982 (1931) (brief bibliography, with some significant additions to Wenstrom's); I. E. Mouromtseff and H. V. Noble, "A new type of ultra-short-wave oscillator," Proc. IRE **20**, 1328–1344ff (1932); F. E. Fay and A. L. Samuel, "Vacuum tubes for generating frequencies above 100 megacycles" (Abstract), Proc. IRE **22**, 679 (1934). For an illuminating general survey of this state of affairs where experimental knowledge has outstripped theory, see M. J. Kelly and A. L. Samuel, "Vacuum tubes as high-frequency oscillators," Electr. Eng. **53**, 1504-1517 (1934) (bibliography; see especially Refs. 6, 25, and 26), also published in Bell Syst. Tech. J. **14**, 97–134 (1935). An attempt was made to place the theory of vacuum tube oscillators at ultra-high-frequencies on a sound basis. The study was inaugurated by Benham in 1928 when he explored the effect of electron transit time on the behavior of a diode at moderately high frequencies. Cf. W. E. Benham, "Theory of the internal action of thermionic systems at moderately high frequencies," Part I, Philos. Mag. **5**, 641–642 (1928), and Part II, Philos. Mag. **11**, 457–517 (1931). An extension of the Benham treatment to include triodes with a positive or negative grid was made by F. B. Llewellyn, "Vacuum tube electronics at ultra-high frequencies," Philos. Mag. **21**, 153–173 (1933), also printed in Bell Syst. Tech. J. **13**, 59–101 (1934). See also his "Note on vacuum tube electronics at ultra high frequencies," Philos. Mag. **23**, 112–127 (1935), and "Operation of ultra-high-frequencies vacuum tubes," Bell Syst. Tech. J. **14**, 632–665 (1935). O. J. Pakker and G. DeVries, "On vacuum tube electronics," Physica **2**, 83–97 (1935), advanced a now generally accepted theory of transit-time effects within a triode, and confirmed this theory experimentally, showing for the first time a negative resistance caused by finite transit time, at frequencies low compared with the reciprocal of the transit time.

29. See below, p. 91.

30. B. J. Thompson and G. N. Rose, "Vacuum tubes of small dimensions for use at extremely high frequencies," Proc. IRE **21**, 1707–1721 (1933). See also B. Salzberg, "Design and use of 'acorn' tubes for ultra-high-frequencies," Electronics **7**, 282–283 and 293; B. Salzberg and D. G. Burnside, "Recent developments in miniature tubes," Proc. IRE **23**, 1142–1157 (1935).

31. H. B. Jackson, "On some phenomena affecting the transmission of electric waves over the surface of the sea and earth," Proc. R. Soc. London 70, 254–272 (1902). This is the paper in which Jackson first demonstrated the role of the Earth in the transmission of radio waves.

32. Taylor, *op. cit.* Note 19, p. 10.

33. The effect was looked for shortly after by L. D. Grignon, S. M. Hudd, and Frank C. Jones, who made some pioneer observations on 3 m. They did not find the effect as large as expected. See Frank C. Jones, "Pioneer short-wave work," QST 7 (5), 8–14 (1923).

34. W. J. Brown, "Ultra-short waves for limited range communication," Proc. IRE 23, 1129–1143 (1930), especially pp. 1137–1138; L. F. Jones, Proc. IRE 21, 362–363 (1933); Wenstrom, *loc. cit.* Note 9.

35. Carl R. Englund, Arthur B. Crawford, and William W. Mumford, "Some results of a study of ultra-short-wave transmission phenomena," Proc. IRE 21, 1 (1933).

36. Guglielmo Marconi, *op. cit.* Note 15, p. 237.

37. "*Naval Personnel*," June 1943 (unpublished), p. 10. A copy of this report is in the Naval Research Laboratory, File C-267-5 Radar 1.

38. W. J. Brown, *op. cit.* Note 34, p. 1142.

39. In the 17th century, Père Marin Mersenne measured the velocity of sound by what was tantamount to a crude pulse-echo method. He used a pendulum to time the echo. Similarly Fizeau's measurement of the velocity of light in 1849 is an application of the pulse method, in reverse (as it were). His method consisted of transmitting a beam of light between the teeth of a rapidly rotating toothed disk, and returning it from a mirror through the toothed disk once more, where it was observed in the eyepiece of a telescope. The disk acted as a high-speed shutter that "pulsed" the transmitted light, allowing it to pass through on its return only if the light pulse had traversed the distance to the mirror and back in the time required for the tooth to be replaced by the adjacent groove (or an integral multiple of that time). Since the distance traversed could be readily measured and the elapsed time was known from the angular velocity of the disk, the velocity of light could be determined at once. The Fizeau apparatus was an optical "radar" system in which the distance was known and the velocity of light was unknown.

40. The results are summarized in English in Tech. Phys. USSR 4 (10) (1937).

41. This literature is touched upon in C. D. Tuska, "Historical notes on the determination of distance by timed radio waves," J. Franklin Inst. 237, 1–20 and 83–102 (1944). An excellent survey of the problem is to be found in the article by K. Dziewior in *Luftfahrtforschung*, No. 16, 326–338 (June 1929). I am indebted to Mr. E. M. Deloraine of the International Telephone and Telegraph Co. for calling my attention to this article and for putting in my hands an excellent unpublished survey of the literature on distance measurement and obstacle detection: Note Technique No. 32, Documentation sur la Detection d'Obstacles, Les Laboratoires L. M. T., January 1940 (unpublished).

42. Eaton, U.S. Patent No. 1,854,122, filed 16 March 1929; Ewald (assigned to Telefunken), U.S. Patent 1,965,632, filed 23 September 1929.

43. A number of different patents, perhaps the earliest being H. Löwy, Brevet Français No. 574,969, filed 20 November 1922.

44. Patterson, U.S. Patent No. 2,022,517, filed 25 June 1931.

45. Jenkins, U.S. Patent No. 1,585,591, filed 18 April 1924; Espenshied, U.S. Patent Nos. 2,045,071 and 2,045,172, filed 29 April 1930; Bentley, U.S. Patent No. 2,011,392, filed 10 August 1928; Levy, U.S. Patent No. 1,844,859, filed 1 April 1927 in the U.S. and 3 April 1926 in France; Turner, U.S. Patent No. 1,982,271.

46. H. Löwy, U.S. Patent No. 1,585,591, filed 17 July 1923. There is a discussion of the Löwy patent in Tuska, *op. cit.* Note 41, p. 11.

47. Cf. above, Note 39.

48. E. Wolf, U.S. Patent No. 1,924,174; R. W. Hart, U.S. Patent No. 1,924,156; both filed 19 May 1930.

49. A. E. Kennelly, "On the elevation of the electrically conducting strata of the Earth's atmosphere," Electr. World Eng. **39**, 473 (1902). Heaviside's suggestion that a single layer existed was written three months later and published in December 1902, in the article "Telegraph theory," *Encyclopedia Brittanica*, 10 ed., Vol. 13, p. 215.

50. Leonard J. Fuller, "Positive evidence of ionosphere," Western Electrician (1912).

51. Taylor, *op. cit.* Note 19. A. E. Kennelly, "The conference, in relation to amateur activities," QST, **8** (12) (1924). Sir Joseph Harmor, "Why wireless electric can bend round the Earth," Philos. Mag. **48**, 1025–1036 (1924).

52. John L. Reinartz, "The reflection of short waves," QST, **9** (4), 9–12 (1925). According to R. Jouast ["Some details relating to the propagation of very short waves," Proc. IRE **19**, 479–480 (1931), and "Les ondes tres courtes," Onde Electr. **9**, 8 (1930)], the discovery of the skip distance was anticipated by Colonel Chaulard of the French Army.

53. A. Hoyt Taylor, "Investigation of transmission on the higher radio frequencies," Proc. IRE **13**, 667–683 (1925).

54. A. H. Taylor and E. O. Hulburt, "The propagation of radio waves," Phy. Rev. **27**, 189–215 (1926).

55. R. L. Smith-Rose and R. H. Barfield, "On the determination of the direction of forces in wireless waves at the Earth's surface,." Proc. R. Soc. London Ser. A **107**, 587 (1925), and "Some measurements on wireless wave fronts," Wireless Eng. **2**, 737 (1925).

56. R. L. Smith-Rose and R. H. Barfield, "An investigations of wireless waves arriving from the upper atmosphere," Proc. R. Soc. London Ser. A **110**, 580–614 (1926).

57. Appleton's earliest report, according to a citation by H. R. Mimno, "The physics of the ionosphere," Rev. Mod. Phys. **9**, 1–43 (1937), is to be found in Tijdschr. Ned. Radiogenotsch. **2**, 115 (1925), which I have not consulted. The account I have used is the article of E. V. Appleton and M. A. F. Barnett, "On some direct evidence for downward atmospheric reflection of electric rays," Proc. R. Soc., London Ser. A **109**, 621–641 (1925), which is the first extended account.

58. The method is somewhat too complicated to describe here, but it can be readily understood that interference effects with both transmitter and receiver fixed can only be observed if the frequency is shifted.

59. The general proposal of studying the reflecting layer by radio was due to Breit. Early in 1924 he proposed to his friend Tuve, then an instructor in physics at Johns Hopkins University, that they collaborate in an attempt to transmit radio waves from the Carnegie Institution's Terrestrial Magnetism Laboratory to Baltimore by reflecting them from the ionosphere. It was Breit's original plan to direct short waves skyward by means of a large parabolic reflector of wire netting, about 100 ft in diameter, and have the skywave, as we now call it, received by Tuve at Johns Hopkins. Since the proposal of having the Terrestrial Magnetism Laboratory embark on radio research was something of a novelty, the possibilities were discussed at a meeting, held in November 1934, at which leading radio experts in Washington, like L. W. Austin and A. Hoyt Taylor of the Naval Research Laboratory, and others were called in for consultation. It was during a discussion immediately after this meeting during dinner at Breit's house that Tuve proposed to Breit that they try to use pulses along the line previously suggested by W. F. G. Swann and J. G. Frayne at the University of Minnesota. At the close of the spring term in 1929 at Johns Hopkins, Tuve came to Washington and joined his colleague in these historic experiments.

60. M. A. Tuve and G. Breit, "Note on a radio method of estimating the height of the

conducting layer," Terr. Magn. Atmos. Electr. **30**, 15–16 (1925). See also G. Breit and M. A. Tuve, Nature (London) **116**, 357 (1925).

61. G. Breit and M. A. Tuve, "A test of the existence of the conducting layer," Phys. Rev. **28**, 554–575 (1926).

62. M. A. Tuve, personal communication.

63. A general survey of the progress of ionospheric investigations is found in L. V. Berkner, in *Physics of the Earth*, edited by J. A. Fleming, Terrestrial Magnetism and Electricity, Vol. 8 (McGraw–Hill, New York, 1939), pp. 434–491; and in Mimmo, *op. cit.* Note 57. Some discussion of the various methods used is to be found in Tuska, *op. cit.* Note 41.

64. M. A. Tuve and O. Dahl, "A transmitter modulating device for the study of the Kennelly–Heaviside layer by the Echo method," Proc. IRE **16**, 794–798 (1928).

65. For a description of this circuit, see Chap. 4, p. 81.

66. Georg Goubau, "Eine Methode für Untersuchung von Echos bei des Ausbreitung elektromagnetischer Wellen in der Atmosphare," Phys. Z. **31**, 333–334 (1930).

67. E. A. Appleton and G. Builder, "A simple method of investigating wireless echos of short delay," Nature **127**, 790 (1927).

CHAPTER 4

Early Radar Research in the U. S. Navy

The purpose of this chapter is to describe America's first sustained radio detection development, which was begun in 1930, and which culminated in the production of the earliest successful pulse radar devices to be manufactured in this country. The work we are about to describe took place at the U. S. Naval Research Laboratory, the contributions of whose Radio Division to the development of shortwave radio communication we had occasion to mention in the previous chapter.

The Naval Research Laboratory is in many ways unique as a military research organization, chiefly for the relative independence which its civilian investigators enjoy. In contrast to Army laboratories where the atmosphere is likely to be that of a military establishment, where an officer is assigned to each project and the civilian investigator is subject to direct supervision, at the Naval Research Laboratory there is no participation in the scientific work by the assigned Naval officers. These officers are responsible for the overall administration of the Laboratory, and are charged with the conduct of all official Navy business, but do not control the research policy. In the period with which we are mainly concerned, the scientific policy at the Naval Research Laboratory was laid down in broad outlines by the Bureau of Engineering and executed by the civilian personnel. The Bureau had the power of the purse, for it supported the Laboratory from the funds it received from the annual appropriations. There was, in addition, a small direct appropriation from Congress, not passing through the hands of the Bureau of Engineering.

Founding of the Naval Research Laboratory

Before the First World War the United States Navy had no center of research and development, a fact that was officially deplored by farsighted officers like Admiral Melville and Admiral Bradley Fiske, both of whom pointed to the influence of scientific research institutions upon Germany's strides in naval armament. The idea of a Naval Research Laboratory dates from the First World War and is directly attributable to the initiative and prestige of Thomas A. Edison. In 1915 *The New York Times* published an interview in which the famous inventor expressed his opinion that the government should maintain a great research laboratory jointly under military, naval, and civilian control. This proposal still represents, in broad outline, what many persons still feel to be the ideal solution to the problem of fundamental military research. The suggestion aroused the interest of the Secretary of the Navy, Josephus Daniels, who was being harassed by the flood of ill-digested and well-intentioned proposals that were deluging the bureaus of the Department. Daniels wrote to Edison to enlist both his support and his prestige for a plan of creating a department of inventions and development to which these ideas could be referred.

Out of a visit to Washington of Edison's personal representative, M. R. Hutchinson, and a discussion at East Orange between Edison and the Secretary, sprang the proposal for the Naval Consulting Board. The activities of this body, of which Edison was the nominal head, have been recorded in some detail.[1] The Board was at first asked only to investigate, test, and improve inventions submitted to the Navy; later this was broadened to include inventions submitted to any government department. The Board was composed almost exclusively of industrial engineers who had been proposed by the principal engineering societies of the United States, and included such names as L. H. Beeksland, Hudson Maxim, Elmer A. Sperry, and W. R. Whitney of the General Electric Company.

The greatest achievement of the Naval Consulting Board was the creation of the Naval Research Laboratory, a proposal due to the initiative of Edison himself. After the formation of the Naval Consulting Board, and when Edison had become familiar with the Navy's problems, he revived the idea of a laboratory, this time for the Navy alone, and sent Hutchinson to discuss the matter with Secretary Daniels.

The credit for suggesting the laboratory goes to Edison; but the actual plans as to the nature and scope of the work of the laboratory were drawn up by a Committee of the Consulting Board. This Committee, headed by Edison, visited Congress and in 1916 obtained an appropriation for $1,000,000 for an experimental and research laboratory to explore all phases of naval technology including the "improvement of radio installations."[2] The project was postponed because of the confusion accompanying America's entry into the war; and the Naval Research Laboratory was not opened until July 1923. The original staff of the Laboratory, which was located on the grounds of the old Bellevue Naval Magazine, was drawn from the small radio and sound laboratories which the Navy had operated during World War I. These were formed into a Sound Division headed by H. C. Hayes, and a Radio Division under A. Hoyt Taylor.

The Radio Division at NRL was formed by a consolidation under Taylor of the personnel of three wartime radio laboratories: L. W. Austin's Naval Radio Laboratory, located at the Bureau of Standards and consisting of about a dozen persons working on long-wave propagation; the Radio Test Shop at the Washington Navy Yard, a group concerned with the maintenance and testing of the Navy's ship and shore installations, but which did a small amount of research and development; and the Naval Aircraft Radio Laboratory at Anacostia.

The nucleus of this last-named group had originated in 1917 at the Navy's Great Lakes Training Station.[3] The organizer was A. Hoyt Taylor, who had recently left a position as head of the Physics Department at the University of North Dakota to take a lieutenant's commission in the Navy. Taylor was made District Communications Superintendent at the Great Lakes Station, and on his own initiative set up a small radio research group to deal principally with low-frequency problems. His principal associates were Louis A. Gebhard and Leo Clifford Young.

Taylor was a graduate of Northwestern University; he had taught briefly at Michigan State College (1900–1903) and at the University of Wisconsin (1903–1908), then had followed the migration of American physicists to Germany where he received his PhD at Göttingen in 1909. While at the University of North Dakota, where he taught physics between 1909 and 1917, he acquired a substantial reputation in the field of radio through his work with his amateur station 9IN located at the University. Louis A. Gebhard was employed between 1913 and 1917 by the old Marconi Wireless Telegraph Company on the Great Lakes. The third member of this group, Leo Young, was in the tradition of the resourceful, inventive, gadget-minded small-town American boy with no education beyond high school, but with a great and consuming passion for radio. Also an experienced amateur, Young had operated the Fort Wayne, Indiana, High School radio station in the years before the war. In 1917, after five years in the telegraph department of the Pennsylvania Railroad, he enlisted in the Navy as a radio operator and found himself at Great Lakes joining Taylor in radio research.

The group migrated from place to place in the next few years, but the core of the personnel—Taylor, Young and Gebhard—remained together. Late in 1917 they moved from Great Lakes to Belmar, N.J., where they were in charge of transatlantic radio communication for the Navy. Then they joined the Navy's small Aircraft Radio Laboratory in Pensacola, Florida, which was later moved to the Naval Air Station at Hampton Roads.

Late in July 1918 Taylor was put in charge of the entire experimental division, including aircraft radio experimentation, at Hampton Roads. After the Armistice, the Bureau of Engineering decided that it would be more convenient to have the laboratory moved to Washington; a Navy radio group had already been set up temporarily at the Bureau of Standards. Late in the fall of 1918 Taylor established his group in three wooden barracks at the Naval Air Station at Anacostia. It was here that Taylor and Young did the five-meter work which resulted in the early proposal for a radio detection device.

In May 1923 the Radio Division, numbering 23 civilian employees, began work some time before the Naval Research Laboratory was formally opened. The Sound Division arrived at about the same time. The Laboratory was officially opened on July 1, 1923. The Director—who in those days was not in residence—was Captain E. L. Bennett. Commander E. G. Oberlin was the first Assistant Director, and since he was in residence, it was upon his shoulders that the administration of the Laboratory mainly fell.

As the reader of the previous chapter already knows, the 1922 effort to interest the Navy Department in ultrahigh-frequency communication failed; evidently the suggestions were not taken seriously. At all events the project was dropped in favor of other work in which the Navy was more deeply interested. The main preoccupation of the Laboratory between 1923 and 1930 was the vitally important, if somewhat routine, business of keeping the Navy's standard communications equipment—radio telephones, direction-finders, etc.—abreast of a rapidly improving art, in which new tubes and new circuits and new advances of all sorts were being made daily.

One of the most important early projects was the study of transmitter frequency control by means of piezoelectric crystals. When a quartz crystal of special characteristics is connected to the grid circuit of a vacuum tube, in place of the ordinary tuning element, the tube will only oscillate (and produce rf energy) at the frequency determined by the characteristics of the crystal. This crystal holds the transmitter inexorably to the chosen frequency. In 1924 the first high-power crystal-controlled transmitter was built by Gebhard, Matthew Schrenk, and E. L. White, and for a long time this was used for the Navy Department's communications with the American Embassy in London. Most of the Laboratory's attention was focused upon problems of this sort; but, as we have seen, the workers at NRL were able to make fundamental contributions of great historic importance to the development of long-distance radio communication on short waves and to the study of the propagation characteristics of these waves. Although some of this work had to be done unofficially and after regular working hours it had important practical consequences. The Navy Department at first was not enthusiastic about work that appeared—as radar was later to seem in its early stages—to be something of an impractical plaything, but which led to the design of the first high-frequency radio sets used by the fleet. The Naval Research Laboratory also developed special radio equipment for aircraft and, as previously described, gave important assistance to Breit and Tuve of the Carnegie

Institution in building radio equipment for the first exploration of the ionosphere by pulse echoes.[4]

In the early years of the Naval Research Laboratory's existence, especially under the leadership of Captain E. G. Oberlin who was in succession Assistant Director and Director over a period of eight years, the Naval Research Laboratory developed rapidly. It reported to the Secretary of the Navy, not to a Navy Department bureau, and had only a limited officer personnel.[5] About 1932 it was brought more rigidly into the bureau organization and placed under the Bureau of Engineering which henceforth controlled the expenditures at the Laboratory, and exerted more direct influence upon its research. The Directorship was then brought in line with other Navy assignments, subject to the policy of rotation of duty, and the Director was replaced every two years. At the same time the officer personnel at the Laboratory was expanded; nevertheless civilian control on the actual research level was not seriously threatened.

At the time, the transfer was appealed to the General Board and the House Naval Appropriations Subcommittee, but was allowed to stand by a Secretary of the Navy somewhat indifferent to research. The effects of the transfer to the Bureau, especially the loss of their relative independence, was not pleasing to the research personnel, nor were the results commendable. The Laboratory of course found it difficult to serve all the Bureaus equally. But the chief results were that routine test work for the Bureau of Engineering bulked much larger in the Laboratory's work than it had before, and that a tighter control was exerted over research problems.[6] In theory it had always been the case that no work was tolerated which did not come under a problem assigned by the Navy Department. The investigators were supposed to account for their time in terms of these problems, especially "bread-and-butter" problems offering prospect of some definite success. Understandably, the problem system was if anything, more rigidly enforced after the transfer to the Bureau.

This change had little or no effect upon the internal organization of the Laboratory. The civilian superintendents remained fully responsible for and with full authority over the work within their divisions. They reported only to the Director of the Laboratory. Within the framework of the problem system, the research program in the Radio Division was largely determined by the decisions of Superintendent Taylor, and his principal advisors Gebhard and Young, both of whom became associate superintendents in 1935–36. The only way of breaking new ground under this system was (1) by working overtime or (2) by interpreting a problem broadly enough or loosely enough to permit new work being done under it. With a chronic shortage of funds and manpower, especially during the depression years, it is evident that neither of these methods could be used

extensively. It is necessary to understand these factors in order to appreciate the somewhat unfavorable conditions under which radar was developed at the Naval Research Laboratory.

Echoes from Airplanes

Renewed interest at the Naval Research Laboratory in the possibilities
of radio detection dates from an observation of great importance made by
a young Associate Radio Engineer, L. A. Hyland. This was the detection
of radio waves accidentally reflected from aircraft. A similar observation
in England somewhat later was one of the factors that led to or at least
encouraged the British radar development. The surprising fact was not
that radio waves were reflected from aircraft—the laws of physics predicted this—but that the effect was large enough to be observed. In their
discussions of the 1922 observations Taylor and Young had considered
the possibility of plane detection, but thought that the energy from a plane
would be too small.[7]

In the summer of 1930 Hyland was working on direction finding experiments at ultrahigh frequencies as a member of C. B. Mirick's Aircraft
Radio Section. Using a frequency of 32.8 MHz (9 m) transmitted as a
horizontally polarized beam from the Naval Research Laboratory, Hyland was studying the directional reception obtained with a single-wire
antenna about 15 feet long attached fore-and-aft along the fuselage of an
experimental O2U land plane. The plane was being tested on the ground
about two miles from the transmitter at the compass rose of the Naval Air
Station. Hyland was observing the maximum and minimum signals obtained as he turned the plane around on the compass rose, pointing it now
toward and now away from the transmitter.

In the early afternoon of June 24, Hyland made his important observations.[8] He noticed that occasionally, after the plane was turned so that a
perfect minimum was obtained, this was disturbed unaccountably, and
that for a time a signal would come in irregularly when only a steady
minimum signal should have been observed. After repeated observations
it became apparent that these fluctuations occurred when planes flying in
the vicinity of the Air Station crossed the line between the receiving antenna and the transmitting station at NRL. The phenomenon was so startling in fact and in implication that Hyland dropped the work he was
doing and hurried to the Director with news of the discovery.[9]

The observation was confirmed the following day in the presence of
Mirick, Ross Gunn, and a Lieutenant Rowe. On this occasion a specially
requested plane made flights at various altitudes between 1000 and 4000
ft. Mirick prepared a memorandum that same week for the Superintendent of the Radio Division, A. Hoyt Taylor, who returned from his annual

leave shortly after and instructed Mirick to follow up the work vigorously. A small portable field receiver with a single-wire antenna was hurriedly thrown together and the experiments were repeated at distances of 4, 6, and 10 miles from the transmitter. In all cases the presence of planes was indicated by the periodic variation (beats) of the signal received from NRL.

This observation was very soon interpreted in terms of the earlier Taylor and Young observation of 1922. In a memorandum[10] prepared for Commander Almy toward the end of August, the phenomenon was described by Hyland as resulting from interference between a directly transmitted wave and a wave reflected or reradiated[11] from the airplane.

Similar observations were made during the late summer and early fall with some interesting variations. It was determined that the effect could be observed from airplanes flying as high as 8000 ft. The phenomenon was also studied using a vertical beam antenna which confined its energy to a narrow vertical cone. An airplane flew across this cone from various directions and at various altitudes. An extremely weak ground signal—attenuation was necessary if the ground wave was to beat against the sky wave—was obtained from this beam at a distance of 10 miles. When the plane flew into the cone the usual change in the ground signal was observed. Successful tests were made on a frequency of 65 MHz and the same variations were heard from a receiver placed in a car. It was experimentally demonstrated that the effect could not be obtained when the ground wave was not present. A few calculations were also made to ascertain just what the periodicity of the variations should be for a plane flying at a given speed at certain altitudes between the transmitter and the receiver. The results obtained appeared to agree, as well as could be ascertained without recording apparatus, with the results observed in the tests.

Taylor described the experiments in a detailed memorandum for the Chief of the Bureau of Engineering, called attention to the earlier observations of 1922, and emphasized the potential usefulness of this equipment for detecting moving objects and for determining their velocity.[12] He emphasized the possibility of using the principle on shipboard to avoid collision in fog with an iceberg or other ship, the vessel itself furnishing the motion. A very short-wave beam transmitter would be located at one end of the ship and the receiver at the other. This memorandum was transmitted to the Bureau with an endorsement[13] by Captain E. J. Marquart, the Director of the Laboratory, in which he gave his opinion that the matter was of the "utmost importance", adding that if it could be developed it would be of the greatest military and naval value for defense against enemy aircraft. He recommended that a problem be set up for the development of this apparatus and that it be given a high priority. Shortly afterward,[14] the Bureau assigned the historic project number W5-2S to research on the use of very high-frequency radio waves to detect the presence of enemy vessels or aircraft.

Apparently little work was carried out on W5-2S during 1931 for want of funds and encouragement, although the administration of the Naval Research Laboratory seems to have done its utmost to secure full support. At the end of the year the Director informed the Chief of the Bureau of Engineering that the pressure of other problems prevented the actual prosecution of research on this problem, and that "until greater priority is given to this work and additional time and facilities are made available, progress must be limited to such research as can be made during intervals in the prosecution of other work having greater immediate importance."[15] Almy, then Assistant Director, later testified to Hyland's energy in following up the problem despite the handicaps, and to the time he devoted to it after working hours. In the autumn, some tests were made on the airship *Akron* which had been asked to circle the Naval Research Laboratory in order that high-frequency direction finders might be calibrated. The reflection tests were made with personal equipment belonging to Hyland and Young and at Hyland's own expense, and consisted in picking up energy reflected by the airship from three Washington broadcasting stations.[16]

Although the favorable results obtained were reported to the Bureau of Engineering, there were evidently widespread doubts about the applicability of this principle to Naval problems. On January 9, 1932, the Secretary of the Navy officially informed the Secretary of War of the results obtained during the previous eighteen months. The memorandum included the following significant paragraph:

Certain phases of the problem appear to be of more concern to the Army than to the Navy. For example, a system of transmitters and associated receivers might be set up about a defense area to test its effectiveness in detecting the passage of hostile aircraft into the area. Such a development might be carried forward more appropriately and expeditiously by the Army than by the Navy. Other applications of special importance to the Army may occur to the War Department.

During 1932 attempts were made to determine how well the velocity of aircraft could be determined from the beat frequency of the interference effects. Some scattered observations were made between ship and ship, and between ship and shore, in the course of which it was shown that a tug could be detected a mile away, when it passed between the two stations. Nevertheless comparatively little work was done on such slowly moving objects, for fast moving aircraft showed the phenomenon much more readily. Airplanes were detected at distances as great as 50 miles from the transmitter.[17]

In the letter to the Secretary of War, mentioned above, the Secretary of the Navy confided that "the phenomena upon which it is based are not known in any other country, or outside of our own military organizations in this country, except in the Patent Office, where applications for patents are being held in a confidential status."[18] Considerable excitement therefore reigned at the Naval Research Laboratory when on January 12, 1933—nearly a year after the transmittal of the letter to the Secretary of War—Bell Laboratory engineers, Carl R. Englund, Arthur B. Crawford and William W. Mumford reported upon and discussed in an open meeting of the Washington Section of the Institute of Radio Engineers their independent discovery that radio waves were reflected from aircraft. The account subsequently published by these men, under the heading "Field Fluctuations from Moving Bodies" is as follows:

> It is well known that the motion of conducting bodies, such as human beings, in the neighborhood of ultra-short-wave receivers produces readily observable variations in the radio field. This phenomenon extends to unsuspected distances at times. Thus, while surveying the field pattern in the field described above, we observed that an airplane flying about 1500 ft (458 m) overhead and roughly along the line joining us with the transmitter, produced a very noticeable flutter, of about 4 cycles per second, in the low-frequency detector meter. We then made a trip to the nearby Red Bank, N. J., airport, distant about $5\frac{1}{2}$ miles (8.8 km) and observed even more striking reradiation phenomena. Nearby planes gave field variations up to two decibels in amplitude, and an airplane flying over the Holmdel Laboratory and toward this landing field was detected just as the Holmdel operator announced 'airplane overhead'. These were all fabric wing planes...This airplane reradiation was noticed at various subsequent times, sometimes when the airplane itself was invisible.[19]

On January 19, 1933, E. D. Almy, now Director of NRL, wrote the Chief of the Bureau of Engineering describing the disclosure at the Washington meeting and urging that the detection problem be removed from confidential status, and that the patent disclosure be placed in the hands of the Judge Advocate General's Office at once for prompt prosecution of the patent.[20] The Bureau accordingly took the appropriate action and a patent application on the beat method was filed on June 13, 1933, in the names of Albert H. Taylor, Leo C. Young, and Lawrence A. Hyland.[21] Patent No. 1,981,884 for a "System for Detecting Objects by Radio" was granted them in 1934.[22] (See Fig. 4-1.)

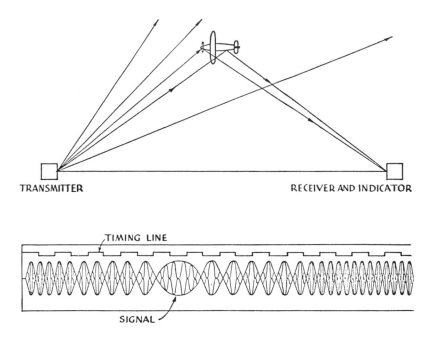

TRANSMITTER   RECEIVER AND INDICATOR

TIMING LINE

SIGNAL

FIG. 4-1. Taylor, Young, and Hyland beat system of detection. U. S. Patent No. 1,981,884.

Very little progress seems to have been possible on aircraft detection by the beat method after 1933. The difficulties of the method were all too apparent. A report sent to the Bureau of Engineering in March 1933 shows that the work had not progressed beyond the exploratory stage. A theory of detection had been sketched in some detail so that the frequency of the beats could be correlated for certain special cases with the path of the detected plane. The theory had been roughly tested by setting up an aircraft radio transmitter and receiver on the ground about a mile apart and observing flights at varying altitudes made over the line by a target aircraft. An oscillograph was arranged to record on motion picture film the receiver output.

This report included a concrete proposal, along the lines suggested to the Army, for transforming the equipment into a practical military device. Although planes could be detected well beyond the range of sound or of vision, with an apparatus similar to that used in the experiments, and though some information on the distance and velocity of the aircraft could be obtained, little idea of position and direction was provided by a single installation. They therefore suggested a chain of transmitting and receiving stations distributed along the line to be guarded. These would report in to a common center where the results would be correlated. Since photographic recording meant an undesirable delay, some attempts were made to record the signals directly with a spark discharge perforating a moving tape.[23]

Early in 1934 preparations were made for a demonstration to the Naval Appropriations Subcommittee of the House of Representatives. A system was assembled consisting of a self-oscillating transmitter operating on 60 MHz with a 500-Hz modulation (this was set up in a nearby field, with a gasoline-driven power supply) and a super-regenerative receiver with adjustable dipole antenna installed in the penthouse on the roof of Building 12, a building then about half completed.[24] This seems to have been the first semipermanent apparatus built specially for beat detection. It was kept operable for several years after the development of the pulse method was undertaken, and was used in a number of demonstrations.

Of considerable historic interest was a demonstration of this beat equipment given to members of the House Naval Appropriations Subcommittee in February 1934 which paved the way for an independent appropriation earmarked for the use of the Naval Research Laboratory, the first such appropriation intended to support a long-term effort. A sum of $100,000 was added to the annual appropriations for 1935 and gave the Laboratory a degree of financial independence.[25] In Congress the man instrumental in securing for NRL the much-needed funds was Representative (later Senator) James Serugham, a Nevada Democrat who was the dominating personality, and later the Chairman, of the House Naval Appropriations Subcommittee. The drive to secure these special funds seems to have originated in the Naval Research Laboratory, and to have been due in the first instance to the efforts of Taylor, Superintendent of the Radio Division, Hayes, Superintendent of the Sound Division, and the Director, Captain H. R. Greenlee. It received strong support from Admiral S. M. Robinson, then the Chief of the Bureau, and somewhat later from Admiral H. G. Bowen, who in May 1935 became the Chief of the Bureau of Engineering, and from Captain Hollis N. Cooley who succeeded Greenlee as Director of NRL.

The Pulse Method: First Attempts and Failure

It is much too easy, now that we are fully aware of the immense military value of radar, to assume that its revolutionary significance must have been fully grasped at the beginning and that the development of a device of such potentialities must have fired the imagination of all persons to whom the idea was disclosed. This was far from being the case. It is an important fact that during the early phases of its development at NRL, pulse radar encountered a great deal of skepticism not only in the Bureaus, where in certain quarters it amounted to outright antagonism, but in NRL's own Radio Division, and that this did not evaporate until the feasibility of the equipment had been demonstrated beyond a doubt. The vicissitudes and obvious weaknesses of the beat method were to some degree responsible for this attitude; in fact, the earlier experience with radio detection may be said to have made it harder, rather than easier, to recognize in embryo the vast importance and the essential simplicity of the new idea.

It is generally conceded at the Naval Research Laboratory, where so many persons had a share in bringing pulse radar into existence, that Leo Young, who had been actively interested in both previous radio detection attempts, was the first to hit upon the idea of using reflected *pulses* of radio energy for the detection of aircraft and other targets. Young has said that the idea occurred to him at the time of the revival of interest in radio detection resulting from the Hyland observation; he recalls having mentioned the possibility to Hyland late in 1930.[26] Be this as it may, nothing was done about the idea for some time because the difficulties seemed almost insurmountable. Extremely short pulses—much shorter than anything being used in ionospheric work—would be required. It would be difficult to produce these short pulses and to build a receiver that would detect them, and still more difficult to design an indicator, or electronic time-base (as the British rather aptly prefer to call it) to display the transmitted and received pulses and to measure the extremely short intervals of time. Before a program was actually launched some calculations were made and the possibility was pretty thoroughly discussed, with Taylor, the Superintendent of the Radio Division, with Ross Gunn, the Technical Advisor to the Director, and others.

Young began the investigation without official approval, working without assistance and after regular hours on the design of a timing circuit to operate a cathode-ray tube with a circular sweep. After some opposition he persuaded his superiors to assign a man to the problem on a full-time basis. Early in 1934, Robert M. Page, who was working under Young in the Research Section of the Radio Division, was transferred to the project, presumably on full time.

Page, the young man to whom fell during the next few critical years the bulk of the early work in the design of the first successful radar, had come to the Naval Research Laboratory in June 1927 immediately after his graduation from Hamline University. At the time of his transfer to the detection problem, Page was still, after eight years, at the bottom of the civil service professional ladder. Up to this time, he had been almost wholly preoccupied with frequency standardization problems and high-precision frequency measurement, matters of great practical concern, since without careful control, the Fleet's vast artillery of radio transmitters, often on closely related frequencies, could turn into a cacophony of mutual interference. Between 1932 and 1934 Page was fully occupied with the design of a so-called decade frequency system, a problem that was canceled by the Bureau of Engineering in February 1934 leaving Page free for his new assignment.[27]

Page began by familiarizing himself with the work that had been going forward at NRL on the beat method of detection, and helped to set up the demonstration apparatus for the Congressional visitors. He noted the defects of the system, chief of which was that the direction of the echo could not be obtained, and listed in his log book the other desiderata for a successful echo system: that it should be possible to locate the transmitter and receiver close together, as for example on the same vessel; that the apparatus should be capable of detecting airplanes at distances up to 50 or 100 miles; that it should be possible to determine whether the reflecting object is approaching or receding from the station or ship and to determine its speed. The pulse system as finally developed during the next four years was to display all these advantages.

The pulse detection problem was tackled on March 14, 1934. The entry recording this event is of some interest:

> Upon failure of the Bureau of Eng. to re-open the decade problem, work was resumed with Mr. Young on the Airplane Echo problem. It was decided to attack this problem in a manner similar to that by which super-sonic depth finding is accomplished.[28]

Page and Young continued work on the indicator problem from the point to which Young had carried it. During April and May, they worked on the sweep circuits to produce a circular trace on the cathode ray tube.[29] They obtained their first pulses by means of a multivibrator circuit (Fig. 4-2); and though this circuit, which, as we have seen, had recently become popular in ionospheric work, did not give as sharp a pulse as had been expected, they found after trying other possibilities (*inter alia* a Van der Pol relaxator) that it was the best for producing very short pulses.

This circuit, which contains resistances and capacitances, but no inductances, is in reality a simple two-step amplifier with regenerative feedback; i.e., the output of the second stage is coupled through a condenser to the grid of the first tube. The combination produces pulses roughly rectangular in form at a wide range of recurrence rates. If the circuit is symmetrical, that is, if the components of the two sides are exactly duplicated, the pulses produced will resemble those in the upper part of Fig. 4-3. Page and Young, anxious to produce short pulses separated by a relatively long resting time, made the circuit asymmetrical by choosing different time constants for the grid constants on the two sides. This produced a pulse of the sort shown in the lower part of Fig. 4-3. The first pulses were about 10 μsec long. This may not have been the first time a multivibrator was built asymmetrically; but it is certainly the first time that pulse keying had been performed at such extremely short pulse lengths.

During the first weeks of the project Page and Young worked together, but later Page worked almost entirely alone without any manual assistance. He benefitted from frequent discussions with Taylor and Young and with the men who shared his Laboratory, Wesley F. Curtis and La-Verne R. Philpott, from both of whom he received encouragement and fruitful suggestions.

The lack of official enthusiam for this project, still conducted under the old project number of the beat system, and with its low priority, is most

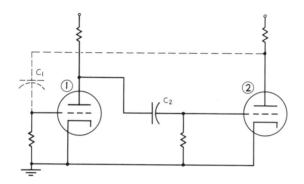

FIG. 4-2. Multivibrator circuit.

Waveforms from symmetrical multivibrator

Waveforms from asymmetrical multivibrator

FIG. 4-3. Waveforms from symmetrical and asymmetrical multivibrators.

convincingly demonstrated by the number of times that Page was borrowed from the echo work and assigned to other apparently more pressing problems. Perhaps half of this first year was spent on such special assignments[30] having nothing to do with radio detection. Some progress of a preliminary sort was made despite these handicaps.

During July, a receiver that could be adapted was obtained from the receiver group at the Laboratory, and Page made a first and unsuccessful attempt to evaluate the time constant (calculate the decrement) of a multistage amplifier for pulse reception. An experimental test oscillator with an acorn tube in a self-oscillating or squegging circuit was built and used for testing the receiver. It was generally agreed that it would be wise to build the earliest radar on relatively low frequencies, even though they were not ideally suited to radio detection, in order to test the principle and establish a basis for the more difficult problem of building equipment on higher frequencies. A frequency of 60 MHz was chosen and in a conference with A. Hoyt Taylor it was agreed to build a transmitter using RK-20's (transmitting pentodes) in a push-pull circuit keyed by the multivibrator. After a long interruption, during which Page was occupied with other problems, the months of November and December were taken up with building the transmitter and installing it on the roof of Building I so that it pointed out across the Potomac, looking slightly upstream, to pick up the aircraft which flew with considerable regularity up and down the river.

The first attempts, late in December 1934 and early in January 1935, to get echoes by the pulse method were hopelessly unsuccessful and a grievous disappointment to all involved. No pulse echoes were observed on the cathode-ray tube, but only beats. The outgoing pulse was so widened by the receiver that it became a big blob or smear on the tube, and when the signal from the plane got into the blob it was undetectable except by its beating against the main pulse. A mood of black despair settled over everyone, even those like Young and Page who had been most enthusiastic and optimistic. In Page's case the mood lasted only a few days until he was able to think through the causes of failure and describe what had to be done to remedy it. He soon realized that a fresh start would be necessary, and that a completely scientific job of redesigning the receiver was essential. It was necessary to discover the transient response characteristics of such a receiver and to determine means for preserving fast response under conditions of extremely high voltage gain, problems not previously encountered with ordinary communication receivers.

The failure was taken more seriously by Page's and Young's superiors. There was some disposition to feel that the project might be a blind alley like the earlier detection effort. Nevertheless during 1935 the problem went forward under extremely low priority.

Successful Pulse Radar

During January and February 1935, Robert Page devoted what time he could to a consideration of the pulse receiver problem. A propitious start was made when he ran across a theoretical paper from the results of which he was able to work out the relationships that should obtain in a pulse receiver. Contrary to the doctrine then current at NRL, it emerged that the bandwidth and gain relationships in such a receiver are independent of frequency. Page's first receiver had been designed for communication purposes; and it now became clear that the proper design criteria for pulse receivers were quite different than for communication. Communication receivers had been designed for high selectivity at high gain, whereas a radar receiver should not be selective, since it must pass a band of frequencies included in the envelope of the pulse. The criterion of stability in a communication receiver was merely that it should not oscillate, whereas more rigid conditions had to be met by a pulse receiver. To design a receiver capable of withstanding and recovering from the effects of the transmitted impulse it was necessary (1) to overcome the blocking effect on the receiver input tubes of the strong transmitted signals; and (2) to reduce the feedback to zero to eliminate the ringing in the receiver which so distorted the received signal.

Following the theoretical analysis, careful measurements were made on the individual components to measure their capacitances to see where they could be reduced, the purpose being to raise circuit impedances as much as possible, and extreme precautions were taken to eliminate regenerative feedback. At Taylor's request, the receiver was modified in several respects to make it suitable for high-frequency communication, which would provide a hedge should the radar development end in failure.

Because of the low priority, it required 5–6 months for the new receiver to emerge from the shop. In September and early October, though Page was once again assigned to another problem which required most of his time, he was able to make a few measurements on the receiver. During the autumn a series of confidential monthly reports, each only a paragraph or two in length, were transmitted by the Director of the Laboratory to the Chief of the Bureau of Engineering. They contained requests for additional personnel and showed that a year and a half after the project had been begun only two persons were associated with it, Page "part time," and L. C. Young "very small portion of the time."[31] On November 22, a young engineer, R. C. Guthrie, was assigned to the problem under the direction of Page.[32] From the addition of this full-time assistance real progress in the radar program can be dated.

Guthrie began work on a low-power experimental model of a new transmitter built along lines suggested by Page's associate Philpott. The new transmitter used what the British call a "squegging" circuit, a self-interrupting, plate-quenched circuit, that made the transmitter self-keying and eliminated the multivibrator. The device appeared to promise pulses as short as 2 or 3 μsec,[33] and provided its own keying power, whereas the earlier transmitter with its modulator had a high average power drain, steadily drawing power that was only used during the pulse.

The first model was built with acorn tubes and gave less than a watt of peak power and pulses of about 20 μsec. It was Guthrie's next assignment to build a transmitter using this design with commercial transmitting triodes.[34] It was proposed to operate it on 28.3 MHz—despite the obvious objection that equipment in that frequency could never be used on shipboard with the size of the directive arrays that would be needed—because of the need to demonstrate the principle and also because a directive array on this frequency was in operation for communication purposes at NRL.

After Guthrie had assembled and tested the transmitter it was moved from Building 12 to a field house, then located underneath the big transmitting array and between the two towers that supported it. The receiver which Page had improved and trimmed to fit the new frequency was installed in the penthouse of Building 12, with a half-wave antenna on the penthouse roof. Tests on the finished receiver showed that it was extremely fast compared to anything then in existence and had been designed with a wide margin of safety. Feedback had been eliminated, and complete recovery from saturation by signal levels of several hundred volts was obtained in a matter of 2–3 μsec.[35]

The first receiver, which had shown such serious regenerative feedback, had obtained all its gain on a single frequency (10 MHz). To keep down feedback, the gain per frequency was held down in the new receiver and the total voltage gain of 100 million from the first grid to the last plate was obtained by using two intermediate frequencies (with two local oscillators) and rf amplification ahead of the first IF stage. A linear sweep was substituted for circular sweep on the indicator.[36] Figure 4-4 compares the performance of Page's first and second radar receivers.

It was with this experimental equipment that the first radar echoes were successfully observed at NRL on Tuesday, April 28, 1936, using pulses 5 μsec in length. Planes were picked up at random, since scheduled flights had not yet been arranged, out to the not very staggering range of 2½ miles. Since the beam was stationary, it was not possible to follow planes that did not stay in the beam; and the beam was so narrow that it was difficult, during later scheduled flights, for a plane to stay in the beam intentionally. On the following day they raised the plate voltage of the transmitter from 5000 to 7500 V, and on April 30 a plane on a scheduled flight was tracked out to 8¾ miles.[37] These first successes were observed only by Page and Guthrie. On May 6, a plane was followed out 5–9 miles

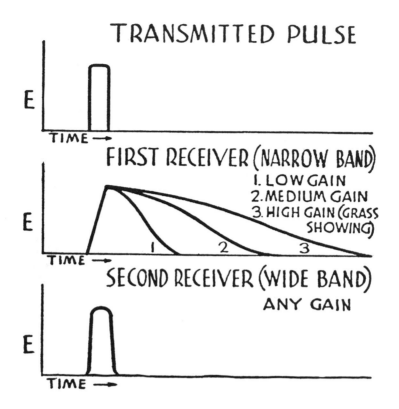

FIG. 4-4. Receiver characteristic of Page's earliest pulse radar.

during a demonstration given to Captain Cooley, the Director of the Laboratory, and to Taylor and Young. Later the same day, after some fruitful tinkering with the receiver connections, they tracked one plane out to 11 and another to 17 miles. It was in the course of this demonstration that Taylor, enthusiastic over the success, but constantly aware of the conditions that had to be satisfied aboard ship, urged that the next step was to investigate the possibility of going to higher frequencies and stressed the importance of designing a duplexing system to allow a single antenna to be used for transmitting and receiving. The type of solution that was later adopted was suggested shortly afterwards by Leo Young.

Tests were made almost daily wth the 28-MHz equipment through the month of May and early June. On June 3 Page and Guthrie were able to report a range of over 25 miles, even under conditions of bad static. The equipment was successfully demonstrated to Bureau of Engineering officials headed by Commander W. J. Ruble, head of the Radio Division, and on June 26—miraculously, considering the crudeness of the equipment, without breakdown—to a group of visiting admirals that included Admiral H. R. Stark, the Chief of Naval Operations, and Rear Admiral Harold G. Bowen, Chief of the Bureau of Engineering, who became one of the warmest supporters of the radar development in higher quarters, and who in 1940 became the Director of the Naval Research Laboratory.

The immediate consequence of these successful demonstrations was that the value and importance of the new equipment was established beyond a doubt among those immediately concerned with the Navy's research and development program. At the request of Admiral Bowen, the problem of radio detection was given the highest possible priority in the Naval Research Laboratory and raised from a confidential to a secret category. In a letter to the Director of NRL, Admiral Bowen expressed satisfaction with the successful demonstrations and urged that work now be centered upon "providing for shipboard use of equipment operating at the highest frequency consistent with obtaining the required power and providing in a single device both detection and ranging."[38] At this time an analysis of the state of the radio detection art was prepared by Page and transmitted to the Bureau of Engineering. In this document the pulse method was compared with other known methods of aircraft detection. While admitting that there were serious unresolved engineering difficulties, the reported concluded:

> ...the discontinuous signal consisting of short pulses recurrent at a fixed frequency is overwhelmingly superior to all other methods. Indication is visual, continuous, fully automatic, direct reading, unaffected by signal variation, never false or ambiguous, shows all reflecting objects in its range separately and simultaneously, relates each reflecting object unerringly to its own distance, and can give no indication of a reflecting object unless that object actually exists at a distance shown by the indication, while the accuracy of measurement is inherently greater than that of any other known method.[39]

Although attention was promptly turned to developing improved equipment on various higher frequencies, the 28-MHz detection apparatus was kept in operation for demonstration purposes with Guthrie in charge. During the remaining months of 1936 the transmitter power was raised and a number of refinements were introduced.

An able campaign of high-pressure salesmanship within the Navy on behalf of radar was conducted by Admiral Bowen and by Captain Cooley, the Director of NRL.[40] Captain Cooley was a man of infectious enthusiasm and great ability who possessed an extraordinarily large acquaintance within the Navy, and who button-holed a succession of influential Naval officers as they passed through Washington and persuaded them to visit the Laboratory to see demonstrations of the new device. The record of successful operation during demonstrations was so extraordinary as to seem almost a violation of the laws of nature. There were however several close calls: on one occasion the equipment was inoperative until a few minutes before it was to be displayed, but worked perfectly throughout the test; on another occasion serious and inexplicable interference was encountered just before an important demonstration. At the very last minute the case was found to be a faulty commutator in one of NRL's own motor generators. In the fall of 1936 the equipment was demonstrated to Admiral A. J. Hepburn, Commander-in-Chief, U.S. Fleet. In February 1937 it was shown to Charles Edison, the Secretary of the Navy, and to Admiral William Leahy.

The continued success of these demonstrations was an important factor in marshalling the necessary support for the radar development throughout the Navy. Equally important were the efforts of A. Hoyt Taylor, the Superintendent of the Radio Division, of the Director of the Laboratory, and of Admiral Bowen at the Bureau of Engineering, to convince the representatives of the Navy Department of the importance of radar, and to obtain from Congress adequate funds for the support of an expanded program.

Before the end of 1936, expressions of interest in radio detection equipment began to arrive from the Fleet. In December Admiral Hepburn forwarded to Captain Cooley an interesting letter that had originated in the Fleet in connection with tactical exercises held off the Pacific Coast in early October.[41] The Commanding Officer of Destroyer Division Sixty-One had written a comment on these exercises and called attention to the need of a device to penetrate fog and "to increase the tactical and fire control efficiency of the fleet and to reduce the dangers of navigation." The writer was evidently unaware that work of this sort was proceeding at NRL, and the proposal seems to have been promoted by the publicity concerning the *Normandie* apparatus.[42] Brief endorsements were made on the document's journey to CINCUS, all of which strongly supported the request in the original letter, and several of which (in order to indicate the feasibility of the idea) called attention to the published accounts of beam detection or referred to the work at NRL. Admiral Hepburn forwarded this letter to the Chief of the Bureau of Engineering and with it another from the Commanding Officer of Destroyer Division Sixty-One commenting on tactical exercises held 16–18 November 1936.[43] The comment read:

> Fog again played a major part in the tactical exercises–delaying the sortie from San Diego, interrupting exercise three and forcing abandonment of exercise five. In each of these instances radio gear of the "Normandie" type would have been of great assistance.

This comment was doubtless dispatched with considerable vehemence, for the letter also reported that exercise 4 had been "completed before *Aylwin* (DD355, the Flagship of the Destroyer Division) and *Dale*, delayed by fog, had joined the group screening *Ranger*."

Equipment on Higher Frequencies

Very soon after the first successful tests, work was begun on a series of higher frequencies. During the rest of 1936 Guthrie attempted to duplicate on 50 and 60 MHz the results that had been obtained on 28.[44] Early in the summer Guthrie modified the field house transmitter to operate at 50 MHz and connected it to a 16.8 MHz curtain array of which it used a harmonic. When it was tested against aircraft in July, it was found that the curtain gave multiple lobes; on one of these beams, however, very good results were obtained out to 15 miles.

The transmitter had used a pair of Gammetron 354 tubes operated in push-pull. Tests were now made to determine the frequency limit of these tubes; and it was discovered that they could operate with good efficiency up to 80 MHz. In mid-September Guthrie began work on an 80 MHz beam array which was suspended between two of the towers so as to direct the beam in a southerly direction down the Potomac River. After Page's receiver had been transferred to the field house and attached to the array, excellent results were obtained up to 30 miles, using the newly developed duplexing switch, shortly to be described.

Attempts on these frequencies were soon abandoned because of success on a frequency that subsequently was adopted for the first shipboard radar, namely 200 MHz.

Early in May the problem of developing a 200 MHz system that would operate on a single antenna when either transmitting or receiving was entrusted to a new engineer assigned to Page's group, A. A. Varela[45] who was put to work independently on the task of designing a transmitter and perfecting the duplexing system suggested by Young and Page.

Varela's transmitter circuit consisted of two tubes in push-pull with cathodes, grids, and anodes all tuned by resonant parallel lines. The tubes were Western Electric 304-B tubes. While this was being designed, work went forward on possible antenna structures for use on 200 MHz. A novel all-metal array, consisting of six Yagi antennas connected in parallel and mounted without insulation, was built for this new frequency. The 28 MHz receiver was used for 200 MHz reception as an IF amplifier; a modification of a Bureau of Standards concentric-line receiver was used as a preselector and frequency converter to convert 200 to 28 MHz. The engineering of a new 200-MHz receiver was carried out under Page's direction by John Gough and G. E. Pray who were borrowed from the NRL receiver group for a brief period.

The 200 MHz system, from which the Navy's first shipboard equipment was developed, received its first echoes on July 22, 1936,[46] less than 3 months after the start of the project. This was a remarkable achievement especially when it is remembered that the first duplexer was developed by Varela in the same period.

The performance of the equipment was encouraging but not impressive. During tests against airplanes and a blimp, no echoes were returned beyond 10 or 20 miles because of the low transmitter power and poor gain of the Yagi array, which did not give as narrow a beam as the large curtain arrays used at the field house. Reflections from buildings and trees were stronger than with the 28-MHz equipment, and the echoes of planes were mixed with ground echoes or ground "clutter".

A day or so after the first 200-MHz tests, the transmitter was moved inside the penthouse next to the receiver, and the duplexing equipment built by Varela was tried with complete success. This device is deserving of some attention.

A. Hoyt Taylor had been the first at NRL to agitate for a device to permit the transmitter and receiver to operate on the same antenna. For this to be possible, some sort of switching device is necessary to protect the receiver (especially its input circuits) from the effects of the transmitted pulse and to isolate the transmitter during the time the echo is received, so that the weak echoes will not be wholly or partially lost by following the transmission line back to the transmitter. Since there were no practical mechanical switches which could open in a few microseconds and at the repetition rate of a radar set, electronic switches were clearly necessary.

In the course of an early conversation between Young and Page on the possibility of putting transmitter and receiver on the same antenna, Young suggested all the essentials of the idea which was later adopted: namely, using resonant lines with a spark gap or gas tube located at the high-impedance point of the resonant circuit. The essential feature of this device is that the spark gap or glow tube breaks down into a conducting path of low impedance (i.e., virtually a short circuit) when ionized by a sufficiently high voltage such as that applied during the building up of the transmitted pulse, but that the gap or tube presents a high impedance to a low voltage such as that of a received signal.

Page and Varela nevertheless went forward on what seemed a simpler tack, namely to use the resonant lines coupled directly to the input triodes of the receiver. When used with a low-power transmitter this scheme proved satisfactory; although the life of the input tube was understandably short, it held up sufficiently for tests to be made. After the tests on the 200-MHz system this was used successfully for 30-MHz equipment.

Several other important developments took place during the course of the 200 MHz development. During the summer studies were made on the effect of a plane's aspect on signal strength, and on the possible polarizations of the transmitted energy. They found that the strongest reflections were obtained when the plane was head-on, tail-on, or broadside; it appeared that many more echoes were received when using vertical instead of horizontal polarization, because of defects in the antenna system. Subsequent work was accordingly done for some time with vertical plane polarization.[47]

It also appeared in the course of the early tests that the indicator sweep circuit they were using gave too short a timing line with neither sufficient range nor resolution. One of Page's assistants, H. E. Reppert, was given the assignment of making a double-line sweep, one fast and the other slow. They ran into difficuties building stable circuits of this sort and found it simpler to design a multiple-line sweep, produced by repeating on the face of the tube a single line with high sweep frequency. This gave high resolution and long range simultaneously. Marker pips were introduced at 2-mile intervals.

The winter of 1936–37 was spent assembling improved 200 MHz equipment, and in particular building a stacked array with a rotatable support.

In the spring of 1937—at the insistence of Commander M. E. Curtis, the energetic executive officer at NRL, and over the objections of Page who felt that the equipment was not ready for such a test—the 200-MHz equipment was installed aboard the destroyer *Leary*, then at dock in the Washington Navy Yard. Two trips, lasting about a week each, were taken during April with NRL engineers aboard, down Chesapeake Bay to Lynnhaven Roads. Except for demonstrating the weaknesses of the equipment in this early stage of development—it gave a peak power of about 2 or 3 kW and a maximum range of only 18 miles—and giving the engineers their first opportunity to observe radar performance in the absence of ground clutter, it is not clear that the development was greatly furthered by the expedition.

Although many and varied improvements went into the equipment during the next two years, without doubt the most important was the development of a high-power transmitter. This involved the adoption of both a new transmitter circuit and of new transmitting tubes.

During the fall of 1937 the effort was concentrated on the development of a transmitter using four tubes connected in a ring circuit, a "series-parallel" arrangement consisting of two pairs joined in the filament circuit. The tubes used were still the Western Electric 304-B's, which, although satisfactory for their intended use, could not withstand the extreme voltages required for radar. With this transmitter, reflections from planes were still very disappointing.

The most important step, the capstone in the design of the first radar, was the development of a six-tube transmitter in a ring structure.[48] This oscillator was produced by the combined efforts of Robert Page, Guthrie, Varela and I. N. Page. Of equal importance was the decision to use Eimac 100TH tubes in this ring oscillator.

The Eimac tubes were designed and manufactured by Eitel–McCullough, Inc., a small but very dynamic and skillfull concern in San Bruno, California, manufacturers of transmitter tubes for the radio amateur. This market required tubes of great durability and power, designed to withstand the sort of abuse often given tubes by amateurs. Eitel–McCullough were aware that the "ham" was likely to run his transmitter tubes well above their declared rating; unlike more conservative companies, they made proper allowance for this enthusiasm. In November 1934 they produced a tube known as the 50T which was extremely successful and which was found to have plate dissipation capabilities far beyond the original 50-W rating.[49] In November 1936 this tube was redesigned in certain respects, to profit by this characteristic, and was designed as type 100T. This was produced in two models, 100TL and 100TH, indicating respectively a low and a high amplification factor. The first tubes of this design shipped to NRL were a pair of 100TH's sent from San Bruno in December 1936. In the same year, as a result of a request from the Signal Corps, work was begun by Eitel–McCullough to modify the 100TL in various minor ways. The result was the 100TS which became, after later improvements, the Army's VT-127 which was used in the SCR-268 radar.

NRL's new transmitter with the Eimac tubes was completed in January 1938, and marks the first really successful transmitter design at 200 MHz. On January 28 it was attached to the antenna; two days later, when the transmitter had been properly matched to the array, it was determined that 6 kW of peak power were being delivered to the antenna. On February 1, 1938, planes were followed out to 40 miles and giant signals appeared on the indicator for targets 30–37 miles away; this equaled the performance of the earlier 80-MHz equipment. On February 14, in a demonstration of the equipment, a plane was followed out to 50 miles, the limit of the sweep.

These tests were made with separate transmitting and receiving antennas, for it was discovered that the duplexer used in the earlier equipment was unsatisfactory with the higher powers now being obtained on 200 MHz. A new and improved duplexer was built, using the spark-gap technique originally proposed by Young. On February 17, 1938 the system was finally operated on a single antenna with satisfactory results. Page recorded this in his notebook: "This completes the entire 200 Mc development of radio echo equipment. This development was started on the 8th day of May 1936, and completed on the 17th day of February 1938."[50]

Development of the XAF

The crucial decision that had to be made in the light of the successful conclusion of the 200 MHz work was whether to build an engineered version of this prototype for service test and possible manufacture or whether to bend every effort towards the development of more compact equipment on still higher frequencies. Experimental work was already well advanced on 400–500 MHz, and the engineers were prepared to shift the emphasis to this new problem. They were ordered, however, by Captain Cooley, to keep the 200 MHz equipment working until equal or better performance was available on higher frequencies.

The decision to give the highest priority to carrying through the 200-MHz equipment for use in the fleet seems to have been made by the Bureau of Engineering in consultation with the Director of NRL and A. Hoyt Taylor. In reply to a request from the Bureau for tentative performance specifications and for a report on the present status of the equipment "with a view to the possible incorporation of this equipment in new battleships," the Director of NRL submitted the information that the development of a practical apparatus for shipboard use, at least to serve as a preliminary model for service test, could be completed within a year. In an important conference at the Bureau of Engineering on February 24, 1938, attended by Captain Cooley, Commander Curtis, and Taylor from the Naval Research Laboratory, the need was expressed by a number of officers for early service tests of the 200 MHz equipment.

By the end of February it had been agreed that the Laboratory would build a partially engineered model for installation on some vessel in the fleet. Accordingly the main effort of the Laboratory during the next eight months was put on designing and building such a system, to which was given the designation XAF. The program was placed under the general supervision of L. A. Gebhard, and redesign was carried out by Reppert, Guthrie, Varela, and the two Pages. The total cost of the apparatus— which included the manufacture of the antenna support structure by the Brewster Aeronautical Corporation—came to something over $16,639. In the construction of the XAF no electrical improvements were introduced, for the design had been frozen as of February 15; but changes in mechanical design were made to fit it for shipboard installation. The antenna size was somewhat reduced with a slight loss in range.

While this development was in progress, there was much debate and correspondence as to the proper location of the XAF aboard a ship and the appropriate vessel for the installation. It was not until early December, however, when NRL announced that it was ready to ship the equipment, that it was finally announced that the equipment would be installed aboard the USS *New York* during the vessel's current overhaul at the Norfolk Navy Yard and would be tested during the maneuvers in which the ship was to take part. The choice of the *New York* was largely the result of the recommendations of Leo Young who the previous winter had been aboard her to make propagation tests on experimental high-frequency communication transmitters and receivers.

The XAF was shipped from NRL on December 8, and during the rest of the month it was being installed aboard the *New York* under the direction of Reppert and Guthrie. Some delay was occasioned during the installation by the inability to weld the base of the antenna support while the vessel was loading ammunition in anticipation of the maneuvers. The XAF's unwieldy rotating antenna structure was installed in place of a 20-ft range finder, just forward of the foremast and directly over the navigation bridge, at a point just 80 ft above the waterline. This location had previously been selected by Young. The control apparatus was installed in the antiaircraft station directly behind the antenna.

Almost simultaneously it was arranged to have an experimental radar system called the CXZ, which had been hurriedly built by RCA, installed on the *Texas*. This experimental set was the result of a disclosure to RCA by the Bureau of Engineering of the successful results at the Naval Research Laboratory.

Disclosures of Pulse Radar

It is of some interest to learn at what stage the highly secret pulse radar development at NRL was disclosed to the U.S. Army Signal Corps and to the industrial concerns that were later to be participants in the radar program. In view of the scarcity of funds and manpower that plagued the development in both Navy and Army laboratories, it may also be of some interest to discover the extent of the cooperation between the Army and Navy and the manner in which the burden was shared.

The Signal Corps Laboratories, which had long since been informed of the beat system of detection developed at NRL, first learned of the Navy's pulse radar on the occasion of a visit of one of its engineers, W. D. Hershberger, to the Naval Research Laboratory in January 1936. Besides being given a demonstration of the beat equipment, Hershberger learned from Robert Page of the proposed pulse method and of the work then being undertaken to design a satisfactory pulse receiver. As is made clear in the succeeding chapter, the beginning of pulse radar work at the Signal Corps Laboratories was a direct consequence of this information. In April, Page paid a return visit to the Signal Corps, and, with the approval of the Director and under the guidance of Hershberger, was shown the laboratories and shop at Fort Monmouth and given the opportunity of discussing the detection problem. He learned of the microwave research that was being carried on, and found that a pulse transmitter for the 2–3-m band, keyed by a multivibrator, and a super-regenerative receiver were both being assembled. At Hershberger's request the brief monthly progress reports from NRL were forwarded to Fort Monmouth. The first indication reaching Hershberger that NRL had achieved success with the pulse method seems to have been provided by the sudden and unexplained cessation of these reports when the project was changed by the Navy from a confidential to a secret status.

Early in January 1937, a demonstration of NRL's pulse equipment was put on for the benefit of the Chief Signal Officer; and a month later, at his request, the equipment was demonstrated to a delegation from Fort Monmouth composed of Captain Rex V. D. Corput, radio engineer Paul L. Watson, and assistant radio engineer R. I. Cole. Echoes from airplanes out to 35 miles were demonstrated to the visitors using the 30-MHz equipment. The work in progress on other frequencies was displayed and a full discussion of the principal technical problems ensued. Another demonstration to a Signal Corps group—this time of the 200 MHz equipment—took place in March 1938. In September of that year an inspection trip to Fort Monmouth was made by Page and one of the officers on duty at NRL; a month later Leo Young and E. C. Hulbert paid a similar visit to the Fort Monmouth and Fort Hancock laboratories of the Signal Corps.

These formal demonstrations, and brief, widely spaced visits—during which complete disclosures were made on both sides—continued to constitute the mode of collaboration between the services. There were no prolonged visits, no exchange of engineering personnel, no coordination of programs, no jointly planned attacks on any of the basic problems. The radar development of the two services progressed along parallel, but essentially separate, lines.

The beat method of detection had been publicly patented; and in 1935 the Bureau of Engineering decided to give to the General Electric Company all the essential details on this type of detection for use in connection with contracts that company had with the War Department. But until mid-1937 no disclosure to any commercial concern was made of the pulse method. In the summer of this year the administration at NRL withdrew its objection to what it considered premature disclosure of pulse radar, and full information on this new device was made available to the Bell Telephone Laboratories, whose engineers paid visits to NRL in July and

November 1937. Although Bell seems to have been reluctant to embark at once on a program of radar systems development, RCA, to whom a disclosure was made at about this same time, began work on an experimental prototype that was given the Navy designation CXZ. The system had been designed at RCA without help from, and for a time without the knowledge of, the Naval Research Laboratory on the basis of information suplied through the Bureau of Engineering. In the early fall of 1938, when the CXZ was nearly completed, it was inspected by Robert Page and Commander Pierce of NRL during a visit to the RCA Laboratories.

The CXZ was installed on the *Texas* at Norfolk Navy Yard on January 4–6, 1939. The speed of this installation was made possible by the fact that the equipment was mobile and its largest unit consisted of a transmitter weighing about 1000 lb and mounted on four large rubber-tired casters. The antenna structure, a horizontally polarized stacked array which could be slowly rotated by a motor, was mounted above the transmitter. Transmitting unit and antenna were installed on the deck above Flag Plot in the *Texas* and the receiver was mounted below on a table in Flag Plot. The transmitter itself used four RCA 834 tubes (similar to Western Electric 304B's) arranged in two separate push-pull units with the filaments coupled. It gave a peak power of about 1 kW. The system operated at 385 MHz, a higher and less thoroughly mastered frequency than that used by the XAF. Varela joined the RCA representatives aboard the *Texas* to follow the operation of this system during the maneuvers as NRL observer.

Radar is Tested in the Fleet

The Navy's first radar was given exhaustive and rigorous tests on fleet maneuvers carried out in the Caribbean area during January and February 1939. This was the occasion for finding out the true merits and the real limitations of the equipment. The XAF on the *New York* was operated and supervised, and the tests were conducted, by R. M. Page, assisted by Reppert and Guthrie. Careful studies were made of the directivity of the beam, and of the possible interference between the XAF and the ship's communications and radio direction finding systems. Observations were made of the value of the equipment in station keeping and in range spotting during firing exercises at surface targets. The XAF demonstrated its ability to pick up shell splashes and to follow 14 in. projectiles in flight. A reliable range of 20,000–24,000 yards was obtained on battleships and cruisers, and a range of 14,000 to 16,000 yards on destroyers. Special tests were carried out on the ability of the equipment to pick up submarines when surfaced and when partially submerged, and it was discovered that signals were obtained at 3000 yards as long as the conning tower was above the surface of the water.

On two occasions the XAF demonstrated its ability to frustrate night destroyer attacks, picking up the darkened destroyers and giving their bearing to the battleship's searchlights. During the fleet problem, XAF was used to detect aircraft for the Black Force, and kept an accurate report on the friendly aircraft. During the actual "battle", however, XAF was of little value because it could not distinguish between friendly and enemy planes. The important conclusion was reached that an airplane recognition system operating through the XAF would be of inestimable value.

In general the trials were extraordinarily successful as far as the XAF was concerned, exceeding in nearly every respect the most sanguine expectations of Page and his associates. The gear was subjected to severe operating conditions, and ran continuously 16–24 hours a day during the actual fleet problem, in high winds, through rain storms, and during gunfire. Failures were minor and readily repaired. Although the system had been designed primarily to detect and locate airplanes at long range, and every factor in its design had been considered with this end in mind, the other applications for which it was tested on the cruise, and which were suggested during the tests, were byproducts; and it was recognized that each of these additional duties could be better performed by using the same principle in apparatus designed particularly for these other applications.

Unfortunately, aboard the *Texas* the performance of the CXZ was not nearly so favorable; extremely low ranges, unreliability of performance, and inability to withstand the effects of gunfire from the 14 in. guns led the Commanding Officer of the *Texas* to describe the equipment as being in a highly experimental stage, and unsuitable for war.

The officers who had observed the performance of the XAF and personally watched it in operation were unstinting in their praises. The Commanding Officer of the *New York,* Captain R. M. Griffin, expressed himself as satisfied with the reliability of the equipment and impressed with its manifold uses and its undeveloped possibilities. The device, he reported officially, would have most far reaching effects on tactics; with such equipment standing guard against surprise, there would be no necessity for widespread formations and ships could cruise in tight formations, stronger tactically, and in a position to render mutual support against all forms of attack. The more compact formation would also be more difficult for the enemy to locate and communications within the formation would be facilitated. Captain Griffin strongly recommended that the device be installed on all aircraft carriers and as soon as practicable on other vessels, and that no reduction in size at the expense of range should be required, especially for CV. "The device looks big, but really caused very little inconvenience. After all, we can't expect to get something for nothing. It is well worth the space it occupies."

From Admiral A. W. Johnson, Commander Atlantic Squadron, came an equally eloquent testimonial in favor of XAF. "Commander Atlantic Squadron," he wrote, "considers that the equipment is one of the most important military developments in radio since the advent of radio itself. Its value as a defensive instrument of war and as an instrument for the avoidance of collisions at sea justifies the Navy's unlimited development of the equipment." The test indicated that it would protect a task force against surprise attacks by aircraft and by surface vessels in darkness or conditions of low visibility. He strongly recommended that the equipment be made a permanent installation in all cruisers and carriers.

An extremely important meeting was held on May 1, 1939 in the Office of Naval Operations, to discuss the next steps in the procurement of radio echo equipment of the XAF type. It was attended by an important list of officers representing Operations and the various Bureaus. From NRL there were present the Director, Captain Cooley, his deputy, Lieutenant Commander Pierce, and three technical men, A. Hoyt Taylor, R. M. Page, and L. A. Gebhard. Page gave a brief description of the tests made on the *New York* and described the limitations of the new device and the possibilities of improving it. In the discussion that followed it was proposed to recommend that procurement of ten of the devices in their present form with only minor changes be got under way at once.

The technical features of the XAF were disclosed in detail to engineers of the Bell Telephone Laboratories and RCA to enable them to bid for the manufacture of the preproduction sets. A contract for the manufacture of six of those sets, exact copies of the XAF to which the Navy gave the designation CXAM, was awarded to RCA in October 1939. The first of these sets was delivered by RCA in May 1940 and was installed on the USS *California* at Puget Sound in the summer of 1940. Five other CXAM sets were installed on the cruisers *Northampton, Pensacola, Chester, Chicago,* and on the old carrier *Yorktown.* At this point it was decided to modify the design to include an improved antenna and amplidyne instead of thyratron control. Fourteen units of the set, called CXAM-1, were ordered and these were installed on the battleships *Texas, Pennsylvania, West Virginia, North Carolina, Washington;* on the heavy cruiser *Augusta;* on the carriers *Lexington, Saratoga, Ranger, Enterprise, Wasp;* on the auxiliary seaplane tenders *Curtiss, Albermarle;* and on the light cruiser *Cincinnati.*

Serial No. 1 of the CXAM's was removed from the *California* after the raid on Pearl Harbor and installed as a search set at the Army Base, Oahu, T. H. In the summer of 1942 it was removed and installed on the USS

Hornet and subsequently lost when that historic carrier was sunk at Guadalcanal.

The Naval Research Laboratory took an active part in these early installations. For a period of about six months L. C. Guthrie supervised the actual installations, and went to sea on the several ships for brief periods to observe the performance of the equipment and to show the officers and men how to operate it. The response to the new equipment aboard the vessels varied from cool indifference to great enthusiasm. Gradually, however, the extraordinary utility of radar for station-keeping, and for navigation in fog and darkness, brought its value home to all but the most skeptical.

NOTES

1. Lloyd N. Scott, The Naval Consulting Board of the United States (Washington, DC, 1920).

2. HR 15947, Naval Appropriation Bill, 1916.

3. The information is based on "Naval Technology in Today's War," a report of the U.S. Naval Research Laboratory, Washington, D.C. September 1942, and conversations with A. Hoyt Taylor and L. C. Young. For a number of details, I am indebted to Dr. Taylor for permission to consult in typescript a chapter of a work prepared on the early history of radio. Some valuable facts have also been gleaned from the articles of John H. Hightower of the Associated Press which appeared in the spring of 1943 and were later reprinted as Senate Document No. 89, 78th Congress, 1st Session. Though an extraordinarily competent piece of reporting in matters of detail, in view of the complexity of the story, the emphasis is in a number of respects somewhat misleading.

4. See above, pp. 57–59.

5. The information in this section is drawn from personal conversations with men at the Naval Research Laboratory and from a somewhat acid account of the relations with the Bureaus given by an ex-member of the Laboratory's research staff: F. Russell Bichowsky, *Is the Navy Ready?* (Vanguard, New York, 1955).

6. It is noteworthy that in the recent reorganization by an Executive Order (Oct. 1, 1945) the Naval Research Laboratory was removed from the direct control of the Bureau of Ships and placed under an Office of Research and Invention, in the Office of the Secretary, headed by Rear Admiral H. G. Bowen. A previous decision to set up NRL under the Secretary's Office was reversed by a ruling of the General Board (April 1, 1941).

7. Interview with Leo G. Young, August 14, 1944.

8. C. B. Mirick, "Memo for Superintendent, Radio Division," 27 June 1930, NRL files C-F42-1/67 (E).

9. L. A. Hyland to the Honorable Frank Knox, via Captain E. D. Almy, June 26, 1941. Hyland errs slightly: Almy, then a Commander, though he seems to have been the officer in charge, had the title of Assistant Director of the Laboratory. He became Director with the rank of Captain about January 1931.

10. L. A. Hyland, "Memorandum to Cmdr. Almy, Subject: Radio-Super-Frequency Phenomena," 19 August 1930, NRL file F42-1167 (E).

11. In this early radio detection work the phenomenon is often spoken of as "reradiation" as well as "reflection". Though there is no physical distinction between the two, reradiation may have seemed a natural term with wavelengths nearly corresponding to the linear dimensions of the aircraft.

12. Memorandum from Naval Research Laboratory (A. Hoyt Taylor) to Chief of BuEng, 5 November 1930, "Subject: Radio—Echo Signals from Moving Objects." Report on NRL files C-F42-1/67 (E 4222).

13. 1st Endorsement, 11 November 1930 (NRL files C-F42-1/43). See also 1st endorsement by the new Director, E. D. Almy, 16 January 1931, which said device appeared to have great promise, though "No estimate of its limitations and practical value can be made until it has been developed," and recommends that a "problem of high priority be set up for the development of the method and apparatus in order that its limitations and value may be established."

14. BuEng, Radio Division, Problem Specifications: Problem No. B1-1, 25 Nov. 1930; Problem No. W5-F2, 21 January 1931.

15. E. D. Almy wrote later: "The problem was carried on with extreme difficulties due to the fact that the resources of the Laboratory were extremely limited in funds and in personnel, and failed to get official support of this problem. However, it was continued as an active project during the remainder of that tour of my duty." To the Secretary of the Navy via Director of NRL from E. D. Almy, First Endorsement, 27 June 1941 to memorandum of Hyland to Secretary Knox (NRL files), in which Hyland not only complained of inadequate support and no funds but asserts incorrectly that no problem was set up while he was at NRL (he left the Laboratory in 1932 to start his own business).

16. Director of NRL, E. G. Oberlin, to Chief, BuEng, 2nd Endorsement of C-F42-1/67 (4574) 30 December 1931. Evidently drafted for the Director by A. Hoyt Taylor. Secretary of Navy to Secretary of War, Subject: Radio—Use of Echo Signals to Detect Moving Objects, January 9, 1932. NRL files, Eng. C-F42-1/43 (1-4-WS).

17. NRL files, WF-2 Confidential—Investigate use of radio to detect the presence of enemy vessels and aircraft, n.d." (ca. July 1, 1932).

18. See above, Note 14.

19. C. R. Englund, A. B. Crawford, and W. W. Mumford, "Some results of a Study of Ultrashort-wave Transmission Phenomena," Proc. IRE **21**, 465 (1933).

20. Director, NRL to Chief, BuEng, "Subject: Radio—Means for the Detection of Moving Objects Through the Employment of Radio Equipment," 19 January 1933. NRL files C-S67/43 (4972). Also BuEng to Director NRL, 1 February 1933, NRL files C-A12-a (9-29-1920-W8).

21. H. G. Bowen, Assistant to Judge Advocate General, "Subject: Radio—Means for the Detection of Moving Objects Through the Employment of Radio Equipment," 3 March 1933.

22. Official Gazette of the United States Patent Office, Vol. 448, No. 4, Nov. 27, 1934, p. 821, and U.S. Patent No. 1, 981, 884, 1934.

23. Director, NRL (E. D. Almy) to Chief of BuEng, Memorandum 28 March 1933. "Subject: Radio—Use of Radio to Detect Enemy Vessels and Aircraft (BuEng Problem W5-F2)" NRL files C-F42-1/67 (5039).

24. R. M. Page's Official Log. Vol. III, 24 August 1932 to 14 January 1936, Register No. 171.

25. Contrary to the statement spread abroad by the Government's official radar release, the sum of $100,000 was not specifically for radio detection, and particularly not for pulse radar which was not taken really seriously until late in 1935 or early in 1936.

26. Interview with Leo C. Young, August 14, 1944.

27. Interview with Robert M. Page, October 6, 1944. Page held the civil service rating of P-1 from 1927-1935.

28. Page's Log. Vol. III, p. 79, entry March 14–31.

29. A start was made using 1000 Hz as the sweep frequency. This was amplified by two independent amplifiers in parallel, giving a relative phase shift of 90°. The output of each amplifier was put on one pair of deflection plates of a General Radio cathode-ray tube oscillograph to give a circular pattern.

30. Page's Log, Vol. III, *passim.*

31. NRL files, C-A9-4/EN8 (W5-2S) Confidential, Director NRL to Chief of BuEng, 9 September 1935, 4 October 1935, 1 November 1935, 2 December 1935.

32. Interviews with Robert Page and with R. C. Guthrie; Page's Official Log, Vol. III, pp. 144–5. Guthrie's appointment is not mentioned in the report of 7 January 1936. It is however referred to on 1 November that the "new engineer" had been selected and steps were being taken to complete the necessary papers.

33. Perhaps as a means of stirring up interest and support, the reports from the Director of NRL to the Chief of the Bureau mention in October that "several of the current magazines" had disclosed the fact that some activity was under way in Germany of radio detection and remarked that they "appear to have put considerable thought and work into this problem and have worked out a system for area protection, making use of centimeter waves." The following month the report quoted the recent news items concerning the *Normandie's* device for detecting obstacles in her path. (See Chap. 8, p. 225.) Monthly Report of 7 January 1936. NRL files C-A9 4/EN8 (W5-25).

34. Gammetron tubes made by Heinz and Kaufman Co. in California.

35. Memo from L. C. Young and R. M. Page to Judge Advocate General, 15 June 1937, NRL files, S-A13-2 (1).

36. Monthly Report of 7 January 1936. NRL files C-A9-4/EN8 (W5-25). Later, a new sweep circuit was built to give a logarithmic time axis to give greater accuracy at close range without sacrificing long-range measurements.

37. Monthly Report of 5 May 1936. NRL files, C-A9-4/EN8 (W5-25) Page Log Book Vol. III, pp. 9–19.

38. Memorandum from BuEng (H. G. Bowen, Chief of Bureau) to Director NRL, "Subject: Radio—BurEng Problem W5-2S–Aircraft Detection and Ranging," 12 June 1936. NRL files C-S67/36 (6-10-W9).

39. Memorandum from Director NRL to Chief of BuEng, "Subject: Radio—Use of Radio to Detect Enemy Vessels and Aircraft–Special Report on BuEng Problem W5-25," 11 June 1936. NRL files S-S67/36.

40. H. M. Cooley, Director NRL, to Admiral A. J. Hepburn, 7 January 1937. NRL files C-S67/36.

41. Commanding Officer, Destroyer Division Sixty-One to Commander Scouting Force, 22 October 1936 and endorsements. NRL files A16-31DD355.

42. See Chap. 8, p. 225.

43. C. O. Destroyer Division Sixty-One to Commander, Destroyer Force, 24 November 1936 and endorsements. NRL file DD355/A16-S (253).

44. Interview with Guthrie, October 6, 1944; Page's Log Book Vol. IV, pp. 27–8; R. M. Page, Memorandum to the Director NRL, 29 January 1937. NRL files, S-S67-5, "Radio Detection of Moving Objects." Secret, 6-36-6-39.

45. Page's Log Book, Vol. IV, pp. 10-13; see also Monthly Report on BuEng Problem W5-2S, dated 29 May 1936.

46. Page's Log Book, Vol. IV, pp. 33-4.

47. The prototype XAF, the Laboratory's first complete radar system (1938), used horizontal and not vertical polarization.

48. Page's Log Book, pp. 58–68.

49. This information was kindly supplied by Mr. R. L. Norton of Eitel–McCullough, Inc. in response to a request.

50. Page's Log Book Vol. IV, p. 73.

CHAPTER 5

Radar Research
By The Signal Corps

In the course of 1928, the Army's detection research at the Frankford Arsenal with infrared rays mentioned in Chapter 3 shifted to what seemed a much more promising line of attack. As a result of the accidental discovery that a red light reflected from a white shirt could be detected by the thalofide cell[1] used as the sensitive receiving element in these experiments, tests were undertaken at once to investigate the feasibility of detecting aircraft by a two-way method using reflected infrared energy emitted from a source on the ground. Late in the year reflected signals from aircraft were obtained in this manner at a slant range of 5000 yards.

In 1930 the infrared problem was taken over by the Signal Corps; and in February 1931 a project concerned with "Position Finding by Means of Light" was set up at the Signal Corps Laboratories, Fort Monmouth, New Jersey. At the time of inauguration of the detection program, this institution, an outgrowth of the Signal Corps Radio Laboratories established in 1918 on the site of Camp Alfred Vail, was housed in nine wooden buildings of World War I vintage and its personnel numbered 6 officers, 23 enlisted men, and 66 civilians. Its principal responsibility was the improvement of the Army's radio and wire communication and its meteorological service. Since June 1930 the Laboratories had been under the direction of Major William R. Blair.[2]

Some years before, as Officer in Charge of the Research and Engineering Division, Office of the Chief Signal Officer, Major Blair had called attention to the inadequacy of sound detection for fast-flying modern aircraft; and before a joint meeting of the technical committees of the Ordnance Department and Coast Artillery Corps, he had recommended the allocation of funds for the development of detection by means of heat or high-frequency radio. He was not able to pursue the matter until it came under his jurisdiction as Director of the Signal Corps Laboratories.

Infrared Research

The infrared detection project was assigned to the Sound and Light Section of the Laboratories, headed by a civilian physicist, S. H. Anderson. Tests performed during the next two years with reflected infrared rays revealed that to track even such a favorable target as a Navy blimp to

93

a distance of 32,000 feet would require a considerably stronger source of radiation than could be provided by a battery of four 60-in. searchlights. During the fiscal year 1933, the reflected infrared method was abandoned, and research was simultaneously concentrated on (1) the old method of using heat radiated from the target, and (2) reflected radio waves.

It would depart too far from our purpose to describe in any detail the work done during the next few years on the "one-way" method of heat detection. The difficulty with "background temperature", such as the heat radiated from clouds, which had caused so much trouble at the Frankford Arsenal, was partly overcome by using a compensated thermo-pile designed to cancel out the background temperature. In April 1934 such a thermopile backed by a 60-in. parabolic metal reflector was installed in a searchlight mounting at Battery Halleck, Fort Hancock, the military reservation on Sandy Hook, overlooking the entrance channel to New York Harbor. During the next few months ships of all kinds, includ-ing several of the most famous luxury liners, were picked up and tracked on the outbound ocean course. Similar apparatus was tested in the sum-mer of 1935 from a site at Navesink Lighthouse near the base of Sandy Hook, from which location it successfully picked up large tankers and freighters leaving or entering Ambrose Channel some 13,000 yards away. On one occasion the 79,000-ton *Normandie* was tracked on a clear after-noon to a distance of 30,000 yards. Tests with this equipment led to a largely speculative article in *The New York Times* in the summer of 1935 headlined, "Mystery Ray Sees Enemy at 50 Miles."

Meanwhile, unknown to the Signal Corps, another infrared project was in progress within the Army. As a result of exercises held at Fort Knox, Kentucky, in May 1933, revealing the hopeless inadequacy of sound loca-tors, the Corps of Engineers placed a contract with the General Electric Company. The duplication of effort was uncovered in the summer of 1934; and early in 1936, the Signal Corps equipment was shipped to Fort Monroe, Virginia for a secret comparative test with the General Electric Company's heat detector. On the basis of these tests the Coast Artillery Board pronounced both sets to be roughly comparable, both showing promise against marine targets, but neither adequate for the detection of aircraft.

The Use of Radio for Detection

In 1932–33 a program of radio detection research at the Signal Corps Laboratories got under way under the auspices of Anderson's Sound and Light Division. For a brief time two men were assigned to the work: Floyd Ostensen and W. D. Hershberger, the latter recently transferred from the Sound Division of the Naval Research Laboratory; after 1933 the work was carried on for several years by Hershberger alone with whatever occasional assistance he could obtain from his colleagues in the other divisions.

Although there is good reason to believe that Blair came to the Signal Corps Laboratories with some notions of his own about the possibilities of radio detection, it is hard to escape the conclusion that the setting up of a program was in large measure influenced by the successful experiments at the Naval Research Laboratory. As early as December 1930 the Navy's beat experiments were demonstrated to Army representatives from the Signal Corps, Coast Artillery, and Air Corps. In January 1931, and again in July, information on the Navy system was forwarded from the Chief Signal Officer to Fort Monmouth. To each of these communications Major Blair replied that the Navy's method was only suitable for general area protection, and not for accurate location of aircraft as required for antiaircraft fire. He called attention to an earlier analysis he had made pointing to the advantages of much higher frequencies.

Under the aegis of Major Blair, whose training as a physicist and whose personal research experience doubtless influenced him in this decision, the Signal Corps Laboratories embarked on a program of detection experiments using microwaves. Theoretically, it is true, these provided the ideal solution but practically, in view of the existing state of the art, they led the Signal Corps investigators down a blind alley which the Navy was fortunate enough to avoid.

We shall take up the Army's research on microwaves in more detail in a later section. It began with a survey of the current possibilities of generating, detecting, and amplifying waves less than a meter long, which in turn led to transmission tests made with Barkhausen–Kurz tubes and split-anode magnetrons. The experiments covered a range of wavelengths from 75 to 9 cm. In one of the early field tests on 9 cm the magnetron transmitter and the receiver were placed 40 ft apart, and by means of the interference phenomenon employed at the Naval Research Laboratory a Ford truck was detected at 250 ft as it moved down the perpendicular bisector of the line joining transmitter and receiver. Early in 1934 a detection system was built on 9 cm using an RCA split-anode magnetron, giving about a watt of amplitude-modulated cw radiation. This was a simple detection system using pure reflection, with the crystal audio receiver carefully shielded from the transmitter by a building—in one instance the corner of Navesink lighthouse. This did away with reliance on interference effects.[3] The tests were made during July and August in the presence of Irving W. Wolff and E. G. Linder of RCA, in equipment being placed on the parapet of Battery Halleck, Fort Hancock, and beamed at shipping in the harbor. Good reflected signals were received from a 500-ton boat 1000 yards out, but the apparatus was not powerful enough to detect reflections from larger boats passing through Ambrose Channel three miles away. These results were not very encouraging; much better results were being obtained by Zahl and Anderson with their infrared equipment. Nevertheless as late as July 1935 Blair showed no sign of abandoning the project. In his annual report of that date, he wrote:

> To date the distances at which reflected signals can be detected with radio-optical equipment are not great enough to be of value. However, with improvements in the radiated power of the transmitter and sensitivity of the receiver, this method of position finding may well reach a stage of usefulness.[4]

The Army's Adoption of the Pulse Method

Blair's faith in the potentialities of microwaves was not shared by the men most closely concerned with the project. Hershberger, in particular, became gradually convinced that the state of the microwave art was not such as to promise successful radio detection on these frequencies in the immediate future. He several times expressed his belief that the detection experiments should be done on longer waves where more powerful equipment could be built, as at the Naval Research Laboratory. This conflicted with the policy at Fort Monmouth which confined the Sound and Light Division, where the detection experiments were being carried out, to work below one meter, leaving work above this wavelength to the Radio Section.

Hershberger shared Blair's disapproval of the Navy's beat method of detection where precise location was not possible. The cw direct reflection method was also inadequate, for the receiver had to be effectively shielded or widely separated from the transmitter in order to receive the reflection.

In the fall of 1933 Hershberger proposed a pulse method closely resembling that described in the Löwy patent,[5] and first used by Breit and Tuve[6] and differing in certain respects from the method finally adopted for radar. Like Löwy, Hershberger had described his proposal[7] as one for an electrical Fizeau wheel, or for electrical shuttering. He proposed to interrupt the transmitted beam into short bursts or pulses and synchronously block and unblock the receiver. The shuttering frequencies applied by a multivibrator to both the transmitter and receiver were to be of the order of broadcast frequencies (i.e., pulses lasting 2–10 μsec). As Hershberger has described it, this provided for alternate operation of the transmitter and receiver so that the receiver is blocked while the transmitter is operative and the transmitter is silent when the receiver is operative. This proposal differs from radar in that the pulse length and the interval between the pulses would be of approximately the same duration, whereas radar is characterized by short pulses separated by relatively long intervals, of the order of 1000 μsec. If one attempted to put the reflected signal from Hershberger's projected apparatus on a cathode-ray tube with this high pulse recurrence frequency, only objects nearer than two miles would be received unambiguously without being blanked out by the succeeding pulse. It was not, however, Hershberger's intention that the system should give range data as it stood, or that it should be used with a cathode-ray tube. It was expected that the output of the receiver would go to a meter which would integrate a succession of the received responses. Ranging would be theoretically possible by varying the shuttering frequency, but not obtained by direct reading as in radar.

That the Signal Corps Laboratories intended to investigate the Hershberger proposal is indicated by a paragraph of Major Blair's Annual Report, dated July 1934, in which the Director expressed his opinion that a new approach to the radio detection problem was essential and remarks that consideration "is now being given to the scheme of projecting an interrupted sequence of trains of oscillations against the target and attempting to detect echoes during the interstices between the projections."[8] Although Hershberger in 1934 made unsuccessful attempts to pulse or key his 50-W, 50-cm Hollman tube by means of a multivibrator, he devoted most of his time to developing detectors for microwave frequencies, and no serious efforts were made to push the new detection proposal. Outside stimuli brought a revived activity in this direction early in 1936.

During 1935 the Signal Corps Laboratories received each month the brief and cryptic reports from the Naval Research Laboratory reporting on the work of Page and Young on pulse receivers. Late in the year Lieutenant Colonel Roger B. Colton of the Office of the Chief Signal Officer called the attention of Colonel Blair and his subordinates to a recent proposal for radio detection which had reached the Office from the United States Bureau of Standards. At a conference held at Fort Monmouth on December 5, 1935 Colonel Colton requested that Hershberger write an appraisal of this Bureau of Standards method and accompany it with a general statement as to the best solution to the problem of detecting aircraft by radio means. It was also proposed that Hershberger should visit both the Bureau of Standards and the Naval Research Laboratory in the near future.

Colonel Colton's drive and initiative gave a strong impetus to the detection work at Fort Monmouth. Hershberger's two short studies were completed the day after the conference. He reported that the proposal of Diamond and Dunmore at the Bureau of Standards was fundamentally the same as the beat or interference method used at NRL; and he used the occasion of the more general memorandum to call the attention of his superiors to the principal reasons for the failure up to that time of the Army detection program: inadequate personnel—the detection program was still only a one-man affair—and the preoccupation with microwaves. He strongly urged expanding the personnel and advised that attention be concentrated on the wavelength range from about 50 to 150 cm.

Hershberger compared the two methods of detection under consideration at the Signal Corps Laboratories: the method requiring wide separation or shielding of receiver and transmitter, and the electrical shuttering method which promised to permit the transmitter and receiver to operate in close proximity. He went on to describe the carrying out of the second scheme as "a tremendous undertaking", and concluded:

> If the problem of aircraft detection is to be prosecuted with any vigor, more personnel is needed for experimental work in building tubes, developing circuits and apparatus, and conducting field tests. It is not a one-man task.[9]

In response to Colton's request, Hershberger spent January 15–18, 1936 in Washington, during which time he devoted a half-day to the Bureau of Standards and a day and a half to the Naval Research Laboratory. His visit to the Bureau added nothing except the discovery that no experimental work on radio detection had been done by Diamond and Dunmore.

At the Naval Research Laboratory, Hershberger witnessed a demonstration of the Taylor–Young–Hyland interference device and included in his report a lengthy description of this equipment. At the close of his report, he mentioned the Young–Page investigations on the pulse method which had not reached the stage in which echoes were received. He described the method as "somewhat akin to the 'pulse' method of measuring the heights of ionized layers in the ionosphere" and remarked that "a method somewhat similar" was under consideration at the Signal Corps Laboratories as a promising approach, and would probably be tried if time and facilities were made available for the purpose.

Hershberger's report was forwarded to the Chief Signal Officer the second week in February by Colonel Blair, who strongly indicated his preference for the pulse method and concluded that with the necessary directive and facilities, the Signal Corps Laboratories could produce within a short time tangible results in the development of a one-station radio locator capable of yielding range as well as direction.

In February 1936 the War Department, by a directive of the Adjutant General, put an end to the jurisdictional confusion into which the radio detection had fallen within the Army and assigned the further development unequivocally to the Signal Corps, emphasizing their desire "that the development of these detectors be given the highest priority practicable, with particular emphasis on the development of the detector of aircraft". The directive was accompanied by military characteristics—a theoretical preliminary to all procurement of new devices—for a heat detector to be used against marine surface craft and for a detector of aircraft. Both were thought of primarily as searchlight directing devices. The aircraft detector must reach 20,000 yards under average atmospheric conditions, and 10,000 yards with rain, mist, smoke, or fog. It must point a searchlight within 1° in azimuth or elevation.

Although this was the official go-ahead signal, funds were not readily forthcoming to back it up. In April, the Chief Signal Officer requested that the War Department make available $40,000 for the project during the coming fiscal year, obtaining it, if necessary, by a special request to Congress. The reply, by order of the Secretary of War, deserves to be quoted in full:

The funds set up for research and development of Signal Corps projects are considered sufficient in view of the many other pressing needs of the Army. The development of an efficient means of detecting the approach of aircraft is considered of such vital importance to all branches of the Army that it is considered essential to place it in the highest priority. It is therefore directed that the Chief Signal Officer provide the additional funds required for the development of the detection of aircraft by a reduction in the amounts now set up in the fiscal years 1936, 1937, and 1938, for less urgent projects.

The dilemma of the War Department, aware of the importance of the new project but incapable of finding the necessary funds, could hardly be more evident. The Chief Signal Officer, vigorously supported by the Chief of the Coast Artillery, renewed the attack, but the War Department reaffirmed its earlier position.

The Army's First Radar

Work began on the Army's first pulse device in the spring of 1936, the effort being concentrated on building a satisfactory transmitter operating in the neighborhood of 100 MHz. In April, Hershberger was joined by Robert H. Noyes, who had been transferred from the Aircraft Radio Laboratories at Wright Field. Together during the summer and early fall, they assembled a transmitter operating on 100 MHz,[11] a 20-kHz master oscillator to establish the pulse repetition frequency, a multivibrator keying unit that gave pulses 2 or 3 μsec long, a phaser or phase shifter, and a super-regenerative receiver. The indicator tube was a commercial RCA cathode-ray oscilloscope.

In October 1936 Hershberger left the Signal Corps Laboratories to complete his doctorate in physics at the University of Pennsylvania. The following month the detection project was transferred to the Radio Section under Paul Watson. Two officers, whose names are closely associated with the Signal Corps development of radar, entered the picture at this time. Lt. Colonel Roger B. Colton, whose interest in the radio detection program as a representative of the Chief Signal Officer has already been mentioned, came to Fort Monmouth in August 1936 as Blair's executive officer. The transfer of the detection project to the Radio Section put it under the supervision of Captain Rex V. D. Corput, Jr., officer in charge of radio projects, who later became the first Director of the Signal Corps Radar Laboratory.

In the Radio Section, after Noyes' transfer, there were 15 people in addition to Watson himself. In the months that followed, about half of the engineering time of this group was devoted to the development of what we now call radar. For a while there was no sharp allocation of time or manpower between the detection and communication work. On January 30, 1937, Watson divided the Radio Section into a Receiver Group, a Transmitter Group, and an Engineering Group. The Engineering Group was not particularly concerned with detection, but in the Receiver Group headed by Ralph I. Cole, three engineers, John Hessel, James Moore, and A. C. Prichard, were particularly active in the development. In the Transmitter Group headed by William S. Marks, work was carried out by all the engineers: Noyes, Melvin D. Baller, and John J. Slattery.

In the late fall, after the successful development of a multivibrator keyer which produced a 300-V pulse powerful enough to operate a 100-W transmitter, the principal remaining difficulty was to produce a receiver that would recover rapidly enough from the outgoing impulse to receive the weak returning echo. The first scheme tried was to have a quenching frequency from the master oscillator applied to the super-regenerative receiver at the moment of the "main bang," allowing it to recover its sensitivity before the return of the echo signal. The trouble was, of course, encountered in devising a receiver with the proper recover time. The alternative approach was to protect the receiver as much as possible from the effect of the main impulse. An obvious though clumsy method was to place the transmitter some distance away from the receiver, using two separate antennas for transmitting and receiving. Even for this purpose, an improved receiver was required. In the earliest test, the separation of transmitter and receiver was employed, and use was made of a superheterodyne receiver with low-Q circuits to ensure a sharp damping and rapid recovery time. It is noteworthy at this period that there was no detailed exchange of information with the Navy on the common problems of component design, and no exchange of visits by engineering personnel after Page's appearance at the Signal Corps Laboratories in April.

The Army's first successful tests with pulse radar date from December 14–15, 1936. Some rather hastily assembled equipment built with the components described above was installed early in the month in two trucks, one carrying the transmitting components, another carrying the receiver. The transmitting antenna was a Yagi, only slightly directional. The receiving antenna used a simple dipole. The trucks were driven to a point near Princeton Junction, New Jersey, where it would be easy to pick up commercial aircraft flying on the Newark–Camden airway. The receiver was set up approximately 1 mile from the transmitter, screened by a grove of trees from the transmitter's radiation. The observers were shortly rewarded by picking up an aircraft signal on the cathode-ray tube. James Moore, who had built the superheterodyne receiver being used, was the first to observe the signal. The range indication showed the plane, whose presence was confirmed visually, to be about seven miles away. The tests were repeated without difficulty on the following day.

During the early months of 1937, the effort was concentrated on designing directive antenna systems for use with both transmitter and receiver. An initial attempt was made to design parabolic cylinders of wire netting, with a half-wave dipole along the focal line. Such reflectors for the wavelength in question were found to be prohibitively large, so that investigators rejected them in favor of broadside arrays. In this work they received considerable assistance from the literature which had accumulated in the previous decade, especially from the papers of G. C. Southworth and Edmund Bruce, and by direct consultation with these and other Bell Laboratories engineers. During the winter months of 1937, they built a series of dipole arrays. The size limitation was still important; this they circumvented to some extent by building separate antennas: broad antennas to give a sharp beam in azimuth, narrow antennas to give a sharp beam for elevation. The arrays were horizontally polarized and backed by a parasitically excited reflector. These antennas were tested against planes specially dispatched from Mitchell Field, and against blimps from the Naval Air Station at Lakehurst, New Jersey. By February they had built azimuth antennas, for transmitter and receiver, two dipoles high and six wide, with which they were able to track a B-10B bomber up to 23 miles. The dimensions of these arrays were later changed to four dipoles high by eight wide. Similar antennas with vertical directivity were then developed. These were also parasitically excited reflectors and were pivoted on a frame so that they could tilt in elevation while being rotated in azimuth.

In preparation for a scheduled demonstration of the equipment to take place in May, in the presence of the Secretary of War, attempts were made to develop a more powerful transmitter. One was finally developed by

Marks and Baller using a pair of RCA 806 air-cooled tubes operated well above their rated voltage. The development of an electrical phase shifter for precise measurement of range took place at about the same time. This was an accurate potentiometer with two banks of resistors fed 90° out of phase, and so arranged as to deliver a constant voltage of variable phase to the oscilloscope. A pointer on the potentiometer arm indicated the range to the target. A sequence of preliminary tests made during February and March showed that with the improvement of components the range at which tracking was possible was steadily increasing.

While this development of the 110-MHz equipment was progressing, the Signal Corps Laboratories undertook the development of equipment on 240 MHz which promised to yield a detector of much more compact proportions. A transmitter employing a pair of RCA 834 tubes rated at 100 W was constructed by Slattery for this frequency, and at the same time a super-regenerative receiver was built by Hessel. Smaller arrays made it possible to have only two antennas, one for transmitting and one for receiving, eliminating the need for separate antennas for azimuth and elevation. In May 1937 this equipment was used in a pioneer effort to track meteorological hydrogen-filled balloons to see whether this might offer a method of determining wind conditions aloft, under conditions where an optical theodolite was inadequate.

Two historic demonstrations of this primitive Army radar took place late in May 1937. On May 18 and 19, preliminary tests were conducted in the presence of the Chief Signal Officer, Major General James B. Allison; the Chief of Coast Artillery; the Assistant Chief of the Air Corps; and representatives of the Ordnance Department and the Corps of Engineers. The performance was repeated the following week for the benefit of the Secretary of War, the Honorable Harry H. Woodring; the Chief of Staff, General Malin Craig; and other dignitaries, including members of the Military Affairs Committees of the United States Senate and House of Representatives. On this occasion, both the 110-MHz and the still more experimental 240-MHz equipment were demonstrated. The most important feature of the show was the combined and coordinated operation of the 110-MHz pulse detector with searchlights and with the much improved heat detector. The radio detection apparatus, with its greater range and searching ability, was used to pick up the test planes and when they had come within range to hand over the work to the thermal detector. This, in turn, with a greater directional accuracy than the radar possessed, was used for accurate pointing. The demonstrations were carried on both day and night. In daylight, the target planes were followed visually with telescopes controlled by the radar and the thermal detector. In the nighttime tests, the planes were illuminated by searchlights controlled by the thermal detector. In the first demonstrations, weather conditions prevented the use of the thermal detector; the radio detector was used alone, and the results were only moderately successful.

In the May 26 demonstration, weather permitted the use of the complete system and the score was perfect; on each of the four night approaches the planes were successfully illuminated.

Despite this success, it was recognized that the radio equipment did not quite satisfy the military characteristics set down in Febraury 1936. The information was not continuous; the error in azimuth was considerably larger than the 1° that had been specified; although the maximum range requirements could be met under average atmospheric conditions, the minimum range performance was poor; at a slant range of less than 4000 yards, accurate readings were not possible. Lastly, the equipment was sure to be vulnerable to jamming.

The demonstration brought more than a flattering letter of commendation from the Secretary of War. Secretary Woodring made haste to inform President Roosevelt of the success of the demonstration. In June the Chief Signal Officer inquired of Col. Blair whether $200,000 was an adequate estimate for the cost of the research in the coming fiscal year.

The presence of the Chief of Air Corps at the tests led to a request for the development of a long-range detector with sufficient range against enemy bombers to allow interceptors to take off. Military characteristics were shortly afterward drawn up for such a long-range detector. In August, the Signal Corps Technical Committee revised its military characteristics for coastal detection of ships to include the possibility of radio detection, and altered the specifications for what became the SCR-268.

The demonstrations of May 1937 brought to an end the initial phase of the Army's radar program. The Signal Corps Laboratories embarked on the nearly simultaneous design of the SCR-268, a set developed directly from the experimental equipment just described, and intended for searchlight control and for use with the M-4 gun director, and of the long-range detector, which, in its mobile and stationary versions, was known, respectively, as SCR-270 and SCR-271.

Development of the SCR-268

Although the experimental work at the Signal Corps Laboratories had been conducted with great secrecy, the proximity of the Laboratories to so many other activities at Fort Monmouth disturbed the Chief of Staff who directed the Chief Signal Officer to have the secret equipment moved to a more isolated location for all future work. This led to the establishment during the summer of 1937 of a field station on Sandy Hook within the military reservation of Fort Hancock, about 20 min away by road from Fort Monmouth. During the first winter, work at the field station was done in the open without shelters from the weather. In 1938 antenna shelters were erected of all-wooden construction, buttressed from the outside by heavy timbers. Some time later, a temporary building of steel construction was erected to house shops and storerooms. In order to test radar on more elevated sites, the Signal Corps was given limited facilities by the United States Lighthouse Service at the Navesink Lighthouse Reservation, Sea Bright, N.J., where some of the early microwave tests had been conducted.

When the outbreak of war in Europe brought an expansion of American military preparations, the base facilities at Fort Monmouth became overcrowded, and the administrative as well as technical activities of the Radio Position Finding work was transferred to the Fort Hancock field station. This area gradually filled up, especially after the production of the Army's radar equipment led the commercial concerns which were producing the equipment to build their own antenna shelters on the reservation. Early in 1942 the Signal Corps Radar Laboratory, which later became the Camp Evans Signal Laboratory, was established on an extensive tract of land at Belmar, New Jersey, a few miles inland from the coast.

The Laboratories remained under the direction of Colonel Blair until his retirement in October 1938, though during the last months ill health forced him to delegate the responsibility to others. Colonel Colton was director from June 1938 to August 1941, when he was succeeded by Major Corput, who continued as director until January 1942. The civilian head of the Radio Position Finding Section was Paul Watson, later director of the Camp Evans Signal Laboratory until his death in 1943, at which time he held the rank of Lieutenant Colonel.

In the echelon above the Director of the Signal Corps Laboratories, during the critical period from 1937–1941, was the Research and Development Division, Office of the Chief Signal Officer. This was headed by men whose names we shall frequently encounter in our story: Lt. Colonel Louis B. Bender (later retired with the rank of colonel); and until November 1941, Colonel Hugh Mitchell. In August 1941 a new echelon (the Materiel Branch) was set up above the Research and Development Division and other procurement divisions. Colonel Colton was put in charge of the Materiel Branch, and in November 1941, Colonel Tom C. Rives was made officer in charge of the Research and Development Division. During this period the whole program was under the supervision of the Chief Signal officers who were Major General James B. Allison (January 1935–September 1937); Major General Joseph O. Mauborgne (October 1937–September 1941); Major General Dawson Olmstead (September 1941–June 1943).

The first goal at the Signal Corps Laboratories after the tests of May 1937 was the development of a service test model based upon the experimental equipment used in these tests. Only one unit of this set, called the SCR-278-T1, was completed. It was the first Army radar to employ lobe switching for accurate "positioning" of the target, and it was the last model to include a heat detection unit. After the service tests in the fall of 1938, the thermal unit was abandoned, partly because of its unreliability and its susceptibility to weather conditions, partly because the lobe-switching feature of the radar gave some degree of accuracy in determining position.

At Fort Monmouth[12] the general idea of lobe switching seems to have been first suggested by Colonel Colton, and it was first explored in the summer of 1937. The scheme involves the use of a divided receiving array and can be understood by considering Fig. 5-1. The two-lobed "beam" there shown is not an actual beam of transmitted radiation but represents by a polar diagram the power received by two arrays placed side-by-side at a slight angle to one another. The two directions of maximum sensitivity, corresponding to the axis of the lobe in each case, differ by some angle θ, say about 30°. The shaded area represents the overlapping region common to both beams. Signals from both receiving arrays are displayed simultaneously on an indicator so that they can be compared. Only when the target is dead ahead (along the line indicated by the arrow) will both signals have the same intensity and both pips the same height. If the "crossover" point of the two beams is properly chosen the sensitivity will

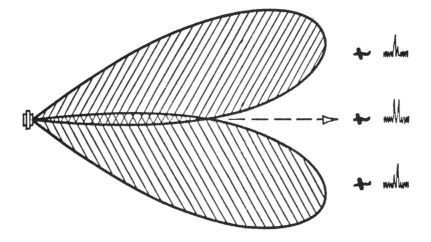

FIG. 5-1. Lobe switching.

be high; for the signal strength varies more rapidly with changes in azimuth along the edges of a beam than at its extremity.

The effect described above was first obtained by a physical division of the azimuth receiving array into two halves, placed side-by-side at an angle to one another. A single receiver was mechanically switched back and forth between the two antenna halves so that signals from both beams appeared simultaneously on the oscilloscope. The first such double array was built in August 1937. By the time the SCR-268-T1 was tested, it was equipped with a double antenna for receiving both in azimuth and in elevation. In most other respects, it did not differ greatly from the experimental equipment used in the May tests, though there were a number of important improvements.

The model of the SCR-268-T1 was completed early in September and arrangements and preparations were under way for service testing in the next few weeks. The schedule was unpredictably disrupted by the famous hurricane which swept the eastern seaboard in late September 1938. The radar set, which was in an exposed location at Fort Hancock, was quite severely damaged; and it was not until mid-October that the equipment was completely restored. Preliminary tests were then carried out by Coast Artillery officers at Fort Hancock, after which the equipment was transported, not without incident and confusion, to Fort Monroe. Here the service tests were carried out between November 30 and December 13.

Extensive observations were made on the ability to pick up aircraft targets. More than 4000 carefully recorded readings—checked, when possible, by optical tracking instruments, and at night by means of searchlights—indicated that the equipment fell somewhat short of the accuracy required, though range performance was pronounced satisfactory. Detailed comparisons with sound locator devices, which were still the regulation means of searchlight control of anti-aircraft batteries, found the radar superior in every respect except for its greater size and for the more numerous and more specialized operating personnel that it demanded.

Some incidental but significant diversions were provided when radar demonstrated its ability to pick up shell bursts, an experiment suggested by Colonel Colton, and—quite incidental to a test conducted for visiting Air Corps officers—obligingly showed that it could guide a friendly plane to safety, the first time, in America at least, that radar was called upon to perform this important role. This, combined with the overall performance of the radar, heightened Air Corps interest in the equipment, especially in the possibilities of long-range detection for air defense.

The board concluded that the device warranted a continuous and energetic development program. The most important consequence, besides the decision already mentioned to eliminate the thermal unit, was the recommendation to classify the SCR-268 as Required Type, Development Type, and Limited Procurement Type, which can be interpreted to mean its adoption as regular army equipment. This official action was approved by the Adjutant General on March 29, 1939.

While the SCR-268-T1 was being built and tested, work had been going forward in parallel on higher frequencies. Early in 1938, though there was considerable progress with experimental equipment on 240 MHz, because greater power was available from commercial tubes on the latter frequency. Sufficient progress, however, had been made in the meanwhile, so that it was decided that the 268's should henceforth be on a higher frequency. The problem was a satisfactory transmitter.

The earliest transmitter on 240 MHz had employed a pair of RCA 834 tubes. A year later a more powerful transmitter was in operation using four of these tubes. The performance, however, was unsatisfactory, and a new transmitter was substituted using four Eimac 100-T tubes. Thus modified, the high-frequency system tracked planes to 10,000 yards in May 1938. In that month, however, work on this project was temporarily discontinued in favor of the SCR-260-T1. Little could be done, in fact, with the tubes then available, to produce a satisfactory set. Specially designed tubes were clearly required. The cooperation of industry in tube development had first been sought in September 1937 when a contract was let to the Westinghouse Electric and Manufacturing Company for the development and manufacture of special high-power triodes at 100 MHz. This resulted in the VT-122 tubes for the SCR-270 and 271 and set a precedent for contracts on higher frequencies.

In the summer of 1938 a thorough survey of available commercial tubes was made to learn their characteristics and potentialities. During July 1938 special transmitters were built on 240 and 205 MHz using four Eimac 100-TL tubes. As a result of these tests the design gradually crystallized around the Eimac 100-TL tubes. The merits of this series of tubes have already been described in the preceding chapter. It was decided to ask the Eitel–McCullough Company in California to produce a higher powered tube on 200 MHz by modifying the Eimac 100 TL's. A contract for the tubes was let in 1938 with this company. This energetic concern responded with models based on the designs suggested by the Signal Corps; these were shipped east by Air Express in the record time of a few weeks after the original request. From this development came the Eimac 100 TS's ("S" for special) which were eventually procured in quantities running into the tens of thousands. A 205-MHz transmitter using these tubes was completed in December 1938.

During the winter of 1938–1939 the design was worked out for a radar set on 205 MHz, eliminating the defects revealed in service testing the SCR-268-T1. It was decided to profit by the higher frequency in such a manner as to obtain both an increase in accuracy and some decrease in the size of the equipment. The design of this new set, the SCR-268-T2, still called for three separate antenna mount units, one for the transmitter and two for the receivers. The design was completed in the spring of 1939, and in the middle of May a model of the SCR-268-T2 was turned over to a detachment from the 62nd Coast Artillery, anti-aircraft, for an extended field test.

The outbreak of the European war in September 1939 suddenly shifted the emphasis from the searchlight and anti-aircraft radar program to the development of long-range equipment. In consequence, the SCR-268-T2 was never officially service tested, and during the lull an improved design emerged for a set to be built entirely on a single mount. This set was also characterized by a feature of considerable importance: electronic lobe switching. This is too complicated to be explained in such a volume as this, beyond saying that by an ingenious phase delay method, the double-lobe pattern can be produced by a single stacked dipole array.[13]

The adoption of electronic lobe switching, which permitted double-lobe performance with a single azimuth or elevation array, was facilitated by the decision to use a single mount for all the antennas. In March 1940 the single-mount SCR-268-T3 began to take shape as much lighter and more compact equipment than earlier models. It was this design which was followed in the large-scale commercial production of the SCR-268.

Persistent attempts had been made to increase the power of the transmitter by the use of two, four, and then six tubes operating in parallel. For the SCR-268-T3 a transmitter using eight Eimac 100-TS tubes operating in a symmetrical ring in push–pull had been designed by Melvin Baller.

The eight-tube ring transmitter was used in the service test model and in the 18 units of the T3 manufactured by the Signal Corps Laboratories in late 1940 and 1941. A 16-tube transmitter was developed during 1940 and was subsequently adopted for the production equipment.

The SCR-268-T3 was service tested April 15–25, 1940 at Fort Hancock by a subcommittee of the Coast Artillery Board. Nine enlisted men from Fort Monroe operated the searchlight and the optical height finder, while other operating personnel were provided from the Signal Corps Laboratories and Fort Hancock. The tests were not as extensive as the previous ones, but the Board noted the improvement in accuracy and the marked reduction in size and greater ease of operation. As a result of the recommendations of the board and the Signal Corps Technical Committee, the set was officially standardized, i.e., adopted as Required Type, adopted type, standard article, and classified as secret.

Before the actual standardization, production began on what later came to be called a "crash" basis. Eighteen sets of the SCR-268-T3, with the eight-tube transmitter, were built by the Signal Corps Laboratories for training purposes out of research funds of the fiscal year 1940. The Laboratories purchased some of the components, built others, and conducted the final assembly at Fort Hancock.

The course of the war in Europe shook loose the purse strings and brought about the abandonment of piecemeal procurement in favor of contracting with a single commercial supplier for the entire set, the supplier being responsible for subcontracting those components which he could not manufacture. This policy was decided upon at a conference in Washington, May 20–21, 1940, of key research, development, and supply officers from the Signal Corps Laboratories and the Office of the Chief Signal Officer. Here it was also decided that because of the novelty and special secrecy of the new equipment, procurement would be handled through the Signal Corps Laboratories rather than through usual channels and the final inspection would be conducted at Fort Hancock.

Bids for the production of the SCR-268 were requested from the Western Electric Company and RCA on July 1, 1940. Western Electric's bid was accepted, and a contract was signed in 1940, calling for the delivery of a total of 212 units, ten of which were to be delivered before February 21, 1941. The order was later revised upward to 418 sets; later, in June 1941, to a total of 520 sets.

The first of the 18 crash units was completed by the Signal Corps in December 1940; the rest were delivered in early 1941. By the end of the fiscal year, June 30, 1941, a total of 85 sets, crash and production, had been delivered and accepted, while 22 others were awaiting test and inspection.

During the fall of 1940, a school of staff sergeants of the Coast Artillery was conducted at the Fort Hancock Laboratory. Instruction was furnished by personnel of that detachment of the 62nd Coast Artillery antiaircraft which had been attached to the laboratory during the development of the SCR-268.

The first installations outside the United States were made in Panama. Early in 1941 a group of Coast Artillery officers, all graduates of the Fort Hancock SCR-268 course, arrived in the Canal Zone and during May and June conducted a theoretical course for personnel of the Panama Canal Department. Practical operational experience was acquired after seven SCR-268 sets arrived early in July. One of these units was installed on Naos Island and another on Perico Island. By October 1941 the seven SCR-268's were being set up at various battery positions in the Panama Department.

Several SCR-268 sets were used during the fall of 1941 in maneuvers of the First and Second Armies. In December 1941 there were 16 SCR-268's assigned to Coast Artillery units in Honolulu, and others were in use by Marine Defense Battalions on other islands of the Hawaiian group. About the middle of September 1941, soon after the first American troops had arrived in Iceland to protect the North Atlantic convoy route, an SCR-268 was installed there on a site called Skeleton Hill.

The SCR-268 remained the Army's standard searchlight radar equipment until January 1944 when, after the introduction of mirowave equipment, it became "limited standard," a polite term for obsolescent. In addition to the searchlight function, the set was used with the M-4 director to point 90-mm anti-aircraft guns for blind firing. This combination was only a stopgap, for the M-4 gun director had been designed to operate with optical equipment and so used target height instead of range. The SCR-268 was equipped with a height-range converter which, by means of a three-dimensional cam, multiplied slant range by the elevation angle to get height. It is clear that the height data were only as dependable as the elevation angle data; and this, because of ground reflections, was distinctly unreliable in the SCR-268.

In spite of its shortcomings, the 268 was used with the M-4 director in various theaters until the more modern gun directors and gun-laying radars, such as the SCR-545 and the SCR-584, became available in quantity in 1944. With careful, skilled handling, good results were often obtained. Considerable success was reported by the Marines who used the SCR-268's to shoot down Japanese planes over Pacific Ocean outposts. It was also widely employed in North Africa. General Colton was informed by Fifth Army personnel that a large percentage of enemy aircraft shot down by our 90-mm AA artillery in the North African Campaign could be credited to the use of the SCR-268 as a gunlaying set. But in Europe its day as an anti-aircraft set was over when the Germans jammed it at Anzio.

Because of the absence of other, more specialized equipment, the SCR-268 was used for a number of auxiliary duties, for which it had not been specifically designed and for which it was not particularly well suited. It was used to give advance warning to AA batteries which then tracked and fired by optical methods; it was sometimes used as a substitute for the M-1 optical height finder when there were severe shortages. The 268's were also used as medium-range warning equipment by Aircraft Warning Units. It will be recalled that twenty 268's had been assigned for this purpose to the Air Defense Command from among the first sets to come off the Western Electric production line. These were modified, by altering the pulse recurrence frequency and raising the plate voltage on the tubes, to produce warning sets having a range of 70 miles. The set thus evolved, with some other modifications, was given the designation SCR-516. Only a small number of these were ever built.

Long-Range Early Warning Sets

Unlike the U. S. Navy or the British Air Ministry, whose work we shall describe in the next chapter, the U. S. Army Signal Corps did not begin its radar development with the design of a long-range aircraft warning set. Nevertheless, the long-range sets, the SCR-270 and SCR-271, were the first sets to reach the troops. The SCR-268 had been designed to satisfy the immediate needs of the Coast Artillery which was intent on finding a substitute for their sound locators used for short-range warning and the direction of searchlights. The long-range sets grew directly out of the early work on the searchlight director. Despite their different function and greatly enhanced power, these sets are more closely related to the original experimental equipment than the SCR-268 itself. They are electrically very similar, although in some ways simpler; they operate on the original low frequency (100 MHz) which was abandoned for the 268; they do not use lobe switching nor give height information.

The earliest formal inquiry about the possibilities of radar for long-range warning was made to the Signal Corps by the Air Corps in May 1937, and was motivated by concern over the protection of the Panama Canal. Later that same month Air Corps representatives went to the demonstration in which the Army experimental equipment was unveiled for the first time to visiting dignitaries from the War Department and from Congress.

Brigadier General H. H. Arnold, then Assistant Chief of the Air Corps, attended the first demonstration on May 18–19. The Chief of the Air Corps, Major General Oscar Westover, was present at the second performance a week later. The success of the demonstration led to a formal request from the Air Corps that long-range equipment be undertaken. This was accompanied by a set of military characteristics which specified a range of 50 miles and—somewhat vaguely—sufficient accuracy to permit interception of enemy planes by our fighters. These characteristics were approved by the Adjutant General in August 1937. The demonstration of the searchlight control apparatus, the SCR-268-T1, at Fort Monroe in the fall of 1940, served further, as we have seen, to arouse interest within the Air Corps in the possibilities of radio detection.

Work began on long-range detectors in the summer of 1937 at the Signal Corps Laboratories. A separate project was set up to deal with the high-power development. Work was, of course, concentrated on the principal problem: trying to find a transmitter powerful enough to meet the requirements. The initial experimentation consisted of carefully testing for this new purpose the various transmitters developed for the SCR-268. The most fruitful step, however, was the letting of a contract with the Westinghouse Company in October 1937 for the development of high-power triodes to operate on 110 MHz.

The tube finally brought forth by the Westinghouse Lamp Divison, after about 1 year of effort, became the heart of the long-range sets. It had many novel features designed to provide the important prerequisites to high output power: strong electron emission from the filament; the ability to withstand the high plate voltages required; and efficient means of dissipating the great amount of heat generated.

The VT-122, as it was called, had a filament consisting of a number of wires shaped into a hemisphere and consequently having a large emissive surface. A mesh grid of similar shape fitted around the filament like a cap. The plate consisted of a still larger copper hemisphere mounted outside the grid and forming the bottom of the tube. The external face of the plate was directly cooled by circulating water. So great was the heat produced that the cooling equipment weighed some 700 lb and thus constituted a component nearly as large as the transmitter itself.

The first four samples of the VT-122 were delivered to the Signal Corps in September 1938. During the next few months, tubes with steadily improving characteristics were delivered; consequently, the development of long-range sets could go forward rapidly in 1939.

Great improvements in engineering design were necessary to handle the increased power derived from the Westinghouse tube. The overall direction of the project, and especially the design of the transmitter built around the new tubes, was the responsibility of Paul Watson, head of the Radio Position Finding Section.

Aside from the new, high-powered transmitter, the novel feature of the SCR-270 series was the system of duplex operation that permitted a single antenna to be used for transmitting and receiving. In the early equipment, recourse was had to using separate antennas for transmitting and receiving in order to protect the sensitive receiver from the great bursts of energy put out in the transmitted pulse. The solution that was adopted resembled that devised at the Naval Research Laboratory, though it seems to have been independently conceived. Though unquestionably important, the duplexer TR switch was less vital to the Army with its fixed land installations than it had been to the Navy or would be later for the development of airborne radar systems.

The principle of the TR switch has already been explained in the preceding chapter. The Army's earliest spark gaps were open structures of metal with electrodes adjustable by hand screws. Much work was done to discover the proper material for the electrodes and to obtain their precise adjustment, to prevent premature breakdown of the gap before the voltage had built up to the proper level, and also to avoid welding together the electrodes. In a later model the gaps were placed in a sealed tube filled with an inert gas.

The TR junctions did not act as perfect switches and let too much energy through to the receiver. Additional protection was provided in a number of ways, by putting a second spark gap at the receiver input terminals, by choosing rugged tubes with a heavy grid for the first stage of the receiver, and by using for the second stage an "orbital beam" tube developed by RCA.

By May 1939, the principal components had been sufficiently developed to permit their incorporation in a complete service test model, the SCR-270-T1. In its first laboratory trials in June, it was consistently successful in picking up and tracking planes flying at 8000 ft with ranges out to 85 miles. Occasional indications against aircraft were reported as far out as 138 miles.

This performance exceeded the original requirements, but did not quite meet altered characteristics that had been substituted in February 1939, and which increased the desired reliable range to 120 miles. It was at this time the long-range radar received its official baptism, the designation SCR-270 being given to a long-range mobile set and SCR-271 to a fixed version of the same equipment.

Two models of the SCR-270-T1 were made and tested during the summer and fall of 1941. One of these was turned over for service testing by the Signal Corps Board which was officially assigned to conduct the tests because Signal Corps troops were destined to operate the long-range sets, even though they would be under the operational control of the Air Corps. In mid-November this second set was installed for testing purposes at West Peak, within the city limits of Meriden, Connecticut, overlooking Long Island Sound.

After extended tests the Signal Corps Board gave the stamp of its approval to the long-range sets, though recommending a number of modifications. It was the board's opinion that the set was able to give satisfactory long-range warning against high-flying planes but that visual observers would probably still be required against low-flying aircraft. In order of importance, the board felt the sets would be of the greatest value in (1) protection of important strategic bases such as the Panama Canal Zone, Hawaii, and Puerto Rico; (2) protection of the continental shores; (3) protection of land frontiers. The board recommended that since for the above purposes most of the installations could be made before the outbreak of war, the fixed SCR-271 sets were better suited than the mobile SCR-270's.

Of the strategic points to be protected Panama was deemed most important. While the first available SCR-271 sets were being completed, sites were picked on both sides of the Isthmus to cover the Atlantic and Pacific approaches to the Canal. The first service test model was completed during the early months of 1940 and was turned over to the Signal Corps Board for service testing. The tests were completed in April and procurement of this set was officially approved. While steps were being taken to launch quantity production at Westinghouse, which had been chosen as the producer, the service test model and another laboratory-built model of the SCR-271 were shipped to Panama in June 1940.

The two locations had been chosen by Colonel Colton on an advance siting survey of the Canal Zone. At Fort Sherman, across Limón Bay from Colón, a 400-ft hill was selected for a site to command the approaches from the Caribbean. The second site, covering the Pacific end of the Canal, was on Taboga Island, about 12 miles from the coast in the Gulf of Panama. About one month was required after the arrival of the sets before they were installed and running. The Fort Sherman installation was completed in September 1940, and the Taboga set one month later. They were handed over to the First Signal Aircraft Warning Company and became operational as an integral part of the defense scheme of the department. Though there was no filter or information center, they reported by telephone to the Department Command headquarters at Quarry Heights. The Fort Sherman set went on the air on October 7, 1940, becoming the first Army radar to become operational as a unit of the country's integrated plan for air defense.

The installation in Panama brought many useful lessons of practical value. The importance of designing radar equipment to withstand the extreme temperatures and humidity of the tropics was brought home to Signal Corps with great force. Problems of siting immediately raised their heads. The first design wherein the SCR-271 had its antenna on a 36-ft tower mounted on the roof of the operating building was only satisfactory at those few sites where there were no obstructions between the set and the sea. This led to designing the next model, the SCR-271A, with a 100-ft tower which could be placed some distance from the operations building.

The policy conference previously referred to, held in Washington at the end of May 1940, made a number of important decisions affecting the long-range radar programs. From funds from the fiscal year 1941 it was hoped to purchase from the supplier 85 long-range detectors, 59 of the mobile type and 26 of the fixed variety. Procurement would be handled through the Signal Corps Laboratories. It was decided to offer bids to the two concerns which had already made important contributions to the set by the vacuum tubes they had developed for it: RCA and Westinghouse. RCA declined to bid, and a contract was awarded to Westinghouse in August 1940. The contract called for 73 mobile sets of the SCR-270-B type and 21 units of the SCR-271's. These figures were later increased. Deliveries began in February 1941. By the end of the fiscal year, June 30, 1941, Westinghouse had delivered 18 of the 270 sets and 28 of the SCR-

271's. As in the case of the 268, the Signal Corps undertook a crash program of five preproduction SCR-270's. These served in an important training capacity while the production sets were being awaited.

A total of 794 long-range sets, both fixed and mobile, were produced between 1939 and 1944, at an approximate unit cost of $35,000 for the SCR-271's and $55,000 for the SCR-270's.

The Beginning of the Aircraft Warning Service

The disposition of the long-range sets and the provisions for their operation and maintenance were determined by the consideration which bulked largest before, and for some time after, the attack on Pearl Harbor. This was the importance of defending the continental United States and of protecting our outlying bases. Closely connected with the plan of defense was the creation of the Army's Aircraft Warning Service (AWS). Although the details of the aircraft warning were the responsibility of the respective departments and defense commands, their proposals were coordinated and reviewed by the War Department in consultation with the Chief Signal Officer, since the Signal Corps had designed the equipment and trained the Signal Aircraft Warning troops that operated it. Within the overseas departments, the preparation of installations was entrusted to the department signal officers and subsequently handed over to the department Air Defense officer.

On March 10, 1939 the Chief Signal Officer was instructed to prepare a comprehensive plan for the organization and operation of the Aircraft Warning Service for the continental United States and the overseas departments. This study was approved in February 1940. This continental part of the study provided for a total of 23 long-range detector stations and 9 information centers.[14]

Late in May 1940 the Commanding Generals of the First, Second, Third, and Fourth Armies were directed by the War Department to revise their plans for the defense of their areas to provide for the use of radio detectors. In March 1941, the organization, training, and operation of the aircraft warning network was transferred to the GHQ Air Forces who made this a responsibility of the Interceptor Commands. By the end of 1941, AWS surveys had been completed for 46 sites in the United States, 10 sites for Panama, 5 for Puerto Rico, 12 for Hawaii, 14 for Alaska, and 5 for the British bases. Surveys had previously been completed for the Philippines.

By early 1942, the Aircraft Warning Service boasted a chain of SCR-270 and SCR-271 stations covering the Atlantic coast from Maine to Key West, and the Pacific coast from the uppermost corner of the state of Washington to San Diego. A filter center for correlating information was established at every second or third station. The interior of the country and the Gulf coast was covered only by visual observers and information centers; this of course made it possible to allocate more radar to the overseas departments.

Among the overseas departments, Panama had top priority and first claim on new equipment. The Panama Department had received the first two preproduction long-range sets. Late in 1940, the first Aircraft Warning Company to be formed was activated in the Canal Zone. By February 1941 it was equipped with a half-dozen SCR-270's.

When the agreement was signed with the British early in 1941 permitting the use of British bases for American Atlantic defense, the Caribbean Defense Command recommended the location of long-range radar stations in Trinidad, British Guiana, Jamaica, and the Bahamas, as well as on American bases. This was approved by the War Department December 4, 1941.

In March 1941, two SCR-271 sets and one 270 were earmarked for Hawaii, one SCR-271-A for Alaska and another for Panama. In subsequent months, as sets of both types became available, they were allocated on a strict priority schedule to the points of greatest danger—Hawaii, Panama, Alaska—and for equipping the Second Aircraft Warning Company and supplying sets to the U. S. Navy.

In July 1941 American task forces sailed for Iceland to occupy the island for the protection of lend-lease supplies and to deny the base to the Germans. This task force included the first Aircraft Warning Company, with three SCR-270's which were mobile but set up as fixed installations.

In Alaska, the Fourth Army was directed early in 1940 to select tentative sites for the location of radar systems. An Aircraft Warning Company, Alaska, was organized in 1940 and reached Fort Richardson, Alaska, on March 14, 1941. Two SCR-271 sets were shipped to Alaska in May 1941, two more in July, and a fifth in mid-August.

In the Philippines an Aircraft Warning Service, employing visual observers, was set up in July 1940 by G-2 for the Philippine Department. Early in 1941 an Air Defense Program was submitted by the Philippine Command which proposed the location of radar systems on three different sites. On May 29, 1941 a radar-trained Aircraft Warning Company, Philippine Department, was constituted at Fort Monmouth with 2 officers and 202 enlisted men. This outfit was ordered to the San Francisco Port of Embarkation in June 1941. One SCR-270 and two SCR-271 sets were shipped to the Philippines in August.

Early in September, the Philippine command submitted a revised report recommending ten detector stations and a filter station. Although this plan was not officially approved by the War Department until November 27, 1941, and the Japanese attack came before it could be carried out, it was reported that by the outbreak of World War II, four SCR-270 sets and two SCR-271 sets had reached the Philippines. The two fixed installations had not arrived in time to be set up for operation; but the four mobile sets could have been, or should have been, in operating condition.

The Attack on Pearl Harbor

In response to a War Department directive issued in October 1939, the Commanding General, Hawaiian Department, was instructed by the Chief Signal Officer to prepare an Air Warning Service plan. Six months later on April 17, 1940 a project was submitted that called for five mobile and three fixed stations, and a filter center at Fort Shafter. This was approved by the War Department late in June 1940.

During the prolonged delay caused by the difficulty of obtaining ground for the three permanent sites in land under the control of the National Park Service the requirements for the Hawaiian department were increased at Washington to provide for six fixed and six mobile radar installations. In the spring of 1941, a Board of Officers was convened in Hawaii to choose sites for the additional stations. Its report, transmitted on October 8, 1941 by that ill-fated personage, Lieutenant General Walter C. Short, pointed out that since most of the vital installations in the Hawaiian Department, including the great naval base of Pearl Harbor, were located on the island of Oahu, sites were accordingly being recommended on the assumption that Oahu was the primary area to be defended.

Time was running out. The additional equipment was not, and can hardly have been expected to be, delivered before the attack. Some progress, however, had been made in carrying out the original plan for the five mobile and three fixed stations. Three fixed-tower sets of the SCR-271 type were received in Hawaii on June 3, but because of delays encountered in preparing the sites, this equipment was not installed or operating by December 7. The Army Pearl Harbor Board found that these installations were not held up by any lack of information or drawings or equipment from the Signal Corps.[15]

No such difficulties were encountered with the mobile SCR-270's which required little or no special construction or advanced preparation. Six of the SCR-270-B type of long-range set were received in Hawaii on the first of August 1941 and were shortly thereafter put into operation.[16] Extensive testing of the sets was carried out in the next few months from installations at Kaaawa, Kawailoa, Waianae and Koko Head, the latter at an elevation above 600 ft.

In November three of the sets were tested in a tactical exercise that resembled in remarkable degree the actual attack of December 7. The exercise began at 4:30 in the morning. Attacking planes were picked up as they assembled near the carrier from which they took off 80 miles away. When they had assembled, the planes headed for Hawaii. This was easily detected on the scope; the pursuit aircraft were notified within about 6 minutes and they took off and intercepted the incoming bombers at about 30 miles from Pearl Harbor.

The hour chosen for these exercises confirms what has already been revealed in the Pearl Harbor investigations, namely that the militarily brilliant Japanese stroke on December 7 not only achieved strategic, but also tactical, surprise. We were not expecting Japan to open up with an attack on Hawaii. Those who were in charge of the Air Defense of Hawaii were moreover fully convinced that the most probable time for an air assault would be from 0400 to 0700. On the morning of December 7, however, the first Japanese planes reached the naval air station at Kaneohe Bay on the east coast of Oahu at 0750; the planes thundered over the Pearl Harbor area a few minutes before eight o'clock.[17]

The air defense of Hawaii was certainly inadequate and, by later standards, primitive.[18] Training of personnel assigned to the sets is reported to have begun only on November 1. The Army Interceptor Command was barely in the first stages of organization. On December 7 the total radar protection of the island consisted of five mobile SCR-270 sets at various points along the coast. As matters turned out, this equipment was perfectly capable of giving adequate warning of the attack that actually took place, had the danger been properly assessed. The following paragraph from the report of the Navy Court of Inquiry contains the heart of the matter:

> Between 27 November and 7 December 1941 the Air Warning Systems operated from 0400 to 0700, the basis for these hours being that the critical time of the possible attack was considered to be from one hour before sunrise until two hours after sunrise.[19]

The role radar actually played on the day of the attack is well known, for it was spread across the record by the Roberts Committee and by the Army and Navy investigating boards. On the morning of December 7, the Opana station, located on the east coast of Oahu, had been on the air from 0400 to 0700. While they were waiting for a truck to arrive to take them to breakfast, two Signal Corps privates, Joseph Lockard and George Elliott, kept the station on the air, so that Elliott could receive some instruction from Lockard, an experienced operator. Just two minutes after closing time a strong echo appeared near the extreme end of the sweep line of the *A* scope when the antenna was pointing north. To an experienced observer, the fluctuations in the signal indicated a large flight of planes at a distance of 136 miles.

At first Lockard was skeptical and checked over the adjustments to make sure the signal was not spurious. When the signals moved in to 132 miles, this skepticism vanished and he called the Information Center by telephone and reported the observation. The center was very nearly deserted, being staffed only by an enlisted telephone operator and an Air Corps officer, with very little radar experience, who was there for training. Having heard that a flight of B-17's was expected at about that time, this officer told Lockard to forget it. Lockard and Elliott continued to track the planes in to a distance of 20 miles from Oahu when they were lost sight of. An advance warning of something over 50 min—a very precious commodity in such situations—was thrown away. At 0755 the first bombs fell on Hickam Field and Ford Island.

Other Signal Corps Radar Developments

It is the intention of this discussion of early Army radar development only to bring the story down to the period when the NDRC radar program must necessarily command our main attention. It is therefore impossible to do full justice to the activities of the Signal Corps during the war years. In June 1940 General Mauborgne, the Chief Signal Officer, informed the Assistant Chief of Staff, War Plans Division, that the fiscal estimates for research and development for 1941 and 1942 were far beyond anything recommended in the past—they amounted, in fact, to $8,338,860—but it was not expected that the Signal Corps Laboratories would be expanded to take care of the new commitments by expansion. Instead, General Mauborgne announced that it would be his policy to place development contracts with commercial firms, as in the previous war.[20]

As it turned out, both were necessary. We have already seen how the expanded radar activity led to the creation first of the Fort Hancock field station, then of the Camp Evans Signal Laboratory (1942). The latter grew to have a peak personnel of 3000 persons. At the same time, the Signal Corps let an increasing number of development contracts with industrial laboratories, and followed attentively the results of NDRC contracts.

It is the fate of military laboratories to have an ever-increasing burden of routine responsibilities thrust upon them during the war. Much of the time of the Camp Evans personnel was taken up during the war with testing equipment developed elsewhere, suggesting improvements to make equipment more serviceable in the field, etc. The proportion of effort devoted to basic research and to the development of new radar equipment consequently dropped off heavily. Perhaps the most important responsibility was the steady improvement in a succession of later models of the long-range sets and of the SCR-268's. Considerable work was also done on the development of several versions of the Mark III IFF (Identification Friend or Foe) for use with various units of Army equipment, and of an IFF trainer.

The most important new program at Camp Evans, however, was launched during the period of preparation for the North African campaign, and continued throughout most of the war. This consisted of the development of most members of the SCR-602 series, which were transportable, lightweight, medium-range, early-warning radar sets. The most successful of these was the SCR-602-T8 (which also received the joint Army–Navy designation AN/TPS-3). The set itself weighed about 100 lb; it had a lightweight parabolic antenna, and operated on 50 cm using a power tube developed at Camp Evans. Even with power units, tent, and spare equipment, the outfit weighed less than 300 lb. About one dozen experimental models were completed by the Camp Evans Signal Laboratory in January 1944. Large-scale manufacture of this equipment was entrusted to the Zenith Radio Corporation. A total of 650 of these sets were shipped overseas before the end of the war.

So far no mention has been made of the work at the Aircraft Radio Laboratory at Wright Field, where during the war the Army's own work on airborne radar was mainly concentrated. In its modern form, this laboratory was created in 1935 when all aircraft radio activities, whether by the Air Corps or the Signal Corps, were assembled in a single organization whose Director reported to the Chief Signal Officer.[21] In 1936 a project described as "Project 19, Priority D, Collision Prevention" was undertaken. This got no further than some preliminary experiments to determine the beam concentration possible using wavelengths of 75 cm and a parabolic reflector about a meter and a half in diameter. This project, which might conceivably have led to some sort of airborne radar experiments, was discontinued as a result of official action in February 1937.[22] In 1940, an attempt was made at ARL to modify the Signal Corps ground radar for aircraft use. This mass of gear was actually flown in the fall of 1940 while the British Mission was in this country. The Aircraft Radio Laboratory soon found itself fully occupied with airborne radar directly and indirectly resulting from this visit. Six months before Pearl Habor, personnel at Wright Field numbered 24 officers and 361 civilians. A radar division was established under Lt. Col. W. Bayer in January 1942. At this time ARL was one of the units under the Signal Corps Air Signal Service (SCASS), headed by Col. John H. Gardiner. His executive officer was Lt. Col. Gilbert Hayden. Col. Hobert R. Yeager was at the head of the Aircraft Radio Laboratory.

NOTES

1. In this cell the sensitive material is thallium oxide fused with sulfur. Exposure to light or infrared rays lowers the electrical resistance of this mixture so that incident radiation is detected by variations in electric current passing through the material.

2. In 1906 Blair had received his PhD in physics at the University of Chicago for work carried out under Professor Michelson and Professor Millikan on the properties of microwaves. Using a Michelson interferometer he had performed a series of experiments with waves 20 cm long investigating the change in phase resulting from the passage of electric waves through thin films. Cf Wm. R. Blair, Phys. Rev. **26**, 61 (1908). Blair later joined the U. S. Weather Bureau and in 1917 was commissioned Major in the Aviation Section of the Signal Corps Reserve. In 1925–26 he attended the Command and General Staff School at Fort Leavenworth; in June 1926 he was assigned to the Office of the Chief Signal Officer.

3. Personal communications from Dr. Irving Wolff.

4. Cited by Harry M. Davis, History of the Signal Corps, Development of U. S. Army Radar Equipment, Part I, 1943, p. 38. This is a Signal Corps document of limited distribution classified "Confidential."

5. See above, page 53.

6. See above, pages 57–59.

7. Personal communication from Dr. W. D. Hershberger.

8. Cited by Davis, Note 4, p. 24.

9. E. F. Hershberger, Detection of Aircraft by Radio-optical Methods, 12-6-5.

10. Davis, Note 4, p. 33.

11. The transmitter used two RCA 834's operated in push–pull with parallel wire lines as tank circuits to give an average power output of 75 W.

12. Mechanical lobe-switching devices were independently being developed by the British for airborne radar equipment at about this same time. See below, page 172.

13. For a fuller explanation see the article in Electronics, September, 106 (1945).

14. Air bases at Chicopee Falls, Mass.; Tampa, Florida; San Diego, Cal.; and McChord Field, Washington were to be protected by radar, and so were vital harbor areas: New York, Norfolk, Charleston, Galveston, and San Francisco.

15. Report of the Army Pearl Harbor Board.

16. See Ref. 15.

17. Walter Karig and Welbourn Kelley, Battle Report, Pearl Harbor to Coral Sea, 1944, pp. 28-9. This seems to be the most complete careful account of the Pearl Harbor attack published so far.

18. The Navy Court of Inquiry pointed out that only a few vessels of the Pacific Fleet had been fitted with radar by December 7, and correctly emphasized that the radar of ships berthed in a harbor surrounded in part by high land like Pearl Harbor could not be relied upon, as of course it was not, for the defense of the Island.

19. Navy Court of Inquiry, Finding of Facts, 1-24.

20. OCSigO 111 (1942) J. O. Maubourgne, Memorandum for the Assistant Chief of Staff, War Plans Division: June 3, 1940.

21. Signal Corps Aircraft Signal Service—History and Activities, compiled as of 1 July 1943.

22. Annual Report, Aircraft Radio Laboratory, fiscal year 1936, pp. 14 and 30. File RD-506.

CHAPTER 6
Radio Detection in Great Britain

To the British, radio detection meant nothing less than survival. None of the later accomplishments of pulse radar in the war—except perhaps its victory over the U-boat—will probably receive as high an appraisal by future historians as its share in winning the Battle of Britain and warding off the invasion of the British Isles in 1940–1941. The British radar equipment of that day was crude, half-engineered, often unreliable; but without it, the war might well have been decided then and there in favor of the Axis. After the disaster of Dunkirk, with its Army stunned and virtually stripped of arms, Britain had three outstanding assets for defending the island: the ships and men of the Royal Navy; the Hurricanes and Spitfires, the Blenheims and Beaufighters, and above all, the pilots of the Royal Air Force; and the defensive screen of its radar warning system. The long-range Chain Home (CH) stations, operated under the most exacting conditions, and often during actual bombardment, by specially picked members of the Women's Auxiliary Air Force (WAAF), were the principal factor in redressing the numerical inferiority of the aerial defenders who were outnumbered about four to one by the attackers. The radar watch permitted a coherent and systematic defense that did not depend upon guesswork, luck, or primary reliance upon standing patrols; it multiplied severalfold the effectiveness of each RAF Fighter Squadron by reducing flying hours, saving pilots, planes, and gasoline, and raising the level of performance of both men and machines. Germany may not have been either psychologically or materially prepared to exploit, as swiftly as the situation demanded, the collapse of Belgium and the unexpectedly swift disintegration of the French armies. Nevertheless, the prime strategic prerequisite for invasion—the control of the air over the British Isles—was never satisfied. The radar warning barrier, probing with its invisible fingers over a hostile continent, and later the radar-equipped night-fighter, able to seek out and destroy the night intruder, gave Britain, if not security, at least security against surprise. To no people can radar be expected to mean as much as to the British; with them it is already a part of legend, inseparably associated with the heroism of the RAF. It stood between them and the enemy at one of the most dangerous hours of their history, just as our carriers and planes did at Midway, and gave them time to recover from a paralyzing defeat.

Even aside from its intrinsic importance, generous space must be ac-

corded the history of radio detection in Great Britain because it is necessary for a clear understanding of the development of American radar
during and just before the war. The first British radar was developed
wholly independently of our own; until the autumn of 1940 neither
country borrowed from nor influenced the other; in fact, until the winter
of 1939–40, each country seems to have been in complete ignorance, except for the vaguest rumors, that the other possessed radar equipment.
But after the formal interchange of secret technical information in the fall
of 1940, and throughout the subsequent war years, cooperation between
British and American scientists was extraordinarily complete and fruitful.
Most important of all, from the standpoint of this account, the research
program of the National Defense Research Committee owed its effective
origin, its early growth, and some of its most important later ideas and
policies to British example, encouragement, and technical genius.

The Character of the British Radar Program

Like the American armed services, the British first developed pulse
radar on long waves (on a frequency of 200 MHz and below) and acknowledge their debt for the fundamental idea to the well-known pulse
method of exploring the ionosphere. But in a number of respects, the
British radio detection development followed a different course from the
American. It began abruptly, without fumblings and with no experimental preliminaries, with a proposal for pulse radar. This was followed up
energetically, and the development proceeded at the accelerated pace
made necessary by the war clouds gathering over Europe. The intermediate evolutionary stage of exploring the beat method of detection—which
the French, German, and American developments all seem to have passed
through—was sidestepped by the British. There were other differences. In
Germany and France, radar was developed by industrial concerns; in
America, the pioneer work was confined to the laboratories of the armed
services; in Great Britain, by contract, the research and development was
mainly entrusted to civilian scientists, drawn mostly from the universities,
from whom a special laboratory was created *ad hoc.*

The speed of British progress in radar before the war—though they hit
upon the radar idea about a year later than we, they rapidly passed us—
was the result of a number of factors. The most important, for it was the
most all-pervading, was the immediacy of the threat to Britain's national
security. This not only imbued all persons connected with the project with
missionary zeal, but made it possible to obtain support and sustain encouragement at the highest official level. Of equal importance was the
novel manner in which the program was organized to short circuit, to a
large extent, the usual official channels, and to avoid entrusting the program to established organizations already burdened with other responsi-

bilities or clinging to deeply rooted and time-honored practices. The extremely high caliber and fresh outlook of the university scientists to whom the work was entrusted left its mark upon the program. The greatest names in modern British physics, among them the leading students of Lord Rutherford, were sooner or later led into radar work. These university scientists brought with them a freshness and boldness of attack, coupled, especially in the first years, with a marked impatience towards the art and procedures of the engineer. Craftsmanship, dependability of performance, and durability were sacrificed for speed and imagination. In British radar there were plenty of both, but it was, in general, less well engineered than the first American radar, which was produced with less profusion and at a more leisurely pace.

The initiation of the British radar program was a direct consequence of the final collapse of the European collective security front about 1934–35, and was not the least important aspect of the rearmament program that the British launched in those years. It is surprising in retrospect to discover how sharply the events of 1934 mark a watershed between the years of peace and the years of undeclared war: the bellicose posturings and growing strength of Hitler and Mussolini; the breakdown of the Geneva Disarmament Conference; the abortive Nazi putsch in Austria which resulted in the murder of Dollfuss; Italy's machinations in the Balkans culminating in the assassination of King Alexander of Yugoslavia; Germany's rearmament in defiance of the Treaty of Versailles; Italy's first threatening gestures towards Ethiopia; Japan's denunciation of the Washington Naval Treaties of 1922.

These events led Great Britain to abandon the policy of disarmament. The Tory opposition, with Winston Churchill baying in the lead, called for a strengthening of British defenses, especially for an expansion of the air arm. The Baldwin government, after several months of vacillation and temporizing, finally came to the same conclusion, and in the summer of 1934 announced a five-year program for expanding the Royal Air Force. The Government was obliged to defend its new policy against the attacks of those who felt that the program of international conciliation was being rushed to premature extinction. But after the now historic debates—in the course of which Prime Minister Baldwin made his celebrated remark that the frontier of Britain was now upon the Rhine—the motions of censure were resoundingly defeated in both houses. The representatives of the people had thus spoken, and the path ahead was clearly indicated.

The Air Estimates which in the years from 1928 to 1934 had hovered in the neighborhood of £17,000,000 rose for 1935 to £26,000,000 and for 1936 to £39,000,000.[1] A similar expansion took place just at this time in government-supported scientific research and development related to national defense. By 1937 nearly half the sum the Government expended in support of research was earmarked for the fighting services.[2] An appreciable portion of the latter must have been accounted for by the needs of radar research and development.

The Organization of Science in Great Britain

In England, as in the United States, by far the greatest amount of pure scientific research takes place in the universities; and such coordination as this research effort enjoys is a result of informal, cooperative action through the specialized scientific societies, and through the British Association and the Royal Society. To understand, however, the environment in which radar was developed, it is essential to know something of the conditions under which so-called applied research was conducted during peacetime in a country where the industrial laboratories are less predominant than in the United States, and where Government scientific agencies are considerably more influential. The emphasis will of course be placed upon radio research.

Scientific research in the British Government is concentrated under the scientific department and in the laboratories of the armed services and under the sponsorship of a government office—the Department of Scientific and Industrial Research (DSIR)—which is an Executive Office responsible to the Lord Privy Seal. The DSIR was founded in 1917 in a frantic effort to compensate for Britain's backwardness in the sort of industrial research from which Germany, like America, had drawn so much of its strength. Its original purpose was the encouragement of scientific research on a cooperative basis within the principal industries; but in addition to supervision of the Industrial Research Associations, which are of little or no importance in the story of radar development, it was made directly responsible for the Government research activities that did not happen to fall under the armed services. Under it were a number of special research boards to coordinate research in various fields of public interest—the Fuel Research Board is only one example—and the two principal nonmilitary government laboratories, the National Chemical Laboratory and the National Physical Laboratory. The former was of very standard scope, and was mainly an analytical laboratory to assist the Board of Trade in standardizing commercial products. The better-known and more important National Physical Laboratory performs much the same function in Britain as our own Bureau of Standards. Like our own institution, it has tended to become so completely preoccupied with its functions of standardizing apparatus and the testing of materials that its contributions to fundamental physics have been limited. It had a radio division of considerable importance.

Government radio research was coordinated under the Radio Research Board. This was constituted as a panel under DSIR and on it were represented the chief Government agencies concerned with radio: the British Army, the Royal Navy, the RAF, the Post Office, the universities, and the BBC. The DSIR and the Radio Research Board jointly supervised a government radio laboratory, the Radio Research Station at Slough.

Each of the three defense services had its own scientific organization, with its own laboratories. The Admiralty and the Air Ministry were each provided with scientific research departments headed by a civilian Director of Scientific Research (DSR) and staffed predominantly by civilians. The radio research in the Royal Navy, under the Admiralty Department of Scientific Research and Experiment, was concentrated in the Admiralty Signals Establishment (ASE) with laboratories at Portsmouth, Haslemere, Witley, and Nutbourne. The British Army had a laboratory at Christchurch, the Air Defense Research and Development Establishment (ADRDE), at which research was conducted on searchlights, anti-aircraft devices, proximity fuses, radio (and later radar), and other devices of importance to the ground forces.

The Air Ministry from the first had made a branch to supervise research and development an integral part of its organization. Since its inception, the Air Ministry had devoted one of the half-dozen major divisions of its organization to a Department or Directorate-General devoted to problems of supply, research, and development. In 1934 this branch was called the Department of the Air Member for Supply and Research. Under this Department, headed at this time by an RAF officer, the famous Sir Hugh Dowding, were a half-dozen subdivisions predominantly civilian and under civilian directors. The two most important were the Directorate of Scientific Research and the Director of Technical Development, which corresponded, respectively, to a department of fundamental research and one of engineering development. Under these two Directors was the RAF's principal scientific laboratory, the Royal Aircraft Establishment at Farnborough. This institution was concerned with all sorts of aeronautic research, and included a Radio Department nominally under an RAF officer, but actually run by the second-in-command, a permanent civil servant, F. S. Barton. The work of this department of the laboratory, like that of its American analog, the Aircraft Radio Laboratory at Wright Field, was concerned with aircraft communication equipment; its chief responsibility was in designing radio sets for service use, and in properly engineering the equipment and testing it for the Royal Air Force. The Radio Department was small; even after three years of war it only numbered some 300–350 persons.

The air exercises of 1934 had drawn attention to the general inadequacy of the protection against enemy aircraft: the ineffectiveness of the observer corps—then the chief source of warning information—and of sound locators; the primitive state of operations rooms; the poor communications. An Air Ministry official in 1934 could discover in the files only 53 folders covering the whole subject of air defense.

The Tizard Committee

The debates on the expansion of the Royal Air Force in 1934 had made the threat of air raids a new and absorbing topic of conversation and the subject of the most morbid speculations. In the course of these debates it was clearly brought out that even a strong air force would not offer complete protection, and that even with the most effective screen of fighters, some of the bombers were sure to get through to their objectives.

Within the Air Ministry there was considerably greater pessimism than in the country at large. There was a growing feeling that there could be no real protection against air attack unless some striking advance was made on the defensive side of air warfare; something as radically new as the discovery of high explosives or the invention of the tank or of heavier-than-air flight; something to redress the balance drastically in favor of the defense. In January 1935, the Secretary of State for Air appointed a committee principally composed of distinguished scientists to "consider how far recent advances in scientific and technical knowledge can be used to strengthen the present methods of defence against hostile aircraft."[4] Except for an announcement late in February in the House of Commons by Sir Philip Sassoon, the Undersecretary of State for Air, giving the composition of this committee and its purpose, the Committee for the Scientific Survey of Air Defence received no publicity and carried on its deliberations in the greatest secrecy.[5]

The idea of setting up such a committee seems to have been discussed in late summer or early fall 1934 by H. W. Wimperis, the Director of Scientific Research, Air Ministry, and his assistant A. P. Rowe.[6] Wimperis drew up a Minute, dated November 12, 1934, in which he called attention to the immense difficulties of defending a great city against hostile aircraft in view of the greater speed and higher ceilings of modern planes, their quieter engines, and the ease with which they could be flown with automatic pilot in cloud or fog. He proposed that a committee be appointed under the chairmanship of H. T. Tizard, F.R.S, Rector of the Imperial College of Science and Technology, and Chairman of the Aeronautical Research Committee, to intensify research into new defense measures, chief among which was the development of a ray of energy capable of putting the engine ignition of a plane out of action, or detonating the bomb load of an aircraft, or of having deleterious or dangerous effects on the human body or upon the metal structure of a plane. He suggested as members of such a committee Professor Patrick M. S. Blackett, F.R.S., distinguished physicist of the University of London, and Professor A. V.

Hill, eminent physiologist, Nobel laureate in medicine, and Secretary of the Royal Society.

This document was transmitted to the Air Member for Research and Development, Air Marshall Sir Hugh Dowding; to the Chief of Air Staff, Air Chief Marshall Sir Edward L. Ellington; and to the Secretary of State for Air, Lord Londonderry. The committee was officially appointed with the addition of the names of H. W. Wimperis, DSR, Air Ministry, and of A. P. Rowe, who served as Secretary to the Committee.

When the Committee held its first meeting on January 28, 1935, with all members present, and Mr. Tizard in the chair, some important spadework had already been accomplished.

Earlier in January, Wimperis had approached his friend and associate on the Radio Research Board, Robert Watson-Watt, who since July 1933 had been Superintendent of the Radio Department of the National Physical Laboratory. Wimperis asked Watson-Watt unofficially, as a friend in whose judgment he had confidence, whether there was anything in the idea of a "death ray"; whether it would be possible to concentrate energy of any sort upon an attacking plane in order to exert powerful destructive effects. Watson-Watt replied offhand that he thought not, but that he would make some calculations and put some thoughts down on paper. After this interview with Wimperis, he discussed the problem of the death-ray with his younger colleague at the Radio Research Station, A.F. Wilkins, and together they did the necessary arithmetic. About a week later, Watson-Watt submitted a paper on the damaging effects of radio beams, demonstrating that as things now stood, it would be impossible to concentrate enough radio energy upon an aircraft to raise the temperature appreciably (which was the only way to cause damage), since the ease of shielding made it pointless to consider interfering with ignition. The paper wound up, "rather pompously" as Watson-Watt himself said of it, with the statement that detection of aircraft was the real problem, and that if asked he would investigate the possibility of using radio waves for detection. When this was written, Watson-Watt had no precise solution in mind; the death-ray calculations suggested, however, that enough energy could be concentrated in the neighborhood of the plane for a detectable amount to be reflected or reradiated.

At this first meeting of the Committee for the Scientific Survey of Air Defence, Wimperis dominated the proceedings. He began by outlining the main purpose of the Committee, and found the members in agreement that the problem of defense was largely one of detecting the position of enemy aircraft, and that because of increased speed of planes and improved silencing methods it was of paramount importance to consider means of detection other than sound. Various methods of detecting aircraft were discussed, such for example as the use of infrared radiation, and Wimperis told the committee that there was some reason to believe that short radio waves emitted from a ground source could be reflected from the surfaces of an aircraft and serve to detect its presence. He reported that Watson-Watt considered that there was some hope of detecting aircraft by this method and that a memorandum prepared by him on this subject would be circulated to the Committee. It was decided that further consideration should be given to this detection possibility when this memorandum had been studied.

During January and February, Watson-Watt and Wilkins worked over the calculations to determine first whether detectable radio energy could be reradiated from an aircraft, assuming the wingspan of a bomber to be a horizontal half-wave dipole 25 m long and knowing the peak power available at 50 m and the sensitivity of a receiver.[7]

Watson-Watt and Wilkins determined the fields of reflected energy to be "about ten thousand times the minimum required for commercial radio communication, so that very large factors of safety are at hand for ranges of the order of 10 miles at flying heights of about 20,000 ft."[8] Watson-Watt proposed taking over intact "the present technique of echo-sounding of the ionosphere," as it was being used at the Radio Research Station. Distances would be read directly on a cathode-ray tube. He believed the time intervals involved, though much smaller than in ionospheric work, would be "quite manageable within the technique, though they involve a very considerable shortening of the pulse durations now used (about 200 μsec)..."

At first Watson-Watt was satisfied to propose floodlighting techniques and the use of three receivers to determine position by time delay measurements. Detection by a Doppler cw method or by Appleton's frequency modulation method were proposed as alternatives if the pulse method "should prove to have unexpectedly great difficulties."

A memorandum containing the above proposals was submitted on February 12 to Wimperis through Rowe; it was accompanied by a letter to Rowe in which Watson-Watt called attention to the unexpectedly encouraging results of the calculations. "It turns out so favourably," he wrote, "that I am still nervous as to whether we have got a power of ten wrong, but even that would not be fatal. I have therefore thought it desirable to send you the memorandum immediately rather than to wait for close rechecking."[9]

A copy of the memorandum was transmitted before the end of February to F. S. Barton, civilian head of the Radio Department at RAE, for his opinion and comments. Barton and his co-workers studied the memorandum and reported to DSR that they agreed in principle with the proposal and that the calculations seemed to give results of the right order of magnitude. They called attention to the observations by the British Post Office, in a confidential report, and to the results published by the Bell Telephone Laboratories, as experimental evidence that a detectable amount of radiation could be reflected from aircraft.[10]

The observation of the British Post Office engineers was nearly identical with that reported in the United States by the men from the Bell Laboratories. Late in 1931 tests were being conducted on a two-way 5-m radio communication link between the Radio Laboratories at Dollis Hill and Colney Heath. During tests being conducted between December 15 and 18, 1931, "it was noted at Colney Heath that beats were being received on the tone transmission from the Dollis Hill transmitter. In all cases an aeroplane was noticed in the neighborhood and in no case were the beats heard when a aeroplane was not present." The only possible explanation was that interference was set up between the directly received waves and those reradiated from the aircraft. On December 17 beats were recorded from a plane that was audible but not visible. The authors observe, "It should be noticed that this type of interference which has been noted on many occasions since December 1931 has been experienced from distant airplanes."[11]

"Stuffy" Dowding, the Air Member for Research and Development, showed a commendable lack of credulity on the subject of Watson-Watt's report. The author of this historic document quotes Dowding as having said, "You know these scientific blokes can prove anything by figures. I want a demonstration." Watson-Watt met Dowding at Farnborough for the first time and agreed to a test, which, in effect, would be an attempt to duplicate under somewhat controlled conditions the results of the Post Office and Bell Laboratories observations.

On February 21, 1935, before they had had a chance to consult the Watson-Watt memorandum, the Committee held its second meeting at the Headquarters of the Air Defence of Great Britain at Uxbridge where they were received by Air Chief Marshall Sir Robert Brooke-Popham, the Commander-in-Chief of ADGB, and by Air Vice Marshall Joubert de la Fertem, who was later to become one of the RAF's most ardent radar enthusiasts, and by other dignitaries. Here they learned from the Commander-in-Chief the main features of the defense plan, and discovered that the tactics for the interception of hostile aircraft by fighter aircraft involved little more than the uneconomical method of using standing patrols near probable enemy objectives, thus reducing interception to a mere matter of chance. There was some discussion of the role of sound locators, for which large sums of money were about to be spent.[12] The outstanding example was a huge concrete sound reflector, 200 ft in diameter, installed soon after the first World War, pointing in the general direction of Paris, with which under favorable conditions ranges of 20 miles could be obtained. Air Vice Marshall Joubert was asked what increase in range of detection off the coast, compared with this figure of the 20 miles, could be considered a definite advance. He replied that detection of aircraft at 50 miles would revolutionize the present methods of controlling fighter aircraft. At this meeting Wimperis spoke of having received Watson-Watt's preliminary memorandum and of his intention of distributing it to the members of the Committee. Arrangements were being made, he told the meeting, for experimental work to begin.[13]

To obtain experimental verification of the earlier observations that it was possible to pick up secondary radiation reflected from an aircraft, Watson-Watt and Wilkins installed a modified ionospheric receiver with a cathode-ray tube indicator in an experimental van, and drove it to a point on a country road near Daventry where the 25-kW overseas 50-m beam came in greatly attenuated. It was arranged that on the following morning an aircraft would be flown into the beam and an effort made to detect the beats. At 9 the following morning Watson-Watt, accompanied by A. P. Rowe as official observer for the Air Ministry and the Committee, drove out to the van. An aircraft, a Weyford two-engined biplane with metal fuselage and fabric-covered wings was flown from the Royal Aircraft Establishment (RAE), and though it occasionally flew out of the beam, the beats came in strongly when the plane was as much as 12 km away. Rowe was favorably impressed. The experiment was carried out with such secrecy that not even the officials at RAE nor the pilot of the aircraft were told the purpose of the flight. In consequence the plane that was used had a long trailing antenna, belonging to its communication system, trailing along behind, which may well have enhanced the effect slightly.[14]

Watson-Watt's memorandum in final form was submitted to the Air Ministry on February 28, 1935.[15]

Meanwhile a search had been begun for an experimental site where radio detection work could go forward in appropriate secrecy and solitude. Someone in the Air Ministry suggested an RAF bombing range and small landing field situated on an island called Orford Ness just off the East Coast, about fifteen miles north of Harwich. The island—actually a spit of land close to shore, separated from the mainland at its northernmost point by only about 50 yards of water—lay directly across a narrow estuary from the village of Orford. The island or peninsula was 10 or 15 miles in length, dismal, wind-swept, and uninhabited. Except for a few broken-down huts and dilapidated hangars dating from the last war, it was a deserted waste. It seemed a possibility, though anything but luxurious. Wimperis and Watson-Watt visited Orford Ness late in February.

On March 4, the Tizard Committee held its third meeting at the Air Ministry. Wimperis reported on the success of the tests with the Daventry transmitter, told the Committee of the visit to Orford Ness, and reported that the site seemed suitable for the experiments and that he saw no reason why work should not begin almost immediately.[16]

Despite this auspicious beginning, the radio detection program might easily have been frittered away, supported parsimoniously, divided up among the existing research institutions, or in any of a dozen familiar ways strangled at birth. That it was given at once the fullest support was in no small part due to the wisdom as well as the prestige of the members of the Tizard Committee and to the confidence reposed in them by Air Ministry officials. It was also due to support by a still more august top secret committee set up at about this same time as an offshoot of the Committee on Imperial Defence. This powerful top level body, called the Air Defence Research Subcommittee, was extremely helpful in obtaining generous funds without interminable delays.

This committee was set up as a result of the initiative of Professor F. A. Lindemann (later Lord Cherwell) who suggested to the Secretary of State for Air that a committee of scientific and service representatives be set up directly under the Prime Minister. Lindemann was informed of the existing Air Ministry Committee and was asked to communicate with Tizard. Still not satisfied that this committee filled the need he had in mind, he induced Sir Austen Chamberlain and Winston Churchill to write to the Prime Minister, suggesting an independent body which would not be a departmental committee of the Air Ministry. Lindemann was invited to join the Committee for the Scientific Survey of Air Defence, but postponed his acceptance until he could find what had happened to his proposal. "I understand the whole question is soon to be debated in Parliament," he wrote on January 5, 1935, "and I think you will agree that in the circumstances it is best for me to defer giving a definite answer for the time being."[17]

On March 19, Sir Austen Chamberlain arose in the House of Commons and asked what steps were being taken to provide defense against aerial bombing. In his reply, the Prime Minister referred to the existence of the Air Ministry committee which, he said, "has already made concrete proposals for a promising line of research for which the necessary facts have been made available and preliminary experiments already started." And he continued:

> The Government have always recognized, however, that the inquiry requires the cooperation of all the defence departments and other public institutions. We have, therefore, decided to appoint also a special subcommittee of the Committee of Imperial Defence through which the Air Ministry Committee will report to the Committee of Imperial Defence itself. This subcommittee will have the direction and control of the whole inquiry, and the necessary funds

to carry out experiments and to make researches approved by this committee will be made available.[18]

The Air Defence Research Subcommittee was composed of the Chairman, Sir Philip Cunliffe-Lister (Secretary of State for Colonies); Ormsby-Gore (First Commissioner of Works); Vice-Admiral Henderson (Third Sea Lord and Controller of the Navy); Sir Hugh Dowding; the Rt. Hon. Lord Weir; Sir Warren Fisher (Permanent Secretary to the Treasury); Sir Frank E. Smith (Secretary of DSIR); Mr. H. T. Tizard; and as secretaries, S/L P. Warburton and Mr. A. P. Rowe. This committee held its first meeting on April 11, 1935, by which time the Air Ministry Committee had been given informal assurances that funds would be forthcoming.

On March 18 Wimperis reported to the Committee that Treasury authority had been received for the expenditure of £10,000 for the detection research.[19] All the appropriate steps were already being taken. The Director of Works and Buildings of the Air Ministry was to visit Orford Ness on the following Wednesday for the purpose of investigating the laying of a cable from the village of Orford under the water to the island in order to provide needed electrical power; and he was also planning to inspect the site for the erection of wireless masts, new buildings, and for possible rehabilitation of the existing buildings.[20]

Provided at last with a formal go-ahead Watson-Watt began during the last weeks in March and the first weeks of April to organize the project at the Radio Research Station.[21] To the detection problem he assigned A. F. Wilkins, who had made the Daventry measurements, Bainbridge-Bell, another worker at the Station, and E. G. Bowen, who came to work at the Station early in April. Wilkins and Bainbridge-Bell were assigned to develop the receiver and the transmitter; Bowen, who since 1932 had been engaged on cosmic ray research in Professor Appleton's ionospheric laboratory at the University of London, was put to work on the transmitter.

No one else at the Station except Watson-Watt's aide, J. F. Herd, had been informed of the new project. These initial preparations were enveloped in the profoundest secrecy. The men kept no data books and drew no circuit diagrams. The preliminary work during April and early May at the Station consisted to a great extent in collecting together as much equipment as possible. At first the project was short of ready cash, and for the most part the equipment was scraped together from cast-off components available in various corners of the Station.

Watson-Watt appeared before the Tizard Committee for the first time at its sixth meeting held on April 10, 1935.[22] He began by saying that his proposals grew out of fundamental radio research which had been undertaken at the Radio Research Station, Slough, on the location of atmospherics, on the measurement of the angle of elevation of short pulse signals received from the U.S.A. on 20 m, and on exploring the ionosphere by the pulse method. These results led him "to consider it practically certain that, by employing 'floodlighting' with radio waves, secondary radiation would be received from aircraft that could be located," [23] and he believed that, although the use of continuous wave excitation might extend the maximum range of detection, the pulse technique offered the best prospects for detection plus location. As an example of the small amount of energy needed for detection, Watson-Watt told the Committee that a Virginia aircraft had recently been detected at a range of a mile or so, using a 20-m beam transmitted from across the Atlantic. [24]

The possible developments and applications of his device Watson-Watt put into four categories. First came long-range detection of aircraft, without accurate location. Although range would be found incidentally with the pulse method, it would not give azimuth or elevation. "A range of 100 km," he said, "does not seem impracticable." Second was moderately accurate location at ranges between 10 and 50 km. Third was accurate location at small ranges (less than 10 km) for anti-aircraft fire control. Fourth, requiring less accurate location than the previous case, was the use of the device to put home fighter aircraft in contact with enemy bombers through radio-telegraph. Watson-Watt requested that a decision be made on the relative priorities to be assigned these four objectives, and the Committee agreed that such priorities should be determined.

Watson-Watt then predicted that work should begin at Orford Ness in a month's time, which later proved to be very nearly accurate. Preliminary study of the transmitter had been concluded and the specifications drawn up, and it should be ready in about a month. The receiving and indicating equipment should also be ready at that time. They were planning to use 50 m at Orford Ness, with horizontal polarization, although better results might be expected at shorter wavelengths, as for example at 7 m. There were several persuasive reasons, however, for using wavelength of 50 m. In the first place the pulse technique and the necessary components were more highly developed in the neighborhood of 50 m. Second, direction-finding techniques for azimuth determination would be easy at 50 m but difficult at 7 m. His last argument was based on the proximity of the research site to a possibly hostile continent. The use of 50 m would permit greater secrecy, for if pulses on this frequency were received abroad, ionospheric research would provide a sufficient explanation. Orford Ness could, if necessary, be associated openly and deliberately with ionosphere work. Seven meters would doubtless have to be used for close range work with anti-aircraft guns.

In response to a question posed by Wimperis, Watson-Watt replied that there should be no difficulty in applying his technique to the detection of ships at sea. He said he thought ships could be located, and their courses plotted, at ranges of the order of 20 miles. In this work vertical dipoles would probably be used.

By the middle of May, the power cable had been put through from Orford to the island and some arrangements had been made to house the workers in the village and to shelter their newly assembled equipment on the island. On May 13, 1935 a historic caravan, consisting of a passenger automobile and two RAF 10-ton lorries loaded with equipment, set out in the early hours of the morning from Slough. By 10:00 A.M. it had arrived at Orford village. Aided by ten RAF ground crewmen, and with the help of the indigenous island transportation—an antiquated Ford car and a fire truck—they moved their bulky equipment through the rain and hail of a vicious East Coast day to their primitive headquarters on the island.

The Year at Orford Ness[25]

With no technical help and with only one other occupant of the island, a watchman maintained there by the RAF, to lend them an occasional hand with the heavy moving, the three men—Wilkins, Bowen, and Bainbridge-Bell—set up the receiver and transmitter units in the two newly prepared huts about 300 yards apart. Wilkins's antenna was virtually nondirectional, a single 75-ft dipole with a twisted wire feed supported at the two ends by two 75-ft wooden towers that the Air Ministry had provided and that were placed on either side of the hut. This antenna sent one rounded lobe out to sea in the direction of Ostend and a similar back lobe over land, in the direction of the RAF field at Martlesham Heath, some 15 miles away. The receiver, designed by Wilkins and Bainbridge-Bell, was a broad-banded affair that fed the signal into a cathode-ray oscilloscope to give what came later to be referred to as an A-scope indication. This provided a 50-mile range scale.

The transmitting equipment was housed in a large frame 8 ft high and 4 or 5 ft square. In the lower half was a bulky thyratron modulator and some power supply units. This fed a pair of power triodes arranged to operate in push-pull, located in the upper half of the compartment. The transmitter tubes chosen to produce the high power that was required were British Navy NT-46's, tubes about 12–15 in. long and surrounded by a pure silica envelope. They were rated to operate at 5000 V and to give 6 kW of cw. The whole apparatus was later described by Bowen as "a marvel of crudity." Many of the circuit elements were thrown together with whatever materials were found at hand. Bowen recalled with special relish an awkward variable condensor made by fitting two ordinary metal cans of different sizes one inside the other!

In general, the specifications which the workers were seeking to meet were summarized in the following table which compares these new objectives with what was achieved in ionospheric work:

	Old	Watson-Watt
Peak power	2 kW	100 kW
Frequency	6 MHz (50 m)	less than 6 MHz
Receiver bandwidth	10 kHz	50 kHz
Pulse repetition rate		50/sec
Pulse length	100–200 μsec	10 μsec

During the next three or four weeks, the workers made steady progress in the installation. The transmitter power was successfully raised from 25 kW to 50 and finally to 75. The pulse length was shortened without trouble to 25 μsec, but difficulty was experienced in going to anything shorter. The bandwidth of the Wilkins' receiver was 50 kHz as required, and the pulse repetition rate was fixed by the 50-Hz frequency of the main power supply from the Orford cable. Soon the separate components were working satisfactorily and being steadily improved, but for some time no actual experiments were undertaken.[26]

Watson-Watt took little part in the actual experimental work, though he kept in close touch with the island. He assumed as his special task that of defending and promoting the project on the level of the Tizard Committee and the Air Ministry. He held long and frequent discussions with Rowe, whom he found to have a special gift for scientific statesmanship and intense missionary zeal for the radio detection program. Both men were anxious to secure the maximum support for the work without sacrificing independence or allowing it to become entangled in service red tape. They spent long hours discussing the military applications of the new device.

Just a little over four weeks after the work had begun at Orford Ness, it was decided to attempt the first detection experiments and to have them coincide with a grand opening, attended by the entire Tizard Committee. While tuning up the system in expectation of the Committee's visit, the investigators at Orford Ness, according to the official record, saw their first signals.[27] A Singapore flying boat which happened to be cruising off Orford Ness was followed out to a distance of 25 km and back again, this being the maximum range apparently reached by the plane. In addition, a Virginia aircraft was followed from Orford Ness to Martlesham, along a track unfavorable for observation, and the indications were less clear than for the Singapore. Though there is some question about these first observations, there is no doubt that the equipment had made great progress. The transmitter was giving about 60 kW of peak power. The duration of the transmitted pulses had been reduced to 25 μsec and though the received pulses, as recorded on the oscillograph, were widened out by the

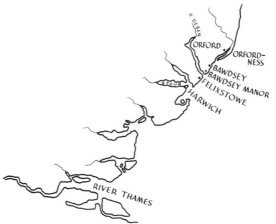

FIG. 6-1. Locations of radar agencies in Great Britain.

receiver response to about 75–80 μsec, this was not deemed excessive at the time.

The Committee spent a pleasant June weekend at Orford Ness. The members arrived at the island, shepherded by Watson-Watt, on Saturday morning, the 15th. They were shown about and listened to an explanation of the equipment.

That afternoon, the first formal experiment was attempted. An old Air Force biplane, a Valencia troop transport supplied especially for the test by the RAF station at Martlesham Heath, flew in from the landward side through one lobe of the antenna beam. Although both the transmitter and receiver seemed to be working perfectly, for some reason that was never discovered, unequivocal signals were not observed.[28] On the next day, Sunday, June 16, another flight with the Valencia was made, and this time the tests failed completely. Conditions for radio reception of any kind were appallingly bad. Nothing resembling a signal was ever apparent, and the screen was confused with bad atmospherics and "noise" interference.[29] The failures were never explained, and Bowen in retrospect has called it "just one of those things." The Tizard Committee, he recalled, did not seem in the least dismayed by this initial setback.

On Monday, June 17, the day after the departure of the Committee, the first real signal was accidentally picked up in the presence of Watson-Watt, who had remained behind. It was a dramatic moment. The signal was strong and first appeared on the range scope at 17 miles. The plane that caused it turned out to be a Scapa flying boat, a twin-engined biplane with 70–80-ft wingspread that came from the Marine Aircraft Experimental Establishment on the coast just below Orford. When it was observed, it was engaged in practice bombing runs at sea, but it flew close enough to the island to be accurately identified. Watson-Watt reported to the Committee later that the plane was continuously tracked to and from a maximum range of 46 km (29 miles), and that the signal responses at the maximum range reached by the aircraft were of such magnitude that a much greater range could have been obtained.

During the summer, the station's performance steadily increased, not by virtue of any startling innovations, but through improved details of construction, better matching of lines, and so forth. In the next few weeks the range went up step by step from the original value of about 25 miles to 40 miles and then to 50. The earliest photographic record, taken July 24, 1935 (which is probably also the earliest A-scope picture ever taken) shows a signal at 54.6 km. The power output soon exceeded the original specifications: by the end of the summer, it had been raised to 150–200 kW by dint of increasing the filament emission of the pair of power tubes. The Admiralty Signals School Laboratory at Portsmouth developed at

Bowen's request triodes with double the filament wattage without being told the use to which they were to be put.

In the same period, a series of external events led the investigators to shift their frequency of transmission. Early in the summer, the 50-m pulses began to be badly jammed, to all appearances deliberately, by commercial broadcasters. By the end of July, it proved impossible to work on 50 m, so the Orford station halved this value to 25, then added another meter to be sure of avoiding interference with stations at 24 m. By the end of September on this new frequency and still using the 75-ft masts, they had observed a maximum range of 90 km,[30] and they were soon getting 60 to 80 km as a regular thing.[31]

While this work was going forward with great secrecy, the British were treated to a demonstration of the inadequacy of their obsolete methods of defense. Mock night air raids were held about the middle of August against naval installations and shipyards. It was reported in the press that 21 bombers had scored hit after hit in the Southampton area and that the defense had proved helpless. Expert opinion was openly pessimistic and the London *Times* issued a solemn warning.[32]

Until the end of the summer, there had been no definite ideas at Orford Ness on how to obtain accurate direction finding, for all their energies had been directed towards obtaining a greater maximum range of detection. The first idea that was suggested for accurately locating a plane was simultaneous detection by two stations and subsequent triangulation. This suggestion was promptly rejected by Bowen who pointed out the difficulties that would ensue when there were two or more planes to be located. Instead, he suggested the use of crossed dipoles and a goniometer which would give an accuracy in azimuth of about 5°.[33]

Thus by the end of the summer of 1935, the main features of an RDF warning system had been blocked out. It was now time to consider setting up a series of warning stations. As early as September, it was decided to give coverage to the Thames estuary by establishing five stations, situated about 25 miles apart. This plan was based on tactical information provided by Air Ministry officials, which assumed the probability of the invasion of the lowlands but which left out of account the possibility that France might be occupied by the enemy.

Plans for the new stations were elaborated during September but it was some time before work was begun on them. By early fall the first new recruits were beginning to arrive at the station, both university men and commercial people. It was not planned to make any basic alteration in the system before making it the prototype of what soon became called the Chain Home Stations or CH Stations. The same wavelength was to be employed, and in order to get increased range, the height of the towers was to be raised from 75 to 250 feet. The first of these 250-ft towers were erected at Orford Ness in the autumn of 1935, and the system at Orford Ness became the first of the historic British Warning Chain.

The first efforts to decrease the wavelength materially were made at Orford Ness from August to October with tubes that operated at 8 m. This first attempt met with no success.[34] Later in the autumn two tubes arrived from the Admiralty Signals School that were capable of operating at half that wavelength, so work was promptly begun at 4 m. Substantial success was reported on this wavelength early in December with the system giving ranges in the neighborhood of 25 km. It was believed possible that they might obtain 100 km on 4 m within a year, a hope destined to be dashed.[35]

By the early autumn it was apparent that a new and more convenient site would be necessary for the expanding project. Watson-Watt, Bowen, and Wilkins drove up and down the southern coast of Suffolk looking for an appropriate location. Just south of Orford a site was finally discovered, a large country estate, Bawdsey Manor, situated on a 90-ft bluff at the edge of the estuary of the River Deben. The estate was the property of a certain Sir Cuthbert Quilter, and steps were immediately begun for its purchase by the Air Ministry. It was planned to turn part of the manor house into laboratories and the rest into residential space for the staff. Other laboratories were to be constructed on the grounds.

The process of moving from Orford Ness to Bawdsey was slow and painful. By the end of February 1936, Bawdsey was only partly occupied, and the moving was only completed in May. Just about this time, someone in the Air Ministry gave the equipment the general designation it was to have for about eight years, that of RDF, presumably meaning "Radio Direction Finder." In May, just at the time of the final move to Bawdsey, Watson-Watt, Bowen, and Wilkins and their colleagues were officially transferred to the Air Ministry Research and Development staff, Watson-Watt being given the title of Superintendent, Bawdsey Research Station.

Bawdsey Research Station

The move to Bawdsey Manor of this handful of pioneers—there were probably no more than a half dozen of them—marks the creation of the Air Ministry's radar laboratory. At first this was called simply the Bawdsey Research Station, though later it became AMRE, meaning Air Ministry Research Establishment. The late spring and most of the summer months of 1936 were devoted to recruiting new personnel. The new men were drawn, according to the prevailing philosophy, largely from the ranks of young academic physicists. Watson-Watt scouted about, visiting the various physics laboratories in different parts of the country requesting, demanding, or wheedling the release of high-grade research men. By

August, the number of workers had risen to about 15 or 20, and as people began to join at an increasing rate, the count soon reached 50.

At Bawdsey Manor the surroundings were beautiful and impressive.[36] The site was about 70 miles from London and distinctly inaccessible, being located across the estuary of the River Deben from the seaside resort of Felixstowe and some distance from any main highway. The estate occupied a high bluff bounded on one side by the Deben and on the other by an artificial cliff of concrete looking toward the North Sea. Below the cliff, a public right of way ran along the beach. The house itself was a handsome red-brick building located in spacious, well-wooded grounds, graced by some fine old shade trees, and with attractive gardens and a greenhouse. On one side the house opened onto a terrace and a bricked-in garden; beyond this was a splendid lawn about a quarter of a mile in circumference.

Although the men at Bawdsey were all under Civil Service, they enjoyed at first a unique and beneficial laxity in the enforcement of the usual rules. The group resembled an Oxford or Cambridge college more nearly than a Government research laboratory: for about six months they kept hours of their own choosing and were pleasantly oblivious of Civil Service red tape. It was Watson-Watt's own idea to turn the manor into pleasant residential quarters as well as a laboratory, where the men—nearly all of whom were young bachelors—lived together and had their meals together in a large wood-paneled hall. Some of their best ideas and most fertile discussions took place in true Common Room style around a fire in the evening.

Part of the manor house served as living quarters, another part provided laboratory space—it was only later, in 1938, that laboratory buildings were erected on the extensive grounds—and still another part continued, for what seemed to everyone a very long time, to shelter the person of Sir Cuthbert Quilter, whose departure was regularly postponed. The manor received its food and other supplies by automobile from the nearby village of Woodbridge. Those who were married lived in nearby Felixstowe and daily took a ferry across the Deben to work, but a lucky few lived in cottages on the estate. Watson-Watt, the Director, lived with the scientists in the manor house until his appointment as DCD (Director of Communications Development) at the Air Ministry. Under Watson-Watt, and serving as deputy, was E. T. Paris, a scientific officer of the War Office who had been brought in to head a group of army workers at the laboratory. When in 1937 Watson-Watt was made DCD, Paris ran the laboratory for a few months until A. P. Rowe, the Secretary of the Tizard Committee, became Superintendent in November 1937.

In good weather Bawdsey Research Station had many of the attributes of a summer camp. The men would stop for a swim in the middle of the

morning; and after hours some worked in the manor garden to keep themselves, and incidentally the garden, in shape, or played cricket on the
magnificent lawn. Despite their lack of routine, or because of it, a great
amount of work was accomplished, and often one could find half the
laboratory working away long after midnight. Under the Watson-Watt
dispensation, no effort was made to enforce the strict 9 to 5 hours or other
regulations of the Civil Service. Gradually, however, this idyllic state of
affairs came to an end. When visitors began to arrive in increasing
numbers even Watson-Watt saw the necessity for stricter compliance with
the regulations. Under A. P. Rowe, a still more rigid system was adopted
and men were forced to sign in and out in true Civil Service style.

During the year 1936, energies were concentrated on developing the
Bawdsey Research Station, on perfecting the early warning system on 26
m, and on establishing the first of the five Thames estuary stations which
had been officially authorized the previous December.[37] These stations
were soon referred to as Chain Home or CH Stations. Trouble of various
sorts slowed up the development. Bowen and Wilkins were at first the
heads of the two main groups, one concerned with transmitters and the
other with receivers and antenna arrays. Very soon, however, Bowen set
up an air-borne group and relinquished the direction of the transmitter
work to another key man, Mitchell.

In the summer of 1936, two 40-ft masts for a Chain Home Station were
being built some distance away from the manor house overlooking the cliff
and plainly visible to persons walking along the beach. Preparations were
under way to give a demonstration of the effectiveness of a CH station to
the Secretary of State for Air and other high officials of the Air Ministry.
A demonstration was, in fact, long overdue, for the new device had so far
been sponsored without convincing proof of its effectiveness. The date for
a demonstration had been several times selected and postponed, and it was
felt that to postpone it further would be to lose the confidence and the
support of the Royal Air Force. Preparations were hurriedly made, but
nothing was really ready and in proper shape. A system was devised to
transmit bearing and range data from Bawdsey to the Fighter Command
Headquarters in London.

The tests took place about the middle of September. A series of mock
raids were made by a squadron of flying boats coming in over the North
Sea. The exercises were disappointing, if not a complete failure; and no
echoes at all were obtained on the first and perhaps the most important
day of the demonstration. In a frantic effort to get some sort of performance before the period of the trials ended, Bowen was called back, and
with Gerald Touch built a new transmitter in 48 hours with which the
attacking planes successfully were picked up at 90 miles. This did not
serve to modify the original poor impression, and an extremely unfavor-

able report was received from the Air Ministry. Watson-Watt read it with appropriate comments to the assembled group.[38] He told them that they had been given just a few month's lease on life and that there would then be another test which must succeed or the Air Ministry would withdraw its support of the project.

There was nothing really wrong with the equipment except that Mitchell's transmitter, which was supposed to be of improved design, was only incompletely developed at the time of the first test.[39] They settled down at the point to improve the system in the months of grace at their disposal. Everyone worked at top speed and there was steady improvement. Watson-Watt himself pitched in, and on one occasion climbed up one of the towers in a howling gale to fix the antenna structure. Bowen was taken off the air-borne project and assigned to work on the ground stations once again because of the importance of the forthcoming trials. Bowen and Touch daily drove the 10 miles from Bawdsey to Orford Ness, which continued in use as a sort of field station, to work on the improved transmitter. The trials were repeated in April, 1937, this time to everyone's satisfaction.[40] With the improved 26-m transmitter, they got a range of a hundred miles on planes that flew out to sea from Martlesham Heath.

Meanwhile, a group under H. Dewhurst had started work at Orford Ness on 13 m and 4 m with the hope of building a mobile set. They got no results on 4 m; but a 13-m system, complete with direction finding and height finding, was working by January, 1937. The main CH stations, operating at 26 m, shifted over during the course of that year to 13 m to avoid the interference and jamming they were encountering at 26 m. A mobile set on 13 m was actually built during the late winter and spring of 1937, installed in a truck and provided with extendable wooden towers of the sort used by the fire departments of some English cities. This system was demonstrated in the summer of 1937 at Dunkirk, but although it had a range of 30–40 miles and had both direction-finding and height-finding features, no interest was expressed in this system.

During 1937 work began in earnest in setting up the chain of early warning stations. By August of 1937, CH stations had been set up at Bawdsey, Dover, and Canewdon and the first experimental filter center for rapidly collecting and centralizing information was established at Bawdsey. In the same month the third air exercises were held using the three RDF stations. By the end of the year or by the beginning of 1938, the first five had been set up.[42] In September 1938, all five went over to 24 hour operational watch with RAF personnel entrusted with operation and maintenance. The filter room was transferred to HQ Fighter Command. Meanwhile the new radio warning technique was gradually integrated into the British defense system. During 1938 it was arranged to extend the chain of stations along the east coast about the Thames estuary as far north as Edinburgh and south of the Thames as far as Portland. Again the stations were to be placed so that the beams just overlapped, that is, about 20–25 miles apart. By the spring of 1939 the whole of the eastern chain had been completed, as well as a few of the stations south of Dover. In the summer of 1939 a CH station went on the air at Scapa Flow.

Once the stations were in regular operation, some of the consequences of sidetracking the radar program and keeping it out of the hands of the regular service establishments became evident. The equipment, from an engineering standpoint, was not up to operational standards, and break-downs were numerous. This early radar was an engineer's nightmare, imperfectly designed, even crude, and very fragile and temperamental. Much engineering redesign had to be done after the stations were in regular operation, and scientists were detached from Bawdsey and assigned to the RAF to help with this work.[43]

These initial engineering weaknesses were a consequence of Watson-Watt's deliberate policy of leaving the work to young physicists with no industrial or engineering experience. This decision, however, probably produced equipment sooner than it could have been got in any other way. Apart from his original suggestion of radar, Watson-Watt's main contributions in this early period can be summed up as follows: he organized and put across the project with great skill; he defended it ably in high quarters; he seems never to have lost faith in its ultimate success; he built up radar research with men of the highest scientific caliber; he very early pointed out that a chain of RDF stations was essential, that a single station would be nearly useless; and he early proposed to link the stations by means of a Filter Center.[44]

Description of CH Stations

The CH or Chain Home Stations were large complex installations operating interchangeably on any one of four frequencies in the band between 23 and 50 MHz (6–13 m).[45] The stationary transmitting arrays were placed on or between 360 foot towers, and consisted of vertically stacked individual dipoles (not stacked arrays). This served to limit the beam to low angles, though it produced a main lobe nearly 60 deg wide in azimuth. Gaps in the vertical coverage were filled by a second stack lower down the tower, usually consisting of four dipoles. Both arrays were backed by tuned reflectors or curtain reflectors to minimize backward radiation. The receiving antennas, also stationary, were mounted on 240-ft wooden towers, and consisted of a double stack of crossed half-wave dipoles at the top of the tower and a similar "gap-filling" antenna located some distance below. Bearing was determined by means of the goniometer used in conventional radio direction finding. The method depends upon the fact that the signal set up in an antenna varies in intensity with the direction from which the incident wave approaches the antenna. The signals from the crossed dipoles are fed into corresponding coils placed at right angles to one another in the goniometer. The alternating current in each coil system sets up a magnetic field, and the resultant field indicates the direction of the incident waves. The field is explored by a search coil, and the current induced in this coil is fed into the receiver. The search coil is rotated until the signal disappears from the scope, and the angle through which it has been rotated is equal to the angle at which the incident waves are striking the antenna.

Two methods were available to determine height with the CH stations, but only one seems to have been widely used. Much the more accurate was the one which switched the goniometer to two antennas on different heights in the mast to determine the angle of elevation of the target by comparison of the reading with a previously calibrated curve of goniometer readings. The other, and more common method, was the familiar one of relying on a previous calibration of the beam and estimating the height of the aircraft by observing when it first enters or leaves a lobe of the beam.

Extremely good range was obtained with this set which would dependably track reliably out beyond 100 miles and usually to 150 miles. Ranging on single planes out to 175 or 180 miles was by no means uncommon. A range accuracy of 1 mile, a bearing accuracy for strong signals of 5°, and height measurements on high-flying planes at 70 miles accurate to 1000 ft: this seems to have been the expected performance of the Chain Home Stations. The stations were placed inland a few miles from the coast where the ground was flat. Because, by later standards, the bearing and elevation data provided by the CH stations were often grossly inaccurate, especially when enemy formations were still far from the stations, Filter Centers, which had been first used to correlate the reports of visual spotters, were set up to assess the value of received plots from several stations and smooth them into the most probable track. The filtered track and other information were passed from the Filter Room to Group and Sector Operations Rooms where they were used in bringing about visual interceptions.

The First Airborne Radar

Late in 1935, one of the members of the Tizard Committee pointed out the necessity for a small radar set that could be carried in an aircraft. In November or December of that year, Bowen was assigned this problem. Although he was still trying to improve the transmitter of the early warning system at Orford Ness, and could only work at the new problem part time and without any assistance, he nevertheless undertook the assignment.[46] After a series of discussions, Watson-Watt and Bowen agreed upon the characteristics that such a set should have. Obviously it had to be very light and compact, and apparently also it had to have a greatly improved, that is a much shorter, minimum range. This last was the most difficult problem and involved the greatest jump so far attempted. In going from the ionospheric work to plane detection, they had passed from a 50 mile minimum range to one of 5 miles. The airborne system was still more exacting and would require a minimum range below 1000 ft. In February, 1936, they were talking of a system on 1 m, using four aerials and a 1-μsec pulse. The maximum range would be about 5 km and the minimum range about 150 yards.[46]

Bowen's first experimental work on the new project began with an attempt to develop a 1-m receiver using RCA acorn tubes, still only available from America. The receiver was finally completed at Bawdsey during the spring of 1936. Encouraged by this success, Bowen organized a small group within the laboratory for the development of airborne equipment. In the fall of 1936 this consisted of Bowen, P. A. Hubbard, S. Jefferson, and A. G. Touch. They tackled the problem of a 1-m transmitter, but immediately ran into the obvious difficulty that there were no tubes in England or America, so far as they were able to determine, that could give a reasonable power output at 1 m. Accordingly, they abandoned work at 1 m and chose instead a wavelength that was possible, e.g., 6 m.

On this wavelength, before the end of 1936, they performed a series of experiments on aircraft targets. In these experiments, they got their first positive result in which one aircraft successfully detected another aircraft. The experimental plane had only a receiver in it built by S. Jefferson and the target plane was illuminated from the ground by a transmitter which had a 9-ft dipole and silica envelope transmitter tubes operating in push-pull, a power output of about 40 kW. By late December 1936, they were flying a Heyford aircraft, borrowed from Martlesham Heath, that had a receiver aboard and a single receiving dipole between the wheels; and they were getting aircraft echoes at 10 and 12 miles. Bowen's suggestion that work proceed on this wavelength was vetoed by Watson-Watt.

Several expedients were tried in the course of the spring to lower the wavelength. An attempt was made to have the silica tubes modified at ASE so that they would work at 3 m, but this was not very successful. Sometime during this period, they made a completely fresh start and investigated the possibilities of a wavelength of about 1 m for which the newly developed Western Electric "door-knob" tubes appeared suitable, at least for experimental purposes. At first there was only a limited supply of these tubes available and they were dependent upon American importation. Gerald Touch was the first of Bowen's group to try them out as transmitter tubes.

Part of Bowen's group was still working on 6 m and some attempts were still going on to get power at 3 m. While the major effort was going forward on 1.5 m, still another project was undertaken, an outgrowth of the earlier Heyford tests. This was currently referred to as RDF-1$\frac{1}{2}$ because it was a sort of cross between RDF-1, the CH system, and RDF-2, the as-yet-unborn aircraft set. This was to operate on 13 m, and to have the transmitter on the ground and the receiver in the plane. A system was actually assembled, but the work was dropped in favor of the 200-MHz-equipment.[47]

Touch built a system on about 300 MHz, with a "door-knob" tube transmitter that gave a scant 50–100 W, and a superheterodyne receiver using a television IF with RCA acorn tubes as mixer and local oscillator. Under Bowen's supervision, these components were turned into an operating system, which was pointed out to sea and got ship echoes. In August they installed their primitive system at Martlesham Heath in the first of two Anson aircraft.[48] The receiving antenna consisted of a length of wire hung along the inside of the fuselage of the Anson, a resonant quarter-wave piece of wire. The transmitting antenna was a half-wave dipole supported on a piece of wool below the bomb bay. Touch flew with the equipment while Bowen was on his annual leave, and consequently had the good fortune to be the first British scientist to fly with airborne radar, and the first to see a ship from the air on a radar scope. On their A-scope type of indicator they picked up a ship, a 2000-ton freighter near Harwich harbor, at a range of 4–5 miles and circled round it.

It was obvious that the system was still lacking in many essentials. It had no method of determining bearing and had hopelessly inadequate transmitter power. When they flew with another plane as target, they got extremely short ranges, 3000–4000 ft. At this time the idea of using a stabilized self-modulating or "squegging" circuit for the transmitter was hit upon, and the early tests indicated that this would be successful. At about this time also they hit upon a satisfactory transmitter tube for the desired wavelengths.[49] Three systems were built using the improved transmitter, which gave a peak power output of 1 or 2 kW at 200 MHz with a 2- or 3-μsec pulse. The overall transmitter efficiency was greater than they had attained before. The improved set was installed in a specially screened Anson; one antenna was a long wire from wing tip to the rear of the fuselage, giving the main lobe abeam; another was a Yagi on top of the wing pointing in the same direction. There was no method of direction finding.[50]

They were severely handicapped by the strict requirements the Air Ministry had specified for airborne interception equipment, namely, that the whole equipment should not weigh more than 200 pounds, that it should occupy a maximum of 8 cu. ft, and that it should draw at most 500

W. Needless to say they were a long way from satisfying these requirements. In fact, later developments have shown that the specifications were unnecessarily stringent. Moreover, they had not been able to meet the performance proposed by Watson-Watt to the Tizard Committee: a minimum range of 150 yards and a maximum of 5 km. The performance of the AI system[51] in December 1937 was such as to give a minimum range of about 1000 ft and a maximum range of 4000 to 5000 ft.[52]

The first flights that Bowen and his co-workers made with this equipment showed that it gave very strong ground reflections but very little sea return.[53] Ship echoes were so strong and clear with the air-borne system that they saw the immense advantages that might be derived from developing a set designed to detect shipping from the air. The priority on airborne interception equipment promptly dropped off as a result of this observation, and all energies were directed toward the detection of surface vessels during the rest of 1937 and most of 1938. It made little difference, as Bowen pointed out, for the equipment was fundamentally the same. Anything they developed toward ship detection would help in the development of an AI set.

The First ASV Equipment

In the spring of 1938, they were able to test the equipment quite extensively. On March 28th, they took the set to Lee-on-Solent—the strait between the Isle of Wight and South Coast of England—where the RAF Coastal Command was cooperating with the Royal Navy in some fleet exercises. Bowen and Touch spent three or four days demonstrating the equipment in one of the Ansons and explaining its tactical possibilities to Admiralty officials and representatives of the Fleet Air Arm and the Coastal Command. The tests were very successful, although the set had no direction-finding feature. Bowen flew the first day, demonstrating the set to Admiralty and Fleet Air Arm representatives. Touch flew with the system on one of the other days, and recalls that he was favored with a brilliant, sunlit day; ten ships of the Fleet in the Solent, including the ill-fated Hood, were picked up clearly with maximum ranges of the order of 10 to 15 miles. The first airborne scope pictures were made on this occasion, Touch having built a crude photographic recording device just a few hours before the demonstration by pasting a strip of paper to block out the trace so that only the top of the main bang and the tops of the pips could show.[54]

The interest of both the fleet and the Coastal Command was aroused by this demonstration, and the result was a request from the Admiralty that six aircraft be fitted with ASV equipment for service trials. The Navy listed some requirements for ASV and information supplied by Dr. A. G. Touch specified that the equipment should provide an all-around search.

The set was to be used for scouting and following, and the Navy insisted that the beam should describe a circle around the aircraft. During the spring and summer months of 1938, Bowen and his men struggled to satisfy these requirements. They adopted Yagi arrays placed above the wings and the tail, which gave them ranges up to 35 miles. At this time work was being resumed on AI.

In September, 1938, important fleet exercises, lasting 48 h, were scheduled to take place in the North Sea. The whole of Coastal Command was to be let loose to track down a group of several ships, the aircraft carrier *Courageous*, protected by the battleship *Nelson*, two cruisers, and a small group of destroyers. The Bawdsey airborne group was invited to participate.

The exercises were scheduled to begin at 7:30 A.M.[55] One of the Bawdsey Ansons, equipped with ASV and with Bowen aboard, took off from Martlesham at 7:15 and flew over the nearby flying boat station of Coastal Command before any of its planes had left the base. They chose a carefully selected course. It was obviously impossible to cover the whole North Sea area, and they preferred to miss completely if they missed at all. They decided to try only the southern half of the North Sea for this appeared, all things considered, the most logical place for the maneuvers to be held. As it turned out, this was well calculated. At 8:15, just an hour after departure from Martlesham, they picked up the task force on the ASV equipment. They followed the signals in to the target and found the *Courageous* surrounded by its protective array of ships. Their appearance brought all the planes off the deck of the carrier. Soon weather closed in on the ships and the other planes searched in vain during the succeeding hours. Bowen's plane returned to the base before noon and the Air Ministry was informed by telephone of their success. They did not communicate with the Coastal Command, whose planes proved unable to find the ships until the 48-hour period had already expired.

These results only increased the Admiralty's impatience to have this equipment available. Bowen tried to get the Air Ministry to order six sets and to provide six aircraft, with the engines properly screened, for use in the Admiralty trials.[56] The Air Ministry gave them no encouragement. In consequence, practically nothing was done to engineer a system for ASV until the outbreak of war. Once again, the emphasis had shifted back to AI, and the Tizard Committee and the Air Ministry were both exerting great pressure upon Bawdsey to accelerate the AI development. When the war broke out, England had no planes equipped with ASV, nor even a set suitable for airborne search, though the techniques developed during the intervening period for AI proved readily adaptable to this purpose.

The Development of AI

Work on an AI system was resumed in the summer of 1938. A "squegging" transmitter, a dipole transmitting antenna, and an improved receiver were installed in one of the Ansons. On a flight at 15,000 ft to see what performance the system gave, they got a range of 10,000 ft on the broadside aspect of a Handley–Page bomber. This was regarded at the time as promising well for the development. It was soon thereafter followed by a study of the reflection behavior of aircraft, and the systematic determination of reflection coefficients for the various aspects of target planes.

Having now in their possession a set with sufficient range they concentrated on developing a method of direction finding by lobe switching. The four antennas (to give up–down, right–left information) were installed in one of three Fairy Battles which were single-engined, all-metal medium bombers that the Air Ministry had sent to Martlesham Heath toward the end of 1938 for use by the Bawdsey AI people. A pair of quarter-wave dipoles on either side of the nose served as azimuth receiving antennas, and there was a pair of elevation receiving antennas on the top and bottom surfaces of one wing. A half-wave transmitting antenna was located on the other wing. Separate cathode-ray tubes were used for presenting the azimuth and the elevation information. This system was put together during the winter of 1938–39, and elaborate tests were run to check the elevation and the azimuth pattern. The measurements were made from plane to plane and involved hundreds of hours in the air.[57]

The Battle, equipped with an AI system not unlike the later Mark IV AI, was first flown in June 1939, and in most respects it lived up to expectations with a maximum range of about 2 miles and a minimum range of 1000 ft. At the time this seemed adequate for a night fighter, a role for which the RAF had already decided to use the Battle.

The AI equipment itself was the crudest sort of laboratory model, but during June it was demonstrated to a succession of high Air Ministry and RAF officials, to Churchill's personal scientific advisor, Lord Cherwell (Professor Lindemann), and on June 20 to Churchill himself, who was already heir presumptive to the premiership of Neville Chamberlain. A demonstration at about the same time to Sir Hugh Dowding was a great success and had important consequences. They gave him quite a show, took him up through the clouds and ran a mock interception. Dowding was very favorably impressed by the performance—even though there was a minor crisis when he inadvertently put his finger on the high voltage terminals of the transmitter—and he was afterwards heard to remark, "We must have these things in the RAF." A consequence of Dowding's visit was a request from the Air Ministry that 30 aircraft be fitted with AI equipment as night fighters before September 1. This was an extraordinary request. The system had scarcely been engineered at all, the design was wholly embryonic, and no plans had been made for production; much of the required material was not readily procurable. Results of any sort were only possible with a prodigious amount of work.

There is some disagreement, even among the participants, as to the conclusion that could fairly have been drawn from the demonstrations to Churchill and Sir Hugh Dowding. They seem to have proved only that it was possible to detect one aircraft from another when the persons in the radar-equipped plane had fairly exact foreknowledge, as they did in these tests, of the general location of the target. They did not prove that it was possible to intercept an enemy attacker under operational conditions, for no steps had been taken to work out a means of directing the defending fighter to the immediate neighborhood of an attacking plane. As it turned out, a vast improvement in the airborne radar equipment, a long and arduous training program, and the development of a special set for GCI (Ground Control of Interception) all had to be gone through before the AI could prove operationally useful.

Unaware of these obstacles, the air-borne group at Bawdsey worked at a terrific pace, spending perhaps 90 hours a week on the project. In July, the first airplanes arrived for the AI work, but they turned out to be Blenheim bombers instead of the Battles that had been promised. At Martlesham a small group headed by R. Hanbury Brown was in charge of the installation, while at Bawdsey, Bowen and Touch worked long hours on the components and negotiated and conferred with the companies that were quickly brought into the picture. Metropolitan-Vickers was commissioned to manufacture the transmitter, and the Pye Radio Company performed prodigies, making radar IF strips from television receivers. According to E. G. Bowen, they produced 70 receivers in ten days. In August the rest of the Blenheims arrived, long before the equipment was ready. As complete systems became available, they were tested by Touch and Waters at Bawdsey and sent out to Martlesham where Hanbury Brown and his assistants put them in the planes, the transmitters in the nose of the Blenheims and the rest of the equipment in the front cockpits.

About six Blenheims were equipped at Martlesham before the outbreak of war. They were flown from Martlesham to Northolt aerdrome, just south of London, where they were assigned to 25 Squadron RAF. One purpose was to have the planes near Sir Hugh Dowding's headquarters during the period of testing and training. Hanbury Brown was transferred with the equipment to Northolt to supervise the maintenance and to train the radar operators. The first two weeks of September were devoted to a series of more or less controlled tests under the direction of RAF personnel. They tested the range, the azimuth and elevation performance on daytime runs from Northolt to Gloucester. Dowding was a regular and most interested visitor; he visited the field nearly every day, pored over the results of tests, asked many questions, and made a number of suggestions. In the first ten minutes of his first visit, he ordered that the indicators be changed from the front to the rear cockpit.

During September, according to Bowen, the rest of the 30 installations were completed, and the operators got a reasonable number of flying hours, though not against the enemy; enough, however, to demonstrate the weakness and crudity of the equipment and to indicate the tremendous obstacles that would have to be surmounted before this new device could be used against the enemy.

After the completion of the initial tests, some daylight mock interceptions were tried at Northolt. Despite the temperamental qualities of the equipment (which used to bother the pilots by occasionally sparking over and catching fire!) and the lack of experience of the operators, they found that interceptions were, in fact, possible. Consequently, in early October, they staged a practice demonstration at that airfield. An attacking "enemy" bomber was reported by the sector organization and the radar-equipped night fighters were sent out under radio control from the Northolt sector. It was discovered, incidental to these tests, that the visual information from the Observer Corps plot came in too late to be of any use.

The sets did not perform as well as in the demonstration earlier that summer and gave only a range of 9000 ft. But some basic lessons were learned in how to use the equipment, as, for instance, that it was essential to provide better communication and improved teamwork between the pilot and the operator. After the completion of the tests and the interceptions, Hanbury Brown returned to Bowen's group, which, as will be related below, the advent of war had caused to move to Perth. The Blenheim aircraft were transferred to Martlesham Heath, where the sets were used for training AI operators.[58]

Up to this point, no mention had been made of efforts to have the AI interceptions directed by the CH stations. No attempts were made to try this out until the late autumn of 1939 when Sir Hugh Dowding began to discuss the possibilities of this sort of fighter direction. At Bawdsey, the experimental CH station remained in operation, after the departure of the laboratory personnel, and was in the hands of an RAF crew. During the fall it was used for the earliest tests on directing fighter interception. On one of these an enemy flying boat was shot down, but although there was AI in the interception it was not working and did not contribute to the final kill. The sole purpose of these early interceptions was to test the ability of the CH stations to do fighter direction, and presumably its deficiencies in this direction were soon apparent.

The Move to Dundee

It was clear to everyone that, in the event of war, Bawdsey Manor was too exposed a location for such an important laboratory as the Air Ministry Research Establishment. Some preliminary steps were taken to move the laboratory to Dundee, but no actual arrangements had been concluded, and the outbreak of war took everyone by surprise.

On September 3, the day of the declaration of war, indescribable confusion reigned at Bawdsey. At luncheon the Superintendent announced that Bawdsey was sure to be bombed until not a stone was left standing, and that everyone must leave before nightfall.

Upon arriving at Dundee, they were flabbergasted to discover that they were not expected and that no space had been set aside. It was necessary to convene the Senate of the University in emergency session to make the necessary arrangements. At the same time Bowen, who was at Martlesham with 20 people and 35 aircraft, received orders to evacuate his whole outfit at once to Perth, the nearest available airfield to Dundee. With considerable difficulty, all the planes and the personnel were transferred before the end of the day; but upon their arrival, they found at Perth a situation resembling that at Dundee.

The best was made of an extremely bad situation. A small amount of space was somehow provided for Bowen's group at Perth. Presented with a *fait accompli,* the Dundee administration made space for the laboratory in a Teachers' Training College associated with the University. It was a good building about four stories high, well-heated and with large well-lighted rooms. But the lack of preparation and planning was manifest in the single fact that there was no alternating current available in the Training College.

Dundee and Swanage

The move to Dundee, coinciding as it did with the outbreak of the war, resulted in the partial disintegration and reorganization of AMRE. A number of men were detached from the Laboratory and scattered as civilian personnel assigned to the Chain Home Stations. The army group which had been working at Bawdsey was split off, at this time or somewhat before, to form the radar section of the Army's ADRDE (Air Defense Research and Development Establishment) at Christchurch.

In the late fall of 1939 there were about 200–300 persons attached to the Laboratory at Dundee. There were about a dozen persons of real brilliance, and somewhere in the neighborhood of 50 persons who were extremely able without conspicuous qualities of leadership. Most of the workers were young academic physicists of tremendous loyalty and energy, but without much engineering experience.

The outbreak of war, though it took away some of the original Bawdsey workers, led to a considerable expansion of the radar work and of AMRE. A. P. Rowe, the Superintendent, and the Assistant Superintendent, W. B. Lewis, the distinguished nuclear physicist from the Cavendish Laboratory, bent every effort to bring in new personnel of high quality, to expand AMRE's facilities, and to convince any high officials who would listen to them of the multifold potentialities of this new weapon of war. The two men were constantly traveling back and forth to London for conferences. Rowe, in particular, was outlining the new projects involving radar that should be undertaken at the first opportunity, the protection against night raids and the use of radar against the submarine. He had drawn up an elaborate chart in which all the possible functions of radar had their proper place and which it was his custom to show to visitors and those newly hired by the Laboratory.

The activities were already expanding in many directions. The principal work at Dundee was centered on the ground stations, on their steady improvement, and on expanding the chain. An obvious outgrowth of this work, and increasingly important, was the development of anti-jamming features and means of identifying the aircraft that the station detected.

One of the most interesting and significant developments, the PPI or Plan Position Indicator, was under development in the autumn of 1939. The work was done by G. W. A. Dummer, a commercial engineer before the war who soon became one of England's outstanding indicator experts. The earliest PPI was an electrostatic one, but this was quickly followed by one with a mechanically rotated coil about the neck of the tube. By January 1940, the PPI was in a fairly advanced state of development and gave all the definition the British radar sets were capable of at that time.

The development of new systems which had begun at Bawdsey continued at an increasing rate at Dundee and later at Swanage. Apart from the airborne systems, of whose development we shall speak further, some new systems had been begun during the three years at Bawdsey. About 1937, some preliminary work was done at Bawdsey on the development of the first British ship detection system operating on 1.15 m, and representatives from the Admiralty were given some training in radar problems before they launched their own work on ship-borne radar at the Admiralty Signals School at Portsmouth.[59] In similar fashion, the first radar gun-laying system, the GL Mk. I and Mk. II, were developed by the army group working at Bawdsey. Another interesting system, the MRU (Mobile Radio Unit), was developed by a group under Preist during 1938, and just as war broke out and the Laboratory was moved to Dundee, this system was going into production at Metropolitan-Vickers. The first installations overseas—on Malta and in Egypt—were completed in the spring of 1939. The MRU's were not only mobile systems but were im-

proved CH stations, and were intended to cover the same band, from 13 m
to 5.4 m. A number of uses were envisaged for these systems: they were
(1) to replace or supplement the existing chain stations; (2) to establish
overseas chains throughout the empire; (3) or to serve as mobile warning
stations wherever they were needed. The production MRU sets were not
ready at the outbreak of the war; in fact, they did not come off the assembly lines until after the fall of France. When they appeared, they consisted
of two large self-propelled vehicles, one for the transmitting unit and one
for the receiver. Telescopic masts for these units were fitted on trailers,
towed behind the other vehicles.[60]

One of the most important projects at Dundee during the winter of
1939–40 was the CHL (Chain Home Low) system, the low-coverage
version of the Chain system.[61] Work was begun in the fall of 1939 under
great pressure from the Armed Forces when it was discovered that German planes could get past the CH warning station by flying in low beneath
the beam. There was great concern in the Air Council and the result was a
tremendous push and excitement—what the British vividly describe as a
"flap"—about developing a system that would remedy this defect. A
group under J. A. Ratcliffe and J. D. Cockcroft, both of the Cavendish
Laboratory, set to work at Dundee and at Cambridge. By the first week of
December, 1939, there were three operating CHL stations, one guarding
the Thames estuary at Walton on the Naze, another placed at the northeast corner of Kent, and the third at Anstruther on the Firth of Forth to
protect the important Naval base. These installations were close to the
water's edge, whereas the CH installations were always *inland*, often as far
as 5 miles. One of the Cambridge men, K. G. Budden, who had joined the
Laboratory at Dundee late in December, was assigned the job of looking
for sites for a chain of CHL station along the East Coast. Such was the
excitement that he was ordered to choose all the sites in three days!

The setting up of these new warning stations began in January and
February 1940, and constituted the main activity at Dundee. By May, the
East Coast chain was nearly complete. In the months that followed, the
chain was extended toward the south, and in June and July, after France
had fallen, there was a new panic to complete the coverage of the South
and Southwest. The new stations were fitted promptly into the British
defense system. The plots from the CHL systems were transmitted to the
nearest CH station and from there they were passed on to a filter room.

In June, after the Laboratory had moved to Worth Matravers, the first
PPI was installed, by way of experiment, in one of the CHL systems that
Ratcliffe was steadily trying to clean up and improve. The new combination was working in the first days of June and was surprisingly, even
startlingly, successful. There was no evidence of the great loss of sensitivity that had been widely predicted for this type of presentation, and con-

trary to all expectations, the device performed admirably at a wavelength as long as a meter and a half. When France was overrun the new system, like everything else available, was used operationally. Many scope pictures were taken, the first in an interminable series of PPI pictures. Later that summer, the PPI on the CHL was demonstrated to Air Marshall Sir Philip Joubert who, as representative for a special device on the Air Staff, was one of the Laboratory's most loyal supporters. Immediately there was a loud clamor that this new indicator be installed on all units of the CHL chain; this was agreed to and, in due season, undertaken. The importance of the PPI—later to be so useful for air-borne search systems—for warning and control systems cannot be exaggerated. One of the principal delays in the CH stations and the first CHL's came from the time it took to convert range and azimuth data to plan position information. The PPI gave this information directly, hence the name that was already becoming generally adopted.

Before the successful test with the CHL systems, it was generally believed that the development of the PPI required, for its proper employment, the concomitant development of a radar system with a highly directive beam. During the Bawdsey years someone had written a proposal for what was termed a "radio lighthouse, " that is, for a radar system with a narrow pencil beam. With a high-directivity beam of this sort, it would be possible to make azimuth determinations of great accuracy without recourse to beam-splitting devices. In addition, the use of a pencil beam would permit the sort of "painting-in" of the target area that the PPI presentation required. To get really good directivity, it was obviously necessary to go below the $1\frac{1}{2}$ m which was the shortest wavelength hitherto tried. The problem was assigned to D. M. Robinson, an experienced industrial engineer who joined the workers at Dundee in December, 1939.

Robinson had an experimental system on 50 cm operating in February, 1940, and there was hope that a search set on this wavelength might be working with the PPI by the following October. In May came a brief interruption when the whole Laboratory was transferred to Worth Matravers, near Swanage on the South Coast. Here the set was installed on a cliff to look for ships, but neither the transmitter nor the receiver had been sufficiently improved to give range on ships greater than a few miles.

Two developments combined to pronounce death sentence on the 50-cm work. One was that the PPI worked so well on the CHL stations. Since it was evident that the 50-cm system could not compete in power with the CHL at $1\frac{1}{2}$ m, Robinson's project lost much of its importance. Moreover, at the same time the work on microwaves which had just started in Robinson's own group began to show great promise. By means of the newly developed resonance cavity magnetron, real power could be produced at 10 cm and everything which 50 cm promised in the direction of narrow beams and high resolution could be done, and done better, at 10 cm. Nevertheless the development continued, at a reduced pace.[62]

The move to Dundee and Perth in September 1939 had been only a stopgap. Neither place was satisfactory. At Perth, Bowen and his airborne group, after cooling their heels for a week or so, were finally given space at a nearby civilian airfield where they took over one small hangar with some adjacent offices. It was extremely crowded. The workers found accommodations as best they could in Perth itself during their brief stay. They left in November, 1939, to install themselves at an RAF field at St. Athan in Wales. This was a very large field, with an important maintenance station, several miles outside of Cardiff near the town of Barry. The move was dictated by several considerations: the hope for better and more spacious accommodations; the fact that at Perth they occupied a civilian field and ran into security difficulties; and lastly, the fact that the maintenance unit at St. Athan had been designated to make the installations of AI and ASV equipment, on which some progress had been made at Perth. Bowen's group remained at St. Athan until the early summer of 1940 when, following the move of the main AMRE Laboratory to Worth Matravers, it was transferred to nearby Christchurch.

By the spring of 1940 Rowe and the Air Ministry officials had agreed upon a more suitable and apparently permanent site for AMRE. Worth Matravers was on the south coast of England, a short distance from Swanage, the well-known summer resort nearly across from the Isle of Wight. A number of considerations dictated the choice of this location: it was a pleasant district in which to live and work; there was an airfield not too far distant; it was at the seaside (for a long time this was deemed an essential prerequisite for a station of this sort); and it was not too far from London. This last was an important consideration, because it meant that it would be easier for important visitors to reach AMRE, and easier for the laboratory administration to keep in touch with headquarters, than had been the case at Dundee. This proximity to London made possible those frequent and informal discussions between the laboratory personnel and the representatives of the armed forces to which Rowe gave the name of "Sunday soviets." Moreover the South Coast was not yet an operational area, and at the time it seemed reasonable to believe that all German air attacks would come from the east, rather than the south.

On May 6 the entire staff of AMRE and all its equipment moved to the South Coast. A great deal of apparatus had to be transferred, and this alone required a special train of 54 cars. Before the Air Ministry took it over, Worth Matravers was the smallest of villages, situated on a plateau only a few miles back of the town of Swanage. The creation of the laboratory completely changed its appearance. About forty special buildings were erected of brick and stucco, carefully camouflaged, and widely separated from one another. There were also nearly a dozen huts with CHL equipment. Most of the workers lived in Swanage itself and traveled to work either in their own cars or by bus along the rough, badly surfaced narrow road that led up to Worth Matravers. It was a delightful place to live, though much of the time, it seemed to them, it was rainy and muddy and there were no adequate arrangements for feeding the large and growing staff at midday. Somewhat later, after the first German air raids, they took over a girls' school about a mile and a half from Swanage, and then a nearby boys' school. These buildings were old, filthy, and depressing, but had the advantage of proximity to Swanage for those who wanted to lunch with their families, and work late in the evenings. The real object was to remove the personnel of the laboratories from close proximity to the operating stations which were the immediate targets of the enemy planes.

The move to Swanage coincided almost exactly with the German attack on the Low Countries, launched on May 10, 1940. While the workers at Worth Matravers were setting up their apparatus, adapting themselves to their new surroundings, and taking up the interrupted threads of problems, they watched, with hypnotic fascination, the incredible speed of the German advance through Holland, Belgium, and France. By the middle of June, France had fallen and the enemy was established in the Channel Islands and had occupied the port of Cherbourg directly across from them on what, just a few weeks before, had been a friendly coast.

The proximity of the Germans and the widespread fear of invasion led the personnel at Swanage to suggest that they be allowed to form a Home Guard unit and have regular drill every day at 5 in the afternoon at the end of formal working hours. A. P. Rowe strenuously opposed the suggestion and said in substance to his men: "I don't want any mock heroics here. If the Germans land on these beaches, the duty of the scientists is to run like hares. I don't want you to waste any time on drilling. Stick to your work benches and get on with research." Rowe followed this up by getting an official opinion from the Air Ministry that in the event of invasion the scientists were to be evacuated inland and were to be given a first alert, telling them to be ready to move within 24 hours. During the months that followed, the men at Swanage were alerted in this fashion on three separate occasions.

The laboratory remained at Swanage just a little over two years. When finally they betook themselves inland, the threat of a full-strength German invasion had long since passed. In the summer of 1940 the responsibility for the laboratory was taken over by the recently created Ministry of Aircraft Production under Lord Beaverbrook, and was given the cumbersome title of MAPRE (Ministry of Air Production Research Establishment). In the autumn of 1940 this was changed to the name it has had ever since, TRE, meaning Telecommunications Research Establishment, and it began to issue a series of official progress reports under that title. During this two-year period, they were constantly plagued by air raids, obliged to spend many hours sitting in the dispersed shelters, and disturbed night after night by alerts that sometimes came as often as 3 or 4 times a week. Southampton, not so far away, was severely hit. The radar warning stations along the South Coast were severely bombed, but the Laboratory itself was not touched. It was not the air attacks, however, that brought about the decision to move, but the fear of German commando raids. Britain had been successful on the night of February 28–March 1, 1942 in capturing a German radar in a commando raid on Berneval on the French coast. It was fear of retaliation and the danger of compromising important experimental equipment that led to the decision to move inland on the last day of May 1942, to its new location at Great Malvern in Worcestershire. By the time of the transfer of TRE to Malvern, the laboratory had grown to be an institution of about 1500 persons.

The Night Fighter Program

During the stay at Perth, the remainder of the original 30 AI sets were installed in Blenheims. Six of these were supplied to 600 Squadron RAF. Meanwhile the tests at Northolt had revealed one major defect in the equipment. The transmitter was literally a "bread-board" creation: it was mounted on an uncovered wooden board, with the result that, because of the light from the transmitter, the approaching night fighter was clearly visible at considerable distances. A modification of the original set with an enclosed transmitter was given the formal designation of AI Mark II. A small number of these were produced, probably between 20 and 50, and a few were actually installed. The AI Mark II was never seriously considered, because a new and improved system, the Mark III, was designed at Perth in the space of two weeks. That had a completely new Pye receiver, and the whole system had a new box and, for the first time, was decently put together. But the minimum range was still not short enough. The set was not sufficiently well engineered, because of the speed with which it was put together, to be successfully used in operations. It was introduced into the field, as one of its designers later put it, "in a fair state of squalor."

At St. Athan the principal job was to install AI Mark III and ASV Mark I in Blenheims of the Royal Air Force. The introduction of the night fighter equipment into service was poorly handled. Not only was the equipment incapable of doing what was promised, but the training of maintenance men and operators was inadequate, and the new sets were introduced without adequate spares and with no test equipment. All told, perhaps two squadrons of Blenheims were equipped with AI Mark III. A few sets were tried out operationally, but without any notable success. The equipment was very poorly received and was harshly criticized by the RAF for its inadequate minimum range. It has been seriously advanced that it was a mistake to try to introduce the equipment with such speed and in such a primitive state of development, for it gave the equipment a bad name in the Air Force and did the night fighter program more harm than good. It is possible that a plane equipped with AI Mark III may have shot down an enemy raider, but no authentic instance has been brought forward.

At St. Athan, where the main attempt was made to introduce AI Mark III and ASV Mark I into the RAF, the working conditions were extremely difficult. There were two unheated hangars, one for research and the other for installation. Three or four portable oil stoves were scarcely enough to take the edge off the cold. The men worked in overcoats and scarves and put down plywood and duckboards to protect their feet from the cold concrete floors. Besides the difficulties of the improvised working conditions, which had a serious effect on morale, the St. Athan arrangements had further disadvantages: they were too widely separated from the main research group, and the small group was concerned with research, prototype development, installation, and flight checking. But they ran into enormous difficulties in getting the equipment in the air. The sets ran beautifully on the ground, but the serviceability of the sets in the planes was very poor. The installations proceeded slowly. Radar took a back seat because the production of the aircraft had all the priority in those critical days. Aircraft for installation purposes were scarce; and, in succession, a number of different planes were suggested or tried for the installations. There was no clear understanding by the Air Ministry or the RAF personnel, perhaps not by Bowen's own group, of how much would be involved in making equipment as complicated as a radar system work under operational conditions.

After some earnest discussion on all levels as to what should be done to salvage the AI program, the decision was reached that a completely new design was necessary, one that would provide the necessary minimum range.[63] A great effort was made at AMRE in Dundee to hit upon a successful design. Under the direction of W. B. Lewis, the scientists at Dundee came forward with a temporary solution of the minimum range problem. At about the same time the EMI (Electrical Musical Instruments Co). was given a contract for the development of an improved modulator, to give the requisite short pulses. The Royal Aircraft Establishment (RAE) which had not yet been called upon in the radar program, was brought in during the spring of 1940 and contributed its engineering experience both to the AI and the ASV program. RAE played an important part; its engineers helped with the overall design, adopted a packaging suitable for installation in aircraft, and helped with the design of the EMI modulator, which provided the final solution for the minimum range problem.

The AI Mark IV, the first successful radar for night fighting, made its appearance in the autumn of 1940. The essential features of the AI Mark IV are as follows: the transmitting tubes were the "micropups," triodes specially developed by GEC which gave about 10 kW at 200 MHz; the receiver was a superheterodyne with one rf stage; there were two pairs of antennas and a mechanical switch to change over the lobes of the two pairs; there were two 3 in. indicator tubes with exponential sweeps. The set had a maximum range of 20,000 ft and a minimum range of 400 ft. An important difference between the Mark III and the Mark IV was the shift from horizontal to vertical polarization. Originally the AI had been simply copied in this respect from the first ASV; some systematic tests at the Fighter Interception Unit showed, on the contrary, that vertical polarization would be greatly superior for the interception work.

One of the great factors in the successful introduction and improvement of the AI Mark IV was the creation of a special, experimental unit of the RAF, the so-called Fighter Interception Unit (FIU), created for the express purpose of expediting the introduction of AI radar into the RAF. It was formed early in the year to a large extent on E. G. Bowen's initiative, and was placed under the command of a distinguished RAF officer, G. P. Chamberlain. This experimental unit was based at Ford in the south of England. Early in the spring of 1940 Hanbury Brown was sent to FIU as a technical representative from AMRE. Serious work on the AI development, from the standpoint of testing and operations, began at this time. The first job was to test the available and competing modulators and to determine which should be used in the Mark IV. The decision to adopt the EMI modulator was the first achievement of FIU. All the subsequent stages of testing and experimenting with the Mark IV AI were passed through at Ford. In the autumn of 1940, the Mark III's were torn out of the planes in which they had been installed, the Mark IV's were substituted, and exhaustive tests were made on how to use the new equipment properly and what further modifications were needed. Most important of all, perhaps, the FIU squadrons began to evolve a tactical and operational doctrine for the use of the equipment, based on the experience acquired, not only in practice interceptions, but also in action against the enemy. At FIU they learned by doing, with the enemy helping with the instruction. It was none too soon.

The fall of France seemed nearly to have brought England to her knees. Even apart from the threat of invasion the occupation of France by the Germans required a change in Britain's entire defense preparations, both against air attack and against the danger to her far-flung supply lines. It also limited the offensive operations in which Britain could engage. In short, England's radar program had to be drastically overhauled. The chain of early warning stations had to be extended, as swiftly as possible, all around the coast. AI and ASV equipment became still more vital with enemy air fields located just across the Channel, and with submarines based directly on the French Coast. AI became more important in everyone's mind when the Germans began the night bombing of Britain.

The Luftwaffe began its assault on Britain early in August, as soon as they had established their new bases in France. These were for the most part daylight attacks. In September, the Germans launched their big assault on London with the first large-scale night raids and continuing daylight attacks. The Germans were guided to their target by radio beams crossing over the particular city chosen as the objective of the particular night. One beam came directly across from the Continent, the other apparently from a transmitter in Norway. There were various modifications of this basic idea, various codings of the beam, and some skillful use of the beam—once the Germans knew the British were aware of its existence—for deception.

The earliest instance of an enemy attacker being shot down by AI Mark IV came in November 1940, when an FIU pilot, Flight Lieutenant Ashfield, who was later killed, had the unit's first success. Installation and tactical experimentation training proceeded through the late fall and early winter. Great progress was made in developing a tactical doctrine and in perfecting the necessary teamwork between operator and pilot. The first really effective results came from the 604 Squadron RAF in the spring of 1941. Although there were scattered successes all during the winter, the night fighters only became genuinely successful at the very end of the Battle of Britain, perhaps even after it had been clearly won. Real effectiveness was only possible when it was used with special GCI equipment. This was being developed during the summer of 1940 and was in operation, successfully teamed with the Mark IV AI, in March or April 1941.

During this period, two types of planes were sent against the invaders' night fighters manned by men with especially good night-time vision, but no special equipment, called "catseye" squadrons; and squadrons provided with AI sets, almost certainly Mark IV in every instance.[64] The eight squadrons of AI-equipped night fighters, flying mainly Beaufighters, with some Blenheims, Havocs, and Hurricanes, are compared in the analysis given below with eight squadrons of catseye planes. The latter included duty night squadrons and day squadrons flying on special night duty. The figures show that the AI planes made a total of 200 contacts with the enemy, destroying 102 enemy planes in these four months. The catseye planes made 120 contacts in the same four-month period and destroyed 56 enemy planes. The operational analysis of the performance of these sixteen squadrons showed two important facts. In the first place, there was great individual variability in combat teams of radar-equipped planes, indicating how important training and cooperation was in such work. In addition it was shown that the great superiority of radar-equipped planes over catseye planes lay in their ability to make visual contact on nights when the pilots were not aided by moonlight. The AI squadrons were able to make 39 contacts on nights when the moon was below the horizon as compared with 8 such contacts for the ordinary squadrons.

The results for the entire four months are summarized in Table I.

Table I. Night Fighter Record—March 6–June 1941.

Squadron	Types	Combats	Combats (moon below)	Enemy Destroyed March–June
AI-equipped				
25	Beaufighters	28	5	16
29	Beaufighters	14	2	7
68	Blenheims	1	1	1
85	Havocs and 5 Hurricanes	29	7	6
219	Beaufighters	43	9	26
604	Beaufighters	71	13	40
600	Blenheims and Beaus	12	1	5
PIU	Blens, Beaus, Havocs	2	1	1
	Total	200	39	102
Catseye				
87	Hurricanes	10	–	1
96	Defiants and 5 Hurries	19	1	5
141	Defiants	12	–	7
151	Defiants and 5 Hurries	31	4	19
255	Defiants and Hurricanes	17	1	9
256	Defiants	11	–	5
264	Defiants	11	–	8
307	Defiants	9	2	2
	Total	120	8	56

Ground Control of Interception[65]

Early in 1940 it was already apparent that a special set, different in many respects from the CH stations, had to be devised for successful vectoring of planes to an interception. The CH stations had been designed to cope with massed daylight raids, not with night bombings, which had not been envisaged on any scale. In the daylight interceptions with CH stations, the radar information was passed into a filter room where all information was coordinated. The fighter interceptors were controlled by VHF (Very High Frequency) radio from the filter room. When this technique was tried on night raids, it failed miserably; the procedure was inaccurate and slow. By July or August of 1940, it was recognized that the interceptor planes must be controlled directly from the radar station and the RAF requested the development of a set that would serve this purpose.

The first actual attempts at controlling interceptors from a radar station were made early in September at Worth Matravers with a CHL station. Practice interceptions, using Blenheims equipped with AI, were made against enemy planes, and some kills were actually made. The equipment used was Ratcliffe's experimental CHL station with the first PPI, and since it had no height-finding feature, heights were obtained from a CH station in the same enclosure.

It was decided to modify the CHL station, incorporating, among other features, height-finding. A method had just recently been developed by K. G. Budden and his co-workers from Dundee at a little field station at Arbroath, in Scotland.[66] Budden and his men were brought to Worth Matravers where work was begun on the first GCI in October or November, 1940. The equipment was installed as an operational station at Durrington, a point about a hundred miles east of Worth Matravers and nearly due south of London.

It might be well to list the points of similarity and difference between the CHL station and the GCI. Both were on 200 MHz and both had separate antennas for transmitting and receiving, and the antennas were hand-cranked. The GCI's abandoned the lobe switching of the CHL's, largely because they were given priority on the first PPI's that were completed. The CHL was fixed, at least at first, whereas the GCI's were built to be mobile equipment. The GCI was run by a crew of 12 for each watch. There were two cathode-ray tubes, an A scope for height finding and a PPI giving both range and azimuth data. The controller, an RAF officer, stood or sat (sitting was a later refinement when he got used to the equipment) at the PPI. The station had a plotter stationed at a map and provided with a small telephone switchboard. All told, five or six vehicles were required for the GCI equipment proper, with two additional ones for the VHF equipment. In January 1941, a sudden push was authorized to produce 10 mobile GCI's, exact copies of the Durrington equipment.[67]

The ASV Program[68]

The lack of official interest in ASV prior to the war has been emphasized above. With the outbreak of war in September 1939, this situation suddenly changed. A number of factors doubtless contributed to this reversal of opinion, principally and fundamentally the realization that in this war, as in the last, Britain's principal strategy would be to try to seal off the German fleet in her North Sea and Baltic ports and at the same time to protect her own sea lanes from German submarines and German surface raiders. There seems to have been a more specific incentive. Just before the outbreak of war both the Germans and the British carried out some discrete aerial reconnaissance over each other's territory. On one of these flights the British had observed a large number of units of the German fleet collected together in the harbor of Wilhelmshaven. With the declaration of war, some 50–60 British bombers took off to sink this quarry. It was a hazy day. When the planes reached Wilhelmshaven the fleet had vanished and—in the absence of airborne search radar—they were unable to locate it. This led to an urgent request for ASV equipment.

The first ASV equipment, the ASV Mark I, was designed at Perth in record time. In many respects it resembled the first AI and was equally tinny and unreliable. The transmitter consisted of two micropup tubes in push-pull, and was made by E. K. Cole, Ltd.

The receiver was built from a Pye television chassis, like the receiver of the AI Mark I; the indicator circuit had a single tube and gave an exponential time base with a single sweep. All told, there were only nine tubes in the receiver-indicator unit, not counting the cathode-ray tube itself and the tubes in the power supply.

The receiver and indicator were manufactured to fit into a Hudson two-motored medium bomber, an American plane widely used by the British for various purposes at this stage of the war. Only a few of these sets were ever built; some were installed at St. Athan. There they fitted two squadrons of Hudsons for the Coastal Command, and built an experimental radar beacon. When Hanbury Brown returned from Northolt to Perth, his first assignment was to prototype a Hudson to receive the ASV Mark I. After about two weeks, the outfit moved to St. Athan. But the work progressed slowly, in part because of the time it took to have the engines properly screened. In January, 1940, these squadrons were sent to an aerodrome at Leuchars in Scotland and the planes patrolled the North Sea as far as the coast of Norway. This was the first operational use of the ASV. The first operational radar beacon was installed at Dundee in January 1940, for use by the Leuchars squadron. The systems were supposed to be operational, but like the first AI's, they were supplied without test equipment. Operators for the system had been given some brief instruction at Perth in the form of lectures, but their training was very inadequate.

Like the first AI systems, this earliest ASV was far from suitable for service use. No record has yet been turned up as to how the Leuchars squadron performed, but it is not certain that they ever detected an enemy vessel. Hanbury Brown has described this early gear as being thoroughly unreliable and possessing "every fault under the sun."

The first successful ASV, like the first successful AI, resulted from a collaboration of AMRE workers and the experienced engineers of the Royal Aircraft Establishment. In fact, RAE was first brought into the picture for ASV work when, early in the winter of 1940, various persons at the top felt that the ASV equipment needed to be carefully engineered and specially designed for fitting in aircraft, and that the experience of Barton and his associates would be invaluable at this point.

It was decided in February, 1940, that a new ASV should be developed. Touch paid his first visit to RAE to attend a conference on this ASV Mark II. In March or early April, RAE was definitely brought into the airborne radar picture. This was accomplished by a transfer of some engineers from RAE to AMRE and vice versa. Touch left St. Athan for Farnborough with a half-dozen or so assistants. Their specific first assignment was to re-engineer the ASV equipment in collaboration with the RAE engineers and to put the development on a firm basis.

Most of the component development work was done in the companies. The Pye Radio Company did the receiver and indicator and the E. K. Cole Company developed the transmitter which used the new "micropup," the tube specially developed for the AMRE airborne group by the General Electric Company at Wembley. Later the E. K. Cole Company also took over responsibility for a receiver-indicator unit. Touch traveled about a great deal, coordinating the development and advising more particularly in the case of the transmitter and indicator. The RAE engineers were responsible for packaging. They had just designed a standard-sized box for aircraft radio equipment, and the ASV Mark II was the first set for which it was used. The prototype assembly was done at RAE, and the first of these sets began to appear in October 1940. Real production started somewhat later, and the separate, completed components converged on the old maintenance center at St. Athan, where J. W. S. Pringle, who had taken over the remains of Bowen's group, was supervising the ASV installations.

The ASV Mark II was planned originally for the aircraft of the Coastal Command of the RAF: the Whitleys, Wellingtons, Sunderlands, and Hudsons. Some were also supplied to the Fleet Air Arm with minor modification of packaging and a slightly altered transmitter. The Mark II had many advantages over its predecessor, and this helps explain why it became the first successful airborne search radar. It was a completely engineered set; it had reliability and longer life; the transmitter was more powerful; the receiver had an additional rf stage and a cathode follower; it had a 6-in instead of the 4-in cathode-ray tube of the Mark I; finally, the Mark II had a linear time base with three range sweeps for 8, 36, and 90 miles.[69]

Long Range ASV[70]

The type of ASV we have just described had a forward-looking antenna with lobe switching, and was sometimes referred to as the "homing" type of ASV. While the airborne group was still at Martlesham Heath, another type of ASV, which came to be called the "beam type" or "long range" ASV, was built with a special antenna designed to send a beam away from the side of the aircraft at right angles to the path of the plane. This was tried out at Martlesham, during one of the periods of low AI priority, with flights out over the North Sea using as a target a Danish butter boat which left Harwich on a daily schedule. Except for the antenna system—a Sterba array along the fuselage and a Yagi on the wings—this early system used the same components as the homing type of equipment.

At St. Athan in February or March 1940, after having completed the installation of the Mark I ASV, Hanbury Brown was asked by E. G. Bowen to determine whether this set would be useful against submarines. They took the prototype Hudson, installed a laboratory system, and went to a point on the South Coast. Here they made experimental runs near Ventnor on a small submarine that had been supplied for that purpose. They determined the range at which they could detect it under different conditions, and made a study of the amplitude of the signal for different aspects of the submarine. The results were not very encouraging. The maximum range on the submarine was only about $7\frac{1}{3}$ miles, or even less. It convinced them that the beam antenna would give better results, longer ranges and less sea return.

Bowen persuaded the Coastal Command to provide a Whitley aircraft, a long, box-like bomber, for the installation. After the normal delay, a Whitley, somewhat old and weather-beaten, reported to St. Athan. Brown and his fellow workers went to work on the plane, boring holes and installing brackets with great energy. A short time later a brand new, shiny Whitley made its appearance, and its pilot formally handed it over to Bowen for the ASV installation. They had gone to work on the wrong aircraft, and in consequence were obliged to effect a trade, receiving a broken, air-weary plane in exchange for a fine new one! The work was interrupted by the move to Swanage in May, and the Whitley was sent to RAE to be fitted with the special antenna. The plane did not return from Farnborough until August or September, 1940.

The Long Range ASV used standard ASV Mark II components with the exception of the antennas. As its name implied, it gave much longer ranges than the homing type, either Mark I or Mark II. In the fall of 1940, the German submarines began to operate from their newly acquired French bases. A squadron of planes equipped with Long Range ASV was based on Northern Ireland and used for the North Atlantic patrol, while the ASV Mark II homing equipment was used in the North Sea, the Channel, and the Bay of Biscay.

Radar with the BEF in France

During the first years of the war, to summarize what has gone before, radar was gradually coming into wide use in the British Isles. The Chain Home Stations and the low-coverage CHL Stations were serving as a warning system. Mobile ground stations and stations for the ground control of interception were, respectively, in production and under development. Airborne equipment was being introduced into the RAF, in the form of airborne interception (AI) equipment, and equipment for airborne detection of surface vessels (ASV). A long-wave gun-laying set, the GL Mark I, developed by the Army had been introduced in the service before the outbreak of the war; and a modified version, the GL Mark II, was under development.[71] It seems appropriate to ask at this point what radar equipment was taken to France in 1939–40, with the British Expeditionary Force.

When plans were being prepared to send an Expeditionary Force to France, it was agreed that some sort of mobile radar equipment would be of great value.[72] The idea, which seems to have emanated from a Group Captain Leedham, attached to the Air Ministry, was under discussion two or three weeks before the war. Preist's MRU (Mobile Radio Unit) would have filled the need precisely, but it was only just going into production. Some sort of stopgap equipment had to be devised and Preist himself seemed the logical man to undertake the job. On September 6, after taking part in the frantic hegira to Dundee, he was commissioned into the RAF with the rank of Flight Lieutenant and was put under Leedham's orders. He was ordered back to Bawdsey Manor from which nearly everyone had departed, and scraped together enough equipment to build a transmitter on the spot in great haste. The equipment was then shipped to Kidbrooke, an RAF field in the south of England, where Preist was assigned to 2 IU (Interception Unit) RAF under Squadron Leader J. W. Rose. The set was speedily assembled and tested in the space of five days by dint of working through a large part of every night.

On or about September 15, Preist took the set to the South Coast and sailed with it for France. Preist was in command of an outfit consisting of 18 enlisted men and noncommissioned officers and a convoy of seven trucks. Poor planning and organization led them to Amiens by a most circuitous route, by way of Brest! After vexatious delays and a slow ride across France in the company of another half dozen trucks—driving at 15 miles per hour and only between 9 A.M. and 6 P.M., and camping out each night—they reached their destination, Wing Headquarters of 60 Wing, at Amiens, on or about September 29.

On Preist's arrival, a conference was promptly held to decide what to do with the RAF's only radar warning station on the Continent. The original use envisaged for the set was to have it cover the advance of the British Army across the Somme, because it was expected that the enemy would try to strafe and dive-bomb the troops during the advance. The delays had been such, however, that when Preist's detachment arrived at Amiens the operation had been completed without incident and the British Army had already crossed the Somme. It was decided to set up the equipment somewhere on the North Coast of France under the supervision of the Air Commodore of the Field Force with Headquarters at Arras. Preist led a siting expedition along the North Coast and finally decided upon a location about three miles east of Calais, where the set would watch for enemy planes coming down the coast and through the straits, a regular German routine in their mine-laying and attacks on British shipping.

This early mobile radar set worked on 40 MHz. Somewhat later it was supplemented by units of a somewhat different species, operating on 70 MHz. These were also to a large degree handmade units, "a lot of gear thrown together," Preist later described them, from factory-made components taken from the GL Mark I and the GL Mark II, the transmitter being made by Metropolitan-Vickers and the receivers by Cossor of London.

The six Ground Mobile or GM units, as these sets were called, were brought to the Continent one after another in charge of six RAF junior officers. It was decided to set up the six sets as a sort of defensive chain against Northern France. Preist did the siting for the first two sets, driving over much of Northern France, until, as he put it, he knew that region better than most parts of England. The first of the new sets was placed about five miles east of Lille to give protection to a fighter field at Seclin. A second unit was placed on high ground near Boulogne, south of that town on the Abbeville road. Counting the original set brought over by Preist, this meant that there were three radar warning stations in France by the end of November. Two additional GM sets came over soon after and Preist was preparing to site those as well when in December he was called back to England. He remembers, however, that a conference was held at Calais just before he left, attended by himself, Watson-Watt, Air Marshall Joubert, and some high-ranking French officers, on what to do with the new stations. It was agreed at this meeting to put the new sets at points on the Belgian border, and though he left before the stations were actually installed, Preist believes that the sets were actually placed there.

The only other radar taken to the Continent with the BEF were 17 GL Mark I sets.[73] These were assigned to various anti-aircraft units scattered at various British positions in Northern France. Most of them seem to have been taken into Belgium when the British and French advanced to meet the German invasion of the Low Countries. At Dunkirk, the anti-aircraft defenses were under the command of Colonel F. C. Wallace, who later came to America with the British Scientific Mission and still later was put in charge of Canada's radar research laboratory. All the GL Mark I AAA units that had been scattered about the front converged on Dunkirk, and their officers reported to Colonel Wallace. He lined the sets up on the beach and blew them up before he himself finally escaped on a destroyer, with the unenviable distinction of having destroyed more radar equipment than any other man.

Colonel Wallace has testified to the extraordinary results achieved with the few GL Mark I sets on the Continent. The 17 sets were responsible for shooting down a total of 400 enemy planes, with the extraordinary average—the result of the great skill and fine training of these early units—of 200 rounds per kill. A memorandum of August 1940, records that the GL Mark I batteries that met the early German raids on England about the middle of August claimed a figure of 375 rounds per kill. This same memorandum laments the tragedy of having to destroy precious radar units at this most critical moment in British history.[74]

NOTES

1. C. G. Grey: *A History of the Air Ministry* (Geo. Allen & Unwin, Ltd., London, 1940), p. 258.

2. J. D. Bernal: *The Social Function of Science* (Macmillan, London, 1939), Appendix II, p. 422.

3. *Ibid.,* p. 258.

4. Minutes of the Committee for the Scientific Survey of Air Defence (S.S.A.D.), 1st Meeting, January 28, 1935.

5. The *London Times,* Thursday, February 28. On March 19, 1935, in reply to Sir Austen Chamberlain's question in the House of Commons as to whether any further steps were being taken to provide defense against aerial bombing, the Prime Minister referred again to this committee.

6. Personal communication from Sir Robert Watson-Watt, and papers put at my disposal by him; also "Draft of RDF. (Radar) Narrative for Official Historian," put at my disposal by the author, A.W.H. Longstaff, at the request of Sir Robert.

7. "Detection and Location of Aircraft by Radio Methods," revised note by Mr. Watson-Watt.

8. *Ibid.,* p. 2.

9. Watson-Watt to A. P. Rowe, letter of 12 February 1935. Watson-Watt's patent application, dated 17 September 1935, is for an invention which "relates to the detection and loca-

tion of material objects by means of secondary electromagnetic radiations received from the object when submitted to primary radiation." It includes the determination of range by the use of very brief pulses, 5–25 μsec.

10. Personal communication of F. S. Barton.

11. Post Office Engineering Department, Radio Report No. 223, "The Further Development of Transmitting and Receiving Apparatus for Use at Very High Radio Frequencies." Case 952. March 6, 1932. The experiments were carried out by F. E. Nancarrow, A. H. Mumford, F. C. Carser and H. T. Mitchell. This document was circulated for official use only.

12. The Admiralty had even been approached about establishing a screen of trawlers, equipped with listening devices, not less than 20 miles from the coast.

13. Minutes of S.S.A.D.

14. Personal communication from Sir Robert Watson-Watt and Dr. F. S. Barton.

15. It bore the date of the previous day, February 27th.

16. Minutes of S.S.A.D.

17. Draft of R.D.F. (Radar) Narrative (see Note 6), pp. 4–5.

18. *Ibid.*

19. The figure given in the Draft, p. 15, is £12 300 for the first year, and is probably more nearly correct.

20. Minutes of S.S.A.D., March 18, 1935.

21. Interview with Dr. Edward G. Bowen, April 27, 1943.

22. Minutes of S.S.A.D., April 10, 1935.

23. By "floodlighting" was meant the steady illumination of the target area with a radiation pattern fairly broad in the horizontal plane, as constrasted with "scanning" or "sweeping" the area with a narrow beam. The main purpose of floodlighting was to facilitate the search for aircraft over a wide area, and to enable a general picture of aerial activity to be obtained in the shortest possible time. Cf. C.H. Stations (Brit. Doc.) 26 May 1942.

24. This observation was evidently made at Slough with Wilkins' receiver sometime subsequent to the Daventry experiment.

25. This section is based largely on information supplied verbally by E. G. Bowen and Sir Robert Watson-Watt.

26. One of this small group upon whom an important share of the credit must be bestowed for the speed with which this development took place is Joseph E. Airey, the ship foreman at Slough, who very soon joined the men at Orford Ness, and was the senior member of a small group of technical assistants who were eventually assigned to work at Orford Ness.

27. Minutes of S.S.A.D., "Notes on a Visit of the Committee to Orford Ness on the 15th and 16th of June 1935." In telling of this early work, Bowen made no mention of these first signals. Either he forgot them, which is probable, or he doubted that they had ever been observed. Watson-Watt was sanguine and hard to discourage.

28. The Minutes report that the Valencia was tracked "with only a small measure of success." Bowen recalled that Watson-Watt was the only one who thought he saw signals.

29. Minutes of S.S.A.D., "Notes on a Visit of the Committee to Orford Ness on 15th and 16th of June 1935." See also Interview with Dr. Edward Bowen, April 27, 1943.

30. Minutes of S.S.A.D., September 25, 1935.

31. Minutes of S.S.A.D., October 12, 1935; December 5, 1935.

32. The*London Times,* Friday, August 16, 1935, p. 5: 1–2.

33. Minutes of S.S.A.D., September 25, 1935. In this connection, in June or July 1935, E. G.

Bowen sent Watson-Watt a draft suggestion for a device resembling closely the Plan Position Indicator. Transmitting and receiving aerials are rotated synchronously, and the received signals are applied "to the focusing cylinder of a cathode-ray tube. The tube has a linear sweep which is rotated in synchronism with the aerials about the point of origin of the outgoing signal." "The image of the face of the tube now becomes a central dark space corresponding to the ground signal, and in the case of a single reflection a circle whose radius gives range directly, and whose intensity varies with azimuth." E. G. Bowen: "An Attempt at Instantaneous DF and Range-Finding," copy date 22 August 1940.

34. Minutes of S.S.A.D., September 25, 1935, October 12, 1935.

35. Minutes of S.S.A.D., December 5, 1935. According to Bowen, the results on aircraft were steadily improved until the 4 m equipment was outperforming the 26 m set in some respects and a final range of 60 to 70 miles was obtained. (The Minutes indicate 82 to 87 km by February 25, 1936.) At this point the transmitting tubes gave out, and with new and allegedly improved tubes, it was never possible to repeat the results and the project was dropped. It was later discovered that these tubes were being blithely operated near their frequency limit. Small changes made in the new tubes to improve the emission made them inoperable at 4 m and the work was brought to a stop.

36. Information on the first year at Bawdsey was supplied verbally by E. G. Bowen and A. G. Touch.

37. It was planned to carry out, by Treasury Letter S-26350102 of 19 December 1935, service trials with these stations and in the light of extending the system to cover the coastline.

38. "To say that I was disappointed on Thursday is to put it very mildly. As you put it yourself, you have to face the fact that very little progress in achievement has been made for a year. Unless very different results are obtained soon, I shall have to dissuade the Air Ministry from putting up other stations. The Secretary of State must have got a very bad impression." Tizard to Watson-Watt, letter of 20 September 1936.

39. The explanation of this failure was simple enough. Bowen's original success had been attained with equipment that was not engineered or in any way serviceable. He ran his transmitter tubes far above their rating and cared little if they burned out. After he left work on airborne equipment, his successors set about designing a transmitter that would hold up. In their effort to get life out of the tubes, they operated them more conservatively and did not get adequate power. The development was in this state when the test took place.

40. Minutes of S.S.A.D., May 1, 1937. The experiments are described as having been so successful "that the 'Defence' Plans Policy Subcommittee of the Committee of Imperial Defense had decided that the organization of the full chain of 20 stations should be proceeded with as quickly as possible, subject to sanction being obtained through the Treasury Inter-Service Committee." Air Ministry Memorandum No. 139 (S-35982(f-5)).

41. Information supplied by Mr. Hanbury Brown.

42. Interview with Dr. Edward G. Bowen, April 27, 1943. The stations were at Bawdsey, Dover, Canewdon, Dunkirk, and Bromley.

43. Later stations were more thoroughly engineered and ruggedly built by British manufacturers under contract with the Air Ministry. In July 1937, production versions of 20 CH Stations were ordered from Cossor and Metropolitan-Vickers.

44. The cost of a CH Station was in the neighborhood of £120 000. (Watson-Watt to Sir Frank Smith, May 31, 1940.)

45. Personal communication of E. G. Bowen.

46. Minutes of S.S.A.D., February 25, 1936.

47. It was intended to be an operational set, and in fact the idea was revived on 200 MHz in 1939 for use against low-flying aircraft. The project was dropped a second time at TRE in 1940.

48. The Ansons were two-motored, reconnaissance aircraft provided, together with pilots and crews, by the RAF station at Martlesham Heath. These were the first two experimental planes actually assigned to Bawdsey.

49. This tube had a tungsten filament with a graphite anode, and was either a Western Electric 304A or a closely similar tube.

50. Information supplied by A. G. Touch.

51. The designation AI, for "air-borne interception" radar, and ASV, meaning "aircraft to surface vessel," were just coming into use at Bawdsey in the spring and summer of 1938. ASV equipment was first called AS, "aircraft to surface vessel," but it was discovered that the designation AS was used by the Admiralty to mean "anti-submarine," so the radar designation was changed to ASV.

52. Information supplied by E. G. Bowen.

53. In all their work they used horizontal polarization. Upon trying out vertical polarization, just to see what would happen, they discovered that the sea return was about 10 times stronger than with horizontal polarization. The two polarizations were compared, using the same equipment, by diving the Anson and then standing it on its tail in a stall.

54. Information supplied by A. G. Touch.

55. Personal communication from E. G. Bowen.

56. Special screening of aircraft engines were necessary at that time, for there was much interference from the plane's ignition system, and no attempt had been made to avoid this difficulty by appropriate changes in engine design.

57. The material for this section came from the three men most closely associated with the AI development: E. G. Bowen, R. Hanbury Brown, and A. G. Touch.

58. Flights with this early equipment revealed that ground echoes from built-up areas were stronger than those from the open countryside. It was therefore wondered if equipment could be designed to utilize this effect as an aid to navigation and bombing. "Town to countryside" ratios were measured for wavelengths of 1.5, 4, and 10 m. It was concluded that the longer the wavelength the better the contrast. This question was dropped and was not revived until later.

59. This later became the Admiralty Signal Establishment (ASE).

60. The MRU's saw wide service during the war. They were used in India, Egypt, and North Africa. They were also installed as an important part of the British early warning system. Preist recalls that he made a tour of the existing stations in the spring of 1940 and found that some MRU's were already being installed. By the autumn of 1940, 50% of the chain stations were fitted with mobile MRU's. The sets helped especially in the rapid extension of the chain to S.W. England after the fall of France.

61. Interview with Dr. K. G. Budden.

62. The development continued for two years and finally culminated in a 50-cm set for search and GCI called the AMES (Air Ministry Experimental Station) Type II, built by TRE in England and in Canada by Research Enterprises, Ltd. It did not come into wide use.

63. Considerable pressure was being exerted from above and a great "flap" was started in February or March to make AI serviceable, because of a desire to start night bombing of Germany. This was not deemed safe unless Britain had perfected her defense against retaliatory raids.

64. Operational Research Section, Report No. 225: "A Classification of Combats during Hours of Darkness by Success of Pilots and Crews." Period, March–June 1941.

65. This is based on information supplied by Dr. K. G. Budden.

66. The site had been chosen because there was a cliff just the right height above the sea.

Three of them began working in April, 1940, looking for a method of height finding that would be satisfactory. They finished their work in August, having developed a method, based on a divided antenna, that was working satisfactorily.

67. About January, 1941, a member of General Arnold's staff, Major (later Brigadier General) Gordon Saville, visited the Durrington station to inspect the equipment. As a result of his visit, the General Electric Company in America was given complete information and the Air Corps hoped to have it copy the English development. Because of opposition from the Signal Corps, the project was held up. Budden was sent over in August, 1941, bringing with him one of the British mobile GCI's. As a result of Saville's pressure, but only after Pearl Harbor, the Signal Corps gave G.E. a contract for GCI. The result was the SCR 527.

68. Information largely supplied by Gerald Touch.

69. These are nautical miles, and their peculiar values are the result of a misunderstanding. The original ranges specified had been 10, 40, and 100 English statute miles, for the Coastal Command representative had insisted that, unlike the Navy, they used statute miles. When it turned out that, instead, they did in fact use nautical miles, the results had to be translated.

70. This section is based almost entirely on the information supplied by Hanbury Brown.

71. S/L F.V. Heakes and Dr. J. T. Henderson: "Electrical Methods of Fire Control in Great Britain—Preliminary Report." Sent from Great Britain, April 1939, copied Ottawa, October 1940.

72. The material in this section was supplied by Mr. Preist.

73. Information supplied by Colonel F. C. Wallace.

74. Files of the National Research Council, Ottawa.

CHAPTER 7

British Disclosures to Her Friends and Allies— French Radar

Disclosure to the Dominions

It was probably inevitable that, despite the most thorough security precautions, some hints of the British RDF development should leak out. Some stories appeared in the American press as early as 1938 which, though garbled as to technical details, correctly revealed that the British Air Minstry was setting up a chain of special radio detecting stations along the East Coast.[1] The French were aware quite early of the existence of the British Chain Home stations, some of which were clearly visible from the French coast; but it was not until 1939 that British radar was officially disclosed to them. As for ourselves, we seem to have remained in complete official ignorance of British developments until after the outbreak of the European war. American Naval attaché reports from London from the fall of 1939 disclosed that from various inadvertent remarks by well informed officials it was apparent that the British were in possession of an aircraft detector that could pick up aircraft up to 70 miles away and give continuous information, accurate enough to direct fighters to the interception. The principle was not clear, though it was presumably electrical, working on the principle of reflected pulses. Six months later rather more complete information was transmitted to Washington.[2]

Before the war Germany seems to have been reliably informed as to the nature and function of the chain stations, and even perhaps to have had more than an inkling of the existence and location of Bawdsey Research Station. At least the British were convinced that this was the case. A story which the British radar people delight to tell concerns the visit of a high Luftwaffe officer to inspect British aircraft developments. He was shown the less secret developments, with of course no mention being made of radio detection. When he found himself alone, for a moment, with a somewhat callow British junior officer, he leaned over and said in a conversational tone, "Tell me, and how are things going at Bawdsey?" Be that as it may, in the summer of 1939, just before the outbreak of war, a German plane flew leisurely the length of the British chain, to the bewilderment of the operators and the embarrassment of officers who could not very well order it shot down.

Similar information filtered out to the British Dominions, at least by 1938, though radar was not officially disclosed to them until early in 1939. In Canada, for example, Major General A. G. L. McNaughton, who was later to command the Canadian Expeditionary Force but who was still at that time President (i.e., Director) of Canada's chief government research laboratory, the National Research Council in Ottawa, instituted preliminary inquiries about British work in radio detection.[3] At the same time the possibilities of the new device were discussed between the Canadian Department of National Defence and the National Research Council. General McNaughton's inquiry was at least contributory to the decision made the following spring by the British Air Ministry to release radio detection to the Dominions. On March 10, 1939, the Chief of Air Staff, Air Vice-Marshall Groil, asked the President of NRC to delegate a physicist to visit England to investigate these important developments. Similar invitations were sent to Australia, New Zealand, and the Union of South Africa. The National Research Council chose the head of its radio section, John T. Henderson, to be Canada's scientific representative. He sailed on March 18, 1939. Upon his arrival he took part in a joint program prepared for the benefit of the delegates, which included conducted visits to the various laboratories, factories, and service installations. In April he sent back a long report (written in conjunction with an officer of the RCAF) giving a general description of RDF, the various types of sets, and the possible applications of radio detection to the defense of Canada. The report stressed the importance of radar to a country faced with having to defend two extensive widely separate coasts. The CH type of station—so valuable to Britain—might not be equally useful to Canada; but ASV systems were sure to be important and "we have suggested that 15 sets be purchased or built for each coast."[4]

Radar Research in Canada

The Canadian National Research Council in Ottawa was the Dominion's chief peacetime research establishment; inevitably it became the principal center of Canada's wartime research.[5] It was supervised by a Committee of the Privy Council and was directly administered by a President and several Advisory Committees. Before the war the President was Major General A. G. L. McNaughton who relinquished this post in the fall of 1939 to command and organize the Canadian Expeditionary Force, and was succeeded by Dean C. J. MacKenzie. The NRC occupies an attractively landscaped building of moderate size situated on the bluff of the river on the outskirts of Ottawa. In peacetime, work was conducted in a variety of scientific fields. Radio research was centered in a small Radio

Section of the Division of Physics and Electrical Engineering. The Radio Section was headed by J. T. Henderson.

In April 1939, General McNaughton informed the Canadian Government that NRC was willing to undertake the special scientific work that might be required by the Canadian Department of National Defence.[6] His request for expanded personnel and increased appropriations was not granted. Radar research was undertaken in the fall of 1939 without special government support by the diversion of NRC funds intended for a proposed ionospheric investigation. With only this slight encouragement the Radio Section expanded from only about three persons in September 1939 to 11 persons by January 1940.

The first assignment of the radar group at NRC came from the Chief of Air Staff in September 1939, and was a request to design an ASV set. At Henderson's suggestion they undertook to modify a Western Electric altimeter so that it would operate by pulses instead of frequency modulation. By the end of November, they had an experimental system in operation that gave about a kilowatt of peak power and could give echoes over short distances from buildings and nearby aircraft. Flight trials in June 1940 were not promising; and after comparison with a second set operating on 1.5 m it was recognized that with the dubious transmitting tubes then available on 67 cm it would be better to follow British procedure and go to longer waves. It was agreed to obtain a British set to study and to copy, and steps in this direction were taken shortly after, but there was a long delay before they received any sample ASV equipment from Great Britain. In fact, no further work was done on ASV at Ottawa until after the arrival of the British Mission in September 1940.

Simultaneously with the beginning of the ASV work it was decided to try to build a coast defense or CD station. The same principle was adopted, that of using a modified Western Electric altimeter. After a winter of experimenting, mainly in an effort to build a suitable antenna, they had not gone very far. During this first winter of 1939–40 nearly all the experimental work on the CD aerials was done in an open field, with only tents for shelter. In February, the Acting President, Dean MacKenzie, authorized the erection of a frame building to serve as a Field Station, but this was not completed until May and was of no use that first winter. The men who did this work recall with feeling the bitter cold and the difficulties of work under these rigorous conditions. In June, it was decided to make the CD set, like the ASV, on 200 MHz. An experimental system was set up at the Radio Field Station that summer with a large rotating aerial array which was intended to be duplicated later at the field site. They worked on this system throughout the winter of 1940–41 and it was demonstrated to staff officers of the Canadian Army in February. A replica of this system was finally installed in an operational site at Halifax in the summer of 1941.

In March 1940, the Radio Section was requested by the Navy to develop some simple form of radio harbor control device to supplement the harbor magnetic loop. A scheme was adopted consisting of a simple guard beam on a relatively long wavelength, the transmitter and receiver to be supplied with billboard or bedspring arrays. The system, called the Night Watchman, was working satisfactorily in the laboratory toward the middle of June and was installed for the protection of Halifax harbor in July 1940. During August and September three naval ratings were trained at NRC to operate and maintain the Night Watchman equipment. In December 1940, Colonel Mitchell and Colonel Coulton of the U. S. Army Signal Corps, who had inspected the set, ordered a duplicate. This was completed and in U. S. Army hands by the end of February 1941.

During the period until the arrival of the British Mission, the staff increased slowly. In July 1940, there were 30 persons; by November 1940, there were 62. Although the support of NRC had never been lavish, by American or even British standards, money began to be made available.[7] About $60,000 was granted from NRC funds at the beginning of the new fiscal year, in April 1940. In September 1940, a sum of $330,000 was appropriated by the newly formed Committee on War Scientific and Technical Development. This grant sufficed to carry forward the initial stages of the development work described above, and other work that will be described in a later section. By the end of July 1941, Dean MacKenzie got a grant directly from the government of about $600,000 that was deemed adequate until the end of that fiscal year, i.e., to 31 March 1942.

French Radar Developments

Until the summer of 1939 there was no exchange of information between the British and French on the subject of radar. In the late spring of 1939, Watson-Watt and A. F. Wilkins crossed to France to see what sort of developments were in progress in the UHF radio field. Radar was not discussed with the French, though the first French experiments with a high-power pulse system had been conducted shortly before. Instead they were only able to report as being of possible interest some UHF direction finding equipment developed by Gerard Lehmann of the Sadir Company.[8] In the summer months a group of French Navy and Army officers visited England in mufti and were shown some or all of the radar equipment in service use. This disclosure apparently did not extend to the French civilian and commercial concerns which were involved in radar. Even after the French commercial laboratories had been nationalized and their men mobilized into the Army at the outbreak of war, they were unsuccessful until early in 1940 in getting information from British sources.[9]

Radar in France was known by the initials D.E.M., standing for *détection électro-magnétique*. The different systems were evolved between 1934 and 1940 and were tried out successively by the French armed forces. Examples of both were in operation at the time of France's military collapse. The French development was similar in general outline to the American. The French did not hit upon the pulse method all at once, as Robert Watson-Watt did in England. Instead, like the Americans they experimented for some time with a continuous-wave type of detection device, based on the phenomenon of interference. In fact, the schemes were nearly identical, although the French carried their early development along a bit further than the men at NRL thought profitable.

The first system of D.E.M. was invented by Pierre David, Chief Engineer of the National Radio Laboratory.[10] David was one of that group of pioneers in very high frequencies who had worked under General Ferrié in the last war, and who continued well into the thirties important work on communication at these frequencies.

The David system, as the French called it later to distinguish it from the pulse system, was proposed about 1934.[11] It used the phenomenon of interference between a direct wave and a wave reflected from an approaching aircraft, which was almost certainly discovered by accident in the course of his experiments on communications at very short wavelengths. The wave reflected from the airplane, altered in frequency by the Doppler effect, was combined with the ground wave to produce beat frequencies in the audio region which were automatically recorded on a moving tape. The system was studied very secretly by David, a few officers of the Corps of Engineers, and the French School of Anti-Aircraft Artillery at Metz; and in 1937 the National Radio Laboratory issued a report giving the theory of the system and the results of the first experiment in the field. About 20 sets built by the Army, the Thomson Company, and the Sadir Company were incorporated into a network that was tested during aerial maneuvers in July and August of 1938.

The equipment was extremely simple and light and could be transported by only a few men. The antennas were nondirective dipoles and the transmitter power—though higher than the Ponte system to be described later—was still very low. In order to set up a warning system that would convey some idea of the location of the approaching planes it was necessary to combine the stations into a fixed chain.[12] The chains were so laid out that a single observer compared the receivers of two adjacent elements of the chain, the information from the observers being fed into a filter station. The system was expected to serve both for early warning and for ground control of interception. A chain of this sort operating on 30 MHz was set up near the city of Rheims in the summer of 1938 and tested during the maneuvers. It was used to direct night fighters to intercept "enemy" bombers. The data from the various observers was telephoned at intervals to the filter room which in turn directed the night fighting planes to the enemy by means of ultrahigh-frequency radio telephone.

The same system was studied by the Naval Laboratory in Toulon, and a few stations were set up along the Mediterranean coast and on one or two small ships. In 1939, a chain protecting the Naval bases along the British Channel, the Atlantic Ocean, and the Mediterranean Sea was put into service. It was composed of stations operating at 30 MHz, with each receiver–transmitter pair separated by a distance of from 20 to 60 miles.

Paralleling the David development, experiments were going forward at the SFR Company (the *Société Française Radio-électrique*) on the Ponte-Gutton obstacle detector. In its final form this was a detection system using pulses, but at so low a power that it appeared to have little if any military value.

The Ponte pulse method had undeniable advantages for a military device, but the system gave such little power that the French armed services rejected it in favor of the David system, crude though it was, which gave considerably more power and range. Early in 1939, a serious effort was made to improve the David system, particularly to give it some directivity. The Army asked Gerard Lehmann of Sadir whether he could not combine in some fashion his direction-finding device, mentioned above, with the nondirectional David detectors. The method developed by Ponte and the SFR engineers was by that time widely known in France. It was this knowledge that led Lehmann, after a brief consideration of the whole problem, to suggest using pulses at or near the frequency of the David equipment.

The development of a high-power detection system using pulses was started in May 1939, in the laboratory of the Sadir Company in Paris under the direction of Gerard Lehmann. By August an experimental station was operating in Paris. In October an experimental system was demonstrated to the Navy near Toulon. This system operated at 50 MHz, using conventional tubes, and gave 20–30 kW peak power. As a result of these tests the French Navy, in December 1939, ordered the building of six pulse systems giving 20 kW peak power at 150 MHz. Cathode-ray oscilloscopes 14 in. in diameter were used for the indicators. The units were specially intended to provide lower coverage over water than was otherwise available and to detect the low-flying German planes that were busy laying magnetic mines in the coastal waters. The rf components of these sets were made by Sadir and the 14-in. oscilloscopes and the video circuits by the *Compagnie des Compteurs*. Before the armistice of 1940 three of these sets were completed and operating in Tunisia. They were used with some success against the Italians during the attack on Bizerte. After the armistice they were sent to Dakar, beyond the jurisdiction of the German Armistice Commission. One of the sets was installed aboard the battleship *Richelieu*.

During the summer of 1939, the Services' technical group headed by General Julien decided to have built a long-range high-power pulse system for use by the French Navy in coast and harbor defense. The possibilities were discussed with E. M. Deloraine and his engineers at the Paris laboratories of LMT (*Le Matériel Télégraphique*), the organization responsible for the television equipment which transmitted from the Eiffel Tower at 46 MHz under the supervision of one of their engineers, Emile Labin. It was Deloraine's opinion that a powerful pulse radar system could be developed by using as far as possible established television tubes and circuits. Work was begun at once on an experimental pulse station.

In September 1939, the LMT laboratory—like the other industrial installations of military importance in France—was nationalized and taken over by the Army. Deloraine and his engineers, all of whom were reserve officers in the French Army, were mobilized and assigned to duty in their own laboratory (*affectés sur place*). Under this new regime the laboratory went over completely to war work and great effort was put on developing the pulse transmitter which General Julien's group had asked them to make. Twelve production units were to be built on LMT's design by one of the company's subsidiaries, the *Compagnie Générale de Construction Téléphonique*. The systems used power triodes, directive antennas using three dipoles, an "A-scope" cathode-ray tube presentation, and a separate antenna system for transmitter and receiver. The first of these sets was completed in March 1940. It was in service at the time of the Italian attack in June, ideally placed on an island off Toulon. During tests it picked up planes at 120 miles, and it was on duty at the time of the Italian attack and is reputed to have helped save Toulon from the enemy.

Shortly before the armistice, when everyone expected the Navy to move to Africa, orders came to take down the set. The dismantling process was begun, when a manifestly false order came telling the men not to disturb the set. This was disregarded. The various parts were put in separate boxes and their location was known to only a single Naval officer. The station was put together again after the armistice, but it never worked as well as before. It was dismounted a second time when the Germans crossed the demarcation line after the invasion of North Africa. The other 11 sets originally planned were never assembled, although their component parts were in various stages of completion at the time the war ended for metropolitan France.[13]

What the Germans Learned in France

In view of the sudden and catastrophic collapse of France it is important to summarize the scant material available at this time as to how much the Germans may have learned of French radar (and, through the French, of British radar) as a result of the debacle. It is equally interesting to discover what efforts the Germans made, and how competently, to learn French technical secrets, for this casts an interesting light on the importance which the High Command attached to radar at different periods. There are three ways by which the Germans could have learned directly about French and British developments: (1) by capture of French equipment from the French forces as a result of the speedy defeat in the North; (2) by capture of British radar brought to France with the BEF; (3) by French disclosure to the Germans after the armistice of their knowledge of pulse radar.

If the testimony of the French engineers can be relied upon, it is improbable that any complete French radar system fell into the hands of the Germans, with the possible exception of certain elements of the northern David chain. No French pulse radar was in operation in the North. All the completed pulse systems in the Mediterranean area seem to be accounted for: the three completed Sadir systems were kept out of reach of the Germans and ended up at Dakar; the LMT system was dismantled and the location of its parts kept secret.

It seems equally certain (remarkable though it may appear) that none of the radar brought to France by the British Expeditionary Force in 1939–40 was left behind or was captured. The BEF brought two different radar sets to France. One of these was a mobile version of the CH stations, called the MIU, and the other was the first radar gun-laying set, the GL Mark I.

The MIU systems were distributed at various points along the front to warn against German aircraft. All of these sets appear to have been successfully brought out without loss before the evacuation from Dunkirk.[14]

As mentioned earlier, the 17 GL Mark I's went with the BEF to France were ultimately lined up on the beach at Dunkirk and blown up.

It is less possible to ascertain how much the Germans learned from the French of the state of the British or French pulse radar developments. Interviews with several of the key radar engineers who were in France during most of the occupation indicate that nothing was officially disclosed on allied pulse radar to the Germans and that the latter were neither particularly clever in trying to ferret out information from commercial laboratories, nor particularly interested. It was not until the spring of 1941 as a result of losses sustained in the Battle of Britain that the Germans really became concerned about radar, and dispatched a mission of high-ranking officers to Vichy who asked, apparently for the first time, to be informed completely about French developments in radio detection. The official answer they are believed to have received was that all the research done in France had been with the David system. To avoid sending them away empty-handed, they were given a full description of the David system and copious documentation concerning it.

NOTES

1. For example, the article headlined "Plane Detection by Television Planned for U.S." appeared in the New York *Herald Tribune* for March 21, 1938. The substance of this article was reproduced in the Aero Digest for April 1938. References to closely allied devices, like Ponte's obstacle detector, and the Army's infrared detector, which are discussed elsewhere, appeared in the press as early as 1935.

2. Naval Attache Report No. 981, London, October 7, 1939: Subject: Detection of Aircraft Approaching Great Britain; Naval Attache Report No. 981, London, March 4, 1940: Subject: Detection of Aircraft Approaching Great Britain.

3. John T. Henderson: National Research Council of Canada. Radio Section: Progress Report for Period June 1939 to January 1942, Ottawa, 1942.

4. S/L F.U. Neakes and Dr. J. T. Henderson: Electrical Methods of Fire Control in Great Britain—Preliminary Report on RDF sent from Great Britain, April 1939. (Copied Ottawa, October 1940). See also Henderson: Report No. 2 on Electrical Methods of Fire Control in Great Britain, Ottawa, June 1939.

5. Edward L. Bowles: Canadian Government Defense Agencies. National Research Council. Research Enterprises Ltd., September 20, 1941.

6. This section is based on John T. Henderson's Progress Report, Ref. 3 above, and on conversations and observations during a visit to the National Research Council in the summer of 1943, made through the courtesy and with the helpful cooperation of Colonel (now Brigadier) F. C. Wallace.

7. The National Research Council radar laboratory never became remotely comparable in size to TRE or the Radiation Laboratory. In January of 1942, its staff was still under 300 persons. Money must have been lacking, but according to the director, Brigadier F. C. Wallace, the limiting factor had not been money but the supply of competently trained physicists and engineers.

8. J. T. Henderson: "Notes on D. F. Systems Used in France, 30 May 1939," Report No. 2 on Electrical Methods of Fire Control in Great Britain, Ottawa, June 1939, p. 73—personal communications from Gerard Lehmann. The visit took place in the week preceding Whitsun (May 28). The inventor of the equipment, Mr. Lehmann, arranged a demonstration in

which a plane flow from Villacouble to Chattes, a distance of about 20 miles. The D̄. F. equipment, which used an Adcock aerial with folded dipoles, easily picked up the plane and gave the bearing accurately to within two degrees.

9. This section is largely based upon information supplied by a group of French radio engineers, employed by the International Telephone and Telegraph Company, during a conference held on February 2, 1944, at the IT&T offices in New York City. My chief informants were Mr. E. M. Deloraine, who in 1939 was the European Director of the Laboratories of the International Standard Electric Corporation; Mr. Emile Labin, who in 1939 was in charge of the television transmitter installed on the Tour Eiffel by the Paris Laboratory of the LMT Company (*Le Matérial Télégraphique*); and Mr. Gerard Lehmann, in 1939 the Chief Engineer of the Sadir Company of Paris (*Societé Anonyme des Industries Radiotélégraphiques*). These men had all been important participants in the French radar development.

10. The Laboratoire Nationale de Radio Electricité was an interministerial radio laboratory of about 200 people situated, after 1932, at Baigneux (Châtillon-sur-Seine). It was an outgrowth of the Laboratoire de la Télégraphie Militaire which had been extremely important during the first World War under General Ferrié, but which had been greatly reduced in size after the Armistice. Many of the wartime personnel shifted from one laboratory to the other. The military laboratory continued to exist and was in close touch at all times with the Laboratoire Nationale.

11. This information was supplied Mr. Lehmann, who had first-hand acquaintance with the David equipment.

12. It should be recalled that this was the feature which led the American Navy to doubt the system's suitability for work at sea, and led to suggesting the method to the Army as more adapted to their needs. In France the Navy was entrusted with coastal defense and consequently was fully as interested in the system as the Army. The French Navy even carried out some installations on shipboard.

13. The information given here was supplied by Gerard Lehmann and Emile Labin who were in charge of the system at this time.

14. Information supplied by Mr. Kenneth G. Budden of the British Air Commission.

15. Information supplied by Colonel Wallace.

CHAPTER 8
The Early History of Microwaves

Several allusions have already been made to early attempts made to build detection devices employing microwaves, i.e., superhigh-frequency radio waves with wavelengths below 30 cm. These attempts were foredoomed to failure, not because the wavelengths were unsuited to detection work, but because adequate sources of power were not available. Only after 1940, after the British invention of the resonant cavity magnetron, was it possible to produce microwaves at powers comparable to those available for long-wave radar.

Logic and basic principles were on the side of those who made these early, fruitless attempts, for only with microwaves is it possible to obtain really narrow beams without using reflectors that are forbiddingly large. It is the narrow beams resulting from using microwaves which have made possible many of the most important refinements of wartime radar.

The Importance of Microwave Radar

For all practical purposes range accuracy is independent of wavelength, over the range of frequencies used in radar; but greatly enhanced accuracy in the determination of bearing and elevation results in passing from radar with wavelengths measured in meters to radar using centimeter waves. This increases accuracy results from the single fact that narrow beams of radiation can be produced more readily at these very short wavelengths.

Without exception, radar systems employ radiation that is focused in beams, instead of being sent broadcast as in the case of commercial radio. The more power the radar transmitter can concentrate on the target, the stronger will be the echo. While the strength of the echo varies directly as the transmitted power, it is proportional to the square of the area of the antenna reflector, so that using a large reflector (large, that is, with respect to the wavelength of the radiation used) is often a very effective way of increasing the performance of a radar set. Power considerations, however, are not the chief reason for seeking narrower beams. Greater immunity to enemy jamming; higher accuracy in the determination of bearing; the ability to make direct determination of the height of a target, to send a beam low over land or water without interference from the reflections produced by the ground or the surface of the water, and to discriminate

185

between adjacent targets: these are the chief dividends derived from narrow beams.

It should be evident that if a sharply directed beam of energy is used and the antenna structure is mounted so that it can rotate, it is only necessary to observe the angular position which gives the strongest signal on the indicator in order to determine the bearing with some degree of accuracy. Special methods are invoked where really high precision is required; but, apart from these, the sensitivity of the single-beam method depends only upon the angular width of the beam. The sharper the beam, the more accurate the determination of bearing.

Such simple considerations do not apply in determining the elevation of targets. Only with extremely narrow beams is it possible to determine elevation merely by increasing or decreasing the angle of the antenna until the strongest signal is observed. With broad beams some of the energy hits the ground on its way to the target, for example, an aircraft, and reaches the target after having traversed this longer path. These waves may arrive at the target in phase or to some degree out of phase with those which took the direct route from the transmitter to the plane. The result of these interference effects is to break up the single beam into multiple lobes with gaps between. The distorted pattern is so complicated that it is impossible to determine altitude merely by changing the angle of the antenna. Cumbersome dodges have to be used at frequencies as low as 200 MHz to make accurate height finding possible.

The other advantages of narrow beams are equally important. The ability to direct a beam low over land or water has an obvious military value. "Low coverage" as it is called, makes it difficult for an enemy plane to fly in beneath the radar beam and thereby escape detection. Likewise the greater angular discrimination of narrow beams is very valuable in all phases of the use of radar. It enables the number and disposition of enemy ships and sometimes even aircraft to be determined accurately. With the use of narrow beams, the PPI type of cathode-ray tube indicator, which paints in a map-like representation of the area being scanned, gives an increasingly realistic picture of the terrain under observation as the frequency is raised and the beam is made narrower. Lastly, the freedom from jamming must be distinguished from the initial element of surprise which using a new and unsuspected frequency always possesses. This was an advantage of microwaves not directly resulting from the narrow beams that they provide, but in addition, a narrow beam is inherently harder to jam. Because of its greater angular discrimination, microwave equipment is less susceptible to jamming devices like "window", even when the metal strips are cut to reradiate on microwave frequencies. Similarly in the case of electronic jamming, a microwave set can continue to operate virtually unaffected except when pointing precisely in the direction of the enemy transmitter.

Oscillators for Microwaves

Microwaves have been known, and many of their properties understood, since the days of Hertz and Righi; but until the development of the cavity magnetron by the British these waves were hardly more than scientific curiosities. After the vacuum tube replaced the spark gap transmitter, they could not be produced by ordinary radio techniques. Conventional triode vacuum tubes can be used over a wide range of the radio spectrum, but only by pushing refinements close to the limit can they be used to generate waves shorter than 50 cm. Special techniques must be employed to generate, transmit and receive microwaves. Before 1940 the principal sources of centimeter waves were (1) spark gap transmitters; (2) Barkhausen–Kurz oscillators; (3) the klystron; and (4) the split-anode magnetron, the ancestor of the device developed by the British.

The earliest type of vacuum tube to show promise as a source of ultra-high-frequency radiation was the so-called positive grid or Barkhausen–Kurz (BK) oscillator. H. Barkhausen and K. Kurz described in 1920 the fact that conventional triodes with cylindrical electrodes can be made to produce sustained oscillations above 300 MHz, provided the grid is kept at a high positive potential and the plate at or near the cathode potential.[1] Under these conditions they were able to produce waves as short as 43 cm, though most of their work was in the neighborhood of 1 or 2 m; but they expressed the opinion that waves as small as 10 cm could probably be produced. Much work was done by latecomers in the field, and a large and somewhat discordant literature has arisen on the subject. There is still no completely satisfactory theory of BK oscillations; and our dependable knowledge does not go much beyond the fact that the oscillations originate in the to-and-fro motion of the electrons as they are accelerated through the grid, retarded by the negative field of the plate, and reversed in direction to traverse the grid once more.[2] In this country the behavior of BK tubes was carefully studied at the Bell Telephone Laboratories, where output powers of several watts were obtained in the range between 400 and 600 MHz (67–50 cm).[3] In general, the ordinary BK tube was found not to be suitable for higher frequencies. To generate higher frequencies the spacings must be made smaller and the structure must dissipate a large amount of heat. In France, A. Clavier devised a modified BK tube in which the grid is an open helix surrounding the axial filament, and in turn surrounded by a cylindrical plate. Tubes of this sort were used in 1931 in experimental microwave telephony between Calais and Dover and three years later in the commercial link across the Channel between the airfields of Lympne and St. Inglevert. The tubes gave a few tenths of a watt at 17.4 cm.[4]

The *magnetron* in its primitive form was first described by A. W. Hull of the General Electric Company's Research Laboratory in May 1921, before the annual meeting of American Institute of Electrical Engineers in New York City.[5] As Lee DeForest once described it, the word "magnetron" was a Greco-Schenectady name of a vacuum tube (diode) operated by a magnetic field superimposed upon the electrical field. The most common form studied at General Electric consisted of a cylindrical anode surrounding an axially placed filamentary cathode (as shown in Fig. 8-1). In the first tubes the magnetic field was produced by a solenoid wound directly on the glass of the tube. The device seems to have grown out of an earlier study of the effects of magnetic fields on the operation of the dynatron triode.[6]

Hull conceived of his device primarily as an electronic valve or switch operated by a magnetic field, not as an oscillator, though some experiments were carried on at GE with the tube as an amplifier and as a generator of low- and high-frequency alternating current. In 1925 they reported success in using the magnetron as a generator of low-frequency radio energy, and were able to get 8 kW at a frequency of 30 kHz with an efficiency of 69%.[7]

The characteristic of the tube is that the magnetic field under operating conditions controls the valve. Above a given filament temperature, if a constant voltage is impressed between the cathode and anode, the current that flows through the tube is independent of the magnetic field until a certain critical value is reached, at which point it falls sharply to zero (see Fig. 8-2). Hull's simple picture of what happens to the electrons in a simple magnetron is as follows: in a steady magnetic field the electrons emitted from the filament (cathode) are accelerated toward the cylindrical plate (anode) by the electrical potential; but at the same time they are acted upon tangentially by the magnetic field with a force that depends upon the velocity of the electrons, their charge, and the strength of the magnetic field. The resultant is a curved path which at some value of the magnetic field (the critical or cutoff value) just fails to strike the anode.

The early Hull magnetron was of too low efficiency for the successful generation of anything except low-frequency rf energy. An important

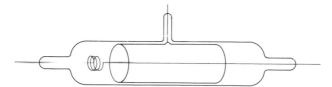

FIG. 8-1. Early General Electric magnetron (Hull, 1921).

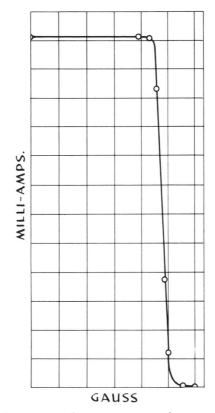

FIG. 8-2. Magnetron characteristic under a magnetic field.

modification, called the *split-anode magnetron,* had to be developed before the tube could evolve into a useful generator of ultrahigh-frequency
waves. This was first described in 1924 by Erich Habann in a Jena disser-
tation, where he showed that if the cylindrical anode of a magnetron were
cut into two hemicylindrical segments, the device would show negative
resistance properties (see Fig. 8-3) and hence when connected to a proper
tuned circuit would be capable of maintaining sustained oscillations.[8] The
common push-pull arrangement for the split-anode magnetron was first
suggested by another German, E. Manns, in 1927.[9]

This early work was done at relatively low frequencies; the next step
was the discovery that the diode magnetron, and, still more easily, the
split-anode magnetron, could be used as a source of ultrahigh-frequency
waves and a possible source of microwaves. The earliest to use the diode
magnetron in this region was August Žáček, who reported in 1924 from
Prague that he had been able to produce waves as short as 29 cm by
operating such a tube in the neighborhood of the critical magnetic field.[10]

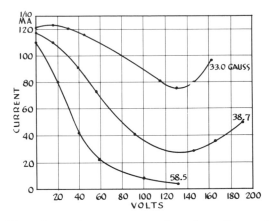

FIG. 8-3. Negative-resistance properties of a split-anode magnetron (Habann, 1924).

Shortly after the work of Žáček, and apparently in ignorance of it, the ability of the magnetron to produce ultrahigh-frequency waves and even microwaves was independently discovered by the Japanese investigators working with Hidetsugu Yagi at Tohoku College of Engineering of the Imperial University at Sendai.[11] Under Yagi, who had studied in Germany with Barkhausen and in America with G. W. Pierce at the Cruft Laboratory, Uda performed a series of interesting experiments on beam transmission at 4.4 m, while Okabe explored the problem of generating superhigh-frequency waves. In 1927 Okabe reported that he had obtained "comparatively intense" electromagnetic waves with a magnetron by keeping the magnetic field strength near its critical value and applying a high voltage to the anode. The shortest waves obtained in this manner were 12 cm long.[12]

Okabe was soon able to report a modification of some importance: he had applied to his tubes the split-anode construction developed for magnetron oscillators at longer waves by Habann and by Manns, and discovered that a greatly increased output was possible.[13] The Japanese investigators found this especially useful for the production of strong oscillation at a wavelength of about 40 cm. Using this equipment for communication purposes with directive antennas and a crystal detector, they received signals up to a distance of 1 km.[14]

The split-anode magnetron was the subject of active, world-wide investigation in the decade following Okabe's discovery. It soon appeared that this tube, when used with a suitable tuned circuit, was a most successful generator of waves from 1.5 m down to about 50 cm, and that with substantial improvement it might be capable of producing still shorter waves with appreciable power.

When the tubes operate in the ultrahigh-frequency range, the frequency is found to depend upon the constants of the external circuit, while in the microwave region the frequency is largely independent of the external circuit and depends upon the transit time of the participating electrons. In the first instance high efficiencies are the rule, as high, for example, as 50%–60%. The sharp decrease in efficiency in going to the microwave frequencies is due largely to transit-time effects; there are also increased circuit losses at the higher frequencies. For producing centimeter waves, it was found that the resonant circuit must be extremely small; and for wavelengths below 50 cm, it was found necessary to mount the whole oscillatory circuit within the tube. This was done by G. R. Kilgore at RCA by constructing the anode segments and the circuit out of a single piece of copper (as shown in Fig. 8-4). In 1932 Kilgore built an internal-circuit transit-time magnetron with which he was able to generate waves of 22 cm with a watt or two of power, and with which somewhat later he was able to get waves as short as 9 cm.[15]

The Kilgore type of internal construction was used for the brilliant experiments performed by C. E. Cleeton and N. H. Williams at Michigan in studying the absorption of energy at microwave frequencies by ammonia gas. With tubes of still smaller dimensions than Kilgore's, he got waves of 1.87, 1.22, and 0.64 cm with enough power to perform his experiments.[16]

Two modifications gave promise of increasing the power of the transit-time magnetron oscillator. E. G. Linder devised a form of split-anode magnetron in which two electrodes were introduced at the end of the plate cylinder. These anodes were kept at a positive potential slightly lower than that of the split plate, and served the same function as tilting the magnetic field, namely, removing electrons from the interelectrode space after they have completed several orbits. This tube gave somewhat greater output and higher efficiency than the simple split plate tube.[17] With this tube they produced waves of 9 cm with an output of 2.5 W and an efficiency of about 12%. This tube was used by Wolff and Linder for transmission and detection tests in collaboration with the Signal Corps in 1934. Four-foot parabolic reflectors were used for transmitter and receiver.[18]

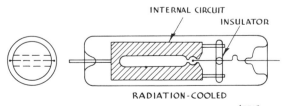

FIG. 8-4. Kilgore's internal-circuit magnetron (Kilgore, J. Appl. Phys., Oct. 1937; see Ref. 15).

In another tube which Linder describes as the "anode tank-circuit magnetron" the anode halves form the resonant line or tank circuit; and the use of a large anode made of refractory material (tantalum) permitted the dissipation of a large amount of heat. With this tube, Linder obtained on the eve of the war an output of 20 W at 3750 MHz (8 cm) and with an efficiency of 22%.[19]

A different approach was followed by H. Gutton and S. Berline of the French SFR Company (*Société Française Radioélectrique*). These workers built a magnetron with 12 anode segments with which they obtained waves of 16 cm with an output of 10 W and 15% efficiency.[20] They used this device (using radiation beamed by parabolic reflectors) to make communication experiments over a distance of 152 km from the Puy de Dôme in Auvergne to Mont Beuvray in the Morvan. They discovered that these waves were propagated without difficulty through fog.[21] It was this equipment which was adapted by Ponte and his co-workers for use as a detection device.[22]

Table I. Magnetron development.

Date	Name	Type of tube	Wavelength in cm	cw output in W
1924	Žáček	cylindrical diode	29	?
1927	Okabe	cylindrical diode	12	?
1929	Okabe	split-anode as transit-time osc.	5.6	?
1932	Kilgore	split-anode mag. with internal cir.	22	1–2
	Kilgore	*ibid*.	9	1
1934	Linder	split-anode, endplate	9	2.5
1936	Rice	cylindrical diode	5	3
1938	Gutton & Berline	split-anode, 12 segment	16	10
1939	Linder	anode tank-circuit	8	20
1940	Randall & Boot	first resonant cavity (6 hole) magnetron	9.8	400

Resonant Cavities and the Klystron

A discussion of electromagnetic resonant cavities introduces one of the novel and characteristic features of microwave techniques. Historically the subject may be said to have come into being in 1936 in connection with the work of Barrow and of Southworth on waveguides[23] (which can themselves be considered a kind of resonant cavity) and in the work of W. W. Hansen on the *rhumbatron*.[24]

In the most general case, a resonant cavity is a region bounded by an electrically conducting shell of any shape; the simplest real cases are those of a sphere, a right circular cylinder, or a rectangular prism made of copper sheeting or other highly conducting material. That such cavities have the properties of a simple tuned circuit, with localized inductance and capacity, was clearly indicated by Southworth, who designed tunable resonant cavities for use in his waveguide transmission studies. These he used as the characteristic feature of a wave meter, in place of a tuned circuit with BK tubes and magnetrons, and as a simple tuned receiver.

At about the same time Hansen at Stanford University had hit upon the idea of using nearly closed resonators of this sort in a scheme for producing high-speed (i.e., high-voltage) electrons. In connection with this work Hansen formulated certain fundamental ideas concerning resonant cavities. He pointed out that any closed conducting surface can contain a periodically repeating electromagnetic field, that only fields of certain frequencies can be sustained in a given cavity, and that the values of the possible frequencies are determined only by the size and shape of the chamber.

Hansen's chief object was to demonstrate the equivalence of such resonators to conventional tuned circuits.[25] Just as the values of L, C, and R of a resonant circuit completely describe the circuit at lower frequencies, so related parameters were found to describe the cavity resonator. These are dependent upon the geometry of the cavity. Hansen calculated the values of the parameters for the cases of a sphere, a cylinder, and a prism. It turned out that the lowest allowed wavelength was about equal to the linear size of the resonator; thus for the case of a sphere, $\lambda = 1.14d$, where d is the diameter. The most striking result came from considering the Q value of such a circuit. This quantity measures the rate at which energy is lost in a resonant circuit, the ratio of the energy stored to the energy lost per cycle, and hence the efficiency of the circuit as a resonator. Hansen gave some experimental confirmation to the theoretical conclusion that resonant cavities are resonant circuits with an extremely high Q.

This knowledge underlies the later development of practical cavities for use in superhigh-frequency work: the rhumbatron, the klystron, and later (after the launching of the microwave radar program) the echo box, the wave meter, the TR box, the crystal mixer, certain local oscillator tubes, and most crucial of all, the resonant cavity magnetron.

The first practical result was the design of an electron accelerator that was shortly after called the rhumbatron. It was conceived by W. W. Hansen and developed in cooperation with Russell H. Varian, a graduate of Stanford, who had recently returned from industry as a research associate in the physics department. The original idea was to activate a cavity made of a cylindrical shell of copper, using one or more acorn tubes as oscillators.[26] The terminals of the tube were attached to an insulated cathode plate and to an anode plate consisting of a perforated diaphragm across the cavity, as in Fig. 8-5. To use the device as an electron accelerator, it was only necessary to modify it so that a stream of electrons could be projected across the cavity by the electrical component of the intense alternating electromagnetic field set up in the resonator. The speed of the electrons can be still further increased by causing them to reverse their direction every half cycle, until the electrons reach a certain speed and are allowed to escape from the accelerating chamber. These reversals of direction are accomplished by causing the electrons to pass at each end of their flight into the field of an electromagnet. By proper adjustment of the fields the electrons can be allowed to escape into the chamber C when they have reached the desired velocity.

Although some experimental cavities were constructed, a new and more important use for the rhumbatron principle was discovered which distracted attention from constructing an electron accelerator. This was the invention of the *klystron.*[27]

The klystron is a velocity-modulated tube in which a beam of electrons is projected successively through two adjacent rhumbatron cavities; the tube converts the dc power of the electrons into ultrahigh-frequency or superhigh-frequency ac power. The reader may follow the process by consulting the diagramatic representation in Fig. 8-6. Electrons emitted from a thermionic cathode (C) are accelerated in the electrostatic field (E) produced by the difference of potential between cathode and grid. The grid serves only to straighten the lines of force of this field and to collimate the electron beam. The electron beam then passes through the field-free space (S), then through a perforated wall, into the high-frequency field of

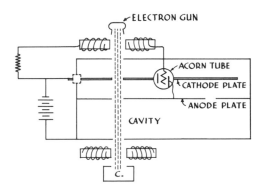

FIG. 8-5. Rhumbatron (W. W. Hansen, U. S. Patent 2, 251, 569).

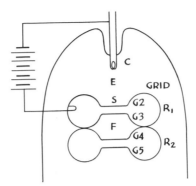

FIG. 8-6. Klystron.

the first rhumbatron (R_1), called the "buncher". The field of this resonator may be very weak, such, for example, as might be produced by a small signal to be amplified. It nevertheless produces a slight modulation of the speeds of the electrons, alternately retarding and accelerating them in accordance with the frequency characteristic of this first resonant cavity. This change of velocity alone is of no practical use until the fast electrons are sorted from the slow ones. There is no appreciable intensity modulation of the current carried by the electrons as they leave the first cavity; but in the much longer flight through the field-free space (F) the differences in their speeds cause them to gather into bunches separated by rarefactions. Thus, before the electron stream enters the second rhumbatron it has been "intensity modulated," i.e., the bunching or sorting process changes the stream from pure dc to dc plus a high-frequency component. When these electrons pass into the second rhumbatron or "catcher" they serve to excite this second resonator which has been tuned to their frequency (that is, to the frequency of the signal voltage). Power may be taken from the catcher by a loop or other device.

In the case just described, the buncher was thought of as driven by an external source of power, such as an antenna receiving radiation, and the klystron is acting as an amplifier. If the buncher is driven by power received through the coupling loop from the catcher, the klystron acts as an oscillator.

The klystron was the result of a search for a source of ultrahigh-frequency or superhigh-frequency energy to be the basis of an aircraft detecting device. The stimulus for such a device came from Sigurd F. Varian, a commercial airlines pilot and the brother of Hansen's collaborator on the rhumbatron.[28]

Unlike his brother, Sigurd Varian had not gone to college. After finishing school at the California Polytechnic Institute, a small technical school at San Luis Obispo, he became a flier, first a barnstormer, then a commercial pilot. From 1929 to 1936 he was a captain with Pan American Airways, flying the route between Brownsville, Texas, and the Canal Zone. Like many other experienced pilots he had given a great deal of thought to the need for some sort of radio beam to enable planes to land through low overcast; with the outbreak of the Spanish Civil War, his speculations began to embrace the idea of a radio detection device to pick up enemy aircraft. He broached his project in 1936 in a long letter to his scientific brother at Stanford, who wrote back that Hansen's rhumbatron might be the starting point for the necessary, but as yet unavailable, source of ultra-high-frequency power. On the strength of this assurance, Sigurd Varian took a leave of absence from Pan American and came to Stanford to work with his brother in the development of a blind-landing and airplane detecting device.

The idea of combining two rhumbatron cavities into the tube that was called the klystron resulted from discussions between Russell Varian and W. W. Hansen of various possibilities for building the desired oscillator. These discussions seem to have taken place in the spring of 1937. The first unit was apparently completed in August of that year, and work was shortly after begun by D. L. Webster, Chairman of the Physics Department, on a theoretical study of its operation. By mid-December, two experimental tubes were in existence, built in removable housings for convenience of experiment, and of course operated on the vacuum pump. Both of the tubes gave energy at a wavelength of approximately 12 cm.

Microwave Waveguides and Horns

It had long been known that fundamentally new techniques were required to provide satisfactory transmission lines to carry rf energy at very high and ultrahigh frequencies. Even at ordinary broadcast frequencies single-wire lines were of limited usefulness because of the low efficiencies arising from the ease with which they radiate energy, and power losses were cut down by using two conductors in a go-and-return circuit, the distance between the two wires being kept very small.[29] If the currents in the two wires are equal and opposite in phase, the field from one wire will cancel out the field from the other, and radiation is inappreciable. Almost perfect shielding and low losses are possible when one conductor completely encloses the other. Coaxial lines, consisting of a central core completely surrounded by a concentric cylindrical sheath, had been in use for some time to provide efficient transmission at high and ultrahigh frequencies. The outer conductor not only serves as one of the two wires of a pair, but also yields perfect shielding. The cross sections of various lines are shown in Fig. 8-7. The insulators shown were usually of rubber or porcelain. The attenuation of well-made coaxial line is proportional to the square root of the frequency, and inversely proportional to the diameters (at optimum ratio). It can be shown that for smallest attenuation there is an optimum ratio for the size of the conductors: the outer conductor should be about 3.6 times the diameter of the inner conductor. Figure 8-8 shows this by a plot of attenuation in a concentric transmission line, which shows the field existing in a hollow tube. That electric waves could be propagated through hollow tubes or pipes without benefit of an inner conductor was first demonstrated in a theoretical paper by Lord Rayleigh in 1897.[30] Although it had been asserted some time before by Oliver Heaviside that waves could not be sent through a hollow tube when a center conductor was absent, Rayleigh showed by application of Maxwell's equations that under certain conditions electric waves could be propagated along hollow tubes. He distinguished two classes of possible vibrations, corresponding to what were later called "transverse" and "longitudinal waves" (Barrow); "E waves" and "M waves" (Carson,

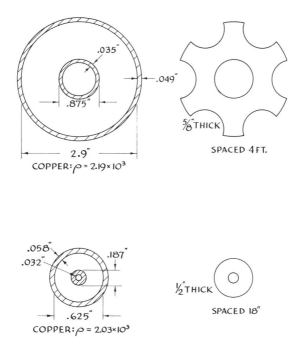

FIG. 8-7. Cross sections, coaxial lines (Sterba and Feldman, July 1932).

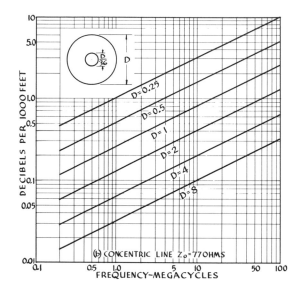

FIG. 8-8. Attenuation of concentric transmission line (Terman, Ra-dio Engineers' Handbook, 1943).

Mood, and Schelkunof); or TE and TM waves (N. H. Frank). Rayleigh postulated a cylindrical dielectric solid, infinitely long, bounded by a per-fect conductor, and he showed that for such tubes, whether of rectangular or circular cross section, there are certain critical frequencies (depending for their values upon the dimensions of the guide) below which the waves are rapidly attenuated, and above which they are freely transmitted.

That real hollow metal tubes showed this property was not experimen-tally demonstrated until 1936 when W. L. Barrow of the MIT Electrical Engineering Department and G. C. Southworth of the Bell Telephone Laboratories simultaneously announced that hollow tubes could actually serve in practice as conductors of ultrahigh-frequency radiation.[31]

The theoretical portion of Barrow's study confirmed and extended the findings of Lord Rayleigh.[32] For an ideal dielectric surrounded by a sur-face of infinite conductivity he proved the existence of a critical frequency above which waves are freely transmitted and found it to be determined by the dimensions of the tube, the critical frequency being inversely propor-tional to the radius of the tube. He treated the special case where the conductivity of the wall is finite but very large (closely approximating that of a copper tube), and where consequently attenuation is important. This was found to be proportional to the three-halves power of the tube radius, so that the attenuation decreases rapidly with the tube size, even when the ratio of wavelength to tube diameter is optimal.

In his paper Barrow presented a graphical comparison of the attenuation properties of the hollow-tube guide with those of coaxial cables (see Fig. 8-9. From these results it appeared that the hollow tube compared favorably with other types of lines in the matter of attenuation; and he went on to indicate that at the very high frequencies at which it would be used, the hollow waveguide "would probably have lower attenuation than any other type of line now in use." The existence of a critical frequency characteristic, he pointed out, would permit such a hollow tube being used with conventional lines and networks and with other hollow-tube arrangements as a very effective high-pass filter.

If a practical transmission system using waveguides were to be possible, then tubes of reasonable diameters must be used, say between 4 and 100 cm. Hollow-tube transmission seemed ideally suited for the handling and conduction of centimeter waves. In his earliest experiments, Barrow made use of a waveguide consisting of a galvanized iron cylinder—actually an old air duct—about a foot and a half in diameter and 16 ft long. His source of radiation was a Barkhausen–Kurz oscillator giving waves about 38 cm in length. These waves were modulated at a low audio frequency and were fed into the hollow tube by a parallel wire feed ending in a probe within the tube (see Fig. 8-10). They were picked up at the other end by a probe connected with a crystal detector, an audio amplifier, ear phones, and a meter.

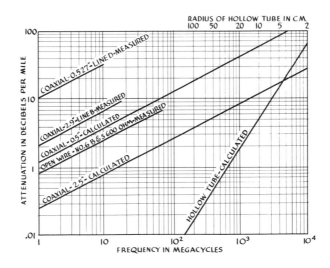

FIG. 8-9. Attenuation in cylindrical copper waveguide compared with measured and with computed attenuation in several coaxial lines and an open-wire line (Barrow, 1936).

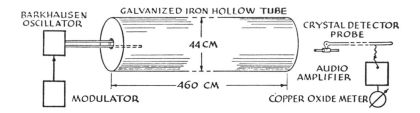

FIG. 8-10. Barrow's experimental waveguide apparatus (1936).

With this equipment, Barrow was able to verify experimentally the critical frequency and the filter action of the tube. Also, by movement of the receiving probe varying distances from the end along the axis of the tube, standing waves were easily detected and gave an accurate measure of the wavelength within the guide. When the guide was suitably terminated (i.e., by its own characteristic impedance) these standing waves were reduced or entirely absent. Barrow reported that speech and other signals were satisfactorily transmitted through the pipe and received at the far end about ten times more effectively than with free-space reception using the same transmitter and receiver.

A number of experiments were performed by Barrow on the waves of the transverse sort (Rayleigh's "vibrations of the second class"). He found that these waves could be excited by using probes inserted transversely across the guide. Since the longitudinal type of wave had been excited by probes along the axis of the tube (see Fig. 8-11) it appeared that the probes excited the electrical field that lay along their length. These transverse waves had an important property: reception was obtained only if the receiving probe was oriented precisely like the transmitting probe; when they are parallel the waves are picked up; if they are at right angles, no signals are received. To Barrow this indicated that a hollow tube could serve as a "multiplex" transmission system. At least three different signals could be sent down the tube without interference if the probes (one for the longitudinal wave and two for the transverse) were properly oriented.

Between 1936 and the foundation of the Radiation Laboratory at MIT in the fall of 1940, Barrow's work expanded considerably. A brilliant Chinese graduate student, Lan Jen Chu, was assigned to the hollow pipe investigation, especially its theoretical side, as a full-time assistant aided by a grant from the Research Corporation. Chu contributed important studies on the properties of wave transmission in hollow pipes of elliptical cross section and in rectangular waveguides, and on the theory of the electromagnetic horn. These results were incorporated into his PhD Dis-

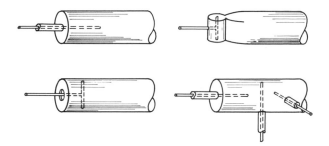

FIG. 8-11. Input circuits for cylindrical waveguide (Barrow, 1946).

sertation in 1938[33] and in various papers that were published singly or jointly with Barrow in 1938–39.[34] An interesting experimental study was published by Barrow and his student, W. W. Mieher, on the natural oscillation in cylindrically shaped cavity resonators.[35]

Much of the attention of Barrow's group was centered, for reasons that will shortly be apparent, upon the development of suitable horn radiators for projecting and receiving directive energy at ultrahigh frequencies. Field pattern measurements were made by various of Barrow's students on rectangular hollow pipes open at the end, on flat horns of rectangular cross section whose sides flare in only one direction (which he termed *sectoral horns*), on the biconical horn applicable to broadcasting work at ultrahigh frequencies, and on horn arrays. At first the field measurements were made in the neighborhood of 40 to 100 cm, with a specially designed oscillator using a Western Electric 316A vacuum tube[36]; the later tests on the biconical and multiunit horns were made at 8.3 cm with a split-anode magnetron.

The Waveguide Work at the Bell Laboratories

The results of the waveguide investigation at the Bell Telephone Laboratories were first published in two separate but related papers, one largely experimental, by G. C. Southworth, and the other by the theoreticians of the Bell Laboratories, John R. Carson in collaboration with Sallie P. Head and S. A. Schelkunoff.[37]

Southworth's papers deal mainly but not exclusively with conduction phenomena in cylindrical waveguides of the sort studied by Barrow. Some space is given to the properties of waves in dielectric wires or cylinders, from which, it would seem, the work on hollow waveguides had derived. Southworth began work on the properties of dielectric cylinders in 1931, his earliest studies consisting of studying the transmission of 100- and 400-MHz radiation transmitted through cylinders of water. The cylinders consisted of a 4-ft column of water enclosed in these copper or bakelite tubes, in some cases 6 in. and in others 10 in. in diameter. The bakelite was sufficiently thin and its dielectric constant so low compared with water that the cylinder could be regarded as only a column of water. His first experiments were concerned with proving the existence of such guided waves by the detection and careful measurement of standing waves. He was able to produce waves of the four principal types which theory had already indicated should be present; and he determined the field distribution and the velocity of propagation (from measurements of wavelengths in air and in the guide) for the two types of water columns (with and without copper shielding) and for the four types of waves which he studied.[38]

Whatever the chain of influence or reasoning, Southworth turned to a study of the propagation of waves in hollow conducting pipes. This meant using microwaves. The program was expanded and was moved to the Bell Laboratories' field station at Holmdel, N.J. Here Southworth used improved apparatus. A Barkhausen–Kurz oscillator gave him energy over the range from 2000 (15 cm) to 1500 MHz (20 cm). Two channels were set up over a distance of 1250 ft across a field. One channel consisted of copper tubing 4 in. in diameter, while the other used slightly larger tubing 6 in. in diameter.

At Holmdel, Southworth and his colleagues[39] explored four of the many modes that may be propagated through cylindrical guides, and which had been studied previously for dielectric cylinders. The proper kinds of "input" or "launching" mechanisms having been worked out to secure pure modes of each kind, and the general disposition of the lines of force characteristic of each of these having been calculated, Southworth and his co-workers verified their characteristics with a probe and crystal detector device quite similar to that used by Barrow. Careful attention was paid to the attenuation properties of these various modes. Figure 8-12 shows the calculated curves for the four principal modes. They show that the attenuation becomes infinite below the critical or cutoff frequency; and that for three of the modes, the E_0, E_1, and H_1 modes, the attenuation drops off sharply just above the critical or cutoff frequency; then as the frequency is still further increased, the attenuation rises once more. The H_0 mode has the striking behavior that for all frequencies above cutoff there is a descending attenuation characteristic. With his Holmdel apparatus, Southworth verified the attenuation curves, in the neighborhood of the cutoff frequency, for the four modes under consideration.

Southworth, like Barrow, showed that the cylindrical waveguide behaves like an ordinary wire line in that standing waves are not produced when the guide is terminated by its own characteristic impedance. He published calculated curves for the characteristic impedance of a 4-in. hollow copper pipe for each of the four common modes, and gave experi-

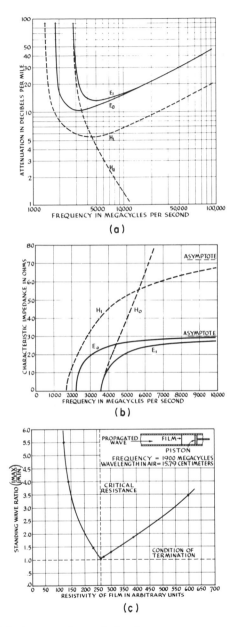

FIG. 8-12. (a) Attenuation of the four common types of waves in cylindrical copper waveguide. (b) Calculated values of characteristic impedance of a 4-in. cylindrical copper waveguide. (c) Typical set of experimental data obtained with various degrees of impedance match. (Southworth, 1936).

mental curves to show how the standing wave ratio (the measure of the magnitude of the standing waves, and hence of the impedance match at the end of the line) varied with the termination of the waveguide.

Southworth also reported, as Barrow had done, that waveguides might play a useful role as circuit elements; and he likewise discovered that radiation is emitted from the opening of a hollow tube, and that by flaring of the ends into a horn this effect is enhanced. An electrical horn of this sort possesses directivity and serves as an efficient radiating load, that is, it is a good termination.

Southworth's apparatus was considerably more elaborate and sophisticated than Barrow's. He made more use of the resonant cavity, which he describes as fundamental to waveguide work. Because it has the properties of a tuned circuit it can serve as a wave meter, or be used in conjunction with either the transmitter or the receiver to increase their efficiency. One of the devices described, the slotted section of waveguide, for use in determining standing waves within the guide, has become an indispensable tool in microwave work. Another of his devices was a coaxial-conductor wave meter, which allowed him to determine accurately the wavelength in free space, and supplement the information obtained with the resonant-cavity wave meter, which merely gives the wavelength within the guide.

Southworth was even more cautious than Barrow in commenting on the possible applications of this new form of radio transmission. Its value appeared to be limited by the fact that the size of the waveguide structure is proportional to the wavelength, and hence it was suitable only for the very highest frequencies. These frequencies were only just being explored, and as yet no suitable source of power was available on these frequencies: "The situation then is that the art at these extreme frequencies is not yet at a point which permits a satisfactory evaluation of practical use."[40]

Microwaves for Radio Detection

Although the U.S. Army Signal Corps was probably the earliest to consider using microwaves for radio detection purposes, work in this direction was also carried out at the U.S. Naval Research Laboratory, in Germany, and in France and in at least two electronic laboratories of industry in the United States. In the United States, microwave detection experiments in the Service laboratories and in industry had been abandoned by about the end of 1937, partly because of the successful development of pulse radar on longer waves. Their place was taken by microwave programs, developed around the use of the klystron, carried on at the Massachusetts Institute of Technology, at Stanford, and (as an outgrowth of both of these groups) at the Loomis Laboratories at Tuxedo Park, N.Y.

It will be recalled that when radio detection was begun at Fort Monmouth under the direction of Major Blair it took the form of investigating the possibilities of using microwaves, or, as they were called at the Signal Corps Laboratories, "radio-optical" waves.[41] Work began in about 1932 in the Sound and Light division, where the infrared detection was being conducted, with a survey of the existing possibilities of generating, detecting, and amplifying waves less than a meter in length.[42] Nearly all of this work was carried out by W. D. Hershberger. The earliest experiments were made with Barkhausen–Kurz tubes giving 19-cm waves, similar to those used for the microwave link only recently put in operation across the English Channel. Some experiments were also made with the modified BK tube designed by Hershberger.[43] This was followed with some experiments using a Westinghouse split-anode magnetron that gave very small power at 9 cm. This was used to build equipment with which the first detection experiments on picking up a Ford truck were performed.[44] During 1934, experiments were made with a German-built Hollman tube, produced in duplicate by x-raying the only available sample. With this tube, which gave 5 W at 50 cm, some transmission tests were performed.

On May 31, 1934, Hershberger paid a visit to the RCA Victor Plant at Camden, N.J.[45] He conferred with Irving Wolff, Engstrom, and others and learned about the pioneer work at RCA on microwave frequencies. In a report to Major Blair he summarized the information he acquired and its possible application to radio detection: the 20-mile communication link on 70 cm, and a novel 9-cm magnetron developed by E. G. Linder and Irving Wolff. He reported that he found the RCA group favorable to his suggestion of a series of tests that summer at Navesink.

The tests with the RCA magnetron, in the presence of the RCA representatives Linder and Wolff took place during July and August 1934.[46] Many of the tests were made with a 75-ft patrol launch borrowed from the U.S. Coast Guard. Transmission tests—always with the ultimate aim of using the equipment for radio detection, an aim shared by the Army and by the RCA men—were performed by setting up the transmitter, mounted in a parabolic metal mirror, on the New Jersey shore at Battery Halleck or Navesink, the two ends of Sandy Hook, and placing the receiver, also with a parabolic reflector, aboard the launch, which would maneuver out into New York Harbor. In this manner the range and pattern of the antenna beam could be easily studied. On one such run, on July 27, 1934, good code signals from the 9-cm magnetron were received on the launch as far as Fort Hamilton, 18 miles away, and on this occasion the workers noted a distinct occlusion of the beam by vessels passing about one mile from the receiver.

In other tests, the RCA magnetron, giving about 1/2 W of power, was placed on the parapet of Battery Halleck, Fort Hancock, and was beamed at shipping in the harbor. The receiver, placed 400 ft away, was shielded by a small building from the direct radiation of the transmitter. With this apparatus, good reflected signals were received from a 500-ton boat 1000 yards out. The apparatus was not powerful enough to detect reflections from larger boats passing through Ambrose Channel three miles away. Much better results were being obtained just at this time by Zahl and Anderson at Fort Monmouth with their infrared equipment. During transmission tests on a 50-cm tube developed in the Signal Corps Laboratories—evidently the modified BK tube with spiral grid and a divided cylindrical plate developed by Hershberger—and set up at Navesink, beats were obtained when the German liner *Bremen* passed between the transmitter and the receiver aboard the experimental launch. The launch was 12 miles out when the 51,000-ton liner passed 300–400 yards astern.

The RCA people had been looking for a useful application for microwave equipment, and became interested in the possibility of radio detection after the tests at Fort Monmouth.[47] In the spring of 1935 they set up a transmission system between the roof of the laboratory at Camden and one of the tall buildings in Philadelphia. Also in 1935 Wolff and his co-workers at the Camden Laboratory built a microwave pulse echo system in connection with a project which they described as "radio vision" intended for aircraft use. The purpose was to form a three-dimensional picture by reflected radio waves. Since they had to determine range and the angular coordinates, they chose the pulse technique in place of frequency modulation because of the possibility of distinguishing the targets in range. The transmitting tube was an end-plate magnetron which was plate modulated to give pulses about 1μsec on 3500 MHz. Double modulation was actually employed: a multivibrator was used to key a 30-MHz amplifier and the resulting 30-MHz pulses were applied to the transmitter. The receiver was not a superheterodyne, and was not very sensitive, and was used only because comparatively stable, noise-free local oscillators were not available. It had a crystal first detector from which was taken the 3000-MHz signal which was demodulated once; the output consisted of pulses on a 30-MHz carrier.

With this equipment, set up on the roof of the Camden Laboratory, they obtained ranges of several hundred yards on buildings in Camden. Somewhat later they used a magnetron detector and abandoned double modulation, and they obtained ranges of approximately $1\frac{1}{2}$ miles, and could pick up buildings in Philadelphia and ships on the Delaware. They worked on the equipment for over a year, and by 1937 had improved it until it could receive echoes from across the river. The chief improvement consisted of a good receiver, a super-regenerator with a magnetron detector. The final apparatus has a Type-B scan (as it was later called), plotting range against azimuth. The set used coaxial transmission line and separate transmitting and receiving reflectors (4-ft paraboloids) tied together. The work was carried on with this system under Wolff's direction until the spring of 1937. Various Army and Navy visitors came to see the equipment, among them Commander Ruble of the Bureau of Ships, but none of the civilian engineers from the Service laboratories.

As early as 1932 Wolff had considered the possibilities of a radio pulse altimeter using microwaves and a cathode-ray tube. During 1935–1936, tests were made with the pulse echo equipment to see what possibilities were offered for getting the short ranges needed in altimetry. It proved to be possible to measure distances as short as 5 ft. The winter of 1937 was one of many aircraft accidents, and this provided the incentive to turn their attention toward echo apparatus to be carried in an aircraft. The management at RCA decided that it would be wise to concentrate on nonmilitary applications, air navigation and ship navigation equipment, and that the existing microwave apparatus should be adapted for collision prevention.

The microwave work was dropped in 1937, in favor of attempts to get results on obstacle detection on 500 MHz (60 cm). Five persons were put on the problem, among them Hershberger, who joined RCA about this time. In the fall of 1937 an echo device, using a 316A doorknob tube as transmitter, was functioning on the roof of the Camden Laboratory. It was demonstrated to a group of Navy representatives headed by Commander Ruble and to Army officers from Wright Field. The Army and the Navy both expressed interest, and at Ruble's insistence the project was made secret. A public demonstration which had been planned for 1938–1939 was abandoned.

The equipment was installed in a Ford aircraft at the Camden airport and flown at the end of 1937 or in January 1938. It consisted of a single transmitter and receiver which by switching to either of two sets of antennas could be used for collision prevention or as an altimeter. Used as an obstacle detector, an antenna in the nose of the plane sent a searchlight beam forward. The equipment was tried out against the Catskills and the Alleghenies, and it was found that the equipment could detect mountains at a distance of 5 miles when flying on a level with the peak. It was necessary to be 2000 ft above the mountain for no indication to appear. With the obstacle detector they found it possible to detect the presence of other aircraft from 3/8 to 1/2 mile ahead. Although the evidence is far from conclusive, Wolff and his colleagues seem to deserve the credit for having built the first pulsed microwave equipment (if not indeed the earliest pulsed detector on any radio frequency) and to have flown what was in effect the first airborne radar equipment in the United States.

Microwave Research at the General Electric Company

In 1936 a paper by Chester W. Rice of the General Electric Company Research Laboratory described some interesting experiments with microwaves.[48] Using BK tubes as detectors and a magnetron of novel design, Rice and his assistants were able to produce and detect 5-cm waves with about 3 W of cw power.

The magnetron is of interest as marking a return to the cylindrical anode type of tube in place of the split-anode construction. It consisted essentially of a long cylindrical copper anode with a short axial cathode. The magnetron acts as a section of resonant line, short-circuited at the output end, and cut to some multiple of a quarter wavelength. It is, therefore, resonant at the applied frequency and supplies its own associated circuit. At the output end, the power goes out over the coaxial transmission line formed by the concentric tube and the filament lead extension. It was a tube of rugged construction, good heat dissipative powers, and a compactly associated resonant circuit. It is of considerable historical interest for having encouraged the later experimenters with magnetrons to adopt some of its general features.

Rice makes a published reference to his attempts to use the microwave equipment for what he calls "radio searchlight problems," that is, the detection of objects by means of reflected radiation. The transmitter and receiver were mounted side by side in a small shack on top of the Research Laboratory in Schenectady, a height of approximately 135 ft. From this vantage point they were able to locate moving automobiles on a neighboring road at distances up to $1\frac{1}{4}$ miles using the beat or Doppler-effect method.

The Navy's Work on Microwaves

Work on microwaves never occupied the center of attention at the Naval Research Laboratory as it did at the Signal Corps Laboratories, but a project of small dimensions was supported for a brief period.

The earliest proposal seems to have come to NRL in the form of a memorandum by a Lieutenant J. N. Wenger forwarded for comment by Captain H. G. Bowen, then Assistant in the Bureau of Engineering, at the end of October 1931.[49] Wenger's memorandum proposed a "radio searchlight" or bearing indicator consisting of a highly directional beam of microwaves and a highly directional receiver. The transmitter and receiver would be training together and the direction indicated by the maximum signal heard in a receiver. The system, it was suggested, could be used for gunnery (determining the bearing of targets in reduced visibility), navigation, scouting, and aviation (where it would help to determine the direction of mountains when flying in fog or at night). In his endorsement,

Bowen suggested that microwaves might be useful for secret recognition systems.

The Wenger proposal was inspected and discussed by the NRL engineers.[50] They pointed out that the originality of the proposal lay not in the use of reflected radio waves, but in the proposal for the use of microwaves which in turn permitted direct determination of azimuth. They recognized that there might be difficulty in rotating the transmitting and receiving reflectors together without direct coupling between them. A report to the Bureau of Engineering at the end of December announced that they felt that the proposal was not practicable at the present stage of development, but that current trends indicated possible future use. They called attention to the similarity of the proposal to that resulting from the Hyland observation, and pointed out that little could be done until higher priority could be given to the work. The possibilities of micro-rays [*sic*] might be better understood if the beat work could be properly carried on.

The proposal of using microwaves was revived in 1933 as a result of the interest and enthusiasm of Lieutenant W. S. Parsons who was stationed as a Bureau of Ordnance representative at NRL between July 14, 1933, and May 31, 1934. During the summer and fall of 1933, Parsons discussed with various scientists and officers at NRL the possibilities of using microwave echo equipment for fire control purposes.[51] Memoranda were drawn up by Parsons for the Director of NRL and transmitted to the Bureau of Ordnance via the Bureau of Engineering in the late summer and early fall of 1933, asking that a microwave group be set up and funds provided. Any effect the memorandum might have had was nullified by the unfavorable endorsement from the Chief of the Bureau of Engineering who emphasized that the progress would depend largely on the results of tube development and might best be accomplished through a development contract with a suitable commercial film.[52]

Parsons returned to the attack with a memorandum of March 20, 1934 to the special Board of Ordnance.[53] This report surveyed the problem quite completely. It pointed out that radio waves about 1 cm should be "practically unaffected by air and its suspended particles," and that wavelengths of about 10 cm would be required to get a sufficiently sharp beam with reflectors of practical size, but acknowledged that there were serious technical difficulties in the way of getting satisfactory ranges with microwaves. Echo work on these frequencies was out of the question. The equipment would give bearing directly because of the narrow beam. The rate of change of target could be obtained by beating stray energy from the transmitter against the reflected radiation. More significantly, Parsons called attention in passing to possible methods of measuring range directly, by a sine-wave modulation of the carrier or by pulse methods, such as he reported were described in the two Submarine Signal patents.[54] Parsons felt that the responsibility rested with the Army and Navy as there appeared to be few commercial incentives for developing in the microwave field.

Despite some support[55] within the Bureau of Ordnance, no authorization for the project was forthcoming from that direction. It was kept alive by interest of another kind. In March 1934, a conference was held at the Naval Research Laboratory, at which the Bureau of Aeronautics was represented by a Lieutenant C. W. Smith, to discuss the problem of an alti-drift meter, i.e., a device to determine absolute height and ground speed.[56] It was agreed at this conference that the Laboratory should investigate the possibility of using microwaves as a solution of this problem, in place of work on a sonic alti-drift meter on which the Laboratory had been working since 1932, and which had been found impractical.

Work was initiated on the microwave project by W. J. Cahill and L. R. Philpott. During the first year they made some preliminary experiments using Westinghouse split-anode magnetrons at 9.2 cm that gave approximately 1/2 W of power. No detection equipment was built. Apparatus was assembled for modulating the 9.2-cm carrier with voice and tone, crystals and vacuum tubes both being used to detect the waves. They observed the quasispecular reflection of the 9.2-cm waves from plane metallic reflectors and verified the theoretical prediction as to the size of beams that could be produced with parabolic reflectors at 9.2 cm. They suggested that the good quality of voice modulation obtained suggested its use for secret point-to-point communication.

It does not appear that the work was continued extensively after 1935. Despite official encouragement from the director of NRL, the project seems to have languished for lack of continued support from the Bureaus. Late in 1937, attention was shifted to 500 MHz. Philpott designed and built an altimeter on this frequency (later called the L band) which, as a result of a suggestion by members of the British Mission in 1940, was transformed into the ASB radar.

European Microwave Developments

The successful IT&T microwave transmissions across the English Channel on 17 cm had caused a great deal of discussion in the United States, and they were frequently cited to give encouragement to work in the microwave field. A growing literature from both France and Germany indicated a widespread activity, on a pioneering level it is true, in microwaves and their properties, including their reflection behavior. In 1935 evidence began to accumulate that work in both these countries was going forward on echo detection using microwaves. A release from the Telefunken Company in Berlin disclosed experiments on radio detection by what was apparently a continuous-wave method using multiple transmitting stations and receivers. The system used 10-cm waves produced by split-anode magnetrons.[57]

At almost the same moment appeared the first notices in the press of a mysterious device for the detection of obstacles which was installed aboard the *Normandie*.[58] This device used microwaves 16 cm in length. These notices were followed in the next few years by more detailed illustrated articles in the popular science and radio journals.[59]

The *Normandie* apparatus was ascribed to M. Ponte, an engineer in the employ of *Société Française Radioeléctrique* (SFR) in Paris.[60] It was a continuous wave detection system operating on 16 cm. Radiation from a positive-grid tube modulated at 7500 Hz was focused into a directional beam by placing the oscillator at the focus of a 30-in. parabolic reflector. The receiving antenna, placed some distance away from the transmitting antenna, was also provided with a parabolic reflector. As shown in Fig. 8-13, the two paraboloids were installed on opposite sides of the ship's bridge. They were linked mechanically, and together they scanned back and forth across an arc about 40° wide, 20° on either side of the ship's heading. The reflected signal was picked up using another BK tube as a detector; the output of the receiver was fed into telephones to the officer's control panel. The transmitter power was low, only a few tenths of a watt, but the directivity of the beam enabled them to detect ships at four nautical miles and channel buoys two miles away. Obstacles could be distin-

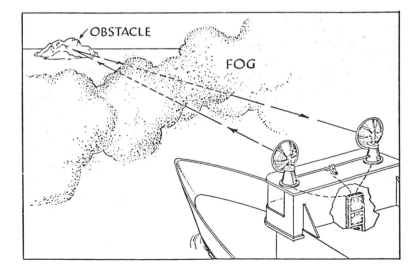

FIG. 8-13. Obstacle detection (from Short Wave and Television, *1937).*

guished in bearing to about 5°. From the patent it is clear that no range-determining methods were available. They proposed to determine range with installations on land by triangulation from transmitter-receiver pairs separated a considerable distance from one another.

A truly important step forward was taken when the SFR engineers substituted a magnetron transmitter and adopted the idea of using pulses. In 1938, Henri Gutton and Sylvain Berline of that laboratory reported success with their split-anode magnetron of 12 anode segments which gave them 10 W and 15% efficiency at 16 cm and with which they made their propagation tests between Auvergne and the Morvan.

An experimental detection station, using the magnetron, was set up by the SFR engineering at Saint-Adresse near the ship channel at Le Havre. From this site the observers were able to follow craft running to Le Havre and Rouen. Superficially, the system closely resembled the *Normandie* installation; fundamentally it was quite different. The magnetron was much more powerful than the BK tube. The paraboloids were larger and gave an angular accuracy of 1° in bearing determination, and greatly improved angular resolution. Targets could be distinguished that were only a few meters apart. More important, the system was really a primitive microwave radar set, for the transmitter gave out pulses 15 μsec long. During the transmission of each pulse the receiver was synchronously blocked to suppress all signals from objects nearer than 200 m. The receiving signal was detected and amplified by a superheterodyne receiver and the output was fed to the plates of a cathode-ray oscilloscope, giving what we have come to refer to as an A-scope presentation. Range could be estimated crudely from the time base to a few hundred meters. This system gave ranges of about 10 km on what the investigators described as vessels of average size (*de dimensions moyennes*).

Detecting obstacles during navigation in fog was not the only use envisaged for equipment. It could equally well serve to detect aircraft, in fog or darkness, and for this purpose it has the great advantage of providing the coordinates with reasonable precision. There was no reason, they felt, why this could not be done with increasing accuracy.

A paper describing the results obtained with this installation was read before the *Société Française des Electriciens* on January 10, 1939, and published soon after in the organ of that society.[62] The authors began with a general discussion of the problems of radio optics and of the advantages of microwaves. Then followed a description of their apparatus. The authors were primarily concerned to call attention to the utility of their invention for collision prevention. They ended their paper, however, with the assurance that detecting obstacles from ships during navigation in fog was not the only possible use for the equipment. It could equally well serve to detect aircraft, in fog or darkness, and for this purpose it had the great advantage of providing the coordinates with reasonable precision. There was no reason, they felt, why this could not be done with increasing accuracy. This appears to be the first published description of microwave pulse radar and in fact, if we except the patent literature, the first published description of pulse radar which all the other countries were by this time jealously guarding as the secret of secrets.

Microwave Detection and Blind Landing at Stanford

The potentialities of the early klystron development at Stanford made it apparent that governmental or industrial support would be required to realize the potentialities of the new tube and to turn it into a usable device. Unaided, this was not possible on a university campus. Efforts to interest the armed services were unavailing, but already in the fall of 1937 interest had been expressed to the Stanford authorities by representatives of the Bureau of Air Commerce and the Sperry Gyroscope Company in helping to support a klystron development because of its potential value in blind landing devices, absolute altimeters, collision prevention apparatus, and aircraft detectors. The Stanford results were disclosed to members of the group at MIT under Professor E. L. Bowles, who were working on blind landing devices in collaboration with Sperry people for the Bureau. Bowles investigated the klystron development at the request of Sperry to see whether it was worth Sperry's while to support the development. On the strength of a very favorable report from Bowles, the Sperry Gyroscope Company, early in 1938, entered into its first contract with Stanford for the support of the klystron program. William T. Cooke was sent West by Sperry to be in charge of the practical engineering aspect of the program.

For a year the work at Stanford remained concentrated in the Physics Department.[63] By the spring of 1939 it was found necessary, in order to accommodate the greatly expanded project, to seek new quarters and to divide the personnel into two groups. A small group, working under Hansen, remained at Stanford where—besides research on the improvement of tubes—experiments were carried out on the airplane detecting device. A larger group, perhaps 30 or 35 in all, was installed in a spacious tool shop that Sperry had acquired at San Carlos, about ten miles from Stanford. At San Carlos, where Cooke was the engineer-in-charge, in addition to work on tube development, the main objective was to design a satisfactory blind landing system based on the use of the klystron. It was here that the Varian brothers spent most of their time. The klystron research continued at Stanford and San Carlos until late in the fall of 1940, when Sperry management decided to transfer it to enlarged quarters in Brooklyn.

Late in 1937, a few months after they had successfully operated the first klystron, Hansen and the Varian brothers began work on an aircraft detection system, using a 10-cm klystron that gave about a watt of cw power and provided a detection range of possibly three or four miles. The earliest antenna reflector was a 16-ft parabolic cylinder mounted on a Sperry 60-in. searchlight frame; this provided a very narrow, high beam, about 15°–20° high and a degree or so wide. The system detected moving objects by the beat method, or, as they preferred to say, by the Doppler frequency of the moving object.

The earliest detection system at Stanford had been a laboratory model which was assembled in the physics building and aimed out of the window at passing cars. The second was a portable system set up in a field about one mile from the physics building. Even when it was used out of doors the detection work was surrounded by no special secrecy or security measures; no guard or barrier protected the equipment in the field. The greatest obstacle proved to be finding aircraft with which to try out the system. There were few planes in the neighborhood, and on the trips they made with their equipment to nearby airports, they were only able to make observations on smaller types of aircraft.

Some detection experiments were also made at San Carlos, even though the system was primarily intended for blind landing. In the spring of 1940, for example, in the course of tests made using a 10-cm klystron and a Barrow 10-cm horn, they were able to pick up automobiles and trains at a distance of a quarter mile, using only a crystal detector with an audio amplifier. As in the case of Hansen's apparatus, the relative motion of these targets appeared as an audio signal of a few hundred cycles per second.[64] In the spring of 1940 distinguished visitors began to turn up at Stanford and San Carlos in increasing numbers to inspect the klystron: Vannevar Bush, General Mauborgne, K. T. Compton, and Alfred Loomis were among the visitors.

Blind Landing and Detection at MIT

For a number of years prior to the establishment of the wartime Radiation Laboratory at MIT, some exploratory work was in progress in the Department of Electrical Engineering at the Institute in the study of microwaves and their applications for aircraft blind landing and aircraft detection. Reference has already been made to the work of Barrow and his students on the basic properties of microwaves. From a scientific standpoint this was the most effective and valuable contribution; but some attention should be paid in passing to the attempts to embody these principles in practical equipment. The program as a whole was under the direction of Professor Edward L. Bowles who was largely responsible for organizing it and securing from a variety of sources the financial support that made it possible.

It is not easy to assign a beginning to this program, for it grew out of earlier research not too closely connected with microwaves. Between 1926 and 1937 Bowles and his segment of the Engineering Department, the Communications Division, were conducting research at a field station located at Round Hill, South Dartmouth, Mass., the estate of Colonel Edward Howland Robinson Green. In the first years attention was centered upon the operation of a shortwave radio communication station, Station XIXV, which made a reputation in amateur circles by keeping in touch with Captain MacMillian's *Bowdoin*, the ship of the Field Museum Expedition that went to Northern Labrador in 1927, and with the Byrd Antarctic Expedition in 1929.[65]

After the first few years the emphasis shifted to attacking various aspects of the problem of aircraft navigation in fog, principally the problem of landing in poor visibility, and various ways—the use of light or heat waves, sound waves, or short radio waves—of conveying intelligence from the ground to the pilot as to his height above ground and his position relative to the landing field.[66] The principal byproduct of this program was a theoretical paper by J. A. Stratton, one of the founders of the Round Hill research program, in which he concluded that "the scattering of radio waves of length greater than about 5 cm is absolutely negligible," but that below this wavelength attenuation would be so great as to preclude the use of infrared radiation through fog.[67] Extensive study on the physical properties of fogs carried out between 1930–1937—which included studies on the transmission of visible light of different frequencies through artificial fog, and careful measurements on the size of fog particles—culminated in a not markedly successful attempt to design a fog dissipator that sprayed saturated calcium chloride solution across the runway of an airfield from a battery of pipes and nozzles.[68]

Colonel Green's support of the research was withdrawn after 1933 as a result of his personal financial difficulties, but the MIT investigators were allowed to continue work on the estate until a year after his death, which occurred in June 1936. From 1934 until the close of the Round Hill field station at the end of 1937, Bowles managed to find other sources of revenue to keep the research going. In particular the fog research was supported in part by three joint contracts with the Navy's Bureau of Aeronautics, the U.S. Army Air Corps, and the Bureau of Air Commerce of the U.S. Department of Commerce. Something of Bowles' talent as a promoter is suggested by the fact that he was able to persuade the Humane Society of the Commonwealth of Massachusetts to contribute a small sum for the fog study in view of its humanitarian and life-saving features.

Late in April 1936, a contingent of the Communication Division, headed by E. L. Bowles, traveled to Washington to hear Barrow read his now historic waveguide paper. Bowles took the opportunity afforded by this trip to discuss with Irving R. Metcalf of the Bureau of Air Commerce the possible continuation of support for the fog research. While they were discussing various schemes for blind landing, Metcalf brought forward an idea of his own, which was later described as the "three-spot system." Metcalf's idea was for a pilot's indicator; it grew out of the fact that it is possible to land an airplane visually on a clear night by means of three light beacons, one marking the center line of the runway and the other two placed at each side of the runway at equal heights above the ground (as shown in Fig. 8-14). The three lights are sufficient to define the plane down which the aircraft must glide, and to indicate the precise glide path down the center of the plane. If the pilot keeps the light equidistant from one another on the same line as he moves in, he will be dropping along the straight line. It was Metcalf's suggestion that the necessary intelligence could be conveyed by radio or infrared beacons substituted for the lights, and that this intelligence could be put on some sort of indicator that would simulate on the instrument face the appearance of the three light beacons. This would be extremely simple for pilots, accustomed as they were to the visual night landings with light beacons.

After rather lengthy discussions and negotiations it was arranged in the summer of 1937 that MIT should undertake this project in cooperation with the Sperry Gyroscope Company. It was to run for a year.

The research was divided into two sections. Development of the indicator went forward under the direction of Professor Draper. The essential feature of the instrument was a cathode-ray tube with a single beam mechanically commutated to produce three separate spots. Into this were fed signals from a Sperry directional gyro and a Sperry turn-and-bank indicator.

The radio phase of the investigation was entrusted to Professor Barrow, who, with his pupils Frank D. Lewis and Donald E. Kerr, were able

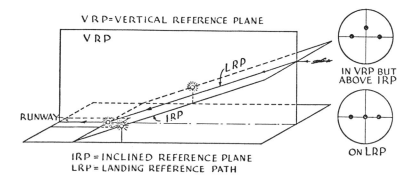

FIG. 8-14. MIT blind landing system.

by the end of the first year to plan the general system and build the principal components. The essence of the scheme was to produce a straight-line glide path by the use of overlapping radio beams from a transmitter located on the ground.[69] As shown in Fig. 8–15, four beams were to establish two overlapping regions, one vertical and the other horizontal, and the landing path was determined by the intersection of these two regions. All four beams were to be on the same frequency, but each was to be modulated with a different audio note. These were to be separated by filters, separately rectified and applied to the cathode-ray tube.

A straight-line glide path was believed to have great advantages from the standpoint of the pilot, for it was assumed that it would be easier for him to control the aircraft when landing at constant air speed and rate of descent than when following the curved path of the earlier landing schemes. The radio components, accordingly, were to consist of a ground transmitter capable of giving directive beams, and an appropriate receiver for the aircraft. The problem of frequency had first to be determined.

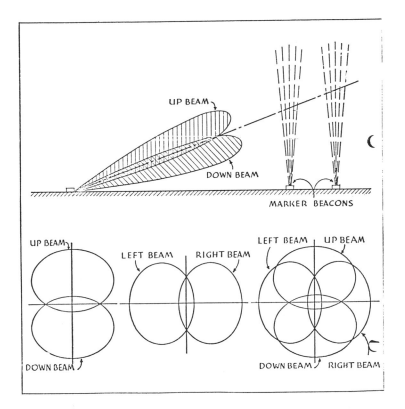

FIG. 8-15. Four-beam glide path.

Since a highly directive beam was desired, it was obviously important to use the smallest possible wavelength. Ten centimeter waves would obviously be best, but because of the state of the art, Barrow decided that a wavelength of 99–100 cm should be selected, for this would permit the use of the Western Electric 316A "doorknob" tubes.

Not all the radio problems were solved in the first year. The accomplishments consisted in exploring the types of receiver and building an experimental superheterodyne for operation at 336 MHz. This included a crystal-controlled local oscillator, a first detector, an IF stage, and a second detector.

As far as the ground part of the system was concerned, Barrow's men concentrated on the antenna problem. No transmitter was actually built and the work consisted in studying antenna patterns, and attempting to define the proper characteristic of the beam required for the vertical indication. The problem of lateral indication was deferred as raising no particular difficulties. A theoretical study indicated that horizontal polarization of the beam should minimize the complications from ground reflections, and so also reduce the variability of the beam as the condition of the earth's surface changed with the weather. When in June 1937 Barrow turned his attention seriously to the blind-landing problem, he and his students had already been making antenna measurements on open-ended hollow pipes and flared electromagnetic horns. These techniques were applied to the new problem. Comparative field-pattern measurements were made using a seven-element broadside array and a simple dipole backed by a cylindrical parabolic reflector. The use of a horn was finally recommended on the basis of these studies.

The contract with the Bureau of Air Commerce expired on June 1, 1938 and the first report was submitted in time to find the Bureau in the throes of formal dissolution.[70] On June 23, President Roosevelt had signed the McCarran–Lea bill abolishing the Bureau of Air Commerce and setting up the new independent agency called the Civil Aeronautics Authority. The new agency opened for business on August 23.[71]

In this winter and early spring of 1939 the activities of the Communications Division were appreciably expanded. By the spring of 1940 Bowles was presiding over a group of some 15 persons working on approximately eight different projects supported from widely different sources. All of these had to do with ultrahigh-frequency problems.

In the spring of 1939 Bowles obtained a grant from the International Telephone and Telegraph Company to sponsor research on dielectrics under the direction of Professor A. von Hippel. This project dealt with a study of the interaction of ultrahigh-frequency radio fields with dielectric materials. The dielectric measurements were made by a hollow-pipe or waveguide method originally suggested by J. A. Stratton and developed by Von Hippel and his assistant Shepard Roberts.[72] The method, in its essentials, consisted of feeding centimeter radiation into a closed section of waveguide ending in a metal boundary, from which the radiation is reflected to form standing waves that can be measured by means of a probe detector moving along a slot. When a section of dielectric material is placed against this closed end of the pipe, the values for the dielectric constant of the material can be determined from the measured value of the field strength at the node and antinode. Associated with this project was the attempt to develop magnetrons as power sources with the hope of generating waves as short as a centimeter in length.

Work on the blind landing apparatus continued until 1941 under the sponsorship of the Civil Aeronautics Authority. Early in 1939, at Bowles' instigation, the radio portion of the system was tried out using a Stanford–Sperry klystron operating on 40 cm as the source of radiation.[73] The tube was flown from the West Coast to Boston in an army plane supplied through the good offices of Wright Field. The klystron and horn unit were tested late in February at the East Boston airport in the presence of observers from Wright Field and from the Office of the Chief Signal Officer. The system showed promise and the Army men flew the beam, with and without MIT men aboard. In the spring the MIT blind landing group cooperated in tests with the klystron-and-horn landing equipment being carried out at Wright Field.

The first flight tests of a complete system (that is, with the cathode-ray indicator plus the radio beam glide path) were made in September 1939. Two of Barrow's horns, fed by separate triode oscillators, were used with the superheterodyne receiver for the first of these tests. A straight-line landing path at an angle of about three degrees from the ground was successfully flown over a distance of somewhat more than 5 miles. Despite obvious deficiencies in the equipment the performance seemed encouraging.

Two investigations were undertaken at MIT by the ultrahigh-frequency group under the sponsorship of the Sperry Gyroscope Company. One of these, the Sperry localizer project, was for the development of a radio beam localizer (a device for locating an aircraft in the horizontal plane) using the same technique of overlapping beams that had been successful in giving an equisignal glide path. Work on this project was carried out during the summer of 1940 under the direction of William M. Hall of the Communications Division. The system used two Barrow horns fed by a single 40-cm receiver that operated a cross-pointer meter. This was later combined with the glide-path apparatus and the three-spot indicator to provide a complete ultrahigh-frequency blind-landing system. Work on this project under the third and last CAA project was not completed until the spring of 1941.

A second—and for this study, a more interesting—Sperry-sponsored project concerned the development of an aircraft detector. The project was undertaken after conferences in the summer of 1939 between Hugh Willis of the Sperry Gyroscope Company and the MIT engineering group. It was to be under the direction of W. L. Barrow and in order to conceal the real purpose of the program it was to be called "The Investigation of Transmitting and Receiving Antennas."

Work on this detection project was started in the fall of 1939. F. D. Lewis, who was a member of the blind-landing team, now temporarily inactive for want of a supporting contract, began with W. W. Mieher and D. S. Pensyl, two of Barrow's students, the development of a crude detection device using the 8-cm split-anode magnetron which had been used for earlier experiments by Barrow and his men.

In December 1939 the future of the program was discussed by Barrow, Willis, and Bowles, and Barrow proposed that the project follow two main lines: (1) the production of very narrow beams; (2) the development of what he described as an "automatic microwave compass." This second objective seems to have been what would now be described as an automatic-following or automatic-pointing apparatus. Barrow emphasized that a great deal hinged on the use of very short radio waves, and that two wavelengths suggested themselves: 43 cm (for which much equipment already was available) and 10 cm. The advantages of 10-cm waves, he pointed out, were so great that the other wavelength should be resorted to only in case of necessity. The magnetron was already available for experimental work at 10 cm, though with very little power, and a sealed-off klystron at that wavelength would soon be forthcoming from Sperry. Barrow also indicated that it might be possible to incorporate a method of range finding. His suggestion, along lines by now familiar to the reader, is as follows:

> Particularly, it appears feasible to use a kind of pulse technique in which the receiver is turned off during the transmission of sharp pulses; the reception of the reflected wave pulses can presumably be made to operate the direction finder and in addition to give an indication of the distance to the reflecting object.[74]

In response to this suggestion Willis expressed the feeling that he would prefer to have this feature left to the San Carlos group for development, for it was understood they were already working on it. Since there is no evidence that pulse methods were being considered at Stanford, this may refer to the various methods being considered by Hansen to obtain ranging data with the cw interference technique.

Lewis, Mieher, and Pensyl working together assembled a crude microwave detection device—at first using the 8.3-cm RCA magnetron as transmitter, later a Sperry 10-cm klystron. The system produced only low-power cw energy and gave them azimuth data only. The maximum range at which a target could be detected by this device was about 300–400 ft. The apparatus consisted of three horns (one transmitting horn and two receiving horns) set up on a movable platform mounted on a Sperry searchlight base. The horns were oriented with their axes about 15° apart and their beams overlapping symmetrically about the pointing axis of the apparatus. The output of each receiving horn was modulated by mechanical means (a paddle rotating in the waveguide). The received signals from the two horns were compared electrically; the driving gear automatically turned the horn carriage until the signals were equalized, that is, until the carriage was pointing directly at the object from which the signal was reflected. With this device the investigators amused themselves by tracking people as they walked across the MIT quadrangle. The project was soon abandoned by the original group who had been working on it when in April 1940, Lewis joined a closely related project at the Loomis Laboratories in Tuxedo Park. Work was continued during the summer of 1940 by a group of four McGill scientists who visited MIT to work on ultrahigh-frequency techniques with Barrow in preparation for war research. They completed the greater part of a duplicate horn array apparatus of the sort we have just described.

Detection Research at the Loomis Laboratories

Early in 1939 still another sponsor was found for the ultrahigh-frequency work at MIT. Alfred L. Loomis, New York lawyer and a trustee of the Institute who for many years had been a friend and scientific associate of many of America's leading scientists, and who had turned a portion of his Tuxedo Park estate into a summer laboratory where he carried out cooperative experiments in applied physics and experimental physiology, became interested in the possibilities of ultrahigh-frequency radio research after conversations with K. T. Compton and E. L. Bowles, who assured him that it was a field of great promise. He agreed to sponsor a program of research, the main feature of which would be a study of UHF propagation to be conducted under J. A. Stratton. This program got under way in the fall of 1939. Donald E. Kerr, who had been an assistant in the blind-landing research, was transferred to work with Stratton "in the immediate formulation of the experimental program, and the exploratory steps to determine the practical range that we can expect to obtain with 50 cm waves which we now have facilities to generate."[75] The next step would be to set up a test link working out of the Tuxedo Laboratory to check the methods. It would then be advisable to apply these techniques to shorter and shorter wavelengths, and to study correlations between meteorological conditions and transmission factors in the centimeter range.

The program did not progress with great speed either at the Institute or at Tuxedo Park. After some delays, the Loomis Laboratories received in November their first 40-cm klystron. At MIT Stratton was conducting a theoretical investigation while Donald Kerr had klystrons in operation and was building a superheterodyne receiver and a klystron-horn transmitter.[76]

Loomis spent the balance of the winter in California working with the groups at Stanford and San Carlos learning all he could of their techniques. Hansen and the Varians demonstrated everything they had to show, and Loomis was especially impressed and interested by the detection work. Upon returning East, Loomis determined to devote his laboratory's energies, not to a long-range nebulous program of scientific measurement, but to a problem that might have valuable implications for the national defense: the detection of aircraft by ultrahigh-frequency waves. He brought back with him some sealed-off 10-centimeter klystrons which the San Carlos people had developed the previous summer and he also induced Hansen to come East to help him get started. At MIT, Loomis made arrangements to have several of the ultrahigh-frequency workers loaned to the Loomis Laboratories for the summer to work on radio detection, and in April 1940 a group that included Hansen, Donald Kerr, and Frank Lewis drove to Tuxedo Park from Cambridge with Alfred Loomis.

By early May some progress had been made in organizing the Loomis Laboratories for microwave work. Although by now the original group at Tuxedo had nearly a year's experience in the ultrahigh-frequency field, a great deal of equipment was still needed. Loomis supported the program generously. The total staff assembled at the Loomis Laboratories for the microwave work consisted of the Director, Alfred Loomis; Charles Butt, who had been at Tuxedo every summer in recent years as research assistant of Professor E. Newton Harvey of Princeton; Philip Miller, who for some time had been manager, machinist, tool maker, and head technician in the Loomis Laboratories; Garret A. Hobart III, a resident of Tuxedo with independent means who since 1935 had been a regular member of the Loomis Laboratories. From MIT were added William H. Ratliff, William Tuller, and Frank Lewis, who became salaried employees of the Loomis Laboratories, and Donald Kerr, who was working on propagation problems with Stratton under the original Loomis grant.

Work was started early in May on assembling an aircraft detector, for it was Loomis' desire to skip all preliminary exploration and to build as soon as possible a locator for actual tests. He felt that they would accumulate useful data much faster if a system were in operation; and in the meantime improved or additional equipment could be incorporated as the work progressed. By the middle of June the system was being assembled and installed in a vehicle of the sound-truck variety bought expressly for the purpose. A second system was set up within the Laboratory and tested with an experimental moving target (Doppler target) that consisted of a row of copper wires projecting from a moving belt.

The apparatus used an 8.6-cm klystron transmitter feeding energy to a novel antenna system.[77] This radiator, called a Tuxedo horn, was a modification of the leaky-pipe linear array proposed by Hansen. The Tuxedo variant consisted of two leaky pipes along two sides of a triangular horn.

The first such radiator produced a fan beam approximately 4° wide and 30° high. Considerable effort was expended to make the radiator mechanically solid and to improve the shape of the beam. Some experiments were also made using a parabolic cylinder as a reflector but this approach was abandoned. The Tuxedo horns were mounted together on a special light platform, resembling a gun mount, projecting from the rear of the truck. This platform permitted the horn array to be rotated a number of degrees in azimuth and in elevation. The transmitting horn could also be rotated about its own axis of symmetry. The receiver consisted of a crystal detector followed by two rf stages; the signal was then fed through other rf and audio stages to ear phones and to the plates of a cathode-ray oscilloscope. The receiving antenna was similar to the transmitting horn and was mounted directly above or below it. A "wave trap" was placed along the front edge of the two horns and between them to prevent the transmitted signal from leaking directly back into the receiver.

The operation of this cw system—like that of the Stanford and General Electric detectors, and like the "long-wave" beat system of the Navy— depended upon interference effects between radiation directly picked up from the transmitter and reflected energy from a moving target. The signal reflected from the moving object is made to beat against the outgoing signal, part of which is fed back directly into the receiving horn. The motion of the object produces an audio frequency resulting from the Doppler effect, and this is received and amplified. The frequency of this note depends upon the radial speed of the moving object.

During July and August the workers at Tuxedo experimented with various targets. The first real outdoor test object was an electric motorboat on the lake, marked by a 90° corner reflector. Balloons were also sent up and used as targets. On other occasions they drove with the truck to points from which they could follow automobiles moving along the highway. They were able to measure the speed of cars with considerable accuracy, the Doppler shift being of the order of 10 Hz per mile an hour. The success with which the observers detected violations of the New York

State speed limit of 50 miles per hour, led one of them to remark "Don't let the cops hear about this."

In late August, the workers drove their truck to the Bendix Airport; and there they followed without difficulty the comings and goings of a number of small planes, Piper Cubs and similar craft. A Luscombe, with its large amount of metal surface, proved to be the best target. Planes were detected at a maximum distance of two miles. Except for the extent of metallic surface, the sensitivity seemed independent of the type of plane. Beyond the distance of two miles, the fourth-power law began to defeat their efforts and the signals were lost in the noise. They made several efforts to follow the Goodyear blimp, but soon discovered that it moved too slowly to give a signal.

Late in the summer a rather unsatisfactory distance-measurement unit was introduced into the truck. A circuit which was afterwards described as being of "extraordinary complexity" was used to detect the phase shift of a signal modulated by an audio frequency. This device was never very successful; and some efforts to use it to make measurements on the movements of the Nyack ferry and other targets were not successful.

The formation of the National Defense Research Committee just at this time gave these efforts a significance that was heightened by Loomis's appointment as chairman of a subcommittee charged with the development of microwave detection devices. Members of the Microwave Committee (Section D-1) and of NDRC itself were among the visitors who came to Tuxedo Park during the summer and early fall to see what was being accomplished. The program was still active when the future prospects of microwave detection development were dramatically altered by the disclosures made at the Loomis Laboratories by the representatives of the British Technical Mission to the members of the Microwave Committee.

The Development of the Cavity Magnetron

The successful development of a source of microwave radiation, giving powers hitherto unrivalled in the centimeter band, was a wartime achievement of British science. The particular stimulus behind the accomplishment was the need to remedy the most serious defect of British AI radar, the lack of directivity of the beam. When the broad, forward-pointing beam of 200 MHz equipment, like AI Mark III and AI Mark IV, intercepted the ground, targets beyond the range of this ground clutter could not be picked up. The operator was powerless to detect enemy aircraft at distances greater than his own height above ground.

This defect of their AI equipment led the British to consider, even before the outbreak of war, the possibility of going to frequencies that would permit them to have really narrow pencil beams. The airborne radar specialists repeatedly urged that a practical form of AI could result only if one used a wavelength of less than 30 cm.[78] This opinion was shared by physicists like J. D. Cockcroft of the Cavendish Laboratory at Cambridge and M. L. Oliphant of the University of Birmingham who were brought into the RDF (radar) program in the summer of 1939. Professor Oliphant strongly recommended that a program be at once undertaken to develop an oscillator that would yield peak powers in the neighborhood of 1 kW at a wavelength of 10 cm. In September 1939, soon after the outbreak of war, the physics laboratories of the University of Birmingham were put under contract by the Department of Scientific Research and Experiment, Admiralty, to undertake a development program leading to the production and detection of radio waves of a frequency 3000 MHz or greater.[79] Somewhat later the British Thomson-Houston Company, Ltd., was sounded out by the chairman of the Committee on the Scientific Survey of Air Warfare to see whether they would be interested in joining the search for a high-power transmitter on microwave frequencies, on the grounds that there could not be "too many people with bright ideas working on the problem."[80] At the same time an attempt was made—not too successfully, as it turned out—to discover the status of the microwave art in America by inquiries through commercial channels. It is highly probable that A. V. Hill was instructed to learn what he could about American microwave developments on the mission he shortly undertook to the United States and Canada.

These other ventures were not needed. By the new year the group at Birmingham had hit upon the solution which was to revolutionize the radar field and exert an incalculable effect upon the course of subsequent events.

According to Oliphant's own account, the investigation began with an examination of the very recent literature and a return to first principles in an attempt to discover what was the chief limitation of existing vacuum tubes.[81] It soon became apparent that the resonant circuit associated with the oscillator was the key to the whole problem, and that the only hope on these extremely high frequencies was to combine generator and circuit in a single unit. It was equally important that this circuit be made of the best possible electrical and thermal conductors. The only tube already in existence which appeared to satisfy these requirements was the klystron, which used high-Q cavity resonators made of copper and therefore possessed high electrical and thermal conductivity. This reduced electrical losses to nearly the minimum value attainable and ensured that energy converted into heat would be rapidly conducted away.

For the first few months the immediate objective of the Birmingham group was to develop an improved klystron oscillator. Before the end of 1939 Oliphant and Sayers had built a tube that could produce about 400

W of cw power at 10 cm. Out of this development during the winter of 1939–1940 came a "double-pulsed" klystron which was delivered to the radar workers at Swanage about the middle of May 1940. This tube was designed to be pulsed in two modes, and could serve alternatively as a transmitting tube and as a local oscillator in a radar set.

Nevertheless, it was quite doubtful that the klystron could ever be satisfactory as the power tube for a radar set. The use of an electron beam necessarily imposed a limitation upon the electron current and consequently upon the power that could be drawn from the tube. The only other tube that offered much promise on microwave frequencies was the magnetron. This tube promised large anode currents not limited by an electron beam; but what Oliphant described as "a very superficial" examination of the existing magnetron schemes showed that the circuits were made of highly resistant materials and in a form where radiative and resistive damping reduced the efficiency. It occurred to the Birmingham group that the solution was to combine the best features of both tubes, and to improve the magnetron by using rhumbatron-like cavities associated with it.[82] The working out of this general idea was brilliantly carried through by two Birmingham physicists, J. T. Randall and H. A. H. Boot.

Randall and Boot had been working at Oliphant's request on Barkhausen–Kurz tubes as detectors for microwave receivers. The two men now turned their attention to this new problem—at first only with the modest objective of building an oscillator with which to test their BK detectors—and in a single afternoon's discussion in November 1939 they blocked out the main features of the resonant cavity magnetron in the form that later became familiar. The type of resonator, the number of resonators and their size, and the nature of output circuit were all decided upon in this initial discussion. Neither Randall nor Boot had any previous experience in this field; and in their first approach to the problem they started providentially from the fundamental physics of the situation, rather than from a preoccupation with the vast and confused magnetron literature.

In deciding upon the shape of the cavity system, the important point of departure was the realization that the familiar type of rhumbatron resonators would not be suitable because cylindrical symmetry in the tube appeared essential. One of the most important decisions was made when with great insight Randall and Boot decided to use the theoretically less efficient cylindrical form of resonator, linked by slots to the central anode cavity. The genesis of the idea seems to have been a consideration of the Hertzian wire loop—fresh in their minds from a recent perusal of the volume of Hertz's collected papers—which constitutes the simplest dc circuit. The solid hole-and-slot structure of the cavity magnetron from this point of view is a three-dimensional extension of the Hertzian loop, as indicated in Fig. 8–16. The most general considerations indicated that at the critical magnetic field strength the electrons would spiral out from the cathode passing the slots at grazing incidence and covering the distance between two successive anode slots in a time equal to half the period of oscillation. To obtain proper performance it would be necessary to decrease this angular transit time between each pair of resonators. To accomplish this it was decided to group a number of cavities around the central anode cavity, six being decided upon. The simplicity of the design would make it easy to construct the whole anode-resonator system from a single machined copper block; and in such a structure heat would be readily conducted away along the webs or segments between neighboring resonators. Since it was not at all clear how much the performance would be affected by small departures from perfect symmetry, to be on the safe side the anode hole was given the diameter of the surrounding resonators.

The wavelength would clearly depend chiefly upon resonator size, but no calculations were available in the literature to give a ready answer as to

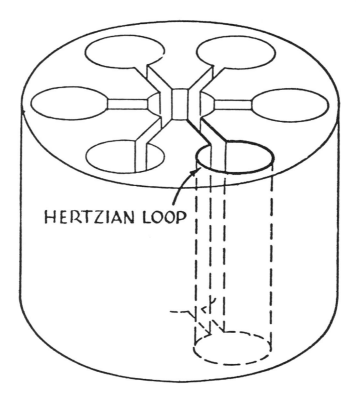

FIG. 8-16. Magnetron as extension of Hertzian loop.

what this should be for 10 cm. An approximation was provided by returning again to a consideration of the Hertzian loop. Hertz's experiments, confirmed theoretically by MacDonald,[83] had shown that the wavelength to be expected from a simple loop of thin wire is 7.94 times the diameter of the loop. It was assumed—and this was largely justified by later experience—that the wavelength to be expected from a single slotted copper cylinder would be independent of the cylinder height and that the unpredictable coupling effect between resonators could be neglected. Applying this relation for a wavelength of 10 cm, Randall and Boot designed the cylindrical resonators of their first tube to be 1.2 cm in diameter, 4 cm deep, with a slot 0.1×0.1 cm. They applied their estimate to the angular transit time in order to determine roughly the operating conditions.

The output circuit used to draw power from this earliest magnetron remained standard for some time to come, until in fact waveguide output circuits were adopted for very short wavelengths and for very high power. This output consisted of a coaxial line, the inner conductor of which terminated in a loop in the central plane of one of the cavities. The magnetic flux parallel to the axis of the cavity passed through the area embraced by the loop, which was chosen to be about half the cross-sectional area of the resonator. The section of transmission line was part of the vacuum system, and was sealed off by a metal glass seal. The inner conductor and the loop were of copper.

The organization of the Birmingham project was only in its early stages as this work was getting under way. There was considerable scarcity of equipment and technical assistance, so that some of the essential apparatus was made by hand; just at the critical stage of this early development matters were thrown into confusion by the necessity of moving to new quarters. Since they had no modulator to pulse the tube, and wished to save time, it was decided to operate it from a dc power supply. A tungsten cathode 0.75 mm in diameter was used as most suitable for cw operation.

The anode block was received from the shop in December. During the next two and a half months it was set up for operation as a demountable, water-cooled tube run on the vacuum pump. The operating conditions were 4-kW input power, 8-kV operating voltage, and a magnetic intensity of 1100 Oe.

The cavity magnetron was operated for the first time on February 21, 1940 by Randall and Boot. It made quite an impressive spectacle with sizzling corona from the output leads. Oliphant, the Director of the Laboratories, was brought in within a half-hour to view the sight. The first impression was that there was too much power being produced for the radiation possibly to be on 10 cm. Randall and Boot spent the first half-day convincing themselves and Oliphant that it was not radiation in the neighborhood of 5 m that was coming from the tube. Within the next day or so they succeeded in determining that the tube was radiating about 400 W at 9.8 cm. The wavelength was measured by means of Lecher wires and the power output was estimated from the light emitted by tungsten-filament lamps and tubular neon discharge lamps. The next few weeks were spent in attempts to repeat the first success in the face of a series of difficulties with the vacuum system.

Despite the initial success there was considerable skepticism at Birmingham and elsewhere at the possibility of getting satisfactory results with a magnetron. It was widely expected that there would be the serious frequency instability characteristic of most other types of magnetrons, and many people failed to appreciate the extent to which the frequency was controlled by the cavities. It took some time for this prejudice to evaporate, for the true merits of the new development to be appreciated, and for the klystron to be relegated to second place. Experiments made in July 1940 showed that the variation in frequency of the cavity magnetron with anode voltage and magnetic field was small and difficult to measure.

The success of the Randall and Boot experiments was kept a closely guarded secret and news of it was not widely disseminated among radar people—for example, only Rowe and Lewis at TRE seem to have been informed of the results. The first visitors to arrive at Birmingham were Admiralty representatives, two radar men from Admiralty Signals Establishment and on another occasion C. S. Wright, Director of Scientific Research for the Admiralty. In April 1940 contact was officially established between the Birmingham group and the Research Laboratories of the General Electric Co., Ltd., at Wembley, the latter being asked to help with the design of a sealed-off, manufacturable version of the tube.[84] E. C. S. Megaw of GEC Laboratories visited the Birmingham laboratory soon after.

Until June of 1940, despite the success of the magnetron development, Randall and Boot were the only men at Birmingham engaged in this work. In June, S. M. Duke, who had been with GEC for many years, joined the laboratory at Birmingham and helped with the task of engineering the tube. Some indication of the shortage of scientific manpower in Great Britain is indicated by the bald fact that, despite the importance of the development, it was not until the late winter of 1940–41 that a third man, W. T. Cowhig, was added to the group.

Besides the completion of a sealed-off tube to which both GEC and Birmingham turned their attention in the late spring, the principal steps that followed naturally upon the preliminary experiments were (1) the operation with a pulsed power supply; (2) the introduction of large cathodes to provide the high anode current necessary for high-power operation; (3) the investigation of improved methods of power measurement; and (4) the investigation of other anode designs to establish the generality of the principle being employed for other wavelengths and for different numbers of associated resonators.

Two successful sealed-off tubes were produced by June 1940. Although it later proved to be possible to assemble tubes without it (as for example by brazing), the key to this swift success was the adoption of a vacuum-tight gold washer technique which had been developed at GEC some years before by Le Rossignol and Boyland.[85] The large glass envelopes and fragile seals of the experimental on-the-pump models were discarded and a simple practical construction was introduced that could be easily manufactured.[86] The anode block was made of tellurium copper into which the central anode hole and resonator holes were drilled. Side stems from the filament leads and output were turned from copper bar and silver-soldered to the anode. The end plates which finally seal off the tube were turned from arsenic-free high-conductivity copper sheet to minimize the possibility of leaks. These were sealed in place by the gold washer technique. Gold wire is inserted between the two copper surfaces and subjected to high pressure and baked at ordinary degassing temperatures. After clamping on the end plates over the gold washers, the tube is sealed to a mercury diffusion pump, pumped out, and baked at 500 °C for an hour and a half. Using oxide cathodes of the optimal size, these sealed-off tubes, designated as E.1189, gave a peak power output of 5–10 kW.

Two steps of great importance in developing the full potentialities of the magnetron—steps, incidentally, which the Russian investigators who had earlier hit upon a similar design had failed to take—were to pulse the tube at unheard of voltages and to insert heavy oxide cathodes of high emissivity. This suggestion about the large cathodes came to the Birmingham group in the form of a paper that had appeared from SFR in Paris on the superior merit of large, oxide-coated cathodes. The Birmingham group decided to try it. It was found, after experiments made late in the spring of 1940 with various cathode sizes,[87] that the power output was raised by increasing the cathode diameter to an optimum value of 6 mm. The use of the large oxide cathodes in magnetrons at 10 cm enabled peak powers of 10–50 kW to be obtained at efficiencies of 10%–20% within a few months of the initial experiments.

The other problems were explored. Modulators were built and the tubes were studied under pulse operation. The early measurements of rf power were succeeded by attempts to use bolometer-type instruments, but these also gave rise to various difficulties. The most successful of the early British instruments for this purpose was a resonant-cavity water-filled calorimeter designed by Sayers. The question had presented itself: How general was this phenomenon? Would it apply equally well to tubes designed for different wavelengths and with different numbers of resonators? By September 1940 (1) an 8-hole, 5-cm design; (2) a 14-hole, 5-cm design; (3) a 6-hole, 3-cm design; and (4) a 30-slot, 2-cm design had all been completed and satisfactorily operated, which showed that the principles embodied in the resonant cavity magnetron were of very general application.

Microwave AI at AMRE[88]

While the search for a source of 10-cm radiation was going forward at Birmingham, some thought was already being given at AMRE to the development of radar on these higher frequencies. Late in December 1939, A. P. Rowe, the Superintendent of the laboratory, called in D. M. Robinson, an engineer who had recently arrived at Dundee, and told him that plans for developing airborne radar on microwave frequencies were shaping up and urged him to discuss the problem with E. G. Bowen, who was in charge of the AI developmental work at St. Athan. On January 1, 1940 Robinson met Bowen at the General Electric Company, Wembley, which provided a convenient halfway point between Dundee and St. Athan; and after a discussion, in which Bowen outlined to Robinson the proposed characteristics for such equipment, the two men proceeded to St. Athan.

In February, Rowe and W. B. Lewis, who had evidently just been informed of the success at Birmingham, took up the subject once more with Robinson and appointed him head of an as yet nonexistent centimeter group. Robinson was not told about the magnetron and did not take this decision too seriously. His attention was still focused on the development of a 50-cm radio lighthouse, and a solution of the 10-cm problem seemed somewhat remote. It seemed more sensible to get experience with 50 cm before attempting 10 cm; to an engineer the thought of a 5 to 1 jump in wavelength seemed somewhat staggering.

Little was done in the microwave field until spring, though some preparatory work was accomplished before the first magnetron was received, and a number of high-grade men were brought to AMRE and assigned to Robinson's group before the move to Swanage. These included A. G. Ward, a Canadian who came from the Cavendish Laboratory, H. W. B. Skinner, who had been working on Chain Home Stations as one of the group physicists under Cockcroft, and who arrived in Dundee in April, and Atkinson Robinson and his group, who did not hear about the magnetron until late in the spring, though the knowledge that Oliphant was building a klystron that promised to yield some power encouraged them to consider the best kind of antenna for 10 cm and the best way of mounting it in aircraft. Robinson had read the papers of Barrow and his students, and had been impressed by the sharp beams, apparently without sidelobes, that were shown in the accompanying illustrations. He urged that horns be used to produce the narrow beams, but this was overruled by

Lewis and Skinner who felt that paraboloids would be easier to put in the nose of a plane and that their operation could be readily predicted from optical theory. Thus it was decided while still at Dundee to use a 30-in. paraboloid as the reflector. Since there was little reason to doubt that 10-cm radiation would pass through a plastic layer, one of the first tasks was to make arrangements for a transparent nose for the aircraft.

After the move to Swanage, which took place in May, Robinson's group received a sudden access of strength. A team of nuclear physicists from Cavendish who had been working together on anti-aircraft rockets at Exeter left the rocket development, now well along, and came to Swanage to join the microwave program. This group included a number of Rutherford's pupils, men like P. I. Dee, who later became a superintendent at TRE, W. E. Burcham, Dee's right-hand man, and S. C. Curran. Shortly thereafter, Oliphant arrived to explain his plans for the 10-cm work, and inspected the hut where his men had been installed.

The new arrivals paid little attention to administrative formalities and set to work in their own inimitable fashion. They were picturesquely un-disciplined. They worked at off hours and far into the night, filling the hut with tobacco smoke and conversation, struggling without much plan or organization, to lay their hands on the only piece of equipment, the one transmitter tube or operating system, that existed for some months.

This group's first step was to set up the Oliphant and Sayers 10-cm klystron. For a time this tube, which was operated on the pump, was the only source of microwaves. With it they got some pulsed 7.3-cm radiation, without much power, but enough to light a neon bulb. It was June before the strange looking object, the resonant cavity magnetron, arrived from Birmingham. Even then there was a delay in putting it in operation, for there was no magnet available that would give enough flux, and no modulator to pulse the tube. Dee and his colleagues solved the first problem by bringing down an enormous electromagnet from the Cavendish Laboratory; while the second was temporarily solved when Atkinson adapted the big metal–glass seal tetrodes which GEC was developing for the Radio Lighthouse work and used them in a pulser.

E. G. Bowen, who had been the earliest to work out the proper characteristics for the centimeter AI equipment, arrived at Swanage from St. Athan in June 1940 having been relieved of his position as head of the longer-wave AI development. Bowen lectured informally on AI problems to the microwave group, talking freely at the blackboard without notes and presenting his material skillfully. It was from these discussions that the main characteristics of British centimeter radar, and, to a large extent, those of the first American microwave radar, were elaborated.

During the summer and early fall a great effort was made to develop proper components for a set on 10 cm. Triodes and BK oscillators were tried as detectors in addition to the crystals on which the laboratory had been working, following Hollmann's work, since December. These were tried with several different sorts of local oscillators, a Clavier tube and a GEC split-anode magnetron (E1210). Coaxial line was used throughout. A two-parabola set was built on a trailer, and some fixed echoes were observed. In the summer echoes were received from a man on a bicycle and shortly after a Blenheim aircraft was successfully tracked.

In search of a good local oscillator, Lewis and Robinson paid a visit to R. W. Sutton at ASE in Portsmouth, who was known to be a very able designer of vacuum tubes. They aroused his interest in the AI problem and persuaded him to make them a tube. Sutton brought his solution to the problem, his reflex klystron, to the laboratory one day in July. It worked admirably, gave more power than any previous tube they had used, was noise-free, and had a high Q.

During August and September, at Oliphant's instigation, Skinner was attempting to develop satisfactory crystals. He succeeded in producing the first sensitive crystal for 10-cm work, and discovered the importance of "tapping down" the crystal, which meant tapping it until the whisker found the sensitive spot that gave the best crystal characteristic. The earliest crystal was enclosed in a glass envelope filled with wax. Further experiments were carried out at Birmingham where Oliphant succeeded in designing the now well-known capsule type of mounting. British Thomson-Houston was brought in by Oliphant and by the spring of 1941 were manufacturing satisfactory crystals following Oliphant's design.

In the autumn, the ground equipment was moved to a better site some distance from Worth Matravers in the nearby school of Lesson House. It was on a hill about 250 ft above sea level with a clear view over Swanage toward the Isle of Wight and opportunity to study the coverage over land and water. Although AI was still the main objective, and there was unremitting pressure for that equipment, they discovered how well microwave equipment performed over water and how well it picked up ships. In November, to test this performance, the Admiralty produced a submarine which was observed from this site and followed to a maximum range of about 6 miles. A party of Admiralty representatives arrived at Swanage in December, studied the set, made tests on shipping, and built a copy of the Swanage equipment which became the prototype of the Admiralty 271 radar, a two-reflector set which saw considerable service.

The microwave work at Swanage was soon divided into two parts, with P. I. Dee in charge of the development of an airborne system, and with Skinner responsible for components development. It was decided to equip the AI with a spiral scan. Early in 1941 the problem of a TR unit to permit the same antenna to be used for transmitting and receiving, was under-ken at Swanage as a necessary preliminary to putting a system on an aircraft.

By the end of the year a number of successful tests had been made with the set operating on the ground. Experimental transmitter and receiver components had been built both at TRE and by GEC and comparative ground tests were carried out in conjunction with GEC.[89] In these tests a Blenheim aircraft was consistently followed to a distance of 10 miles. As a result of these tests a preliminary agreement was reached between TRE and GEC as to which units should be engineered for experiments in the air.[90]

This was not satisfactorily solved for some months. The first attempt was to have separate transmitting and receiving quarter-wave antennas in the same parabola, separated by a circular disk to which they were grounded.[91] A 5-in. disk gave about 25-dB protection to the crystal and there was some hope that the problem could be solved by this method. The effect on the antenna pattern was such as to throw this device into the discard. After experiments during the winter with a modified form of disks,[92] with the use of spark gaps in concentric lines,[93] and with a so-called hybrid system developed by GEC,[94] it was finally suggested by Griffiths of the Clarendon Laboratories at Oxford that one could use a Sutton reflex klystron (ignoring the electron gun and the reflector electrode) filled with gas at a suitable pressure. This is shown in Fig. 8–17.

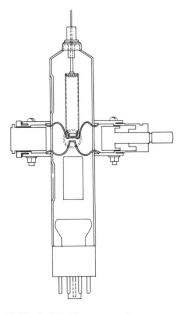

FIG. 8-17. Sutton tube.

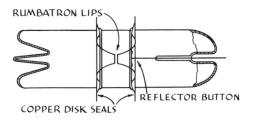

FIG. 8-18. Soft Sutton tube.

Two such tubes were tested in the late winter or spring of 1941, one filled with helium at a few mm pressure, and the other with hydrogen at 9 mm.[95] Both seemed to give good protection and work was carried on actively at both Oxford and TRE on this system which was clearly the most promising. This "Soft Sutton" became standard for 10 cm radar systems. As a result of this development, the British had a single-parabola radar set installed in a Blenheim by the late spring of 1941. (See Fig. 8–18.)

NOTES

1. H. Barkhausen and K. Kurz, Phys. Z. **21**, 1–6 (1920). The paper was submitted in August 1919 as a contribution from the "Laboratorium der Torpedo-Inspection," and may have been the result of wartime research.

2. E. W. Gill and J. H. Morell, Short electric waves obtained by valves, Philos. Mag. **44**, 161–178 (1922). E. C. S. Megaw, "Electronic oscillations," J. Inst. Electr. Eng., Vol. **72**, 313–325, 348–352 (1933). This paper is an excellent review with an exhaustive bibliography of the papers published up to that time. See also the bibliographies of A. F. Harvey, *High Frequency Thermionic Tubes* (Wiley, New York, 1943), and R. I. Saarbacher and W. A. Edson, *Hyper and Ultrahigh Frequency Engineering* (Wiley, New York, 1943). H. E. Hollmann, "On the mechanism of electron oscillations in a triode," Proc. IRE Vol. **17**, 229–251 (1929) [this paper summarizes material that appeared the year before in H. E. Hollman, Ann. Phys. Vol. **86**, 129 (1928)]. See, for example, Hollmann, pp. 241–243, and Sarbacher and Edson, pp. 529–536.

3. M. J. Kelly and A. L. Samuel, "Vacuum tubes as high frequency oscillators," Electr. Eng. **53**, 1504–1517 (1934); C. E. Fay and A. L. Samuel, "Vacuum tubes for generating frequencies above one hundred megacycles," Proc. IRE **23**, 199 (1935).

4. A. Clavier, "Production and utilization of micro-rays," Electr. Commun. **12** (1933), and Onde Electr. **13**, 101 (1934). See also Rene Darbord, Onde Electr. **11**, 53 (1932).

5. The patent applications were filed early in 1920: U.S. Patents No. 1,523,776 (filed 29 January 1920) and No. 1,523,777 (filed 21 May 1920). The first published account was a paper based on Hull's talk, which appeared under the title, "The Magnetron" in J. AIEE **40**, 715 (1921). The first technical treatment is Albert W. Hull, "The effect of a uniform magnetic field on the motion of electrons between coaxial cylinders," Phys. Rev. **18**, 31 (1921).

6. Albert W. Hull, "The dynatron, a vacuum tube possessing negative electric resistance," Proc. IRE **6**, 23 (1918).

7. F. R. Elder, Proc. IRE **13**, 159 (1925).

8. Erich Habann, Z. Hochfrequenztech. **24**, 115, 135 (1924). Negative resistance in an electrical device means that as the voltage across the device increases, the current through it decreases. Any device with a negative resistance characteristic can be used as an oscillator.

9. E. Manns, German Patent No. 450,989 (September 1927).

10. August Žáček, "Über eine Methode zur Erzeugung von sehr kurzen elektromagnetischen Wellen," Z. Hochfrequenztech. **32**, 172 (1928). This is a brief summary in German of Žáček's 1924 paper which appeared in the Czech journal Casopis Peatovani Math. Fys. **53**, 378 (1924), with a French abstract. These results had been cited by J. Sahanek, Phys. Z. **26**, 375 (1935). Žáčak's work is summarized in Czech Patent No. 20,293, applied for on 31 May 1924 and granted 15 February 1925.

11. K. Okabe, "On the applications of various electronic phenomena and the thermionic vacuum tubes of new types," J. Inst. of Electr. Eng. Jpn. No. 463, Suppl. Issue, 13–14 (1927).

12. K. Okabe, "Undamped extra-short electromagnetic waves obtained with magnetron," J. Inst. Electr. Eng. Jpn. No. 467, Suppl. Issue, 61 (1927), and No. 469, Suppl. Issue, 99 (1927). A more complete description of these early experiments, complete with photographs of the apparatus, is given by H. Yagi, "Beam transmission of ultra-short waves," Proc. IRE **16**, 715–741 (1928); it was this paper that inspired Žáček to publish the summary of his own work (see Note 10).

13. K. Okabe, "Production of intense extra-short electromagnetic waves with split-anode magnetron," J. Inst. Electr. Eng. Jpn. No. 476, Suppl. Issue, 33–34 (1928).

14. Yagi, *op. cit.* Note 12, p. 739.

15. G. R. Kilgore, "Magnetostatic oscillators for generation of ultra-short waves," Proc. IRE **20**, 1741–1751 (1932), and "The magnetron as a high frequency generator," J. Appl. Phys. **8**, 672 (1937).

16. C. E. Cleeton and N. H. Williams, "Electromagnetic waves of 1.1 cm and the absorption spectrum of ammonia," Phys. Rev. **45**, 234 (1934), and "The shortest continuous radio waves," Phys. Rev. **50**, 1091 (1936); C. E. Cleeton, "Study of the magnetostatic oscillator and grating theory," Physics **6**, 207 (1935); Neil H. Williams, "Production and absorption of electromagnetic waves from 3 cm to 6 mm in length," J. Appl. Phys. **8**, 655–659 (1937).

17. E. G. Linder, "Improved magnetron oscillator for the generation of microwaves," Phys. Rev. **45**, 656 (1934); I. Wolff, E. G. Linder, and R. A. Braden, "Transmission and reception of centimeter waves," Proc. IRE **23**, 11–23 (1935).

18. See below, p. 219.

19. Ernest G. Linder, "The anode-tank-circuit magnetron," Proc. IRE **27**, 732–738 (1939).

20. This work is summarized in a paper by H. Gutton and S. Berline. Bull. Soc. Francaise Radio (a house organ of SFR Company), **12**(2), 30 (1938).

21. Henri Gutton and S. Berline, "Ondes Hertziennes—Essais sur la propagation des ondes électromagnétiques de 16 cm de longueur," C. R. Acad. Sci. **207**, 325–326 (1938).

22. See below, p. 225.

23. See below, pp. 211–218.

24. W. W. Hansen, "A type of electrical resonator," J. Appl. Phys. **9**, 654–663 (1938)

25. This was elaborated later by Hansen in his lectures to the staff of the Radiation Laboratory in 1941. Cf. S. Seely and E. C. Pollard, "Notes on Microwaves, based on a series of lectures by W. W. Hansen" (unpublished), Chap. 9, "Cavity resonators," pp. 9-4 to 9-11. See also a

summary based on the above by S. Seely, "Group 107. Report on Fundamental Concepts of Cavity Resonators," 25 September 1942 (11 pp.).

26. U.S. Patent No. 2,251,569.

27. R. H. Varian, U.S. Patent No. 2,242,275, filed 11 Octtober 1937, issued 20 May 1941; Russell H. Varian and Sigurd F. Varian, "A high frequency amplifier and oscillator," J. Appl. Phys. **10**, 140, 321–327 (1939). The name "klystron" comes from the Greek verb "klyzein," which expresses the breaking or washing of waves upon a beach, and is also rather unesthetically associated with the English word "clyster."

28. Interview with W. W. Hansen, 17 May 1943. Frank J. Taylor, "The Klystron boys, radio's miracle makers," Sat. Eve. Post, 8 Feb. 1941, pp. 16ff.

29. E. J. Sterba and C. B. Feldman, "Transmission lines for short-wave radio systems," Proc. IRE. **20**, 1163–1202 (1932).

30. Lord Rayleigh, "On the passage of electric waves through tubes, or the vibrations of dielectric cylinders," Philos. Mag. **43**, 125–132 (1897).

31. Both men presented papers on this same subject, with almost identical conclusions, at a meeting of the Institute of Radio Engineers in Washington, DC, 1 May 1936. Each became aware of the work of the other just before the meeting was held. Southworth and his colleagues had already submitted their papers to the *Bell System Technical Journal* for the April issue; Barrow's paper appeared in the *Proceedings of the Institute of Radio Engineers* later in the year. There had been anticipatory papers. In 1934 L. Bergmann and L. Kruegel [Ann. Phys. **21**, 113–138 (1934)] measured the field within a short hollow metal cylinder and the radiation from its open end when a half-wave coaxial antenna was properly excited; but they did not suggest using hollow pipes for the transmission of ultrahigh-frequency radiation. Work on metal-sheathed dielectric wires may be thought of as closely related investigations. Such work goes back to a paper of D. Hondros and P. Debye, "Elektromagnetische Wellen an dielektrischen Drähten, Ann. Phys. (Leipzig) **32**, 465–476 (1910). For the bibliography on the subject, see Harvey, *op cit.* Note 2, p. 208.

32. W. L. Barrow, "Transmission of electromagnetic waves in hollow tubes of metal," Proc. IRE **24**, 1298–1328 (1936).

33. L. J. Chu, "Transmission and radiation of electromagnetic waves in hollow pipes and horns," Ph.D. thesis, Massachusetts Institute of Technology, 1938 (unpublished), 226 pp. and abstract of 4 pp.

34. Lan Jen Chu, "Electromagnetic waves in elliptic hollow pipes of metal," J. Appl. Phys. **9**, 583–591 (1938); L. J. Chu and W. L. Barrow, "Electromagnetic waves in hollow metal tubes of rectangular cross section," Proc. IRE **26**, 1520–1555 (1938); W. L. Barrow, L. J. Chu, and J. J. Jansen, "Biconical electromagnetic horns," Proc. IRE **27**, 769–779 (1939).

35. W. L. Barrow and W. W. Mieher, "Natural oscillations of electrical cavity resonators," Proc. IRE **28**, 184–191 (1940).

36. W. L. Barrow, "An oscillator for ultra-high frequencies," Rev. Sci. Instrum. **9**, 170–174 (1938).

37. G.C. Southworth, "Hyperfrequency wave guides—General considerations and experimental results," Bell Syst. Tech. J. **15**, 284–309 (1936); John R. Carson, Sallie P. Mead, and S. A. Schelkunoff, "Hyperfrequency wave-guides—Mathematical theory," Bell Syst. Tech. J. **15**, 310–333 (1936); W. L. Barrow, *op. cit.* Note 32. The first patent application on the work of Southworth and King (U.S. Patent No. 2,142,159, 3 January 1939) was filed 12 October 1935. The subsequent patents are U.S. Patents No. 2,151,138, 21 March 1939; No. 2,106,771, 1 February 1938; and No. 2,153,728, issued 11 April 1939.

38. These results are summarized in the paper of 1936 (see Note 37), but his detailed description of these earliest experiments was published over a year after he had made known his later and more important work on cylindrical wave guides. See G. C. Southworth, "Some

fundamental experiments with wave guides," Proc. IRE **26**, 807–822 (1937). His early apparatus is carefully described.

39. For other accounts of this work see G. C. Southworth, "Electric wave guides," Bell Syst. Tech. J. **14**, 282–287 (1936), and "New experimental methods applicable to ultra-short waves," J. Appl. Phys. **8**, 660–665 (1937).

40. Southworth, *op cit.* Note 37, p. 299.

41. See above, p. 105.

42. Harry L. Davis, "History of the Signal Corps development of U.S. Army radar equipment." This is a Signal Corps document of limited distribution classified Confidential.

43. W. D. Hershberger, "Modes of oscillation in Barkhausen–Kurz tubes," Proc. IRE **24**, 964–976 (1936).

44. See above, p. 105.

45. Interview with W. D. Hershberger. Memorandum for Officer in Charge, Signal Corps Laboratories, subject: Information on radio-optics obtained on a visit to the RCA Victor plant at Camden, N.J., May 21, 1934. For this magnetron, see above, p. 204, and Wolff, Linder, and Braden, *op. cit.* Note 17. See also Linder, *op. cit.* Note 17.

46. Davis, *op. cit.* Note 42, Part I, pp. 37–38. Photographs of the equipment showing the parabolic reflectors and the mode of mounting the magnetron behind the dish are given by Irving Wolff and E. G. Linder, "Transmission of 9-cm electromagnetic waves," 1935. This paper has been reprinted in *Radio at Ultrahigh Frequencies* edited by Lewis M. Clement *et al.* (RCA Institutes Technical Press, New York, 1940), pp. 421–429.

47. Interview with Irving Wolff.

48. Chester W. Rice, "Transmission and reception of centimeter radio waves," Gen. Electr. Rev. **39**, 368–369 (1936).

49. Confidential Memorandum from Lt. J. N. Wenger, USN. to BuEng, "Radio Bearing Indicator," 21 September 1931, with endorsement by H. G. Bowen, 31 October 1931. NRL files C-S67/69 (9/21-W8). Papers of Lt. W. S. Parsons.

50. L. A. Gebhard, Memorandum for Superintendent Radio Division, "New applications for superfrequencies," 9 December 1931; via L. A. Hyland from W. B. Burgess, 29 December 1931. Parsons papers.

51. Lt. W. S. Parsons to Lt. F. I. Entwhistle, letter 23 December 1935. Parsons papers.

52. Director NRL to Chief, BuOrd, "Superhigh frequency radio—possible use of reflected waves in airplane detection and fire control," 2 August 1933, C-F42-2. Director NRL to Chief, BuOrd, "Fire control possibilities of radio micro rays," 15 September 1933, and first endorsement, NRL files C-S71-8 (9-15-W1), and Parsons papers.

53. W. S. Parsons, Memorandum for Special Board of Ordnance, 20 March 1934.

54. See Chap. 3.

55. BuOrd Memoranda from Section L and Section K to Chief of the Bureau, 8 May 1934. Parsons papers.

56. W. J. Cahill, NRL Report No. R-131, BuAer Problem 1-31, 15 January 1935. NRL Files C-P31-1 (3).

57. "Microwaves to detect aircraft," Electronics **8**, 235–238 (1935).

58. *The New York Times*, September 1935 and October 1935.

59. See, for example, Short Wave Televis. **7**, 604, 630 (1937).

60. A French patent (Brevêt Français 800,869) was granted on 20 April 1935 to H. Gutton and M. Ponte as inventors of this device.

61. See above, pp. 204–205.

62. Elie, H. Gutton, Hugon and Ponte, "Détection d'obstacles à la navigation sans visibilité," Bull. Soc. Fr. Electr. No. 100 (April 1939). This article was reprinted in Ann. Postes, Telegr. Teleph. **28**, 721–737 (1939).

63. Interview with W. W. Hansen.

64. W. T. Cooke letter to E. L. Bowles, 3 April 1940. MIT, ELB.

65. Julius A. Stratton, "The Round Hill experiment station," Tech. Eng. News **121**, 137 (1930); E. L. Bowles, "Science at Round Hill," Tech. Rev. **37**, 18–20 (1934), and "Progress report on short wave research, 1926–1928" (unpublished), and "Polar radio studies," Tech. Rev. **21**, 219 (1929). See also John Stevenson, in *The Boston Transcript*, Saturday Magazine section, 30 March 1929.

66. Proposed Round Hills Reorganization Plan, 29 May 1928.

67. J. A. Stratton, "The effect of rain and fog on the propagation of very short radio waves," Proc. IRE **18**, 1064–1074 (1930).

68. H. G. Houghton, "The transmission of visible light through fog," Phys. Rev. **38**, 159–165 (1931). J. A. Stratton and H. G. Houghton, "A theoretical investigation of the transmission of light through fog," Phys. Rev. **38**, 159–165 (1931). H. G. Houghton, "The size and distribution of fog particles," Physics **2**, 467 (1932). H. G. Houghton and W. H. Radford, "On the measurement of drop size and liquid water content in fogs and clouds," Pap. Phys. Oceanogr. Meteorol. **6**(4) (1938), and "On the local dissipation of natural fog," Pap. Phys. Oceanogr. Meteorol. **6**(3) (1938).

69. E. L. Bowles, "Instrument landing research at Tech.," Tech. Eng. News **19**, 162ff (1938); Donald G. Fink, "3 spots and a horn," Aviation **37**(9), 28ff (1938); E. L. Bowles, W. L. Barrow, W. M. Hall, F. D. Lewis, and D. E. Kerr, "The CAA-MIT microwave instrument landing system," Paper 40–44, recommended by the AIEE Committee on Communication and presented at the AIEE Winter Convention, New York, N.Y., 22–26 January 1940, and released for final publication 20 March 1940.

70. Report to the Bureau of Air Commerce, Department of Commerce, on Blind Landing Apparatus and System, by Massachusetts Institute of Technology, 1 June 1938. MIT/ELB.

71. Henry Ladd Smith, *Airways, the History of Commercial Aviation in the United States*, (Knopf, New York, 1942), p. 310.

72. S. Roberts and A. von Hippel, "A new method for measuring dielectric constant and loss in the range of centimeter waves," Publications from the Massachusetts Institute of Technology, March 1941; L. J. Chu, "Wave guides with dielectric sections," *ibid.*

73. This story has been largely reconstructed from Professor Bowles' papers which he kindly permitted me to consult.

74. W. L. Barrow, Memorandum of Sperry–Barrow research, Conference on program, 21 December 1939, H. H. Willis, E. L. Bowles, and W. L. Barrow, 13 pp., 3 blueprints. MIT/ELB.

75. E. L. Bowles Memorandum to K. T. Compton, 31 October 1939.

76. E. L. Bowles, "A summary of the status of the Loomis research," 10 January 1946.

77. Because of the pressure of events no contemporary report of this apparatus had been preserved. The description of their systems has been reconstructed from descriptions kindly supplied by Frank Lewis, W. W. Hansen, and others.

78. Interview with E. G. Bowen, 29 April 1943. There is a record of Bowen's views in a letter of Sir Henry Tizard to E. V. Appleton, 27 December 1939. Files of the National Research Council of Canada.

79. C. V. D. Report—S.R.F. Department Admiralty, "Magnetron development in the University of Birmingham." Secret. N.d. This is the first published report on the magnetron

development. It was prepared in the summer of 1940 and consists of an introduction by Oliphant and a technical account by J. T. Randall, H. A. H. Boot, and S. M. Duke. Although I have made considerable use of this report, I have been greatly aided in understanding certain facts concerning the development by information kindly supplied by Dr. Randall and Dr. Boot.

80. Sir Henry Tizard, letter to Mr. H. Warren, 18 December 1939.

81. *Op. cit.* Note 79, p. i.

82. Relative claims of Randall and Oliphant. Russian magnetron. (cf. p. 245).

83. H. Hertz, *Untersuchungen über die Ausbreitung der Elektrischen Kraft* (Barth, Leipzig, 1892); H. MacDonald, *Electric Waves: being an Adams Prize Essay in the University of Cambridge* (Cambridge Univ. Press, Cambridge, England, 1902).

84. "Contact had been established with Professor Oliphant's group on magnetron as well as velocity modulation work. The former is certainly relevant to our USW.2 program [an Admiralty radar project], and it is proposed to deal with a request from Birmingham to make a sealed off version of their magnetron as part of their program." General Electric Company, C.V.D. Report No. 3, April 1940, p. 3.

85. British Patent.

86. *Op. cit.* Note 79, pp. 19–20; General Electric Company, C.V.D. Report No. 4, July 1940, p. 4.

87. E. L. R. Webb, Notes on Professor Oliphant's lecture—September 8th 1939.

88. This is based largely on information obtained in interviews with D. M. Robinson and with H. W. B. Skinner.

89. Telecommunications Research Establishment, Worth Matravers: Progress Report for period 16 November–15 December 1940. British Document, Secret, p. 5.

90. *Ibid.* The magnetron used was G.E.C.'s first production effort, the E1198, which showed wide variability in power output. A modulator using a hard valve pentode, the E1155, proved superior to a modulator using an E1191 thyratron. The former tube gave a better shaped pulse and permitted a higher repetition rate. The Sutton tube proved superior to its competitors as a local oscillator.

91. *Ibid.*, p. 4.

92. TRE Progress Report, 16 January–15 February 1941, p. 8.

93. TRE Progress Report, 16 May–15 June 1941, p. 8.

94. TRE Progress Report, 16 January–15 February 1942, p. 8.

95. TRE Progress Report, 16 May–15 June 1942, p. 5.

Section B

The Establishment of the NDRC Radar Program

CHAPTER 9
Inauguration of the
NDRC Program

In an earlier chapter, it was explained that not until the spring of 1939, on the eve of war, did the British Government disclose its RDF (radar) development to the several Dominions, and that the earliest Canadian radar work at the National Research Council was a direct consequence of this disclosure. This, however, did not at once result in close cooperation in radar between Canada and the United Kingdom. It was not until after the outbreak of war that efforts were made to establish closer liaison between the scientific work in the two countries. These efforts were then closely bound up with an attempt by the British to appraise the scientific resources of the United States, especially in the radar field, and to explore the possibilities of a scientific cooperation between Britain and the Arsenal of Democracy. The visit to America in the spring of 1940 by Professor A. V. Hill, Nobel Prize winning physiologist and Secretary of the Royal Society, was intended to accomplish both purposes.

Professor Hill, a close friend of many American scientific men, was attached to the British Embassy in Washington as temporary "scientific attache."[1] It was a roving commission which took him outside of Washington on visits to the leading universities and industrial concerns of the eastern United States and which brought him, in the month of April, to Ottawa. Here he discussed the possibilities of scientific liaison with the leaders of Canadian science and met the "small but active group"[2] working on radar on the basis of information obtained before the war by J. T. Henderson.

The results of Hill's mission were soon felt in London, in the form of steps to establish closer scientific relations with Canada. The Canadians had already taken the initiative by the appointment of Sir Frederick Banting, the co-discoverer of insulin, as a kind of traveling scientific ambassador, and by sending several Canadian scientists and medical men on exploratory visits to England.[3] Hill and the Canadian scientists agreed that "at the present time the greatest need is for someone from England to visit Canada and make himself conversant with our resources and potentialities."[4] Late in June it was decided that Professor R. H. Fowler, distinguished scientist and member of the Committee for the Scientific Survey of Air Warfare, should proceed at once to Canada, where he would be attached to the Office of the High Commissioner to improve the liaison between British and Canadian scientists.

The state of affairs in Canada confirmed Hill in an opinion which had been formed during the discussions in the United States: the need for an overall exchange between England and the United States of scientific secrets of military significance. He pointed out that the Canadian investigators were hindered by secrecy from taking full advantage of the help potentially available in the U. S., and that this could be possible only through an official exchange.

Hill informed Sir Henry Tizard of his proposal before the end of April. The subject was discussed at a meeting of the Committee for the Scientific Survey of Air Warfare on May 2, 1940, attended by a number of persons including Appleton, Blackett, C. G. Darwin, Fowler, G. P. Thomson, Woodward-Nutt, Watson-Watt, and several RAF representatives headed by Air Vice Marshall A. W. Tedder.[5] The idea was received favorably, especially if the exchange were not to be limited to radio and RDF. Tizard reported that the whole plan was being considered by the Secretary of State for Air. After his return to London, on the eve of the Franco–German Armistice, Hill submitted on June 18, 1940 a final report of his mission in which he summarized his discussions with American scientists and government officials and made a point of the widespread sympathy that he found for Britain's cause and the interest that was expressed in a possible interchange of secret scientific information between the British and the Americans.[6] Security considerations prevented any formal disclosure to him of American developments, but he returned with some notion of the existence of American radar, and vague and considerably exaggerated expectations that America had undisclosed secrets in the microwave field.

During the summer of 1940, the diplomatic negotiations were completed on the highest level along the lines and through the channels recommended in the Hill report. The arrangements were embodied in an *aide memoire* signed by President Roosevelt and the British ambassador, Lord Lothian.[7] The proposed interchange was to be very broad, though not entirely complete. It was to be a military interchange, and the visiting British mission would be accredited to the Armed Services. Notwithstanding this restriction, the potentialities and consequences of the interchange which took place in the early autumn were greatly enhanced by the creation in the United States, in June 1940, of a new government agency of civilian scientists, the National Defense Research Committee.

The Establishment of the NDRC

The outbreak of the European War in September 1939 led to a widespread reconsideration of America's scientific preparedness. Although there were frequent discussions during the winter of 1939–1940, nothing of consequence was accomplished until the spring. Vannevar Bush, President of the Carnegie Institution, James Bryant Conant, President of Harvard, Karl T. Compton, President of the Massachusetts Institute of Technology, and Frank B. Jewett, President of the National Academy of Sciences, discussed on a number of different occasions the possibilities of an organization of civilian scientists. Bush and Conant had several opportunities to discuss the problem while serving together on the Committee for Scientific Aids to Learning. They agreed that an *ad hoc* executive agency somewhat along the lines of the NACA was most desirable.

The astonishing success of the German offensive in Norway and the Low Countries in the late spring of 1940 shook Washington with only slightly less violence than London. The collapse of France and the mounting Allied disaster brought swift action.

Having crystallized his ideas in the earlier discussions with his scientific colleagues, late in May, Vannevar Bush had a series of preliminary conversations with Harry Hopkins, who, as Secretary of Commerce, was interested in the related problem of forming an inventor's council. Early in June, Bush outlined his proposals succinctly for President Roosevelt. He presented them to the President in person; when he emerged after a conversation lasting only about 15 minutes, his paper bore the annotation: "O. K. FDR".[8] Bush provided Hopkins with drafts of papers that were required to complete the affair, including a draft of the executive order. In a few paragraphs, the plan called for a new agency, the National Defense Research Committee, to coordinate and to extend research on military devices and instrumentalities, except in the field covered by the National Advisory Committee for Aeronautics. It would be placed somewhat loosely under the Advisory Commission to the Council on National Defense, the body headed by Knudsen and Stettinius, but would be set up "on its own feet" and with its own funds, by virtue of an executive order. Its membership would include the Chairman, a General from the Army, an Admiral from the Navy, the Commissioner of Patents, the President of the National Academy, and three scientists-at-large. As civilian scientists-at-large, Bush proposed the names of Conant, Compton, and Tolman.

On Saturday morning, June 15, Bush conferred with General Watson, the President's aide, and Harry Hopkins at the White House and came away with assurances that the letters of appointment would be written almost immediately and that he should proceed on the assumption that the set-up would be exactly as planned.[9] Then followed conversations with the Chief of Staff and CNO to inform them officially of the new agency to be created and to discuss with them the Army and Navy representatives who might be appointed to the Committee.

General Marshall received with approval Bush's suggestion that General G. V. Strong, an old friend with whom Bush had frequently discussed the problems of scientific research for military purposes, be appointed the

Army representative. Admiral Stark felt that Admiral H. G. Bowen would be the appropriate Navy representative.

It was Bush's impression that both men genuinely welcomed the establishment of the new agency. General Marshall expressed the opinion that, although in peacetime the Army engaged in active research and development, the approach of war meant that the effort in such places had to be directed mainly towards procurement, with research slowed down and in some cases stopped altogether. He expressed himself as much pleased that the NDRC would be in a position to take over a large portion of the Army's research commitments. This view was echoed by General Arnold, with whom Bush also talked, and who requested Bush accompany him by plane to Wright Field the following Monday to look over the situation.

On Monday, June 17, Bush flew with General Arnold to Wright Field, returning the same evening. On the following day—such was the tempo of events—there was held the first informal conference of the National Defense Research Committee in Bush's office at the Carnegie Institution on 1530 P Street. In attendance were all those who, although not yet in possession of their letters of appointment, had been selected for the new committee: Rear Admiral Harold G. Bowen, Brigadier General G. V. Strong, the Honorable Conway P. Coe, Commissioner of Patents, K. T. Compton, J. B. Conant, F. B. Jewett, and Richard C. Tolman.

It was agreed to ask the armed services for their views on what were proper projects for the new organization. In particular, there were three questions to be answered: (1) what military developments were under way, especially what were the programs the services were likely to slow down or abandon in the interest of immediate production; (2) what programs not yet under way the services felt to be worth undertaking; and (3) what current military projects needed to be supplemented. The assembling of this information was entrusted to K. T. Compton.

On June 20, Bush sent a memorandum to the Members of the National Defense Research Committee announcing that lists of research and development projects had been received from the War Department and that similar lists were expected shortly from the Navy. These lists, acquired in the case of the Army through the agency of General Strong, were to be handed over to Compton, who was to summarize the material for the benefit of the committee members.

On Friday, June 21, Bush sent to the scientists on the committee a memorandum embodying "a first shot at an organization chart" intended to help crystallize their ideas before the second informal meeting to be held the following week. His proposal was to divide the Committee into five divisions which in turn should be composed of a number of sections, the true operating units of the organization. He emphasized that in his scheme the section chairmen were the key individuals and should be chosen with care. Revealing that the full extent of the responsibility was then not fully realized, he added that the jobs of the section chairmen would probably be full-time at first, "but it seems to me that they ought to be scientists of caliber and voluntary workers."[10] Each of the five divisions was to be headed by a member of the main Committee.

The Inauguration of the NDRC Radar Program

From the outset there was general agreement that the NDRC would concern itself with some phase of radio detection. As early as his first tentative organization proposals, Bush had planned for a section of Compton's division that would be devoted to detection methods. This was at first conceived in broad terms, for under this heading appeared the words "searchlights, micro-waves, acoustics, infra-red."[11] The list submitted to Dr. Bush by the armed services late in June, embodying projects in which they expressed interest, but for the solution of which they felt their own facilities were inadequate, made precise suggestions along this same line. Although the lists cautiously omitted all reference to research on aircraft detectors, many of the proposals were obviously closely related to this subject. They stressed the importance of general studies of pulse transmission and reception and of basic research in the hyperfrequency field.[12] Furthermore, the Air Corps was especially interested in a number of problems of instrumentation where the solutions undoubtedly lay in the ultra high-frequency or hyperfrequency radio fields. They centered about the problem of fog and haze penetration and the possibilities of reconnaissance or bombing through the overcast; and they hoped for three resulting types of equipment: a precision navigational device to guide or direct an airplane to a point in space by automatic methods or by triangulation; equipment to locate a waterborne target through an overcast with sufficient accuracy to direct bombardment aircraft to that location; and, finally, equipment suitably accurate to permit tracking and blind attack of the waterborne targets. The documents explained with considerable insight: "It will probably consist of radio equipment for transmitting and receiving on a narrow radio beam with an indicating device to accurately indicate in the vertical and horizontal planes the position of the airplane with respect to the target."[13]

The decision to limit the detection work to the field of microwaves was made almost at once. Such a program fulfilled to the general policy requirement that the new organization should undertake long-range projects too speculative for the Service laboratories in time of war. It had the further advantage of leaving the field of longer-wave radar in the hands of the services who had developed it, and who were at present mainly concerned with bringing their new equipment out of the laboratory into practical use as quickly as possible. Furthermore, those civilian investigators who had quietly embarked on radio detection research, at Stanford, MIT, RCA, and the General Electric Company, had begun their work, and acquired some pioneering experience, in the microwave field. The man selected by Compton to head the section on detection, Alfred L. Loomis, New York lawyer and scientist, was one of the most energetic and enthusiastic of these investigators.

Loomis had close personal and scientific connections with Bush and Compton. He had generously contributed to the support of microwave research in the Electrical Engineering Department at the Massachusetts Institute of Technology, and he was at this time conducting some pioneer experiments in cooperation with the MIT group at his own laboratory in Tuxedo Park, N. Y.[14] Bush was following closely the work at the Loomis Laboratories, as previously he had kept in touch with the work at MIT. Late in May, encouraged by reports from Loomis, he had urged the Executive Committee of the Carnegie Institution to allot the sum of $10,000 in support of the microwave detection work.[15]

Loomis chose as his first associates on what came to be known as Section D-1, and later as the Microwave Committee, Ralph Bown of the Bell Telephone Laboratories and two men with whom he had already had contact in his own microwave work: E. L. Bowles of MIT, who became executive secretary of the Committee, and Hugh H. Willis of the Sperry Gyroscope Corporation. These four men held their first meeting as guests of Loomis at Tuxedo Park on July 14, 1940, before any of them, even the Chairman, had received their formal appointments from Bush.[16] The Committee discussed its organization, method, and objectives, and agreed that for effective action it was important to keep the group small. During the next few months, the Microwave Committee filled out its roster with the appointment of the following persons: R. R. Beal, Director of Research of the Radio Corporation of America; George F. Metcalf of the General Electric Company; J. A. Hutcheson of the Westinghouse Electrical and Manufacturing Company; and Ernest O. Lawrence, Professor of Physics and Director of the Radiation Laboratory for nuclear physics at the University of California. With a few exceptions, this original nucleus remained intact during the five years of its active existence, although the Committee grew in size beyond the original expectations of the Chairman. It was a group experienced in the administration of scientific research, well versed in current radio developments, well placed to coordinate the scattered work in this new field, and well fitted to administer the funds soon to be placed at their disposal.

It was agreed in the earliest meetings that the committee members were sitting as individuals, not as representatives of their organizations or institutions, and that there would be full and complete interchange of information between members of the section. The Committee defined its objective in these terms: "So to organize and coordinate research, invention, and development as to obtain the most effective military application of microwaves in the minimum time."[17] The first order of business was clearly to learn what the armed forces had secretly accomplished and to survey the field of microwave research to determine what work was actually in progress in the country and what seemed most promising.

During the summer months, Compton and the members of Section D-1 had officially disclosed to them pulse radar developments of the Navy and the Army. On Wednesday, July 18, Compton and Loomis paid a visit to the Naval Research Laboratory, and following initial talks with the Director, Admiral Bowen, and Commander Briscoe, they were taken by Page, Young and the other NRL scientists to inspect the radio detection equipment.[18]

A few days later Compton and Loomis visited the Signal Corps Laboratory at Fort Monmouth, where they were shown about by Colonel (later General) R. B. Colton and Colonel Corput with every courtesy and with complete candor.[19] However, so great was the secrecy with which the radar developments were shrouded that they became convinced that neither the Army nor the Navy was aware of the work being done by the other service.[20]

The third meeting of the Microwave Committee was held on July 30 in Washington, following a dinner the previous evening at the Wardman Park and an evening discussion in Alfred Loomis' apartment attended by Bush, Compton, and Jewett, at which the section's members were introduced to Admiral Bowen, Commander Briscoe, and Colonel Mitchell.[21] After the meeting, the Committee spent the rest of the day with Admiral Bowen and Commander Briscoe at the Naval Research Laboratory, and inspected the radar equipment.

Late in August, Compton and Loomis went together to attend the Army maneuvers taking place at Ogdensburg, N. Y.[22] They were flown up from West Point in Secretary Stimson's plane, and were present at a test of the Army radar equipment, which took the form of a comparison of airplane detection by radar and by a network of volunteer observers scattered all over the state and reporting in to a large plotting station by telephone. A rival plotting system was set up in parallel, using information supplied by the Army radar equipment which, needless to say, won

It had been agreed at the first meeting that the matter of greatest urgency was to conduct a survey of the work on microwaves actually in progress in the United States.[23] The several committee members contributed to the preparation of a list of all persons in the country engaged, or who had recently been engaged, in microwave work. The survey reflected much important progress: that of Southworth and his associates on wave guides at the Holmdel Laboratory of the Bell Telephone Laboratories; the work of Barrow, Chu, and co-workers at MIT on wave guides and horns; the research of Wolff, Linder, Kilgore, and others at RCA on magnetrons; the klystron work at Stanford and San Carlos under the sponsorship of the Sperry Gyroscope Corporation; the work of H. T. Friis at Holmdel on crystal mixers for microwave work; new tube developments in the velocity-modulation field by A. L. Samuel and his associates at the Bell Laboratories.[24] But in perhaps the most important respect the survey was disappointing; there was no sign that a tube was anywhere in development which might give adequate power on the wavelength which the Committee felt would be desirable: ten centimeters or below. Only two vacuum tubes offered possibilities of giving real power on wavelengths below one meter: the klystron, and another tube being explored on the West Coast, the so-called *resnatron* tube developed by David H. Sloan and L. C. Marshall of the University of California. Neither tube showed promise of giving much power at the frequency specially sought after by the Microwave Committee, but they seemed the only immediate possibilities.

The first working model of the resnatron was completed in January 1940; and by August 1940 the tube was giving an average power output of about a kilowatt in the neighborhood of 45 cm.[25] Since May 1940, the development was sponsored by the Research Corporation, whose President called it to the attention of the NDRC. The subject was brought up at the first meeting of the Microwave Committee by Loomis, who made a personal investigation and submitted a favorable report later in the month at the Committee's third meeting. This was accompanied by a recommendation that the NDRC be urged to appropriate $20,560.00 to the University of California to further this work. This was approved and on November 1, a development contract, NDCrc-25, was signed by the representatives of the University of California and of NDRC. This first microwave contract, made retroactive to September 1, required the development of a high-power version of the resnatron and specified in addition that an attempt be made to lower the wavelength as much as possible.

NOTES

1. He is so described in C.S.S.A.W. Paper No. 24; cf. note 5 for full reference.

2. A. V. Hill, "RDF in Canada and the United States: and a proposal for a general interchange of scientific and technical information, and of Service experience, between Defence Services of Great Britain and those of the United States," London, 18 June 1940.

3. At the time of the Hill visit, a number of these men had already returned from their

missions, while three of them were still in England. Banting was killed on February 21, 1941 in a plane crash in Newfoundland while en route to England on a technical mission (*New York Herald Tribune*, 24 February 1941).

4. MacKenzie to Appleton, ltr. 13 June 1940. Files NRC, Ottawa.

5. Committee for the Scientific Survey of Air Warfare—Minutes of the 11th Meeting held at the Air Ministry, Whitehall, on Thursday, 2 May 1940. C.S.S.A.W. Paper No. 24.

6. See Note 2.

7. Interview with Vannevar Bush.

8. Interview with Vannevar Bush. The original document appears to have disappeared from the OSRD files.

9. The account has been drawn from letters of Bush to Jewett and J. B. Conant dated 15 June 1940.

10. V. Bush, Memorandum to Compton, Conant, Jewett, and Tolman, 21 June 1940.

11. V. Bush, Memorandum to Compton, Conant, Jewett and Tolman, 21 June 1940. See also interview with Dr. Vannevar Bush, 20 August 1940, where Bush says he envisaged the general problem of detection, not merely microwave, and that for some time he had known about American Army and Navy radar and also about British and French developments. "I made it my business to know what was going on."

12. KTC by Irving Stewart to Bush, Jewett, Conant, Tolman, Compton, covering letter of 27 June 1940 in envelope marked "Copies of assignments by KTC to himself, JBC, FBJ, and RCT." Material submitted by services includes "War Department Research and Development Program Fiscal Year 1942" (photostats and charts) and "Brief Descriptive Data on Special Starred Projects Recommended in Budget Estimates FX-1942." These starred projects were not as yet authorized projects in the Signal Corps but were deemed "of the utmost importance at this time" by the Chief Signal Officer. (cf. Memo of J. O. Mauborgne to Asst. Chief of Staff, War Plans Division, Subject: Research and Development Projects, 3 June 1940). On the eve of the foundation of NDRC the Signal Corps had proposed a considerable expansion in research and development. The two-year program for 1941–42 was to amount to $8,338,860, a decided increase over the $3,593,260 previously allocated (and which included a million dollars from the President's Special Fund). This was not to be accomplished by an expansion of the Signal Corps Laboratories but rather by placing development contracts, as in World War I, with commercial firms for the development of the starred items. Major James T. Watson: "Memo for the Asst. Chief of Staff, G-4" 31 May 1940; Mauborgne: "Memo to Asst. Chief of Staff," 3 June 1940. These were consulted in the OSRD files.

13. "Brief Descriptive Data on Special Starred Projects."

14. Cf. above, pp. 236ff.

15. Carnegie Institution of Washington, Meeting of the Executive Committee, 23 May 1940, Confidential Minutes, p. 109. During June, Bush was kept informed about the experiments at Tuxedo. Cf. A. L. Loomis to V. Bush, ltr. of 18 June 1940; "I congratulate you upon the formation of the Committee that you worked so hard to get the President to appoint, and I am delighted to see that you are to be the Chairman... In regard to microwaves, we have got eight people working at Tuxedo at the present time, and this week we are moving all our apparatus into a mobile truck fitted out as a lab, and expect to make tests on automobiles, airplanes, and boats the next three or four weeks."

16. Microwave Committee Minutes, 1940. Meeting No. 1, 13 July 1940.

17. *Ibid.* Meeting No. 1, 13 July and Meeting No. 3, 30 July, p. 5. In an earlier form, the objective included the organization and coordination of "research, invention, development, design, and manufacture." The commercial members obviously felt this was too broad, and Dr. Bown submitted a curtailed and g a atic improved version given above.

18. Notebook of KTC. In these discussions the Navy made known their chief unsolved problems of the moment: how to get a smaller antenna for shipboard search systems on 200–400 MHz; how to get a 400-MHz vacuum tube.

19. *Ibid.* Cf. a conversation with Dr. K. T. Compton, Friday, August 20, 1943.

20. This of course was exaggerated, but there was anything but close cooperation between the two laboratories. Interchanges of personnel were unheard of and visits back and forth were rare and formal. Dr. Bush's testimony is more accurate: "The Army and Navy even kept things from one another. The Navy and the Army had separate developments in fire control; in fact, the Navy refused for a long time to tell the Army about aided-laying." Interviews with Dr. Vannevar Bush, August 20, 1944. p. 2.

21. Microwave Committee Minutes, 1940, Meeting No. 3, 30 July, p. 6; Conversation with President K. T. Compton, 20 August 1943. At this meeting Compton felt that Admiral Bowen revealed little enthusiasm for the disclosures to the British. Bowen, however, used the remarks that were made by the British at the gathering, attributing their radar development to the work of Breit and Tuve, to raise the stock of NRL.

22. Compton Notebook; Conversation with Dr. K. T. Compton, 20 August 1943.

23. Microwave Committee Minutes, 1940, Meeting No. 1, 13 July 1940.

24. "List of Persons and Places Working on Micro Waves", 7/21/40, revised 8/14/40, 4 pp.

25. Interview with Dr. L. C. Marshall, 29 April 1943.

CHAPTER 10

The Foundation of the Radiation Laboratory

The previous chapter has described the rather uncertain prospects that faced the Microwave Committee of NDRC at the time that the British Technical Mission, of whose origins we have had occasion to speak, arrived in Washington and began conversations with the representatives of the Army and Navy in the second week of September 1940. In the course of these discussions, which covered a wide range of technical questions, each nation divulged to the other the details of its secret radar developments.

The British Technical Mission

The chief of the Mission was Sir Henry Tizard, whom we have already encountered as the distinguished chairman of the Committee on the Scientific Survey of Air Defence and member of the top secret Air Defence Research Committee, both of which played leading roles in the early history of British radar.[1] The mission included three military representatives: for the Royal Navy, Captain H. W. Faulkner; for the Army, Col. F. C. Wallace, who had commanded the anti-aircraft on the beach of Dunkirk; and for the RAF, Group Captain F. L. Pearce, who had recently distinguished himself in the bombing of the German pocket battleship *Scharnhorst*. There were three civilians, besides the chief of the Mission: J. D. Cockcroft and E. G. Bowen, the technical advisors on radar for the mission, and a representative from the Ministry of Aircraft Production, A. E. Woodward-Nutt, who served as secretary to the mission. There were also a number of *ex officio* attachés, among them R. W. Fowler, British scientific liaison officer for Canada, and several scientists from the National Research Council in Ottawa.

Most of August was spent in preparations for the Mission. Late in the month, Sir Henry flew to the United States in advance of the rest of the party to pave the way for the Mission. The other members crossed on the *Duchess of Richmond*, an unescorted passenger liner which relied only on its speed to make the crossing safely. Also on board were about a thousand British sailors, brought over to man the first of the over-age destroyers which the United States was consigning to Britain. The *Duchess* landed at Halifax on September 7, 1940, just in time to see the first of the American destroyers enter the harbor. With the exception of Bowen, who remained behind to locate some missing equipment, the Mission at once entrained for Washington.

The opening meeting of the Mission with American Army and Navy representatives took place at 10 AM September 10, 1940. The meeting was presided over by Admiral Noyes. Major General J. C. Mauborgne, the Chief Signal Officer; Lieutenant Colonel H. Mitchell, and Major W. G. Smith were the other American representatives. Tizard, Faulkner, Wallace, and Cockcroft were present at this first meeting, at which the British described in some detail their gun-laying radar, the various types of naval radar that had been developed, and the CHL equipment.

Only the exchanges on radar need occupy us, and these only briefly. On the afternoon of September 12, Cockcroft and Bowen paid a visit to the Naval Research Laboratory at Anacostia and were given a preliminary disclosure of NRL's radar accomplishments. The British visitors described in some detail their newly developed vacuum tubes, and revealed, but apparently did not actually produce, the 10-cm resonant cavity magnetron which was perhaps their outstanding importation. On the following day, in a meeting at the War Department, the British explained the technical details and operational employment of their Radar Chain Stations. On September 14, at the Navy Department, Bowen supplied complete information on the British developments in airborne radar. After a second visit to NRL, and several hectic days devoted to delving into American developments in radio blind-landing, the radar representatives of the mission spent two days at Fort Monmouth, where the Director, Colonel R. B. Colton, showed them all the U.S. Army's radar and communications developments.

About the middle of September, the way was cleared to have the appropriate members of the Tizard Mission meet the section leaders of NDRC.[2] By special arrangment, the Mission was empowered to treat with the civilian scientists on the same terms as with the Armed Services. The first contact between the radar members of the Mission and representatives of NDRC took place at an informal evening conference at the Wardman-Park Hotel on September 19. This was attended by K. T. Compton, Alfred Loomis, and Carroll Wilson from NDRC; Admiral Bowen and Colonel Mitchell; Cockcroft and Bowen of the British Mission, and J. T. Henderson of the National Research Council of Canada (NRC). In the course of the conversation, Loomis outlined the general American position in microwaves, and invited the Mission to visit his laboratory at Tuxedo Park, New York at the end of the month.[3]

During the remaining days of September and the first weeks of October, Bowen and Cockcroft kept to a grueling schedule of visits to the various industrial laboratories in which electronic development was in progress, to the Aircraft Radio Laboratory at Wright Field, to MIT, and to the Loomis Laboratory at Tuxedo.[4]

The first extended talks between the NDRC and the British Mission took place over the weekend of September 28–30 when J. D. Cockcroft and E. G. Bowen of the Mission and J. W. Ball of NRC met with several members of the Microwave Committee and NDRC as guests of Alfred L. Loomis at Tuxedo Park. The visitors were shown the Doppler-effect detection equipment being worked on by Loomis group. A fruitful and historic three-hour discussion was held on the evening of September 29 and on the morning of September 30, during the course of which the British explained their long-range objectives in radar—especially the importance to them of a microwave airborne radar set without the defects of the longer-wave AI—and showed the Committee members the sample cavity magnetron they had brought with them as their prize exhibit (see Fig. 10-1).[5] When the mode of operation of the tube and its capabilities were explained, the interest and enthusiasm of the Americans were raised to the highest pitch. On the following day, October 1, Loomis called a luncheon meeting in New York attended by Hugh H. Willis, R. R. Beal, and Ralph Bown of the Microwave Committee, at which the status of the microwave development in Great Britain was reviewed as it had been related at Tuxedo Park by Cockcroft and Bowen. The item of greatest interest was the resonant cavity magnetron which Loomis described to Beal and Bown.[6]

During the first week of October, the British representatives, with the full approval of the United States authorities, disclosed the cavity magnetron to engineers of the Bell Telephone Laboratories, the organization the British had selected to manufacture the tube in this country.[7] On October 3, a group gathered in the office of Ralph Bown at 463 West Street. Besides Bown, it included Cockcroft and Bowen of the British Mission, a representative from NRC, and five of the most brilliant Bell Laboratories tube experts: A. L. Samuel, C. E. Fay, J. O. McNally, J. R. Pierce, and J. R. Wilson. Cockcroft at first displayed the "micro-pup" tube and then set upon the table a box containing the magnetron; the tube was passed from hand to hand while Cockcroft described the development of the tube by Oliphant's group at Birmingham, and astonished them by describing its rated peak power of 10 kW at 10 cm. Plans were made to have the tube operated as soon as possible for the benefit of the Bell Laboratories engineers. On Sunday, October 6, the tube performed for the first time in this country. Bowen brought the magnetron to the Whippany laboratory of the Bell Telephone Laboratories, where it was run without any attached load in the field of an electromagnet supplied by BTL. The wavelength was checked on the spot and it was estimated that the tube gave about 6.4 kW power. The assignment of making a copy of the British tube was entrusted to V. I. Raned and J. B. Fisk, and the sample tube was left with

FIG. 10-1. Sample magnetron brought to America in September 1940 by Tizard Mission.

the Bell engineers for this purpose. Work began immediately the following day at the Electronics Research Laboratory, and an x-ray photograph of the British tube was made.

Creation of the Radiation Laboratory

Meanwhile, the disclosure of the magnetron had exerted a profound effect upon the members of NDRC and Section D-1. The new tube removed the chief obstacle to a successful development program in the microwave field. A major breach had suddenly been opened in the line, through which the reserve strength of American university and industrial resources could pour. With energetic exploitation of this initial stroke of good fortune, microwave radar could be developed in time to be useful in the war. This was the article of faith—though it was not universally shared—upon which the Microwave Committee, with the prompt and active encouragement of Bush and Compton, based their subsequent decisions.

During the first two weeks of October, a series of important conferences was held in Cambridge, New York City, and Tuxedo Park, in the course of which the main outlines of a concrete program were formulated by the Microwave Committee. E. G. Bowen, who was remaining behind after the departure of the Tizard Mission, took an active and influential part in these discussions, by describing the British Air Ministry's radar research organization, and by helping to lay down the specific objectives for a microwave program.

On the weekend of October 12–13, the future character of the American microwave program was for the first time clearly outlined. A meeting at the Carnegie Institution on Friday, October 11, attended by Bush and Compton, was followed by informal conferences held at the Loomis Laboratories during the next two days. The participants in these important discussions were Bowen and Cockcroft; Loomis, Bowles, and E. O. Lawrence of the Microwave Committee; and Carroll Wilson of NDRC.[8]

There was general agreement that a central laboratory under civilian direction should be set up at once, staffed, in the manner of Britain's AMRE at Swanage, as much as possible by research physicists from the universities of the country. This policy had proved extremely successful in England. It was at first felt when the possibilities were canvassed that the laboratory should be set up at Bolling Field, Washington, D.C., where a large heated hangar with associated laboratories would be erected by the Army.

It was also decided, largely on the basis of British need, to concentrate on three projects, not greatly different from those proposed by the American armed forces and already being considered by the Microwave Committee.[9] Project I, and the project of greatest urgency from the British point of view, was to build a 10-cm AI system. Project II, also to be entrusted to the National Research Council of Canada, was to develop a precision gun-laying radar capable of great accuracy. Project III was to design a long-range aircraft navigation device, one in which no signals were sent out by the aircraft, but which, when a plane was over enemy territory some 500 miles away from its base, could tell the navigator his position within a quarter of a mile.

Definite steps were agreed upon to launch Project I. Bowen, who was England's outstanding authority on airborne radar, drew a block diagram of component parts necessary for a microwave AI system, and from his knowledge of the operational requirements, laid down the specifications the equipment should be designed to meet. For experimental purposes, it was decided to ask the principal electronic and electrical concerns to design and supply a few units of each of these components.

These proposals were officially agreed upon at an important meeting held in Washington, on Friday October 18 at the Carnegie Institution, attended by Bush and Compton and by several Army and Navy representatives, Lt. Col. Mitchell, Major Smith, Commander Briscoe, and Lt. Commander Tucker.[10]

As a result of a last minute decision arrived at two days before, the Microwave Committee at this meeting unanimously approved plans to establish the microwave laboratory at the Massachusetts Institute of Technology. Various factors entered in this decision. Some delays had been encountered in getting matters under way at Bolling Field and—what was doubtless more important—it was pointed out that the NDRC was not empowered to administer its own laboratories but should operate through contracts with existing institutions. Moreover, work on microwaves being already in progress at MIT, the atmosphere should be a congenial one. When Compton arrived in Washington on October 17 he was confronted with the proposal, agreed upon in conference the previous day by Bush, Jewett, Loomis, and Bowles, that MIT offered a better prospect than Bolling Field. He was persuaded to give his approval and to telephone MIT to ascertain if the required space could readily be made available.[11]

It was unanimously voted at the meeting of October 18 that five development contracts be approved by NDRC along the lines suggested at Tuxedo Park, the equipment to be delivered in 30 days. The Bell Laboratories were to deliver five copies of the British magnetron; General Electric was to supply the permanent magnets for use with the magnetrons; pulsers were to be built by Westinghouse and RCA, each company to produce a laboratory pulser using commercial tubes and one or two service pulsers of special design. The Sperry Gyroscope Corporation was asked to build five parabolas and five driving gears (scanners). General Electric was to produce a single parabola and driving gear. The Radio Corporation of America was to supply five 12-in. cathode-ray tubes, as well as five intermediate-frequency amplifiers for use with the Bell receiver equipment, and two complete experimental 3000-MHz receivers. The Bell Laboratories were to deliver five units consisting of a tube detector and tube mixer. Sperry was to produce two crystal mixers and a number of 10-cm klystrons, for possible use as preamplifier tubes. Work was begun at once, in nearly all instances, without waiting for the letters of intent sent to the five companies on October 25, 1941 by Vannevar Bush.

K. T. Compton began negotiations with the Commanding General, First Corps Area, for the use of the National Guard Hangar at the East Boston Airport. A preliminary conference on space was held at MIT in the office of President Compton on the morning of October 24; the meeting was attended by A. L. Loomis, Carroll Wilson, L. A. DuBridge, John

Trump, and E. L. Bowles. After a general discussion, the group was joined by K. T. Bainbridge and L. A. Turner and a tour of inspection of the available space was made by the visitors. The results of this survey were embodied in a letter in which Loomis officially informed Vannevar Bush and the NDRC of the decisions of the Microwave Committee and submitted their recommendations to the central body for the development contracts, totalling $138,425, and for the occupation of the available 11,925 square feet of net space.[12]

On October 25, 1940, the NDRC approved the program as submitted to it by the Microwave Committee. This covered the five development contracts for the components and the contract with the Massachusetts Institute of Technology, which was later signed on February 5, 1941. Under this contract, NDCrc-53, the sum of $455,000 was allocated for the first year of the Laboratory's existence. How modest this appropriation was should be evident from the fact that it envisaged a laboratory of only 50 persons, including technical assistants, mechanics, and secretarial help.[13]

Foundation of the Radiation Laboratory

E. O. Lawrence had actively joined the work of the Microwave Committee early in October in response to an urgent summons from Bush.[14] The importance of his addition to the Committee was evident as soon as it was decided to draw mainly upon American academic physicists in finding personnel for the new laboratory. He alone, perhaps, of the members of the Microwave Committee could readily have enlisted the support of his colleagues in the American Physical Society in a project the details of which could not at first be disclosed.

Names of possible "consultants" were considered by the Microwave Committee in the first week of October, even before the plans for the Laboratory had been completely drawn up or approved.[15] K. T. Bainbridge of the Harvard Physics Department was the first to be brought actively into the picture by Lawrence,[16] and was one of the earliest outside the Microwave Committee to be officially "cleared" by the authorities for work on radar. On October 17, he visited the Naval Research Laboratory and learned of and witnessed for the first time the Navy's radio detection devices; on the following day he attended the historic meeting of the Microwave Committee which launched the program.[17] Bainbridge also took part in the conferences with K. T. Compton, Alfred Loomis, E. O. Lawrence, and members of the Microwave Committee at which it was decided to offer the post of Director of the new microwave laboratory to L. A. DuBridge, Chairman of the Physics Department and, since 1938, Dean of the Faculty of Arts and Sciences of the University of Rochester.

On or about October 15 Lawrence called DuBridge from New York on the telephone and informed him that important defense work was being organized and that he was needed. DuBridge, feeling, as he described it later, "that if Lawrence was interested in the program that was what I wanted to be in," took the train that evening for New York. At a conference at the Hotel Commodore the following day with Lawrence and Loomis, DuBridge was persuaded to accept the post of technical director of the microwave laboratory. Together they looked over a list of American physicists and discussed names.[18]

Immediately after their conference, Lawrence and DuBridge visited L. A. Turner at Princeton, discussed the program in general terms with him, and took steps to have him made a consultant.[19] Shortly afterwards, DuBridge went to the University of Indiana where a group of physicists from several midwestern universities were meeting in what was described as an interdepartmental seminar.[20] A number of the mid-westerners, among them F. Wheeler Loomis, Chairman of the Physics Department of the University of Illinois,[21] were sounded out in general terms by Du-Bridge.

More active recruiting took place in Cambridge by Lawrence, Du-Bridge, and the Microwave Committee during the week of October 28–31 when a conference on applied nuclear physics brought to MIT some 600 physicists from all parts of the country. Most of the future members of the Radiation Laboratory Steering Committee were present, many of them having been apprised in advance by E. O. Lawrence and by DuBridge of important developments. On the afternoon of each day, a seminar talk for the benefit of the physics group was given on nonsecret aspects of ultra-high-frequency work by Barrow and his associates. At a series of conferences—a meeting of the Microwave Committee to which several of the physicists were invited, and two luncheon meetings at the Algonquin Club in Boston—the leading nuclear physicists had explained to them in general terms the proposal for a microwave laboratory and the nature of the overall problem. Among the men taking part in these discussions, besides those already committed to the program, were F. Wheeler Loomis of Illinois, I. I. Rabi of Columbia University, E. U. Condon, J. C. Slater, J. G. Trump, A. J. Allen, E. M. Lyman, and many others.

At the conclusion of the Conference, most of those who had been approached left Cambridge, but a small group remained behind and, strengthened by personnel chosen from Harvard and MIT, began a general discussion of certain key microwave problems and took steps to occupy the space set aside for the new laboratory in the wing of the main MIT building occupied in part by the Department of Electrical Engineering.

On Monday, November 11, there was held a preliminary general group meeting of the Laboratory personnel. At this meeting, and a more formal one held the following day, attended by Alfred Loomis and E. O. Lawrence and presided over by the Director, who had just arrived, the main outlines of the Laboratory organization were agreed upon.[22] Research problems were parceled out among seven technical sections concerned with developing or improving the chief components of the system. These were: Section I, pulse modulators, K. T. Bainbridge, Chairman; Section II, transmitter tubes, I. I. Rabi, Chairman; Section III, parabolas and antennas, A. J. Allen, Chairman; Section IV, receivers, L. A. Turner, Chairman; Section V, theory; Section VI, cathode-ray tubes, W. M. Hall, Chairman; Section VII, klystron test sets, F. D. Lewis, Chairman. A final section, Section VIII, was charged with the technical integration of equipment being manufactured or designed for assembly into a system by the Laboratory.

By the middle of December, the organization consisted of some 35 persons. About half of the promised space had already been occupied and a roof laboratory—a wooden penthouse covered with grey-green tarpaper—had been erected on the roof of MIT Building 6, and a second story was being added. The personnel consisted of about 30 physicists, three guards, two men in charge of the stockroom and the purchase of supplies, and one secretary. The Laboratory was placed under the supervision of an Executive Committee consisting of Loomis and Bowles of the Microwave Committee, L. A. DuBridge, the Scientific Director of the Laboratory, and Melville Eastham, President of the General Radio Company, the business manager.[23] It had already received the name of Radiation Laboratory, selected because it concealed, yet in ironical fashion expressed, the field of activity of the new laboratory. The adoption of the name used by Lawrence's cyclotron laboratory at Berkeley suggested the natural hypothesis that this group of nuclear physicists was engaged in nuclear physics, then deemed a harmless and academic occupation.

In the second half of November, Lawrence continued his effective recruiting for the Radiation Laboratory by visits to Yale and Columbia, and by a leisurely train trip across the continent during the course of which he stopped off at the University of Chicago, Purdue, the University of Wisconsin, the University of Minnesota, and a number of other institutions. E. C. Pollard, R. G. Herb, A. E. Whitford, Dale Corson, and several others were brought to the Laboratory as a result of this trip.[24]

The Bell Laboratories delivered their first five magnetrons to the Radiation Laboratory precisely on schedule on November 12.[25] The first units of the other components began to arrive shortly after. It was recognized from the beginning that these components were only a starting point. Work was coordinately begun on testing and adapting the delivered items, on design of new components suitable for use in an aircraft, and on assembling a first working microwave radar from the equipment available.

On November 17, E. G. Bowen drove up to Boston with A. L. Loomis and E. O. Lawrence and visited the Laboratory the following day for the first time. In the afternoon, to those members who had been cleared, Bowen gave a talk on military tactics and operational use of airborne-interception radar, with special emphasis on the problems of AI-10. On the following day, he held a series of discussions with each of the leaders of the separate components sections.[26]

On the afternoon of December 16, a planning meeting was held at the Laboratory and a schedule to be met was laid down.[27] This provided that by January 6 a microwave system should be working on the roof; that by February 1 equipment should be working in a flying laboratory, a B-18 plane to be supplied by the Army; and that by March 1 a system should be working in an A-20-A attack bomber, which at that time was the most likely choice for a night fighter. A group charged with the assembly of the system was at once set up, and L. W. Alvarez was appointed expeditor to ensure that the schedule would be met.[28]

During the last two weeks of December, the first experimental radar system was assembled in the roof laboratory, largely from the components supplied with such admirable dispatch by the commercial concerns.[29] The system had separate antennas for transmitting and receiving, since the problem of a duplexer, or what the British referred to as a TR (transmit-receive) box, to permit the use of the same antenna for transmitting and receiving, had not yet been solved. This system was first successfully operated on January 4, 1941, two days ahead of schedule, and picked up echoes from the buildings of the Boston skyline across the Charles River. S. N. Van Voorhis has been principally credited with wrestling the first roof system into successful operation.

In order to design a system for use in an aircraft, it was imperative that it operate with only a single antenna. Yet with a single parabola without a duplexing or switching device, or at least some protection for the receiver crystal, the main transmitted pulse would burn out the receiver crystal. While various solutions were being tried, it was discovered early in January by J. L. Lawson that a klystron used as a pre-amplifier tube would serve effectively as a buffer for the crystal. While only a partial solution to the TR box problem, it permitted the roof system to be operated with a single paraboloid on January 10. To the Director, who was in Washington, was sent a cryptic wire informing him of the success: "have succeeded with one eye."[30]

There were many things wrong with this early system; in fact, there were few things that were really right. The rf plumbing consisted entirely of hand-made, bead-supported coaxial line, poorly designed, poorly matched, and subject to frequent breakdown. Tuning stubs were introduced at so many points in the rf system that tuning was distinctly an adventure. The use of the klystron buffer made the signal-to-noise ratio in the receiver very bad indeed. The principal cause of low power had, however, been uncovered and remedied during January. Although Rabi's group had very early discovered that the magnetron could produce considerably more power than the British had suspected, an error had been made in determining the precise frequency at which most power was put out—spectrum analyzers were not yet available—and the receiver was incorrectly tuned to the magnetron.[31]

As yet, the roof system had only received ground echoes. Attempts to pick up aircraft signals had failed and some observers, among them Bowen and E. O. Lawrence, began to doubt that the system was capable of performing this essential feat.[32] Equipment that gave such weak ground signals seemed not to offer much promise of detecting aircraft. The best signals were obtained by pointing the antenna steadily at the target (searchlighting); the use of scanning would require still more power. By frantic efforts and some last-minute improvisation, which consisted of pointing the paraboloid by hand using a crude telescopic sight, aircraft signals were observed at a distance of 2 miles on February 7, 1941, in time to be reported by telephone to a gloomy session of the Microwave Committee, which DuBridge and Lawrence were attending in Washington. The report of this success changed the mood of the meeting which voted confidence in the microwave AI program.[33]

This success had an inspirational effect somewhat out of proportion to its scientific importance. An earlier suggestion of Lawrence and Bowen that more than one system should be in operation at one time was put into effect in the middle of February when the Executive Committee of the Laboratory decided to create a Roof Systems Group under L. C. Marshall for testing components and for the general improvement of microwave system design.[34] A special systems section under E. M. McMillan, assisted principally by J. Halpern and H. Schultz, was formed to prepare the AI installation in the B-18.

Between February 13 and March 5, the system that had been in the mock-up was worked over and modified for installation in the plane.[35] The equipment performed satisfactorily on March 5, its last day in the roof laboratory; it gave a strong signal on the water tower six miles away that had become a standard target. On March 6–7 it was installed in a B-18 plane, equipped with a special Plexiglas nose transparent to hyperfrequency radiation, which had been flown up from Wright Field by an Army crew. The equipment was first flown on March 10 with poor results; but its performance was steadily improved during the rest of the month.

On March 27, there took place a flight in the B-18 with E. G. Bowen, E. M. McMillan, and other scientists aboard which had important consequences for the Laboratory. The equipment performed admirably, and for what was probably the first time an airborne microwave radar was tried out for ASV purposes.[36] Tests on a 10,000-ton ship gave strong signals at 9 miles from an elevation of 2000 feet; and the "sea return," i.e., interfering echoes from the surface of the water, was much less than had been feared. Encouraged by this success, the plane was flown to New London to look for submarines operating near that base. There the pilot made successful runs on a surfaced submarine and the men found that from an altitude of 500–1000 feet strong signals were obtained at a distance of 3 miles.

The first flight of the experimental AI equipment on March 10 can be taken to mark the end of the first phase of the Laboratory's history. On this date the Microwave Committee submitted its first report on the Laboratory to Bush, describing the progress that had been made in the AI development, upon which most of the Laboratory effort was concentrated, and the less extensive results of Project II, gunlaying, and Project III, Long Range Navigation, to be discussed below. The report revealed that during the four months the Laboratory had grown to number 140 persons: including 90 physicists and engineers, about 45 mechanics, technicians, helpers, guards, and secretaries, and six Canadian guest scientists. The organization was substantially unchanged. About 14,000 square feet of floor space had been occupied, including offices and conference rooms. This rapid growth, and the importance of the work on which the Laboratory was engaged, resulted in the authorization by NDRC at a meeting on March 7 of an additional $300,000. It was estimated at this time that an additional $665,000 would be required to continue the work of the Laboratory until November 1, 1941, and that another $1,200,000 would be needed to prolong the microwave work to the end of the fiscal year on June 30, 1942. It was about this time that John D. Rockefeller, Jr. agreed to underwrite the salaries of the technical staff of the Radiation Laboratory between the dates of September 1, 1941 and July 1, 1942. This generous and unpublicized gesture was important in assuring the continuity of the Laboratory's efforts and its steady growth, even though the funds themselves were not required when a new and more liberal contract was signed between MIT and the Office of Scientific Research and Development on September 23, 1941.

Up to the month of March, the main effort at the Radiation Laboratory had been to get a plane in the air with radar aboard, and to meet as closely as possible the schedule laid down in mid-December. This phase of the Laboratory's history was now closed, and there began a period of greatly expanded and diversified effort. The Laboratory's attention was still primarily concentrated upon the design of perfected AI equipment suitable for operational aircraft. This was predicated chiefly upon the successful design of new and improved components. However, there was increased activity in Projects II and III and additional development along two main lines: (1) new applications of the 10-cm AI equipment, especially those involving the use of microwave radar over water; and (2) the development of radically new types of radar.

Research at the Radiation Laboratory

Unquestionably, the most important aspect of the Laboratory's effort during the first year—and almost equally vital in the years that followed—was the development of improved components and the steady growth of knowledge of the properties and behavior of microwaves: how to produce them, carry them along coaxial cables or waveguides, receive them, amplify them, and display them.

A word should be said about the manner of work. The Laboratory was loosely and informally organized, small enough for constant interaction of the various parts, and even for frequent exchange of personnel. The spirit and morale were very high. The lines separating the different sections were anything but formal barriers. Men drifted across them freely, to aid one another in a tight spot, even sometimes to trespass to good effect upon someone else's preserve. It was a picked group, fully conscious of its undiluted strength, as yet untroubled by problems of production and higher diplomacy, unencumbered by administrative routine.

These men shared with the industrial laboratories, chief among them the Bell Telephone Laboratories, the experience of laying the foundations of a new engineering art. The conditions and objectives of research were widely different from what most of the men had been accustomed to. It was applied science; and it was also wartime science. Especially during the first year, the men relied upon empirical investigation of the cut-and-try variety, guided by their theoretical training and insight, but without benefit of much practical experience in radio engineering. The rediscovery, en route, of familiar engineering practices was not an uncommon experience. On the other hand, they were free from a heavy accumulated load of engineering rules of thumb not always adaptable to this new field. The wartime urgency of their work meant that a wholly logical, planned attack on a problem as in peacetime was almost never feasible. Speed was the all-important consideration and there was no time for leisurely theoretical exploration or fundamental research. Most of the knowledge was acquired by building something as quickly as possible and trying it out. Theoretical knowledge grew *pari passu* to be plowed back into the work at a later date. Hence the importance of the various experimental systems soon scattered throughout the laboratory. Experimentation consisted mainly in trying out new components and new ideas as swiftly as possible in the experimental systems on the roof or in the B-18.

Development of Improved Components

The first systems were frankly experimental, intended to educate the laboratory members in the new art, and to help them obtain some general familiarity with the properties of microwaves. All of the earliest systems were assembled almost entirely from the components supplied by industrial concerns under the first contracts. These components in some instances had been considerably modified and changes in them were constantly being made. A rebuilt Westinghouse pulse modulator; a Bell copy of the British magnetron; a Sperry paraboloid and scanning gear; a receiver consisting of a Bell Laboratories crystal mixer, a grounded-grid triode local oscillator, together with an RCA intermediate frequency amplifier; these found their way into all the early experimental systems. The only component in the early systems designed and built entirely by the laboratory was the very important synchronizer unit, designed by E. M. Lyman, of which about twenty were built by hand in the first few months.[37] It was used to provide the triggering pulse to the modulator and blanking pulses to the receiver, and to synchronize the sweep circuits for the cathode-ray tube.

Only the briefest outline is possible here of the complex activity which produced, within less than a year and a half, an operationally satisfactory, if primitive, microwave radar system. Work began on the development of components during the first days of the laboratory.

The essential features of the 10-cm magnetron were not substantially altered.[38] The development of power-measuring techniques and of spectrum analyzers made it possible to understand the potentialities of the tube and to get much more power out of it than the British had been led to expect at first. The introduction of a technique called "strapping," the importance of which the Laboratory learned from the British in the fall of 1941, greatly increased the stability and the efficiency of the magnetron. Although no wholly satisfactory theory of magnetron operation was evolved, much was learned about the modes in which the tube can oscillate.

The most important achievement—and one upon which the magnetron group concentrated from the very first—was the development of a magnetron operating on 3 cm. This was successfully accomplished in the spring of 1941 with the adoption of some novel changes in magnetron design. In both the 10- and 3-cm work the Laboratory was greatly aided by a vacuum-tube model shop facility provided by the Raytheon Manufacturing Company in Newton, Massachusetts. Here a handful of tube experts produced experimental magnetrons, modulator tubes, etc., following suggestions and drawings submitted by the Radiation Laboratory. Raytheon operated under a subcontract from MIT's Division of Industrial Cooperation, MIT being reimbursed under the NDRC contract OEHsr-5.

The pulser or pulse modulator, supplied by Westinghouse and used in the early experimental systems, produced pulses of the required length and repetition rate, but was too wasteful of space, weight, and power to be satisfactory.[39] As early as November 8, 1940, the pulse modulator problem was discussed with a view to developing a unit suitable for aircraft use. Two important developments of the modulator group, both ideas proposed by J. C. Street, laid the foundations during the winter of 1940–41 for the later art of radar pulse modulators. The first of these was the development of a "bootstrap" cathode follower circuit; the second was the adoption of the pulse-forming network. These were important elements in the design of the so-called Service Modulator and Laboratory Modulator, both of which were manufactured during the year by Raytheon. The use of the network did much to improve pulse shape, and later became the basis of the high-power modulator using the pulse-forming line with rotary spark gaps. An important contribution to the work of the modulator group was the development of testing equipment—rf envelope viewers and synchroscopes—which permitted the study of pulse shapes. Special adaptations of the basic circuits were used in designing the modulators for various laboratory systems intended for production. Work was begun late in 1941 on the high-power modulators and on the development of a modulator using oxide-coated cathode output tubes, among them a lightweight pulse modulator manufactured by the Stromberg-Carlson Company, later referred to as the Navy Standard Pulser.

Improvements in antenna design consisted largely in finding the proper way of feeding rf energy to a standard parabolic reflector, and determining the optimum focal length and the proper design and proper matching of the radiating dipoles.[40] The effort was concentrated upon getting a high-gain pencil beam with low side lobes, without introducing very novel reflector dish design.

The receiver problem divided itself into the rf and the IF problems.[41] The earliest IF receiver strips from the Bell Telephone Laboratories and RCA had been designed on the basis of television experience and were only a first approximation of what was required for microwave radar. In its broad outlines, the later receiver development was conservative and there was no departure from the basic superheterodyne principle; yet there were fundamental improvements in circuit design which made it possible to build high-gain and broad-bandwidth receivers with proper transient response. The problem in microwave radar was to build receivers that could tolerate unbounded signals and escape paralysis from the effects of the main transmitted impulse.

The receiver rf problem was part of the broader problem of handling rf energy on these frequencies. The answer to the question as to which first detector, whether a crystal mixer or a grounded-grid triode, was better, hinged in great part upon a solution of the duplexing or TR box problem. The earliest Laboratory experimental systems used a crystal detector, then shifted over to the use of a Bell Laboratories grounded-grid triode, and finally settled on the crystal mixers which became standard for all subsequent microwave radar. This was both because crystals had won out over tubes in the race for greater sensitivity as detectors and because the solution of the TR box problem gave adequate protection to the crystal.

In the spring of 1941, an rf Group was created under W. W. Salisbury.[42] The earliest rf work at the Laboratory, in connection with designing a 10-cm AI system, was centered on three main problems: to design improved coaxial lines and line components such as tuners and rotary joints; to evolve measuring equipment to test the components under development; and to solve the TR box problem. The coaxial line was first radically improved by J. L. Lawson, who designed a beaded line using a particular nonuniform spacing of the polystyrene beads, and finally by adopting the use of brass stubs instead of beads to support the inner conductor, an idea brought back from the Bell Laboratories by J. R. Zacharias. To meet the need of measuring equipment, standing-wave detectors, wavemeters, and wattmeters were developed and improved. In March of 1941, Lawson designed a successful and rugged spark-gap TR box, which was followed later in the year by the adoption and improvement of the British so-called soft Sutton Tube TR. The Lawson TR was introduced into the B-18 system about April 1 and made it possible to eliminate the klystron buffer entirely, with a resulting improvement in signal-to-noise ratio so that the B-18 system in mid-April was able to pick up ships at a distance of 15 miles. A further improvement in the duplexing system was the adoption of "pre-plumbing." This consisted in the preselection of the proper length of transmission line between the magnetron and TR box, so as to ensure, without the use of special tuning devices, the minimum loss in received signal.

With the appearance of the first 3-cm magnetrons in the spring of 1941, the whole art had to be translated to this new wavelength. Waveguide transmission lines were adopted instead of coaxial lines, which would have had to be prohibitively small. The properties of waveguides had to be carefully studied, and a completely new set of components (tuners, rotary joints, waveguide "T's" and angles, flexible waveguide, etc.) had to be developed. Receiver rf components, crystals and local oscillators, and a

TR box for the new wavelength all were needed. The impossibility at this time of "pre-plumbing" the 3-cm magnetrons led to the development of a so-called anti-TR, another glow-discharge device inserted in the line to keep the transmitter from absorbing any appreciable amount of the received signal.

The key problem of the Indicator Group was the cathode-ray tube itself. The earliest indicator tubes used by the Laboratory were those supplied by RCA under the first contract. Although they served a useful experimental purpose, they were recognized to be only a stopgap. They were large electrostatically deflected tubes with a 12-in. face and a screen composed of a single layer of phosphor having only slight persistence. E. G. Bowen had described in general terms the importance of long-persistence tubes, but had been unable to give anything but general information concerning British developments.[43] He explained, however, that the British used a duplex-layer screen composed of two phosphors, an outer layer emitting orange light with a slow decay when activated by the light from the inner screen. NDRC contracts were let to the General Electric Company and to RCA Victor early in 1941 to develop long-delay cathode-ray tubes along these lines. The two research laboratories worked closely together in cooperation with the Radiation Laboratory indicator section, which served principally as a coordinating and testing center for research and later for production control. Delicate techniques were developed for measuring the characteristics of the phosphors in tubes submitted by General Electric and RCA. R. F. Bacher was assigned the task of coordinating this development, which he did with notable success.

By the summer of 1941, what later became standard indicator tubes had been adopted by the Laboratory.[44] These were for general use but were especially suited to airborne installations, being smaller and more compact than earlier experimental models. They were tubes with a flat face, and a duplex-layer persistence screen. Tubes 7 and 5 in. in diameter were designed, in both of which the electron beam was focused and deflected by a magnetic field, instead of by electrostatic means. Large-scale production was begun at General Electric and RCA during this year.

For AI work, these tubes were adopted to produce the so-called Type C scan, with a rectangular image in which elevation is plotted against azimuth, and a Type B scan giving range against azimuth. The improved cathode-ray tubes were also used for the PPI scopes or Plan Position Indicators built in the laboratory during 1941. The work was undertaken on the basis of general information about the British PPI development, but without specific design data.[45] This type of intensity-modulated indicator has a linear sweep that takes its origin at the center of the tube, and which is rotated in synchronism with a rotating paraboloid.[46] The Laboratory's first PPI, which was probably the earliest built in this country, was a magnetically focused and deflected tube with coils that were mechanically rotated (see Fig. 10-2). It was developed for an experimental shipboard system on the USS *Semmes*. An electrostatically deflected tube was built at nearly the same time for the earliest experimental 10-cm ASV system. By the middle of 1942, two types of PPI indicators had been devised: one with a mechanically rotated coil, and a second (electrostatic) type with a fixed coil using selsyns to provide the proper vector compo-

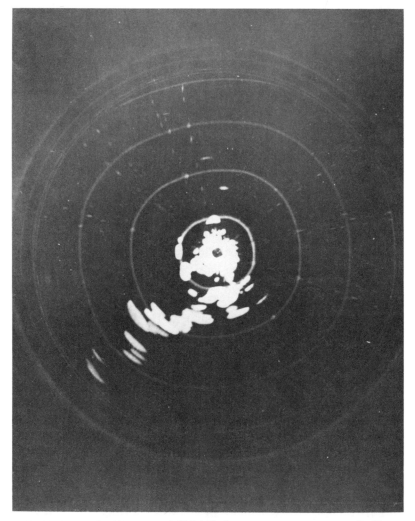

FIG. 10-2. Early 10-cm PPI (1941), from equipment on roof of hangar-laboratory at East Boston Airport.

nents. Although the electrostatic type was never able to attain the precision of the magnetic tube, it drew less power and therefore saved quite a bit of weight, and was thus preferable for use in airborne equipment.

Late in the year, an important change was made in the synchronizer unit. It was incorporated into a single box with the indicator circuits to produce the unit called the control central or indicator central. This important component was the central timing device, the heart of the modern radar systems. It establishes the pulse recurrence rate, triggers the modulator, which in turn operates the magnetron, and produces sweep voltages for the indicator tubes that are synchronous with the transmitter pulses. Much attention was paid to developing circuits for a high-speed 1-mile sweep. At about this same time, circuits were devised to introduce range markers electronically on the sweeps.

What Happened to Project I

In a conference held on January 17, 1941, at Wright Field, attended by Radiation Laboratory physicists and British representatives, the Army spokesmen expressed their doubts as to the desirability of installing AI-10 equipment in the Douglas A-20 attack bomber, as had been tentatively suggested, and instead, made known their preference to have the equipment designed for installation in the Army's new night fighter, the P-61, then in the mockup stage at the Northrup plant.[47] It was agreed that a trial installation in an A-20-A might serve as an intermediate step. In February the Army asked the Radiation Laboratory to provide equipment for 15 experimental P-61's and for one night-fighter version of the Douglas XA-26-A attack bomber.

In April, the B-18-A equipment, improved by general tinkering, and by the incorporation of the Lawson TR box and addition of better indicators, was shown to visiting officers from Wright Field and demonstrated on April 29 to Sir Hugh Dowding, Commander of the RAF Fighter Command. A Canadian Boeing aircraft used as a target was picked up at a maximum range of three miles. During the month of July, this experimental AI system was flown to Wright Field and demonstrated somewhat prematurely to the Army Air Forces, who viewed it critically.

Early in April, soon after the initial flights of this flying laboratory, a so-called A-20 version of the AI equipment was assembled by L. W. Alvarez, T. W. Bonner and their associates in a mock-up in the roof laboratory.[48] Late in May, this system was sent at the Army's request to the Bell Telephone Laboratories in the care of two Radiation Laboratory men, J. Zacharias and J. Cook, who were loaned to Bell for the rest of the year to help engineer a finished set.[49] From this cooperation emerged the first production AI set, the SCR-520, of which 50 were produced by the Western Electric Company before the end of 1942. This set, too cumbersome for serious consideration as an AI and of which only about a hundred were ever produced, was modified shortly after Pearl Harbor into the first production microwave ASV set, the SCR-517 or ASC, which was produced in considerable numbers. Its much improved lineal descendant, Western Electric's SCR-720, in which Bell engineers incorporated the latest improvements in the 10-cm art, actually became America's standard night-fighter radar installed in the P-61's, the much publicized Black Widows. The 720's began to come off the production lines in the spring of 1943 and several thousand had been delivered by June 1944.

A second system destined for an A-20 aircraft was completed at the Radiation Laboratory in June 1941. The plane that was being modified to receive it had not yet been delivered, so the set was taken to Wright Field by Bonner, where it was demonstrated for several weeks in a trailer parked on a nearby hill. The system was finally installed in the A-20 plane and flown for the first time late in September and handed over to the Army for tactical experiments at Mitchell Field. Shortly after the attack on Pearl Harbor, this plane was flown to the West Coast, where it is reported to have constituted America's entire night-fighter protection in the event of an invasion of the Pacific Coast.[50]

In June of 1941, an American AI-10 system prepared for installation in a Canadian Boeing 247D was taken to England by Dale R. Corson of the Radiation Laboratory for comparison with the British experimental AI-10, which had reached approximately the same stage of development.[51] The important discovery was made during these comparative tests that the American transmitter gave much more power than the British, but that the British had developed a more sensitive receiver. The performance of these two systems was therefore roughly comparable. Great improvement resulted when the best features of both systems were experimentally combined on the spot by testing the transmitter with a British receiver. As a result of this experience, the Americans adopted the British-type crystal mixer in place of the tube mixer and brought back the soft Sutton Tube TR box.[52]

During this period of testing, the laboratory began procurement of the components for the fifteen P-61 sets and for ten comparable sets which the British had requested for installation in Beaufighter aircraft for the RAF. However, as the year drew to a close, it was increasingly evident that service interest in "crash" procurement of AI equipment was anything but acute; the production of the P-61's themselves was seriously behind schedule, and it was felt that they could be equipped with production radar equipment when it appeared. The British also showed signs of losing interest in AI, for the last phase of the Battle of Britain had clearly been won. Even before Pearl Harbor precipitated us into the struggle, there were indications that some of the other functions of radar, particularly ASV, were to appear more important.

Microwave Radar Over Water

The discovery in March 1941 that microwave radar performed admirably over water led the Laboratory to explore more thoroughly this behavior of microwave and to design sets to utilize it. In the spring, Marshall's Roof Group expanded its activities to become a systems group for the development of types of radar not in the province of Projects I, II, or III. This group of physicists and engineers built the Laboratory's first microwave shipboard equipment (SSV), its first microwave ASV systems, and the first microwave system for coast defense and harbor entrance control duties. These systems were all characterized by being relatively straightforward adaptations of the AI-10 system, with the incorporation of the PPI the only significant innovation.

In April 1941, the Navy granted permission to the Laboratory to install an experimental ship-search system aboard the USS *Semmes*, a "four-stacker" destroyer of World War I type, operating out of New London, Connecticut, under the command of Lieutenant Commander W.L. Pryor, Jr., and assigned to radio and underwater sound experimental work. Pryor had been detached from the Radio Division of the Bureau of Ships for ship duty; he had a general familiarity with the Navy's radar program, and showed a keen interest in the SSV equipment the Laboratory installed aboard his ship. Installation of the microwave equipment aboard ship began on May 6, and the first A-scope signals were obtained a week later. The group primarily responsible for this installation were L.C. Marshall, J.R. Zacharias, H.R. Crane, D. Corson, E.C. Pollard, and R.H. Schumann.

Four distinguishable models of SSV-10, embodying a succession of improvements, were placed aboard the *Semmes* before the end of 1941. The first model consisted for the most part of AI-10 components. The familiar 30 in. paraboloid was installed on the mast, just below the crow's nest, some 50 feet above the water. The most striking departure of this first ship system from the AI-10 was the incorporation of the Plan Position Indicator. This was the first use of the PPI in any American radar installation. It was of the magnetic-mechanical type and resulted from the combined efforts of E.C. Pollard, S. Seely and J.R. Zacharias. With it the first PPI signals were obtained aboard the *Semmes* on June 5. In August 1941, R.E. Meagher began the engineering of what developed into the first satisfactory rotating-coil PPI, and his units were given experimental training aboard the ship.

Between June 9 and July 1, the *Semmes* made coastwide cruises and a trip to Bermuda that gave excellent opportunities for observing the performance of the system at sea. Land signals were picked up at distances varying from 6 to 19 miles and ships were followed to 10 miles.

During the late summer and fall of 1941, under the supervision of R.M. Emberson, who became project engineer, Models 3 and 4, incorporating innumerable changes and improvements (such as gyro-stabilization of the spinner for roll), were successively installed and tested. By November 4, when the *Semmes* put to sea again with Model 4 aboard, the system was giving 4 miles range on submarines, 8 miles on aircraft, and 26 miles on land.

During January and February 1942, Model 4 was removed and Model 5, new in almost all respects, was installed. This incorporated the results of Meagher's PPI development, an indicator similar to those used on the PBM-1 aircraft installation, the XT-3 truck, and the blimp K-3. The permanent Laboratory crew aboard the *Semmes* during this whole period were R.M. Emberson and C.M. Hammel, who were joined by others when the occasion demanded.

During March and April 1942, there were cruises to Norfolk and to Florida. The system was operating satisfactorily and in the spring of 1942 the Laboratory felt that its responsibility was at an end, although Captain Pryor was anxious to keep the SSV-10 on board, until it could be replaced by production equipment. The set remained aboard the *Semmes* until March 1943, when it was removed to make room for the SF set, but Laboratory personnel were no longer assigned to it permanently.

On a number of occasions, the various models of SSV-10 were demonstrated to visitors from the Bureau of Ships, the Naval Research Laboratory, the Bureau of Ordnance, the Raytheon Manufacturing Company, and other manufacturing concerns. In June 1941, soon after the first model had been demonstrated, the Navy placed its first microwave radar contract with Raytheon to develop, with the assistance of the Radiation Laboratory, a shipboard microwave set, based largely on the experience of the *Semmes* installations. The system that resulted, the SG, proved to be one of the most successful microwave radar sets. This development is discussed in a later chapter.

During the summer and fall of 1941, an experimental 10-cm system, closely resembling the system aboard the *Semmes*, was installed in a truck, the XT-3, to study the possible use of this type of equipment for harbor control purposes and coast defense. On November 18, it was set up on Deer Island, commanding the principal channel into Boston Harbor and the site of a Harbor Entrance Control Post jointly operated by the Army and Navy. So successful was the equipment in aiding the work of the Control Post, by supplying accurate range and bearing on all ships entering or leaving the harbor, that after Pearl Harbor this experimental Laboratory equipment remained on 24-hour duty until replaced by a production set. As a result of visits to the Deer Island installation by Army and Navy officers during December of 1941, the Army ordered a crash production of 50 sets based on the Radiation Laboratory truck system. The production of these sets was undertaken by the Research Construction Corporation, NDEC's factory-sized model shop, which went into operation late in 1941. The first of these production sets, to which the Signal Corps assigned the designation SCR-582, was installed in June 1942 at the Boston Harbor Entrance Control Post. Five of the original crash units reached North Africa in January 1943. Late in 1942, two modified SCR-582's, provided with a larger paraboloid and, in order to get greater range, a high-powered modulator, were sent to the Panama Canal Zone to supplement, by their low coverage, the longer-wave early warning network.

The possibilities of microwave ASV equipment had been mentioned many times in early discussions at the Radiation Laboratory. The first steps were taken in the spring of 1941, shortly after the historic flight in the B-18 flying laboratory, when the Roof Group initiated the adaptation of AI equipment to ASV purposes. An experimental set was built and installed during the summer in the first of many aircraft which the Navy assigned to the Laboratory for experimental work. This was an XJO-3, a Lockheed transport which had been specially adapted to this new job at the Naval Aircraft Factory in Philadelphia by the addition of a plywood nose and other changes. The first airborne PPI was built for this system. The ASV system's performance was carefully tested on a number of flights out of Boston and Philadelphia. It improved so noticeably that on September 26, in an attempt to test the ability of the system to work through the overcast, a ship was picked up from 8000 feet at a distance of 40 miles. The PPI operator guided the pilot until he could see the ship from 2000 feet.

In September, the Bureau of Ships authorized the Radiation Laboratory to carry out further tests on microwave ASV in order to assist the Bell Telephone Laboratories in developing an ASV-10 for the Navy. It was finally decided that a semi-operational installation should be made in a Navy PHM-1, a twin-engine Martin flying boat. A system was assembled during December and first flown on January 3, 1942, on a trip from Boston to Philadelphia. The installation was carefully tested on flights out of Norfolk and from bases in Florida during January and February 1942. At Banana River, Florida, comparative tests were run against a British Mark II ASV. By May 1942, the system had been operated a total of 156 flying hours and was reported to be capable of detecting cargo vessels at 45 miles and submarine conning towers at more than 15 miles.

Although it had originally been intended that the XJO-3 and FBM-1 systems experience should find expression in the ASC, the ASV system the Bell Laboratories were designing for the Navy, the influence was actually felt more directly in a set designed during 1941–42 at the Radiation Laboratory for the British and, independently of the Bell Laboratories, for the United States Navy.

At the end of July 1941, D.M. Robinson of the British Air Commission arrived at the Radiation Laboratory to explore the possibility of acquiring a small number of microwave ASV sets for use by the RAF Coastal Command. These sets were to be installed in Liberator bombers being supplied to Britain under lend-lease. Two specially modified Liberators, known as Dumbo I and Dumbo II because the bulbous radar dome beneath the nose enhanced the planes' already elephantine appearance, were equipped with prototype units of microwave ASV during the winter of 1941–42. The Dumbo I equipment flew for the first time from the East Boston Airport on December 11, 1941, the day Germany and Italy declared war on the United States. It was successfully demonstrated shortly thereafter to British and American officers and was flown to the United Kingdom in March 1942, where it underwent trials in Northern Ireland during April. The second Liberator was rapidly equipped and demonstrated at the end of April to the Secretary of War, General Marshall, General Arnold and other high-ranking officers. These two systems served as prototypes for a crash program of 17 similar systems manufactured by the Research Construction Corporation, of which 14 were for the British. The first of these DMS-1000 sets was handed over to the British representative in August 1942; the remainder were delivered by December 1942. They were able to play a valuable part in the battles against the submarines in the Bay of Biscay.

By the time the British received their first production unit, Radiation Laboratory ASV equipment had already seen service use and drawn blood from the enemy. The story of the ten B-18 ASV equipments hastily thrown together at Army request early in 1942 is of great importance in the history of the Laboratory. Their success gave the organization much needed confidence and a sense of direct participation in the war, and in large measure made up for the disappointing and inconclusive end of the AI program. This story is told in a later chapter.

Early History of Project II: Fire Control and
Automatic Tracking

The development of gun-laying radar at the Radiation Laboratory was much less influenced by British requirements and specifications than AI. This was true, in large measure, because the British Scientific Mission had entrusted the problem of developing a microwave anti-aircraft radar set, along lines already being followed in England, primarily to the National Research Council in Ottawa. At the Radiation Laboratory, therefore, the development proceeded along quite original lines.

Although in the first months of the Laboratory's existence attention was almost exclusively focused on the AI problem, by the end of the year 1940 it was determined that conical scanning should be investigated as a means for the precise determination of target position in a rather vaguely formulated radar gun-laying project. A special group, known as Project II, started work on microwave gun-laying radar systems in January 1941, using this principle. The important idea of wholly automatic tracking in azimuth and elevation was introduced at that time. Once the system had picked up a target, for example an enemy aircraft, it was proposed to have it lock on and follow, the antenna continuing to point at the target despite high speeds or violent evasive action. Data on the plane's three coordinates would be continuously fed to a gun director or searchlight.

In conical scanning, the boom from the antenna is rotated at high speed about the axis of a paraboloid, so that it describes a cone of revolution with its apex at the antenna. This produces the same effect as consecutive lobe-switching in the horizontal and vertical planes; that is, the rotating beam overlaps itself only at the axis of the paraboloid, and produces what is tantamount to a narrow pencil beam along the axis. A constant signal is received from a target at which this pencil beam, and hence the axis of the paraboloid, is exactly pointing. Conical scanning was first experimentally produced by wobbling the entire paraboloid, later by spinning an eccentrically placed dipole. With the proper circuits, the angular deviation of the target from the axis of the parabolic reflector can be detected and converted into an "error signal." By means of commutating circuits, this is changed into a dc voltage which drives the servomechanism and keeps the antenna pointing at the target. This is automatic tracking in angle. Automatic range tracking was not incorporated into production equipment, although it had been begun in the Laboratory before the appearance of production sets.

To profit by the accuracy in range measurement inherent in radar pulses, extremely precise range circuits were devised to produce high-speed sweeps. A special range unit or synchronizer was built to produce sweep generating voltages for the cathode-ray tubes, generate the trigger pulse to the driver unit of the modulator, and provide the range gates designed to eliminate target confusion, one of the principal obstacles to automatic tracking. For example, when two targets are at the same bearing and nearly the same slant range, the antenna may hunt from one target to the other or take up an intermediate position between them. Range gates are designed to remedy this difficulty by confining the reception of signals to a short interval of time, i.e., to a portion of the indicator trace. The "narrow gate" first used in the production equipment, and the still narrower H^2 gate added later as a modification kit, were important features of Radiation Laboratory fire-control radar.

The Project II Group was organized at the end of January 1941 with L. N. Ridenour as Group Leader. L. L. Davenport, I. A. Getting, A. M. Grass, and John Meade were early members. F. F. Hines, A. C. Ruge, and I. F. Woodruff of MIT served as consultants. On February 6, 1941, the first crude tests on a conical scanning system were performed using as the source of radiation a klystron transmitter placed on the roof of the Walker Memorial Building at MIT.[52] This test showed the feasibility of determining direction by means of conical scanning, and even indicated that with this crude equipment an angular accuracy of 1° could be attained. Shortly afterward, Ridenour worked out the theoretical expression for conical scanning, which was used in a later report.[53]

On February 18, the group held a formal meeting at which a block diagram for a complete conical scanning system was drawn up. A range-measuring device and range-blanking system were worked out. Ridenour first proposed a system making use of a mechanical commutator to obtain correct pointing of the antenna. Responsibilities were assigned as follows: Grass, indicator and ranging circuits; Meade and Woodruff, mount and servo; Hines and Ruge, spinner design; Getting, range-blanking system.

Ridenour visited the Naval Research Laboratory and the Bell Telephone Laboratories in search of information on ranging circuits, while Woodruff made a survey at various manufacturing plants of servo equipment and mounts for a spinner. A satisfactory mount was found in the new aircraft 50-in. machine gun turret of the General Electric Company, and two of these were ordered on March 22, 1941 for delivery in six weeks. Sidney Godet and Richard Porter of General Electric visited the Laboratory on April 1 to discuss the proposed system. General Electric was then asked to build an amplidyne control system to operate with an electrical rather than a mechanical commutator.

On April 2, the assembled system was moved to the new roof laboratory and signals were received that day. The delivery by the General Electric Company of the amplidyne-controlled aircraft machine-gun turret late in May made possible the first demonstration of automatic tracking of an aircraft in elevation and azimuth on May 31, 1941.

Although the most successful set to grow out of these early experiments was the mobile SCR-584, embodying all the features described earlier, the first efforts were directed toward using the conical scanning feature, without the addition of automatic tracking, for an airborne radar gun sight and for a ship fire control system.

Project II enjoyed the strong support and personal interest of E. L. Bowles, Secretary of the Microwave Committee, who was largely responsible for bringing the problems of the Navy, the Army Ground Forces, and the Army Air Forces to the attention of Project II, and for giving the program the impetus it needed when so much of the laboratory's effort was concentrated on airborne equipment.

ARS (Aircraft Radio Sight)

The aircraft gunsight program, which was somewhat premature, never went beyond an experimental installation demonstrated at Wright Field in January 1942. The Aircraft Radio Sight, also known as RGS (Radar Gun Sight), was intended to supplement AI equipment in an aircraft by furnishing sufficiently precise indication of direction ($\pm$ 1°) and range ($\pm$ 100 yd.) to permit the gunner to aim and fire his guns effectively in the final stage of an interception. Basically the system, proposed in April 1941, used a conical scanner consisting of an 8-in. paraboloid with an indicator.[54] Otherwise, AI components were employed throughout.

ARS, for which L. J. Laslett served as project engineer, was installed in the B-18 Flying Laboratory and demonstrated to Major F. C. Wolfe in September 1941, after which a 12-in. paraboloid was substituted for the earlier one. Later, a change was made in the receiver from a tube to a crystal mixer, resulting in an increase in range from $1\frac{1}{4}$ miles to 2 miles, but the use of the crystal seemed to give a less satisfactory minimum range. After a demonstration at Wright Field in the B-18 in January 1942, the project was dropped because the Army had no aircraft suitable for such an installation. The program of airborne radar gunsights had to await the development of the lighthouse-tube transmitter, which permitted the design of compact lightweight systems.

Radar Mark 9 and its Derivatives

The Navy program resulted in what was chronologically one of the earliest, though not the most fruitful, production contracts stemming from Radiation Laboratory research and development. Since this represents the Laboratory's first contact with the Bureau of Ordnance, it is perhaps appropriate to say a word concerning the status of Navy fire control at this time.

On the eve of our entry into the war, the need for coordinating the Navy's program of research in fire control devices—largely in the hands of industrial concerns—led to the creation, on April 10, 1941, of the Fire Control Section (Re4) of the Research and Development Division of the Bureau of Ordnance. Lieutenant Commander M. E. Murphy was made

Chief of this section. In October 1941, a Fire Control Radar Subsection (Re4f) was set up with Lieutenant Commander D. P. Tucker as subsection Chief.

At the time of the creation of these sections, the Navy's modern ships had a good optical main-battery fire-control system and an anti-aircraft fire-control system for defense against horizontal and glide bombing. Although fire-control radar was still more or less in the design stage, Main-Battery Radar Mark 3 and Secondary-Battery Radar Mark 4, both long-wave sets, were on order; in fact, the first Mark 4 was installed for testing on the destroyer USS *Roe* in September, 1941. On the other side of the picture, the Navy had no machine-gun director in the fleet or developed, nor any 3-in./50-calibre or 5-in./38-calibre gun directors even under design. The emphasis in Re4 was placed upon an expanded radar fire-control program, and upon the development of lead computing sights and machine-gun directors. The first problem of Commander Tucker's Radar Subsection (Re4f) was to provide radar sets for use with heavy-machine-gun-battery directors Mark 45, 49, and 50, which were under development.

Shortly after assuming his duties as Chief of Re4, Commander Murphy had invited representatives from the Bell Telephone Laboratories, the General Electric Company, and the Radiation Laboratory to attend a meeting at the Ford Instrument Company in Long Island City, New York, on June 13, 1941, to discuss the possibility of adding 10-cm radar equipment to the 1.1-in. AA Gun Director (Mark 45). The director had not, of course, been designed for use with radar; there was no room on it to mount anything except a small paraboloid. There seemed to be wide agreement that only range information could be provided by radar since directional information would require some sort of pip-matching scheme which in turn would mean components too bulky to put into the director. At this point, L. N. Ridenour suggested that by using circuits developed by his group at the Radiation Laboratory, accurate directional information could be provided without difficulty. This proposal was enthusiastically received; Commander Murphy requested a NDRC development project; and the company representatives seemed relieved at not having to commit themselves to what seemed a rather speculative development. A Laboratory model was promised in about two months. Out of this project grew the first microwave ship fire-control radar set, the Mark 9, produced by the Western Electric Company.

Assembly of an experimental model of the GL-10 was immediately begun at the Radiation Laboratory with John Meade as project engineer. The set was to meet the following requirements: range, 500–600 yards; angular precision, 1/2° to 1/4°; maximum paraboloid diameter, approximately 24 inches.[55] No equipment except a small direction indicator and a range data receiver could be put inside the director; no structural alterations in the director, except in the slip ring assembly, were to be permitted, since the construction of the director model was nearly completed before the addition of radar equipment was considered.

A conical scanning radiator consisting of a 24-in. paraboloid with dipole antenna and parasitic reflector was mounted on the director. The rest of the equipment was mounted below deck. The receiver is of interest since it was designed especially for this sytem and was the first 60-Hz receiver in the Laboratory and the first to have automatic gain control. Another special feature was the "T and E [train and elevation] Spot Indicator," designed by John Meade, a type of indicator used later in several Western Electric sets.

During September and October 1941, the experimental system was tested at the Naval Proving Ground, Dahlgren, Virginia. A contract was awarded, in January 1942, to the Western Electric Company for the Radar Mark 9, to be based on the Radiation Laboratory model that had been demonstrated.

The Laboratory severed connection with the project at this point. The pointing circuits, the error scope, and the conical scanning principle were the main features duplicated in the Mark 9. Only about ten sets were built, and they were never installed because the Director Mark 45 was cancelled. The Radar Mark 9's were sent to Navy radar schools, where they were useful for maintenance training.

Twenty sets of a slightly improved version, the Mark 10, intended for the Director Mark 50, were built by Western Electric and installed on ships CL-4 to CL-13. The space requirements in filtering Mark 10 to Director Mark 50 were so difficult to meet that the Mark 10 turned out to be impracticable. Nevertheless, it went through several modifications until it was abandoned when the Director Mark 50 was cancelled.

A further modification, the Radar Mark 19, intended for use with Gun Director Mark 49, can be considered another descendant of the GL-10, since it differs little from Mark 10. A considerable amount of repackaging and other modification was done, but the Laboratory took no part in this development. Besides being repackaged, two details brought about a considerable improvement over the Marks 9 and 10: the use of a crystal mixer and the adoption of preplumbing. Although its range was somewhat increased, the Mark 19 suffered because of the limitation on the size of the dish that could be installed on the director. Twenty-five hundred Mark 19's were at first ordered, but this number was cut to 250 and later to about 100 when the Director Mark 49, in its turn, was cancelled.

The Radiation Laboratory had no part in the later modifications; the principal service of the experimental GL-10 set was to demonstrate to the

Navy and its commercial contractors that a small microwave radar could be made to work in a ship fire-control system.

XT-1 (SCR-584)

The results obtained by the Project II Group with the 10-cm automatic tracking system led the Radiation Laboratory to build a mobile unit, mounted in a truck called the XT-1, to serve as an experimental system for further research. The system had both conical-scanning and automatic tracking. It was not originally intended to be a model of a military weapon, although it eventually became the prototype of the SCR-584, one of the most successful sets derived from Radiation Laboratory research and development.

The project was undertaken without any specific Service request. Considerable interest, however, in the microwave gun-laying had been expressed by Colonel W. S. Bowen, President of the Coast Artillery Board, and also by Major J. E. McGraw, on the occasion of his visit to the Radiation Laboratory in May 1941. Major McGraw was at that time engaged in writing military characteristics of a searchlight control unit to be used with the M4 anti-aircraft gun director for 3-in. AA guns. A copy of these characteristics was forwarded to the Laboratory. According to this document, if such a unit were to serve as a radio position finder for searchlights, the range desired was 15,000 yards (the maximum range of a 60-in. searchlight). If this range could not be obtained, a 5000-yard range would be acceptable for "night operation" of the M4 director. There was also in existence a set of tentative specifications drawn up by the Signal Corps for a GL radar for anti-aircraft use. This was for a dual-purpose system: to act as an early-warning device at vertical angles greater than 20 degrees and ranges from 24,000 to 44,000 yards; as an accurate position finder at ranges from 1,000 to 24,000 yards.

The purchase of a truck was approved by L. A. DuBridge on June 1, 1941,[56] and work was begun under the direction of I. A. Getting in an old hangar, where the Radiation Laboratory's Building 20 was later erected. Getting was assisted by L. L. Davenport, who was then working with the Project II Roof System and did not join the XT-1 group formally until November 12, 1941. David T. Griggs and E. W. Smith assisted in the mechanical design; but Griggs, who but a short time before had been employed by the Laboratory as a pilot to fly his personal plane as a radar target, left to become project engineer for the AGL system. Sidney Godet, of the General Electric Company, worked on the servo circuits and kept in close touch with the IT-1, in the words of the Group Leader, "with energy far in excess of what could have been expected on the basis of a mere contract."[57]

The truck was delivered in the middle of July and the radar components began to be assembled. The 48-in. paraboloid with spinning dipole, the transmitter rf, and receiver components were sufficiently advanced so that the main effort could be put on the automatic tracking and precision range circuits. The General Electric Company machine-gun turret with its auxiliary equipment was received at the Laboratory about September 1 and was used for the next year and a half until the United Shoe Machinery Company was asked to build a pedestal mount specifically designed for radar gun-laying applications. This second General Electric turret was much more elaborate than the first one used on the Roof System. It included an electronic commutator and was mounted on a frame on top of the truck so that it could be raised during operation and lowered for transportation.

The indication equipment in the XT-1 consisted of two 5-in. Type J scopes, one covering 32,000 yards, the other giving a finer indication at 2,000 yards. The Type J, as previously described, is a modification of the Type A indicator in which the time sweep produces a circular range scale on the circumference of the cathode-ray tube; the target signal is displayed as a radial deflection of the time trace.

Since the J scope gives only range indication, it was planned to obtain bearing information from a search set—the SCR-268 was the one envisaged—and by means of repeater dials to set in the azimuth and elevation until the target appeared on the J scope. The design of this type of indicating system was largely the work of A. M. Grass,[58] who had the crystal-controlled range circuits in operation in May 1941. In July, Britton Chance and E. F. MacNichol took over the responsibility of simplifying and further developing accurate timing circuits to measure range so that Grass could put all of his effort on automatic range tracking.

The control equipment of the XT-1 permitted manual tracking by means of an external handsight, automatic angular tracking from the radar signal, or hand tracking by means of pip-matching. By the end of October, the new turret was considered adequate. Tracking was first accomplished with the XT-1 at the East Boston Airport on November 8, 1941.[59] The system was demonstrated to the Signal Corps at Fort Hancock early in December 1941. The later story of Project II—the service testing of XT-1, its adoption by the Army, and the design and engineering of the SCR-584—is told in a later chapter.

Project III: Long Range Navigation (Loran)

The need for a system of navigation independent of weather conditions, and intended in the first instance for the use of long-range bombers, had been among the original proposals made by the American Armed Services to NDRC. It was also one of the subjects discussed in the conversations with the British Technical Mission. Early in October 1940, A. L. Loomis formally proposed to the Microwave Committee that they adopt a scheme in which pulsed radio waves from fixed stations are used to produce a grid or network of hyperbolic lines from which a fix can be obtained by the operator in an aircraft or ship carrying a specially designed pulse receiver. This scheme is identical with the British navigational system called Gee, about which the members of the British Mission were only imperfectly informed. Loomis' suggestion seems to have been arrived at independently of the British, though discussions with members of the mission may have helped to clarify or perhaps crystallize the project.

A simplified case must suffice to explain the principles of this extremely important new method of navigation. If two synchronized transmitting stations several hundred miles apart send out simultaneous pulses, a receiver at any point on the perpendicular bisector of the line between the stations will of course receive the pulses at the same instant. At all other points within range of the two stations there will be a fixed difference or delay in the time of arrival of the two pulses, the value of this time delay depending only upon the geographical position of the receiver. The locus of those points on the Earth's surface having a constant time-difference in the reception of the two pulses is one of a series of confocal hyperbolas described about the two stations. The time difference measured by a special receiver-indicator corresponds to a line of position which can be determined by consulting a prepared chart. When the time delay from a second pair of stations is similarly translated into a second line of position, the result is an accurate navigational fix. Hyperbolas result from the pulses sent out from two Loran or Gee stations.

Loomis' proposal was adopted by the Microwave Committee, and a Coordination Committee for Project III was set up to arrange for the procurement, field installation, and testing of suitable equipment.[60] As in the case of the AI-10 radar, the first equipment was to be supplied by industrial concerns. A frequency of about 30 Mc/sec, almost the same as that of Gee, was first chosen. A transmitter peak power of 2000 kW was specified; but the exact pulse recurrence rates and methods of synchronization were not settled when the original equipment was ordered in December 1940.

After the organization at MIT of the Radiation Laboratory, the technical responsibility was assigned to a special Navigation Group set up in January 1941 under the direction of Melville Eastham. During the spring of 1941, while awaiting the delivery of the Project III equipment, this group concentrated upon a technical review of all aspects of the proposed high-frequency system.[61] This reconsideration resulted in some small-scale experimentation during the summer of 1941 with a system using medium frequencies. The initial tests were so successful that the original high-frequency plan was abandoned early in 1942. This new system became known as LRN (Long Range Navigation) and these letters were later expanded into the word LORAN.

The basic evolution of the Loran system was virtually completed by September 1941, when the present system of receiving and comparing the time of arrival of pulses was evolved. This highly precise time-measuring technique used precision circuits similar to those of the radar range unit, and a cathode-ray tube indicator with a double trace. The pulses of two stations appear on the separate traces, which are carefully marked off by electronic time markers. By superimposing the pulses, the time difference can be read directly from the scale with an accuracy of about five microseconds.

Growth of the Laboratory, 1940–1942

On January 8, 1941, F. W. Loomis, Head of the Physics Department of the University of Illinois, joined the Radiation Laboratory and assumed, in fact though not in name, the role of personnel director. In July 1941, Loomis was made Associate Director of the Laboratory. Largely as a result of his efforts, the staff of the Laboratory had increased by that date to 225 persons, of whom 141 were staff members (i.e., scientists and others of professional grade) and 30 were technicians, machinists, office personnel, guards, etc.

The space available did not grow proportionately. Taken as a whole, the original space in MIT Building 4, the laboratory on the roof of Building 6 (see Fig. 10-3), and the so-called West Laboratory, where shop facilities, laboratories, and offices were located in space supplied by the Institute's Mechanical Engineering Department in Building 3, provided about 19,800 square feet of space. The figure, together with a promise of an additional 3000 square feet, meant an average of 100 square feet of working space per person, less than half what has been considered in industrial laboratories to be efficient working space. Although numerous flights with experimental equipment had been made from the East Boston Airport, it was not until July that the Laboratory was able to take possession of 6300 square feet of hangar space in the National Guard building. The aircraft then available consisted of an Army B-16 bomber and a private airplane.

In the early summer of 1941, further expansion of the Roof Laboratory seemed urgent, but there were numerous objections, for such temporary inflammable structures were causing concern to the Cambridge city authorities and it was feared that the Roof would become dangerously over-

FIG. 10-3. Roof of MIT Building 6 in 1945, showing temporary penthouse roof laboratories. The bulbous structures are plywood "radomes" used primarily to protect experimental radar antennas from wind and weather, secondly for concealment.

loaded. Very serious consideration was given at this time to a proposal that the entire Laboratory should be moved away from Cambridge to a site near Mitchell Field, Long Island. This plan, however, was abandoned, and instead ground was broken on August 1 for a permanent fireproof building to be erected behind the main Institute buildings. Designated as Building 24, it was first planned as a three-story structure to house the main administrative offices and provide some laboratory space.

Even with the new building, it was apparent that more space would be needed, principally because of additional projects taken on at the request of the Navy, and on January 2, 1942, NDRC approved the purchase by MIT of the Hood Milk Company Building located on Massachusetts Avenue within two blocks of Building 24. This was a three-story brick structure which could be easily renovated.

With the acquisition of the Hood Building, 111,000 square feet of space (exclusive of airport space) was available to accommodate the new total of 477 employees, distributed roughly as follows: 208 staff members, 70 technicians; 190 secretaries, mechanics, draftsmen, guards, clerks, and helpers. One floor of the new building was assigned for the temporary use of the Radar Counter Measures Laboratory (later known as the Radio Research Laboratory) until such time as this group could make arrangements to move to Harvard.

The Microwave Committee was one of the earliest groups within NDRC to expand the original notion of NDRC's function beyond that of developing and delivering to the Services a laboratory model or prototype of a new device, and to foresee the desirability of making available several preproduction units of new equipment for demonstration, trial, and, in some cases, actual combat. If the NDRC could supply a small number of units, say 5 or 10, the Services would be in a better position to evaluate the new instrument to detect weaknesses, to suggest modifications, and to estimate more accurately military needs for large-scale production.

It soon became apparent that the Microwave Committee could not very well rely upon commercial suppliers, whose facilities were rapidly becoming saturated by large-scale production, to engineer and produce by hand the small quantities desired. To meet this need, and to relieve the Radiation Laboratory of this responsibility to some extent, the Microwave Section established a so-called D-M Section (the M standing for Model) to operate a Model Shop for the manufacture of limited numbers of copies of developmental microwave radar equipment to be supplied to the Army, Navy, and British. There was strong sentiment against creating a government-owned factory that might compete with commercial firms by producing on a large scale, like Canada's Research Enterprises Ltd.; accordingly it was agreed that a limit of $100,000 should be placed upon any order. The membership of Section D-M was composed of six representatives of concerns manufacturing this type of equipment, plus a representative of the Radiation Laboratory and a chairman who was to be an experienced industrialist not connected with a company having substantial contracts in this field. The Section D-M members were Melville Eastham, General Radio Company, Chairman; L. A. DuBridge, Radiation Laboratory; W. R. G. Baker, General Electric Company; P. R. Bassett, Sperry Gyroscope Company; M. J. Kelly, Bell Telephone Laboratories; W. M. McCurdy, Westinghouse Electric and Manufacturing Company; H. A. Poillon, Research Corporation; E. W. Ritter, RCA; C. G. Suits, General Electric Company.

The Research Corporation of New York, a non-profit organization, devoted in peacetime to the support of scientific research, was chosen as contractor; and upon the recommendation of the Microwave Section, NDRC on September 26, 1941 authorized a $300,000 contract for the establishment of a Radar Model Shop and to cover its operation for a six-month period from October 1, 1941 to June 30, 1942.

The Cambridge Division of the Research Construction Company, Inc., was set up in Cambridge with Ely Hutchinson as manager. Building space at 230 Albany Street, Cambridge, had been acquired by October, and by November 6 an order had been accepted from the Radiation Laboratory for 50 scanners. Orders for various small parts followed and in December there was an order for ten B-24 ASV systems, known at RCC as the DMS-1000. These were normal model shop orders. However, there was violent departure from the original program, and a beginning of the "crash" program policy that was RCC's most useful form of production, when, effective December 17, 1941, the Signal Corps ordered 50 sets of the SCR-582.

The rate of growth of the Laboratory was increasing markedly and the period of sharpest increase was between January and June 1942. Immediately after Pearl Harbor, during the last months of 1941 and the first months of 1942, there was much discussion of the desirable extent of expansion of the Radiation Laboratory. This was indeed the turning point in the history of the Laboratory. For the first time, it was clear to all concerned that the work was not to last a matter of months, but a matter of years. The problem of expansion was actively discussed in the Microwave Committee and in NDRC. There was strong opposition to further growth of the Radiation Laboratory from certain industrial representatives on the grounds that the Laboratory was encroaching upon the legitimate sphere of industrial enterprise and was not moreover properly constituted to be an engineering organization. They expressed a preference for keeping it a small research organization concentrating on fundamental research. Under the leadership of Alfred Loomis, who was a firm believer in the necessity of a "follow through" policy, the Microwave Committee strongly recommended "a severalfold increase in the number of scientists and engineers engaged in its research and development program." K. T. Compton concurred and late in February 1942 wrote an important memorandum to Bush and Conant supporting the Microwave Committee's position.

The attack on the size of the Radiation Laboratory had carried with it the proposal that the Laboratory was becoming too large for efficient operation and that it might be well to decentralize it by dividing the microwave radar work among various other universities; but the Microwave Committee was unanimous, Compton testified, that to subdivide the Laboratory would impair its efficiency in its principal function of developing microwave radar systems.

This, however, was not necessarily true for certain types of fundamental research which could be dispersed without encountering insuperable difficulties in maintaining proper security or coordination. Steps were already under way that would help to allay the criticism that Section D-1 had so far let only two university contracts—one with MIT, the other with the University of California—while having dispersed its microwave activity quite widely among industrial concerns. This would make available to the program the services of technical personnel who could not readily be released from their respective institutions. A committee was set up at the Radiation Laboratory to draw up a list of approved projects which could be "farmed out" and to circularize educational and research institutions that might be interested. A total of 24 separate contracts were initiated with 15 universities and two research institutions. Fifteen of these contracts were entered into in 1942. Although it was originally intended that these projects should be in the Restricted category, the large majority were made Confidential and five of them Secret.

These contracts were mainly for the solution of problems of microwave theory or basic electronic research. They included a study of cathode-ray tubes at Brown University; a waveguide investigation at the California Institute of Technology; studies on electromagnetic theory and the properties of range and computer circuits at Cornell University; the development of attenuators and rf test equipment at the Polytechnic Institute of Brooklyn; crystal research at the Purdue Research Foundation and the University of Pennsylvania; magnetron cathode studies at the Bartol Research Foundation; the development and wide study of dielectric materials in the laboratory of A. Von Hippel at MIT; and contracts with the Franklin Institute, Kansas State College, the University of California, Rensselaer Polytechnic Institute, and other institutions. These were all independent OSRD contracts administered by Section D-1, though they had some of the features of subcontracts of the Radiation Laboratory. Technical liaison with the Radiation Laboratory was ensured by assigning to each project the individual in the Laboratory who was most directly in a position to use the results of the investigation.

In the course of 1942, two major OSRD contracts were let to other universities. One of these was to Harvard for a Radar Counter Measures (RCM) Laboratory (this remained a Section D-1 contract until the creation of Division 15 of NDRC in November 1942). The other was with Columbia University for the operation of a special microwave laboratory, a tube and circuit laboratory devoted to fundamental research in transmitting and receiving radiation in the wavelength region of 1 cm. The Columbia Radiation Laboratory, as it was eventually called, was placed under the direction of I. I. Rabi of the MIT Radiation Laboratory and J. M. B. Kellogg of Columbia. It limited its activities to component development and became the advance research group in NDRC's microwave program.

The expansion of the Radiation Laboratory necessitated a series of revisions in the contractual agreements between MIT and NDRC. On June 11, 1941, contract NDCrc203 was drawn up to supersede and replace NDCre53. It was to run from November 1, 1940, to October 31, 1941, and authorized the enlarged sum of $815,000. This soon became inadequate.

On June 28, 1941, Executive Order No. 8807 from the White House established the Office of Scientific Research and Development in the Office of Emergency Management, Executive Office of the President. This new organization combined under the same top administration the NDRC and the Committee on Medical Research. The contractual responsibilities and obligations of NDRC were transferred to the new body, of which Vannevar Bush became the Director. The Chairmanship of NDRC, which became an intermediate administrative body for coordinating the war weapons development, was handed over to James B. Conant. The organization of NDRC under this new arrangement remained virtually unchanged, with Division D under Compton and Section D-1 (the Microwave Committee) under A. L. Loomis. It was not until December 1942, over a year later, that after widespread consultation NDRC was reorganized and a rigid "vertical" organization composed of 19 Divisions, with each Division Head reporting directly to the Chairman, was finally instituted. At this time, Section D-1 was raised to the status of a full division, called Division 14.

A new contract, OEMsr-61, was signed September 23, 1941, to cover the period between November 1, 1940, and June 30, 1942. This was for $2,920,000 and showed that the new organization meant business, for it was more than three and a half times as large as the previous allocations. This was replaced by a new contract, signed February 17, 1942, OEMsr-262, under which, together with its numerous supplements, the Laboratory continued to operate until the end. The amount of the new contract was

$4,930,000. The amount authorized increased steadily with each supplement until Supplement No. 14 of May 1, 1945, provided for a sum of $109,180,000 under the contract.

The Organization of the Radiation Laboratory (1940–1941)

During the first year of its existence, the Radiation Laboratory was divided into groups of two main kinds. One kind was concerned with the design, construction, and test of completed systems, while the other was concerned entirely with fundamental research and development of the individual radar components such as pulse modulators, magnetrons, etc.

The organization of the Laboratory, however, remained substantially unaltered until the spring of 1942, even though the number of groups of each kind had multiplied. There were six component development groups, a theory group (which consisted at this time more of a panel of consultants than of a full time group associated with the Laboratory), and a so-called engineering group formed late in the summer of 1941 under F. S. Dellenbaugh to develop power equipment (generators, transformers, rectifiers, etc.). Moreover, the number of groups devoted to systems development had increased to eight.

The week after the Pearl Harbor attack, the problem of laboratory organization was taken up for the first time in an informal discussion between the director and several members of the Laboratory. The problem was widely debated among the Group Leaders during the next three months. This brought about a reconsideration of the basic purposes and functions of the Laboratory, especially the extent to which it should participate in the engineering development and small-scale production of microwave equipment with the attendant responsibilities. There was widespread agreement with the policy of the Microwave Committee, inspired partly by British example and strongly encouraged by E. G. Bowen, that it was not sufficient to make a "breadboard" model of a new device to get it accepted by the Armed Services, but that additional preproduction models were useful for demonstration and Service testing and for emergencies in war requiring small quantities of equipment for immediate use. This had already manifested itself in the creation of the Research Construction Corporation.

The outbreak of war clearly meant greatly increased Service interest and still greater pressure from the Services for the acceptance of new projects. It was manifest that the Laboratory organization, already creaking under existing demands, was unsuited to expanding size and expanded commitments. The organization was quite inadequate for handling the Laboratory's internal affairs and was equally unable to deal effectively with outside agencies. There was serious lack of coordination between the groups, no clear lines of responsibility, and poor distribution of available personnel, which was never adequate for the mounting responsibility. A tighter organization was clearly required. As an interim measure, or so it proved, M. G. White was made coordinator of the components groups and R. F. Bacher coordinator of the systems groups.

With the assumption of additional responsibilities, the Laboratory was now clearly committed to expansion and had to perform the following functions.

1. Research and development (including comparative testing of components and sets).

2. Engineering design.

3. Small-scale production (carried out either through such organizations as RCC or through arrangements with industrial companies).

4. Installation.

5. Operations.

6. Maintenance and servicing.

7. Training of Army and Navy personnel in operation and maintenance.

8. Analysis of operational trials and further development of systems based on these results.

The Radiation Laboratory as organized early in 1942 was able to handle effectively only the first of these functions. The additional activities were those in which the Laboratory had engaged on a tentative scale but for which its plant, personnel, and use were not adequate. To cover all of these functions, DuBridge estimated that an adequate establishment would require upwards of 3000 people, including all types of personnel. This was six times the size of the Radiation Laboratory at that time, a size obviously unmanageable under the existing Radiation Laboratory organization. The urgent question then was: what kind of an organization should be conceived to serve as a framework from which this great expansion could be supported for the duration of the war effort?

The Reorganization

Several varied proposals were made for Laboratory reorganization, the most radical of which was E. G. Bowen's proposal (based on the experience of the British Laboratories) for the creation of three large independent laboratories devoted to developing respectively ground, shipboard, and airborne radar systems. This was an extreme form of the proposal for a vertical organization wherein the component groups would be assimilated by the particular kind of system, and research on components and the development of new systems would go on in each division.

The loss of autonomy implicit in Bowen's proposal was stoutly opposed by the components group, which felt that they could function better under a horizontal system which left their groups intact.

The proposal for a horizontal division made up of components groups received short shrift. The principal objection was that so much of the stimulus for research and development work comes from seeing the results of the work go directly into practical use. If a "breadboard" model of a device were to be removed from the research and development group while still in the "breadboard" stage, much of the incentive for the development would be missing. Again, some sort of trials under practical conditions were essential for the real development work. Of necessity, then, the engineers who had participated in the development of the equipment would have to follow it through to its later stages.

Bowen's proposal for vertical organization and the proposal advanced by the leaders of the components groups for a horizontal organization represented the extremes of Laboratory thought on reorganization. The final solution, advanced by DuBridge, partook somewhat of both points of view and was agreed upon in March 1942 when a so-called divisional organization was adopted. This combination of both vertical and horizontal organization brought together the groups working on related components or in basic research into larger units called divisions and brought the related systems groups under a single divisional head. Above the divisions stood the Director's Office and the Steering Committee (see Fig. 10-4), composed of the Director and the Associate Director and the heads of the technical divisions.

The number of divisions finally chosen and the scope of each division's activities were determined, not by abstract considerations, but by the number of men available who were deemed to be of division leader caliber. The Director and the Steering Committee analyzed the available personnel, decided who were the top men, and then built the organizational structure around them. Ten new divisions were set up:

Division 1—Business Office, Channing Turner.

Division 2—Buildings and Maintenance, A. J. Allen.

Division 3—Personnel and Shops, F. W. Loomis.

Division 4—Research, I. I. Rabi.

Division 5—Transmitter Components, M. G. White (modulators, transmitters, rf components, test equipment, power equipment).

Division 6—Receiver Components, R. F. Bacher (receivers, indicators, precision timing circuits).

Division 7—Special Systems, L. W. Alvarez.

Division 8—Ground and Ship Systems, K. T. Bainbridge.

FIG. 10-4. Radiation Laboratory Steering Committee, photographed at one of their weekly meetings in 1944. Seated, clockwise from left: I. I. Rabi, E. C. Pollard, T. W. Bonner, J. W. Hinkley, R. G. Herb, L. C. Marshall, L. N. Ridenour, L. A. DuBridge (Director), J. G. Trump, D. H. Ewing, C. Turner, I. A. Getting, J. L. Lawson, F. W. Loomis, and B. Chance. Standing, left to right: J. R. Zacharias, L. J. Haworth, and D. G. White.

Division 9—Airborne Systems, L. N. Ridenour.

Division 10—Systems Engineering and Production, L. C. Marshall.

The Loran Group under Melville Eastham was listed for a time as a special project outside the divisional organization, but after a few months it became Division 11.

A word or two should perhaps be said about the direction of the Radiation Laboratory, for in many ways the practice in Division 14 was not typical of other NDRC divisions. During the greater part of its history the Laboratory was characterized by a militantly defended autonomy, with respect to both NDRC and MIT. This was never seriously contested. Harmonious cooperation with both was assured by the excellent personal relations which K. T. Compton, himself a physicist and head both of Division D and of the Institute, enjoyed with the Director and other leading physicists of the Radiation Laboratory. Similar close personal ties grew up between DuBridge and Alfred Loomis, Chairman of Division 14.

The Institute never attempted to exert control over Laboratory policy, though in matters concerning building and grounds, plant protection and security, personnel and salaries, and many other allied problems, the Laboratory worked in close cooperation with Compton, J. R. Killian, Jr., the Executive Vice-President of MIT, and N. M. Sage of the Institute's Division of Industrial Cooperation. During the first year of its existence, the direction of the Laboratory, on paper and to some extent actually, was divided between the Microwave Committee and the Director of the Laboratory. Scientific and engineering decisions were the recognized responsibility of the Steering Committee, which came into formal being in the spring of 1941. The Microwave Committee, as represented at MIT by its Secretary, E. L. Bowles, not only exerted general supervision and right of review over the decisions of the Laboratory, but wielded considerable direct influence on the Laboratory's activities, expecially in budgetary affairs and in such matters as the contacts with industrial concerns and with representatives of the Armed Forces. This division of authority proved to be extremely artificial, difficult to interpret, and not generally accepted within the Radiation Laboratory; it was for some time not clear how the respective responsibilities should be construed. In the early months of 1942, at about the same time that the organization into Divisions was adopted, these relationships were clarified, and thenceforth the Laboratory Steering Committee took the lead in formulating policy, as well as in approving projects (after direct consultation of its members with representatives of the Armed Services), and deciding what priority a project should have in the shifting contest for manpower and space among the various Divisions. As time went on, the Microwave Committee took on more and more the character of a Board of Trustees deeply interested in the work of the Radiation Laboratory, but willing to delegate to the Laboratory administration the effective direction of the NDRC radar program.

The Steering Committee of the Laboratory was composed of L. A. DuBridge, the Director of the Laboratory, as Chairman, F. W. Loomis of the University of Illinois, the Associate Director of the Laboratory, and several group or section leaders (later division heads) of the Laboratory: I. I. Rabi of Columbia University, L. N. Ridenour of the University of Pennsylvania, and J. G. Trump, who served at first as K. T. Compton's technical aide and later as Secretary of the Microwave Committee. During 1942, the names of L. W. Alvarez of the University of California, R. F. Bacher of Cornell, L. A. Turner of Princeton, and J. R. Zacharias of Columbia (all Division Heads) were added to the list.

In April 1942, the Secretary of the Microwave Committee, E. L. Bowles, was appointed Expert Consultant on radar problems to the Secretary of War, a position in which his early close association with the microwave program proved to be invaluable. Through his advice and energetic support, the higher echelons of the War Department were kept abreast of the latest radar developments and given the benefit of a small but expert organization in the Pentagon building, many of whose members or consultants were drawn from the Radiation Laboratory. J. G. Trump succeeded Bowles as Secretary of the Microwave Committee, combining this function with his responsibilities as NDRC representative (technical aide) at the Radiation Laboratory. Since the summer of 1941, Trump had confined his attentions to the work of Compton's Division D, Section D-1. His office, which by the time of the reorganization of SRD in December 1942 had grown to consist of five staff members and five secretaries, and had come to be called the Division 14 or Radar Division Office, became a very important administrative unit in the radar program.

During 1942 and 1943, the Radiation Laboratory changed from a small research laboratory to a large, rapidly growing engineering concern. The day-by-day affairs of the Laboratory required the establishment and development of service departments performing the widely varying functions required of a large business. A smoothly functioning personnel department, which also had supervision over a central drafting room and machine shops, developed under the direction of F. W. Loomis. A Building and Maintenance Department was headed by another physicist, A. J. Allen. In the early summer of 1941, K. T. Bainbridge, upon his return from England, took the first steps towards organizing a Document and Reports Office for the Laboratory. This was taken over and somewhat expanded by S. A. Goudsmit at the end of 1941.

No full understanding of the administrative eccentricities of the Radiation Laboratory is possible without recognizing that one of its outstanding merits, in the eyes of its own management, was that it was a physicist's world, run for, and as completely as possible by, physicists. Everything was subordinated to producing an environment for research as free and untrammeled as in a university and to preventing research from becoming entangled or impeded by the growing responsibilities thrust upon the organization. One consequence of this basic article of policy was that physicists assumed a large number of administrative duties, which might perhaps have been assigned to the lay brethren, and carried them out in some instances with extraordinary success, in other cases not quite so effectively. A second consequence was that when service departments were set up, they operated at something of a disadvantage, and not without justice felt that insufficient confidence was reposed in them. On the other hand, no policy could have been better designed to rid the research man of unwise interference, and to give him unsurpassed opportunities for creative work. The Radiation Laboratory came close to realizing a scientist's dream of a scientific republic, whose only limitation was the supply of scientists.

The growth of the service departments reflected the swift expansion of the Laboratory. A Business Office (more properly speaking, a purchasing office, since all other routine business affairs were handled by MIT's Division of Industrial Cooperation) was set up late in 1941, and by the end of 1942 had grown to number about 80 persons, staff and nonstaff.

In January 1941, official attention of the Laboratory investigators had been called to the patent provisions of the contract, and to the responsibilities of the contractor to protect the Government's interest in patentable devices developed under its auspices. A careful system of recording technical developments and scientific data in serially numbered (and carefully dated) notebooks was already instituted and remained in effect throughout the Laboratory's existence. It was not until March 1942 that a small patent office, staffed by three experienced patent attorneys, was set up under the Director's Office and charged with the task of filing patent applications on the results already accomplished and with keeping up with an increasing flow of new inventions and engineering improvements. The activities of this group were somewhat circumscribed by its small size and by the clear directive it received that, like those of other service groups, its activities were ancillary to research and development and should not become a burden upon the scientific personnel. Though a well-intentioned provision, supplemented by the appointment of patent representatives from among the members of each Group, whose duty it was to supply the patent office with the required information, it left the initiative for a time wholly in the hands of men whose attentions were focused upon prosecuting the war.

When, late in 1941 and early in 1942, the Laboratory was clearly entering a new phase of its history, in which production problems were sure to be paramount, the first steps were taken to seek professional advice in such matters. Shortly after the Pearl Harbor attack, a man experienced in business affairs, Armand Herb, was appointed by the Laboratory as "production expediter." His assignment was to follow orders after they had been placed with a view to expediting delivery. This soon took the form of preparing a list of "cleared" companies with which the Radiation Laboratory could place its orders. While conducting this survey, Herb was killed in the crash of a commercial airliner, becoming—despite the globe-encircling travels of Radiation Laboratory personnel during the next few years—the Laboratory's only casualty in the line of duty. In May 1942, a small group was set up in the Director's Office under Rowan Gaither, a San Francisco attorney, to concentrate on these procurement and production problems. This soon grew into a good-sized group that interpreted its instructions as requiring it to do everything possible to aid in the production of radar components, test equipment, and those radar systems produced by the Research Construction Corporation or by the Laboratory. Although the various Group Leaders tended to conduct their own procurement negotiations without outside advice, later in the Laboratory's history the Transition Office, as this group came to be called, emerged from the status of a production fact-finding group to take active part in shepherding equipment through development and into production. It cannot be said that this group was used to the fullest possible extent, largely because the procurement and production were so completely decentralized that the principal Divisions and many of the Groups were, in effect, well-nigh independent business concerns whose activities were coordinated through the Steering Committee.

In the autumn of 1941, a request was made to the Navy Department by K. T. Compton, E. L. Bowles, and L. A. DuBridge that a Navy Liaison Office be established at the Radiation Laboratory to assist the Laboratory in its relations with the Navy. This NDRC request was approved by J. C. Hunsaker, the Navy's Coordinator of Research and Development, and in December 1941, Lt. (later Commander) D. T. Ferrier, USN (Ret.), was ordered to duty at the Radiation Laboratory as Navy Liaison Officer. As the Laboratory expanded its activities, the Navy Liaison Office widened the scope of its functions, though it was never an important means of high-level contact of the Laboratory policy-makers with the Navy. It served primarily as an information service for the Navy Department, and as a base of operations for naval personnel detailed to the Laboratory. Its most important duty was to keep the Navy Department informed of the general course of Laboratory research and of projects that might be of interest to the Fleet; but the familiarity which the NLO had with the workings of the

Navy Department, the Fleet, and the Navy Shore Establishments was extremely useful in the course of the Laboratory's existence. Secret communications for the Laboratory with its personnel in England, Continental Europe, and the Pacific were handled by the NLO through the Navy. At maximum strength, NLO included a permanent staff of about 35 persons with the total number of officers and men who were temporarily assigned to the Radiation Laboratory to follow particular projects sometimes attaining a total of about 200 persons.

A liaison office of the British Air Commission was formally recognized in January 1942 and was staffed by E. G. Bowen, who had played such an important role in the early history of the Laboratory, and by D. M. Robinson, who had arrived in the summer of 1941 to procure microwave ASV equipment for the RAF Coastal Command. In the course of 1942, Liaison Offices of the U. S. Army Signal Corps and the U. S. Army Air Corps were activated. These did not become as large or as influential as the Navy Liaison Office.

One of the most significant developments during 1942 was the launching of an OSRD radar counter-measure (RCM) program and the establishment of the Radio Research Laboratory at Harvard. The Radiation Laboratory was intimately connected with the earliest steps that were taken to organize this program. It is probably not possible to determine who first clearly brought forward the need of a RCM program, for it must have arisen in several different places, especially among persons accustomed to thinking in operational terms. The Naval Research Laboratory, before December 1941, had made considerable progress in designing wideband search receivers for picking up and determining the frequency of enemy transmissions.

In the Radiation Laboratory, it was British influence and example which prompted the start of a counter-measures program. The handful of Laboratory staff members who had visited England in 1941 observed that at TRE radar and RCM were being developed hand in hand. In the fall of 1942, the Director assigned this general problem to L. W. Alvarez, who thought up some schemes of so-called confusion jamming, i.e., jamming where the enemy is unaware that he is being jammed, and began work on defensive counter measures for the microwave equipment, such as possible anti-jamming features in receivers. No work was done on the lower frequencies to be in use by the enemy.

Sentiment gradually crystallized, within the Armed Services, the NDRC, and the Radiation Laboratory Steering Committee, that a separate organization should be set up for the study and development of radar counter measures. It was decided in the weeks immediately following the Pearl Harbor attack that the NDRC should be asked to undertake a project for the development of special radar receivers and jamming equipment in collaboration with the Naval Research Laboratory and the Signal Corps Laboratories.

The Microwave Committee and the Laboratory Steering Committee realized the size of the enterprise and the reason which favored setting it up under separate auspices, instead of within the Radiation Laboratory. In the first place, the RCM program would not be solely or even primarily concerned with microwaves. In the second place, special security considerations entered into the picture. There was no doubt that RCM would have to receive special security treatment, separating it—as the British had done to some extent at TRE—from the radar research. The United States Army and Navy, moreover, insisted upon a complete segregation of counter-measures work, whereas the Radiation Laboratory always opposed compartmentation and internal security barriers, preferring an organization where all could share in the common fund of knowledge. It was urged that the program be set up elsewhere, and that an effort be made to recruit radio engineers instead of physicists, who were increasingly hard to find.

F. E. Terman, head of the Electrical Engineering Department at Stanford University, was selected as Director for the new organization, which soon acquired the name of the Radio Research Laboratory. It was operated under Division 14 of OSRD until the reorganization of that body in the fall of 1942, with Harvard University as the contractor as of March 20, 1942. The Radio Research Laboratory began operations with a small, specially segregated group in Building 24 of the Radiation Laboratory, numbering only three or four persons by the end of February 1942. Shortly thereafter, Terman and his group were assigned the second floor of the Hood Building, which the Radiation Laboratory was just occupying. The Radiation Laboratory furnished a nucleus of three staff members, and for some months the new group received advice from Alvarez, who attended RRL staff meetings to help the new men catch up in the radar field. In July 1942, the Radio Research Laboratory moved to a wing of the Harvard Biological Laboratories, where it remained for the duration of the war.

NOTES

1. This section is based on interviews with E. G. Bowen, Brigadier Wallace, and Carroll Wilson; upon an extremely full diary kept by Bowen during the period September 8, 1940–August 9, 1941; and upon the papers and reports of the British Technical Mission, kindly put at my disposal by the British Central Scientific Office. The most useful items were the reports of meetings, visits, etc., September 9, 1940–December 31, 1940, folders 00-1 to 00-6.

2. Arrangements for contact between the Mission and NDRC were entrusted to Carroll Wilson, Dr. Bush's Technical Aide.

3. Interviews with K. T. Compton and Carroll Wilson; notes by Professor Cockcroft, British Mission papers, 00-1-25.

4. Bowen diary, entries September 20–October 11.

5. Microwave Committee notes by E. L. Bowles; Bowen diary, entries September 25–29; Cockcroft:"Notes on Visit to Loomis Laboratory, Tuxedo Park, 28 and 29 September, 1940", British Mission Papers, 00-1-54.

6. "Events concerning meeting with British Mission and the receipt of the First British Multicavity Magnetron at Bell Telephone Laboratories," August 19, 1943. Information kindly supplied by Dr. Ralph Bown, and derived largely from his personal notes.

7. *Ibid.* Interview with E. G. Bowen.

8. Bowen diary, entries October 11–13; these meetings were fully reported by Bowen in British Mission Papers, folder 00-2.

9. During the discussions at Tuxedo Park, September 29–30, Loomis rather optimistically impressed Cockcroft and Bowen that the Microwave Committee was initiating work on (1) radio navigation for accurate location of bombing aircraft; (2) bombing through overcast; (3) a combined set on 10 cm to perform the function of AI, ASV, IFF, radio navigation, and blind landing: British Mission Papers, 00-1-54.

10. Bowen diary entry October 18; interview with E. G. Bowen.

11. Bowles notes, Monday, October 13, 1940; Bowles Notebook No. III; Interviews with E. L. Bowles, K. T. Compton, and Vannevar Bush.

12. A. L. Loomis to V. Bush, ltr. of October 24, 1940; Bowles desk calendar; K. T. Compton, desk calendar; Bainbridge Notebook #9, p. 1; personal communication of K. T. Bainbridge and L. H. Turner.

13. NDRC–Minutes of meeting of October 25, 1940; in which we read that the chairman "stated that he and the Microwave Section, independently of each other, had made surveys of Government Laboratories, including those of the Bureau of Standards and of the Army and Navy, as well as of commercial laboratories, and had come to the conclusion that no existing laboratory was equipped or manned to carry out the research contemplated under the microwave program. He further stated that the Microwave Section was strongly of the opinion that MIT was the only institution at which the work could be done with the speed which the Armed Services desired. When the matter was presented to the Committee, Dr. Compton stated that he had taken no part in the decision to locate the laboratory at MIT."

14. V. Bush, telegram to E. O. Lawrence, October 3, 1940.

15. Bowles notes, October 10–11, 1940. The term "consultant" carrying with it the notion of advisory duties and no compensation was applied to the first persons appointed to work with the various sections of NDRC. The members of the Microwave Committee and the earliest Laboratory staff members received this appointment. See also Memorandum of Dr. Bush (unsigned) dated 1:25 a.m., October 14, 1940. OSRD files, which asks that Bainbridge, M. J. Kelly, E. W. Engstron, H. W. Beveridge, E. U. Condon, and W. W. Hansen, all of whom were cleared, should be appointed consultants.

16. K. T. Bainbridge, MIT Computation Book No. 9, p. 1.

17. *Ibid.* Cf. also interview with K. T. Bainbridge on May 11, 1943.

18. Personal communication of L. A. DuBridge.

19. Personal communications of L. A. Turner and M. G. White.

20. Personal communication of F. W. Loomis.

21. And no relation to Alfred L. Loomis.

22. K. T. Bainbridge, Notebook No. 9, p. 3; N. F. Ramsey, Notebook No. 5, pp. 2 ff. The earliest laboratory organization list is of November 25, 1940.

23. E. G. Bowen, "Report on Activities of Micro-Wave Section of National Defense Research Committee," December 1940; "General Information" (Hectographed Notice dated

11/26/40.); L. W. Alvarez, Notebook No. 11, pp. 90-91; DuBridge-Neher, Notebook No. 23, p. 3 and p. 90 ff.

24. E. O. Lawrence to L. A. DuBridge, letters of November 25-29 and of December 5, 1940, and of L. A. DuBridge to Lawrence on November 29 and December 3.

25. Alvarez Notebook No. 11; Bowen diary, entry of December 16, 1940.

26. Bowen diary, entries of November 17–19, 1940.

27. Bowen diary, entry of December 16, 1940, I.I. Rabi Notebook, No. 8, p. 15; Alvarez Notebook, No. 11, p. 93.

28. Interview with L. W. Alvarez; Report of the System Group. Special Section F, Report 1, January 30, 1941.

29. *Ibid.*

30. Telegrams in Director's Office file, Folder E. O. Lawrence.

31. Interview with S. N. Van Voorhis and W. G. Tuller; Report of Microwave Sub-Committee, Carnegie Institute (sic); February 7, 1941, dictated by Prof. R. Fowler; RL Report II-4, January 28, 1941.

32. Interview with L. C. Marshall; Bowen diary, entry of February 4, 1941.

33. Interviews with L. C. Marshall and L. A. DuBridge.

34. Radiation Laboratory Report Roof System Report, R-15.

35. Radiation Laboratory Report: A-15: B-18-A Report—February 13 to July 22, 1941; McMillen's RL Notebook.
Pulse Modulator: rebuilt Westinghouse laboratory pulser

Magnet:	GE Ticonal permanent magnet
Magnetron:	BTL magnetron at 9.38 cm
Synchronizer:	built by E. H. Lyman
Receiver:	BTL crystal mixer; RCA IF amplifier, and the 1221 Y local oscillator.
Spinner:	Sperry Model A
Indicator:	RCA standard 5″ oscilloscope for A-scope presentation; and RCA 12″ electrostatic type for Type C
TR:	klystron buffer
RF:	coaxial cable, polystyrene beads, half-wave spacing.

36. The importance of developing a centimeter ASV had been proposed several times by E. G. Bowen. As early as the end of January 1941, Bowen had drawn up preliminary specifications for an ASV system using a PPI, which was expected to give a range of 5 miles on submarines and 10 miles on destroyers. E. G. Bowen, "Copy of Specifications for Centimeter Wave: ASV Systems." January 27, 1941. This was submitted for consideration of the Radiation Laboratory on February 10, 1941. The importance of an ASV program, especially for submarine detection, at the Radiation Laboratory seems to have been first seriously discussed between Alfred Loomis and E. G. Bowen on March 23, 1941 at Tuxedo Park. The following day this problem was discussed in Cambridge with DuBridge and E. O. Lawrence. This seems to have been the background of the flight to New London.

37. Reports on Lyman Synchronizer.

38. Reports of Magnetron Group and Notebooks.

39. Bainbridge Notebook. Modulator reports.

40. Antenna Group reports.

41. Receiver Reports; Tuller, Van Voorhis, and Bell Laboratories.

42. rf Group Reports: Lawson and Salisbury's Notebook. Interview with Salisbury.

43. The first samples of British long-delay tubes were brought to the Radiation Laboratory in February 1941 by Professor Fowler.

44. Indicator section report. Pollard's Notebook.

45. The PPI is mentioned in general terms in E. G. Bowen's report of January 27, 1941.

46. For a description of the PPI see Chapter 2.

47. Bowen diary, entry of January 17, 1941.

48. RL Report A-15, pp. 2–4.

49. Interview with Jackson Cook.

50. A-20 Log Book.

51. Diary of D. R. Corson, interview with Skinner, July 4–16, 1941, *ibid.* July 20–August 19, 1941; Bowen Diary, entries of July 15, August 2, August 4, August 8–9.

52. See above.

53. Radiation Laboratory Notebook No. 37, issued to L. N. Ridenour, pp. 28–29, February 7, 1941.

54. In U. S. Patent No. 2,083,242, Wilhelm Runge, Berlin, Germany, assignor to Telefunken Gesellschaft für Drahtlose Telegraphie; Application Serial No. 3,730, January 28, 1935; Patent issued 8 June, 1937 (issued in Germany 27 January, 1934), the geometrical concept of such a type of scan was recorded, but this was not known to Ridenour at the time.

55. Radiation Laboratory Report B-2, May 14, 1941. "Special Project on Aircraft Radio Sight (ARS)".

56. I. A. Getting, "Section 81 Radar Systems Not Explicitly Intended as Prototypes", April 27, 1943; "Chronology of SCR-584 Development and Design", August 4, 1943.

57. I. A. Getting to L. A. DuBridge, April 20, 1942.

58. Radiation Laboratory Notebook (unnumbered), issued to A. M. Grass, pp. 115–117, "Progress Report" December 1, 1940 to June 24, 1942.

59. NDRC-Division 14-Report No. 230, December 10, 1943. S. Godet, "Two Motor-Driven Gun Turrets". The date given in this report is November 4, 1941. See also Radiation Laboratory Notebook No. 21, issued to I. A. Getting, p. 128, November 12, 1941, "First following of planes by automatic system in RT-1 was on Saturday morning, Nov. 8, (1941) at the East Boston Airport (Sid Godet present...)"

60. Ralph Bown, Bell Telephone Laboratories, the Chairman; Melville Eastham, MIT (General Radio Co.); R. S. Holmes, Radio Corporation of America; L. M. Leeds, General Electric Company; D. G. Little, Westinghouse Electric & Mfg. Co.; Joseph Lyman, Sperry Gyroscope Company; Dayton Ulrey, Radio Corporation of America; F. A. Polkinghorn, Bell Telephone Laboratory, Secretary.

61. In 1941, the group under Eastham's supervision consisted of D. Davidson, Donald G. Fink, Donald R. Kerr, R. E. Lawrance, J. A. Pierce, J. C. Street, W. L. Tierney, R. H. Woodward, and W. Vassers, Jr. In 1943, Eastham was succeeded as Division Head by Donald Fink, later the same year by J. C. Street.

Section C

Technical Developments at
the Radiation Laboratory

Part 1

Radar Systems

CHAPTER 11

Technical Developments in Division 14

The scientific effort at the Radiation Laboratory, where Division 14's work was mainly concentrated, was divided into three interpenetrating categories: the development of new radar systems, the steady improvement of radar components, and fundamental research. To understand the work of the Radiation Laboratory, it is important to grasp the fact that it was an engineering and development plant concerned mainly with the first two of these activities, not strictly speaking a research laboratory, although it was staffed and directed by academic physicists. The great majority of these men were concerned with developing radar systems and components. In 1944 there were about 400 staff members in the components groups and about 250 men designing systems, while only a single division of five small groups numbering in all about 100 staff members was devoted to research.

This last category included exploratory work leading to the design of systems operating on higher frequencies, as well as fundamental investigations, both theoretical and experimental, on the properties of microwaves and the basic principles of pulse-echo operation. Even the most fundamental research—as for example the work of the theory group, composed of mathematicians and theoretical physicists—should be thought of as engineering research, for the aim was to translate the results as promptly as possible into principles of engineering design. Almost no work resembling the fundamental investigations of a peacetime physics laboratory was carried on.

The general lines of policy guiding the distribution of the Laboratory's effort had been agreed upon in the earliest weeks of its existence, and it was accepted that the objective was primarily to get equipment into military use as rapidly as possible. Projects were discouraged which did not appear capable of translation within a reasonable time into a concrete military device. As we have seen, the principle of "cut-and-try" was widely followed, though, of course, it was recognized that there were dangers inherent in this procedure, and that the equipment would doubtless be more nearly perfect, and possible pitfalls would be avoided, if a slower and more methodical procedure could be adopted, and the ground more carefully surveyed by fundamental research. Speed demands risk and compromise.

War research is inseparable from speed: a device which appears too late is of very little use. Speed was one of the outstanding characteristics of the Radiation Laboratory, and the frenetic atmosphere in which work was done—especially by the systems divisions—is hard to convey now that the emergency has passed. The Radiation Laboratory set the pace in most departments of the radar development. As one industrial engineer observed, the Radiation Laboratory "ran several degrees hotter" than other laboratories. The Laboratory's successes and its failures, its speed and initiative, the loud grinding at all points of friction, were all enhanced by this state of thermal agitation.

The pragmatic emphasis was especially strong during the first year of its existence when the men were under pressure to demonstrate to the NDRC and to the Armed Services that they could produce concrete results. Although from the first the Radiation Laboratory had a few loyal supporters among Army and Navy representatives, there was a certain amount of ill-concealed skepticism that a group of physics professors could produce anything useful. Even the more progressive Service representatives felt, with some justice, that the Army and Navy knew their problems better than an inexperienced civilian group possibly could, and had accumulated a certain amount of radar experience, and they certainly felt that to attempt to develop radar on a new frequency was a dubious enterprise when the full potentialities of long-wave radar had scarcely been tapped. There were even a few voices weighty in military councils in 1940–1941 who said that research and new development had necessarily to be curtailed, rather than expanded in wartime when the problems were those of production testing and training. This, it is true, did take place in the Army and Navy laboratories. Like the radar war itself, the Radiation Laboratory passed through a prolonged defensive phase.

Thus, from the beginning, while official policy in the Laboratory sanctioned, nay, strongly pushed, the exploration of new frequencies, it gave only moderate encouragement to other more general varieties of fundamental research. It is significant that this to a considerable extent reflected the opinion and influence of the head of the research division, who, though the Laboratory's most celebrated physicist, was one of the most forceful and insistent advocates of the policy of getting new radar systems into the field.

Nevertheless, the attitude of the Laboratory underwent a perceptible modification in these matters. It was true that great advances could be made in certain branches without benefit of a complete theoretical understanding of a problem. It proved, for example, possible to make extraordinary progress in magnetron design with only a semi-theoretical grasp of the problem and without waiting for the complete and rigorous theory of magnetron operation, which, since it is a problem of great complexity, was never forthcoming. As the scale of operations became larger, and as similar difficulties were more frequently encountered in a number of different groups, the day-by-day utility of fundamental research became more self-evident. It was a great saving, for example, to put rf component design on a coordinated theoretical basis upon which, in turn, sound engineering decisions could be based. Similar contributions were made to receiver and indicator design. The larger the Laboratory became, so it appeared, the more indispensable this type of work became.

Another factor in this basic policy decision was the question of timing. The most important decisions of how much long-range work a war laboratory should undertake and how long-range it should be hinges necessarily on an estimate of the length of the war. In view of the lack of, or perhaps the impossibility of, strategic guidance from the highest levels, this was reduced to pure guess work. During most of 1941, the feeling still persisted that after the solution of a few basic tasks the Laboratory might disband or be put to work on other problems. It was hard to think much more than a year ahead. But after the attack on Pearl Harbor and the unchecked succession of German and Japanese victories during 1942, it was clear that the Laboratory had a major role cut out for it, that the war was not going to be soon over, and that problems whose solution lay a year or more away were sure to be of value. Nevertheless, the length of the war was always underestimated. The Laboratory could not believe that the war would drag on until after the middle of 1945. In this it reflected the optimism of those who were talking of a second front in Europe in 1942 and who fully expected it in 1943. The Laboratory, it can be said in retrospect, adopted a somewhat shorter-range point of view toward many of its problems than would doubtless have been the case had it been possible to know that the Laboratory was to last five years.

The progress of research and development was strongly favored by the generally elastic and decentralized organization of the Radiation Laboratory. The combination of vertical and horizontal features, which left the component groups free and independent, had almost certainly more advantages than disadvantages. Since the components and the systems groups were essentially on the same plane, and headed by men of equal caliber and aggressiveness, even though a large part of the components work was done for the systems people, this resulted in points of friction and disagreement. There was constantly a certain amount of friction that some other organizational scheme might have avoided. Nevertheless, the interaction of the two sorts of groups upon each other was an invaluable source of stimulation and mutual education. There were constant discussions and considerable rivalry between the two groups. The components people in general felt that they had a higher technical competence, while the systems people felt themselves closer to the military and tactical problems and the realities of the war. There was a natural difference in tempo in the two types of work. The systems divisions were more subject to central pressure, more concerned with commitments to companies and the Armed Services, more constantly pressing to turn out equipment at top speed. Had the components groups been administratively subject to the systems groups, it is doubtful whether components would have been improved as methodically as was possible under the existing arrangements. A final advantage was that this organization made the various parts of the Laboratory more completely interdependent; between the two, constant cooperation was inescapable; and by the participation of components men in the work of various systems divisions a new improvement in shipboard radar might be carried over, if applicable, to the airborne projects. With a single exception, the organization as it existed left no division self-sufficient and made it impossible for them to withdraw gradually into complete isolation. The exception was the Loran Division, Division 11, which, because it made its own components, had its own special problems, and had a special security status, became a rather exaggerated case of rugged individualism.

Even with the conditions that promoted consistent interchange between the groups, and the coordination on the level of the Steering Committee and the Director's Office, the separate divisions, as far as their methods of work and other internal affairs were concerned, were so many autonomous republics. Without exception, they were left free to operate—limited only to broad priority directives from the director and the Steering Committee—according to the temperament and preference of the division head and his group leaders. Important problems were voluntarily discussed on most occasions with the director and his advisors.

According to the particular disposition of the division head, and to the scope and nature of the problem assigned to him, the divisions were differently organized and administered. They were in general divided into a number of groups charged with specific aspects of the overall problem. The extent of the subdivision reflected in some cases the complexity of the problem and the size of the division, and also to some extent the disposition of a particular division head to delegate authority or subdivide his responsibility administratively. Therefore, in certain divisions, the group leaders had great independent authority, while in others the division acted very nearly as a unit under the division head. Group independence was much greater in the components divisions than in the systems divisions, partly because of the greater specialization of the problems, partly because of the high caliber of the individual group leaders.

The divisions were not subject to constant supervision in the matter of either expenditures or internal bookkeeping. As far as the service groups in the Laboratory were concerned, all requests made by division heads and group leaders were directly honored without the necessity of formal approval from the Director's Office. Similarly, the divisions could in varying degrees be correctly described as being nearly autonomous in their relations with the Armed Services and with industrial concerns. Some of the most fertile discussions with the Armed Services leading to the design of new equipment took place in direct exchanges between division heads, group leaders, and ordinary staff members and Army and Navy representatives. This was especially true in the early years. This administrative decentralization prevented the Director's Office from becoming a single bottleneck through which everything must pass. The director was frequently consulted, but he was called upon to intervene only at critical stages in negotiations with the Services or when basic policy considerations were involved. The routine contacts were not directly controlled or even minutely supervised.

Perhaps the most important feature of the Laboratory's research picture was that, though it was oligarchic on the sharply defined levels of Division Head or Group Leaders, there was complete and effective democracy within nearly all the groups. Research is no respecter of persons; and as the Germans learned to their dismay, there is no place where the *führer prinzip* is less applicable than in the research laboratory. Ability and originality were the only requirements; suggestions were welcomed from whatever level they were made. It was no crime to contradict or discomfit a Group Leader. Independent initiative in the matter of new projects and the trying out of new ideas was given remarkably free rein, considering the inescapable shortages of manpower and space. Throughout the history of the Laboratory, this was especially true in the Components groups, where a new venture was less likely to require a major effort.

The activities of the Radiation Laboratory were formalized early into approved projects which received various formal numerical designations. When the projects were conducted in response to Army or Navy request, Army-Navy control numbers were assigned to the projects, and these were used by NDRC for their bookkeeping. A separate decimal system was used by Division 14. Formal project status of this sort was particularly important in the case of a system development project which was generally a complicated affair, making greater demands on manpower and space, shop time, drafting room time, flight time at the airport, etc., than a component development. It almost always required extensive cooperation from the components groups.

In the earliest days, little more than the approval of a group leader or division head and the sanction of the director was required to launch a systems project, and it was the director's policy to allow those to flourish as they sprang up and not to try to estimate too precisely the future importance of the various early projects. A priority system was opposed whenever it was suggested, as long as the Laboratory had about half a dozen or a dozen projects. The situation was different when it was a case of 40 to 50 projects. As the pressure of work got steadily greater and the critical items like manpower, space, etc., became more thoroughly saturated because of new projects, the maturing of earlier projects, the requirements of crash programs, and the expansion of field activities, it became more of an affair to launch a new project. Toward the close of 1942, a strict priority system was instituted within the Laboratory which was rigidly enforced during the years following. A priority or precedence list was prepared at approximately monthly intervals and approved by the Steering Committee. This list dictated the order in which the shops, drafting room, and components groups were to proceed in their work on the different systems.

During 1941, the number of systems projects rose slowly from the original three to about 14 at the end of the year. In the spring and summer of 1942, soon after the reorganization of the Laboratory and its rapid expansion in space and manpower, the number of systems projects increased to something above 40. This number remained at between 40 and 50 during the rest of the Laboratory's history, despite the later expansion. That the projects did not continue to increase with the growth of the Laboratory is understandable when it is remembered that many of the later projects were extraordinarily complex, that many of them involved crash production problems, and that earlier projects had given rise to field activities demanding substantial personnel.

The priority system was designed to include all systems projects within the Laboratory that were intended for a definite military use. It was not supposed to cover system research or development work of a general sort carried on with the system divisions, nor projects on component research and development. A given project was placed high on the priority list when it was deemed both extremely important and extremely urgent. The importance of a set was judged by the effect it might be presumed to have on winning the war, and by the extent to which it constituted a major advance in the fulfillment of an important operational requirement. It was considered urgent if it required a considerable effort to meet imminent established production schedules; if a crash program for Laboratory or RCC production of a few sets for immediate operational use had been authorized; if the project had urgent Services backing; if it had excellent prospects of such backing once its possibilities were demonstrated; or if the equipment was destined for specific operational use at a relatively early date.

It was expected that the priority list would be used by the administrative heads of the Laboratory as a general guide to determine where the effort of the various divisions of the Laboratory should be placed. Projects high on the list commanded greater space and engineering personnel; earlier and more extensive use of shop and drafting room facilities; earlier attention or additional personnel in the component groups; and preference from the Transition Office and Business Office in expediting orders or production facilities.

The following paragraphs, as quoted from a memorandum prepared for the Steering Committee early in 1944 by the Director of the Laboratory, explain the significance of the priority list and how it should be interpreted.

It should be recognized that presumably all projects on the Radiation Laboratory Priority List are of considerable importance. The Radiation Laboratory must proceed on a broad and well-rounded program. Thus, groups working on projects which happen to be low on the list should still be made to feel that their project is an essential element in the Laboratory program and is worthy of their best efforts. At the same time, the Laboratory program must be kept sufficiently flexible to give unusual attention where necessary to projects of great urgency. The Priority List should be a general guide to individual groups in the continual restudy of their programs and shifting emphasis, which should always be under way.

To be of use, the Priority List must be revised frequently, taking account of the changing aspects of the war, and particularly of the changing status of individual projects as they prove greater promise or proceed through the production stages. The frequent revisions of the Priority List, however, should not cause equally frequent major upsets in the programs of individual goups. Work which commands a major effort in a group one month, should not be halted the next because of shifting priority ratings. Rather, each group must study its own personnel and facilities, and their distribution in relation to the Priority List, with an attempt to retain flexibility in assignment of personnel, but at the same time, continuity in the important efforts of the group.

The development of a new radar system, as of any new military device, involves decisions and judgments that are both technical and military and the simultaneous application of scientific and tactical knowledge. It is necessary in every case to appraise with great realism what a technical development is capable of at a given time, or what it is likely to yield in the near future. It is equally essential that the tactical conditions under which the equipment is expected to operate be thoroughly understood. So much, at least, is clearly required of the engineers designing a piece of equipment or promoting its use. However, it is also unavoidable that other considerations of what we may call a strategic nature should enter into the decision to exploit a given technical possibility: whether the equipment is likely to be urgently needed at a particular stage of the war; what effect changes in the course of the war will have on the utility of this equipment; whether or not the gains resulting from its introduction are sufficient to warrant, in certain instances, elaborate programs of retraining and replacement.

As we have just indicated, these are factors which, though primarily military, had to be taken into consideration by the Director and the Steering Committee in assigning priority to Laboratory projects. The men were constantly obliged to adjust the program of the Laboratory to the waxing and waning of Service interest in new equipment in response to the changing face of the war. This shifting of Service interest was often the only way in which the broader strategic decisions made on the highest level were manifest on the level of research and development.

The Radiation Laboratory staff members were in a peculiarly strong position to influence basic decisions on radar procurement because they were in possession of both sorts of knowledge, technical and tactical. Since they were in most things the pacesetters of the microwave development, their technical knowledge was only equalled by that of the Bell Telephone Laboratory—though in some departments it was certainly excelled. By a peculiarly effective series of operational contacts, first those with the British, and then later those primarily provided through the Army Air Forces and the Laboratory's own field organization, the Radiation Laboratory was able to combine this technical supremacy with an extraordinarily detailed knowledge of the tactical situation, at least in certain theaters and for certain radar problems. By contrast, where such tactical knowledge was denied it, the Laboratory's contributions were less outstanding.

To the outsider who might expect an institution like the Radiation Laboratory to be merely a passive supplier, filling requests that reached it from the Service procurement agencies, the aggressiveness of a civilian organization in shaping the Service procurement program may come as something of a surprise. A surprisingly large number of the systems projects on which the Laboratory worked were initiated by Laboratory personnel, though eventually all the successful projects had some sort of service backing and were supported by a formal service request. Many projects resulted from informal discussions of Army and Navy representatives with Radiation Laboratory staff members. In these talks, an Army or Navy officer might say: "We badly need something to do thus and so." Or a Laboratory staff member might say: "We think we have something that can do such and such, are you interested?" Especially in the early period of the Laboratory's history, before its capabilities had been demonstrated, it was usually more efficacious for a model to be built and demonstrated. There were, however, many instances of projects stemming entirely from a written or verbal Service request. In particular, most of the projects undertaken for the Navy, especially for the Bureau of Ships, had their origin in this way.

It was thus the recognized function of the systems groups to be fully informed concerning the tactical problems in their respective fields. This they achieved through discussions with Army and Navy officers, especially those back from combat, and with civilians returned from the field, and by means of reports received from a variety of British and American agencies. In the latter category were the reports transmitted to the Radiation Laboratory from the Joint Electronics Information Agency in Washington and the OSRD Liaison Office. The systems men were also expected to keep generally informed on the technical possibilities of radar components design. Thus, as new tactical requirements arose, or as new technical developments were perfected, it was possible for the systems divisions to propose an overall design for a new radar set to meet a specific tactical need.

When a project was approved or under consideration, at some stage of the proceedings the Army or Navy representatives prepared the military characteristics, in varying degrees of detail, stating the general requirements. These usually accompanied the formal Service request. In some cases they were extremely general, in others very detailed. These were sometimes prepared after consultation between technical and military personnel. After consultation of the systems men with representatives from the components groups, the technical characteristics of the set were drawn up. The components groups within the Laboratory were then requested to develop the components required for the new system and to detail men to work with the project. The systems group named a project engineer for each system whose first task was to coordinate the design of the various components. This arrangement was formalized by the appointment of a Project Committee, headed by the project engineer and including the components men assigned to the project. The project engineer was responsible for assembling the components into a final system; for providing the general mounting and assembly of the final equipment, and for its laboratory, field, and service tests; for following the equipment through design into production; for assisting in the introduction of the equipment into the field; and for working on improvements, attachments, and modifications to keep the set adapted to its tactical functions. It was the job of the components groups, on the other hand, to take care of all problems connected with the development, design, engineering, and production of the individual component parts of the equipment.

CHAPTER 12
Airborne Development on Ten Centimeters

Airborne radar systems were, on the whole, the most spectacular products of the microwave development. From the beginning, the emphasis at the Radiation Laboratory had been on the design of airborne equipment, and the largest proportion of the Laboratory's systems effort—there were some who felt it was perhaps too large a proportion—continued throughout the war to be devoted to designing radar to fly in various types of aircraft and perform a variety of functions. The Radiation Laboratory became the principal development agency upon which the Army Air Forces and the Navy's Bureau of Aeronautics relied for new radar equipment. This is readily understandable, for when the Laboratory was founded there was no American airborne radar program of any consequence. In this field the Laboratory had a great initial advantage in the tactical and operational knowledge supplied by E. G. Bowen and other British representatives. Moreover, it was an area of radar development in which the Army and Navy laboratories could not point to extensive previous experience, and for which microwave radar, the Radiation Laboratory's specialty, was preeminently suited. An important offshoot of the Laboratory's early work was the launching of the microwave airborne radar program at Bell Laboratories, the Radiation Laboratory's most substantial rival.

The initial assignment of the Radiation Laboratory, the development of a 10-cm AI (Aircraft Interception) system around the British magnetron, broadened into a complex program of four main types of radar equipment: (1) AI systems; (2) ASV (search) radar; (3) airborne fire-control radar; and (4) blind-bombing radar. The two most widely publicized Laboratory-developed systems, two systems which had a clearcut and important influence on the course of the war, belong in the second and fourth categories. These were, respectively, the "George" radar, which helped to break the back of the German submarine offensive, and the "Mickey" radar, which enabled the Army Air Forces to sustain the strategic bombardment of Germany during the winter months of 1943 and 1944.

The development of airborne radar was enormously influenced by shifting tactical considerations and the changing fortunes of war. The development of microwave AI, principally a defensive weapon, was mainly stimulated by the Battle of Britain, which was won before the development was completed. When the enemy offensive slackened, the need for improved nightfighters faded, and so did the frantic and desperate clamor for AI equipment. After Pearl Harbor the enemy submarine offensive provided a new and grave menace; the result was to stimulate greatly the development of ASV. The use of this weapon marks in a sense the Allies' transition from defense to offense; it protected our supply lines by seeking out the enemy and destroying him before he could attack, and, in so doing, permitted the building up of our forces overseas, both men and materials, in preparation for the great offensives in Africa and Europe, and the air assault upon Germany. The air offensive, begun in 1942, reached its full strength in 1943–1944. This was reflected in the increased amount of time and effort expended by the Radiation Laboratory in the development of a number of increasingly effective radar aids to bombing.

Airborne gunlaying systems are an example of a weapon whose development and procurement was adversely affected by changing tactical considerations. Like the AI, fire-control systems are a defensive weapon, suitable for use either in nightfighters, as a defense against enemy bombers, or in bombers, as a defense against enemy nightfighters. By the time the equipment was ready for production there was no tactical demand for it; on the other hand, when a need arose, as during the invasions of Africa and Italy, no equipment was available, since it had not been put into production in time.

Aside from tactical considerations, the development of airborne radar was strongly affected by the physical limitations entailed by its use in aircraft. Gear for installation in aircraft must be small, lightweight, and able to operate at the reduced air pressures encountered at high altitudes. As a result, airborne radar systems came to consist of a number of small, compact, airtight units. So-called pressurization, by which the unit was maintained at normal atmospheric pressure no matter what the altitude, was a necessary provision in airborne radar. In general, light weight was obtained by compactness, by simplicity, and by the sacrifice of range— low power meant light weight, but also short range. Fortunately, very long range is not, in general, an essential feature of an airborne radar system.

ASV (Aircraft-to-Surface Vessel)

The various 10-cm ASV sets developed by the Radiation Laboratory for submarine search were its first effective contribution to the war. We have seen that as early as the spring of 1941 the Roof Group began adapting the experimental AI equipment for installation as an ASV in an XJO-3 aircraft; and that a similar system for installation in a PBM-1 flying boat was assembled during the fall of 1941 and flown for the first time on January 3, 1942. The first ASV systems that actually saw action were only indirectly the descendants of these early systems. They were hastily thrown together at the Army's request early in 1942.

At a conference between Radiation Laboratory personnel and representatives of the Air Forces held at Wright Field shortly after the Pearl Harbor attack, one of the officers present urged that it would be extremely valuable if a number of B-18 aircraft or some similar type could be equipped on a crash basis with ASV equipment for Pacific patrol work. It was agreed that for lack of aircraft the Laboratory AI program had slowed up beyond resuscitation and that the components intended for the British Beaufighters could be used, with only slight changes, to produce a small number of 10-cm ASV sets.

The Radiation Laboratory AI specialists, principally E. M. Lyman, D. R. Corson, T. W. Bonner, B. L. Havens, and H. L. Schultz, were assigned this new problem, while the Army Air Forces brought together from various parts of the country, under the command of Colonel William C. Dolan, ten somewhat shopworn B-18 aircraft. The planes and their crews began to arrive at the East Boston Airport in February 1942. Working at high speed, the Radiation Laboratory installed equipment during the winter in all ten planes, provided the necessary spares, and equipped a testing and repair truck to service the sets. (See Fig. 12-1.)

FIG. 12–1. One of the ten B-18's equipped with 10-cm ASV, used in antisubmarine strikes in 1942.

As the installations neared completion, it was decided not to disperse the planes, but to assign all the crews, at least temporarily, to Langley Field, where the first planes had already returned, late in March, with their new microwave equipment. Even before all the planes had left Boston, the first operational successes were recorded. On April 2, one of the ASV-equipped B-18's, piloted by Colonel Dolan, and with two Radiation Laboratory staff members in charge of the equipment, shared in the rescue of a Navy observation plane that had been forced down 50 miles at sea near Boston. The floating plane was located by means of the microwave equipment in the B-18, and a destroyer was guided to the rescue.

From their base at Langley Field the first B-18's were speedily acquiring operational experience against the German submarines swarming off the East Coast. On April 1, 1942, one of the B-18's on its first night patrol homed unsuccessfully on an enemy submarine; somewhat later it picked up radar signals from a second submarine, which disappeared before a run could be made; and finally it picked up a third submarine at a range of 11 miles from an altitude of 300 ft, homed on it and sank it. Another kill was made on a flight from Langley Field on May 1. On May 22 five of the crews were ordered to Key West and five to Miami, where they operated until June 19, making one attack, the results of which were undetermined.

In July 1942, the First Sea Search Attack Group was activated under Colonel Dolan's command. This unit, consisting at first only of the B-18 crews, was intended to serve as a development and training unit to try out new antisubmarine weapons and evolve a tactical doctrine. Seven of the B-18's with their Radiation Laboratory equipment—three planes had suffered operational damage beyond possibility of repair—remained with the Group until its inactivation in July 1943. The accomplishments of this unorthodox outfit, set up as a result of the enterprise of Colonel Dolan and of E. L. Bowles in the Office of the Secretary of War, will be recounted in a later chapter.

The effect of these first successes upon the morale of the Radiation Laboratory was immediate and of great importance. It gave the Laboratory much needed confidence and a sense of direct participation in the war; in large measure it made up for the disappointing and inconclusive end of the AI program.

DMS-1000

The British had already shown keen interest in American ASV developments. At the end of July 1941, a technical representative of the British Air Commission, D. M. Robinson, arrived at the Radiation Laboratory to explore the possibility of acquiring a small number of ASV sets for use by the RAF Coastal Command, and to assist in guiding the development.

Robinson found that a good beginning had been made in microwave development at the Radiation Laboratory, and that various component parts of the AI-10 equipment—the modulator and the transmitter in particular—could be used without modification. Antennas with 360° scan and PPI's were being flown. At that time a tube mixer was used in the receiver; but soon after—as a result of the comparison with British equipment—this was rejected in favor of the more satisfactory crystal mixer. Various other improvements were made while the first B-24 plane, later christened Dumbo I, was being modified at Wright Field.

The first experimental system was assembled and put in working order, and the installation in the plane was complete by December. A flight test was made on December 11, 1941, with the equipment operating successfully.

In November, even before the flight tests, the British had been confident enough in the outcome to place an order for 12 sets with the newly established Model Shop, the Research Construction Corporation. The systems installed in the first two Liberators, Dumbo I and the later Dumbo II, served as prototypes for a set given the designation DMS-1000. Work at RCC started in January 1942, under the direction of E. L. Hudspeth, G. A. Fowler, and H. M. Jeffers.

During the early months of 1942 there were various suggestions for installing the DMS-1000 sets in planes other than the B-24; but by June it was decided that the B-24 was the best available aircraft for this purpose. Meanwhile, in March, the Dumbo I had been flown to England. It was successfully flight tested in Northern Ireland, using British submarines as targets. The ranges were better than those obtained with the long-wave British Mark II ASV, and Coastal Command's only comment was "how many can we have, and how soon?"

Meanwhile, Dumbo II flew to Washington, where T. W. Bonner and Jeffers demonstrated the equipment to the Army and Navy. Not much interest was shown in the higher reaches of the Pentagon until the Secretary of War was induced by his radar advisor to fly with the equipment. This he did, before his generals had responded to the invitation, in a plane loaded with depth charges in case an enemy submarine should actually be sighted. Without difficulty the radar located a ship and the plane was able to home on it so that the Secretary could look out of the window and see the results of the pursuit. He was convinced: "That's good enough for me. Let's go home." The next day General Marshall and General Arnold found identical notes on their desks from Mr. Stimson saying, in effect: "I've seen the new radar equipment. Why haven't you?" Dumbo II was shortly afterwards host to many high-ranking officers.

In August RCC delivered its first DMS-1000 set, which was installed in a B-24 at LaGuardia Field. By November the whole order, including three extra sets for the Navy, had been delivered. The British wanted still more; but by that time Philco's large-scale production of the ASG had begun, and British needs could be filled from this source.

These prototype systems were actually used operationally. Dumbo I, which had been returned by the British and Dumbo II were based at Langley Field, Virginia. Dumbo I was credited with sinking a submarine on July 18, 1942.[1] The first DMS-1000 set was used operationally by Coastal Command in early 1943 for submarine search in the Atlantic and the Bay of Biscay. This successful Radiation Laboratory development was pronounced by the British to be the first substantial radar contribution made directly to Britain's war effort by the United States.

ASG

The experimental and preproduction ASV systems so far described had only an indirect effect upon the design of production microwave air search equipment. Moreover, their numbers were too few to have an appreciable effect upon the antisubmarine campaign, although they had shown how effective an antisubmarine weapon aircraft could be. The ASG, by contrast, was a 10-cm ASV set designed by the Radiation Laboratory, engineered in cooperation with the Philco Corporation of Philadelphia, and produced with such speed and in such large quantities that it was able to exert a strong influence on the course of the war against the submarine.

The much-publicized "George" radar, as it was known to the pilots and radar operators, came into being when the Naval Air Station at Lakehurst, New Jersey, expressed an interest, in October 1941, in a radar system for installation in nonrigid K-ships (blimps). Consultations of Navy representatives with Radiation Laboratory members resulted in the decision to make an experimental installation in the K-3 blimp, using Radiation Laboratory equipment recently removed from the XJO-3 airplane.

After operation had shown that the idea was a sound one, a conference was held at Cambridge on February 17, 1942, with Lt. Commander L. V. Berkner and Lt. Commander D. T. Ferrier representing the Navy, to decide on the proper features for a production set for the installation in blimps. The ASC, Western Electric's modification of an AI set for ASV purposes, and very much like the Radiation Laboratory's B-18 ASV, was considered, but dismissed as too heavy and bulky, and with poor chance of speedy production. It was pointed out by L. C. Marshall that the Stromberg-Carlson Company was building a new, lightweight modulator, based on the latest Radiation Laboratory developments and designed to meet Navy specificiations for such components. It was decided to design a new

10-cm ASV set around this modulator and to give the production contract to the Philco Corporation, which had already attracted the attention of the Navy by its speedy performance on earlier contracts. This new set, the ASG, was to have a PPI, whereas the earlier systems, the Radiation Laboratory's B-18 ASV and Western Electric's ASC and SCR-517 (the Army version) had B scopes. T. A. Murrell became the Laboratory's project engineer for this set.[2] Production of the ASG for blimps was speedy. By June 1942 the first preproduction system was undergoing successful trials in blimp K-9. By the time the first production systems were delivered at the end of October, blimps K-3, K-4, and K-9 had begun regular patrol duty out of Lakehurst and South Weymouth, Massachusetts, using preproduction equipment.

Once the practicability of production of ASG was demonstrated, the Navy began to demand ASG for heavier-than-air craft. Considerable thought was put on the matter of adapting the ASG for installation in the XP4Y patrol ship and the PBM-30 patrol bomber, for the Navy's part in the antisubmarine patrol, and for installation in the Army's B-24, which had already proved successful in such work. It was found that only mechanical details would need to be altered to suit the ASG-1 for heavier-than-air craft. Philco accepted a production order for the ASG-1 in June 1942, and ASG-1 sets were among those delivered in October. Since the P4Y was still in the experimental stage, it was decided to try to adapt the construction of the aircraft to the installation of the radar equipment. During the summer of 1942, members of the Radiation Laboratory went to the Consolidated Aircraft Corporation in San Diego to assist in the changes necessary to accommodate the P4Y to the ASG. For the first time an aircraft was designed around a radar set, instead of forcing it to fit into whatever space was available. Radar was thus recognized as a standard weapon of war.

With initial production underway, the Radiation Laboratory concerned itself largely with improvements in components and the possible addition of refinements. Laboratory participation in actual production work slackened, as indicated by the change in status from consultant to advisor, a change that had taken place by the beginning of 1943. In the fall of 1942 the Laboratory had supplied improved designs for various components, particularly the rf lines. These improved components were incorporated in ASG sets produced during the spring of 1943. Meanwhile, various refinements were considered, most of which were incorporated as they reached production. Examples of these refinements are antenna stabilization, sector scan, and the low-altitude bombing attachment (LAB), while tentative antijamming circuits were developed. With wide use by both the Army and Navy, a joint Army–Navy designation of AN/APS-2 (airborne search radar) was adopted, with different models indicated by the addition of a letter. By the end of the war Philco had produced 5547 sets of various types of ASG.

The ASG became the most popular of Navy airborne search radar sets for heavier-than-air craft[3] until the enemy began to use 10-cm listening apparatus, when it became necessary to use 3-cm radar. Blimps equipped with ASG operated on anti-submarine patrol out of Lakehurst, New Jersey, during the time when submarines swarmed off the Eastern Seaboard. The British used ASG successfully during the summer of 1943 from bases in the British Isles for submarine patrol, particularly in the Bay of Biscay.[4] At the time the U.S. Navy was operating in the Caribbean area, with two squadrons of ASG-equipped PBM patrol bombers based on Trinidad. Between June 15 and August 15, 1943, three submarines were probably damaged and one was forced to scuttle. The radar equipment operated satisfactorily 82% of the time and failed only 7% of the time, a successful record.

A chart of the development and production of these 10-cm systems is shown in Fig. 12-2.

Airborne Gunlaying Systems

The development of an airborne gunlaying system (AGL) seemed possible after the first experiments in Project II on the ground gunlaying system, which resulted in the highly successful SCR-584. These first experiments in early 1941 proved the usefulness of microwave radar for target tracking. With minds still focused on the British need for nightfighter operations, the requirements for a radar system to control the guns of nightfighters seemed obvious.

AGL as a Laboratory project was first seriously considered as a result of a request from General Arnold to E. L. Bowles, at the time Secretary of the Microwave Section of NDRC.[5] In compliance with this request Bowles and L. N. Ridenour, then head of the Airborne Division of the Radiation Laboratory, attended a conference at the Douglas Aircraft Company at El Segundo, California, July 8–9, 1941. This was an early example of the cooperation between the Radiation Laboratory and the aircraft engineers that became so common in future airborne development.

Both the Air Corps and the Douglas Company were anxious to have adequate microwave radar equipment for the XA-26A, a new attack bomber designed primarily as a nightfighter. The mockup of the airplane had been based on the presumed use of the AI-10, but Douglas representatives, particularly F. R. Collbohm, wanted a more elaborate fire-control set, which would do justice to the computer which was to be used to

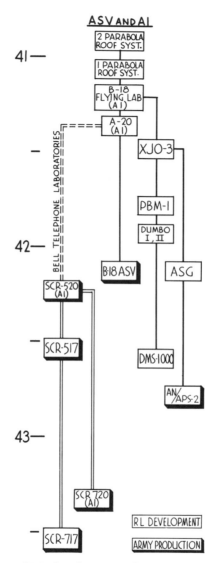

FIG. 12-2. Chart of RL developmental and production systems (10-cm AI and ASV).

permit accurate firing of the guns beyond point-blank range. It was decided that Douglas should plan for the installation of a modified AI-10, but that, for future installations, an improved gunlaying system would be developed. By August the AGL-1 project had been undertaken by the Radiation Laboratory by D. T. Griggs and J. V. Koldam, and in October a

NDRC contract for two AGL systems had been given to the General Electric Company.[6]

The AGL-1 was a 10-cm system which permitted search, automatic tracking, and blind firing against enemy aircraft at night or under conditions of restricted visibility. The operator selected the target manually, after which the system locked on the target and automatically and continuously supplied range, azimuth, and elevation data to the computer that directed the movable gun turrets. The production set had available both B presentation (azimuth against range, primarily used for search) and C presentation (azimuth against elevation, primarily used for tracking). The AGL-1 was originally designed for the XA-26A attack bomber; it was also suitable for installation in the P-61 (Black Widow) fighter.

Flight tests early in 1942 were sufficiently promising to cause the Signal Corps to order 200 sets from the General Electric Company, to be known as the SCR-702A (later as the AN/APG-2). By January 1943, 200 sets to be known as the SCR-702B (later AN/APG-1) were ordered from the Western Electric Company. During 1942 and 1943 various types of airborne gunlaying systems were under development at the Radiation Laboratory and the Sperry Gyroscope Company. These AGL systems operated mostly on 3 cm, and many of them were designed for control of guns in bombers rather than nightfighters. For a development and production chart, see Fig. 12-3.

The only AGL systems ever really in production, however, were the 200 AN/APG-1's built by Western Electric, and a relatively lightweight set for B-29's, built by General Electric.[7] General Electric's contract for the AN/APG-2 was cancelled in October 1944, since production had not yet begun and the 200 sets of the AN/APG-1 fulfilled all Army requirements. Work on the AN/APG-3 began at the end of 1943 and after ground tests in 1945 the Army pushed production of this and a similar system under development by Sperry. These systems represented an attempt to develop a system considerably lighter than the original 400–900-lb systems previously developed; it was part of an intensive program to provide every conceivable form of radar for the B-29.

The AGL program suffered undoubtedly from the fact that it was conceived very early in the Radiation Laboratory's existence and the systems developed remained modifications of a fairly early version of microwave history. The sets were necessarily heavy and relatively complicated, which did not tend to make them popular with the operational branches of the Air Forces. There was never a great deal of enthusiasm within the Laboratory, outside the groups involved. The Army always thought of AGL as a nighttime weapon, whether for use in bombers or fighters, though undoubtedly the daytime accuracy of aircraft gunnery could have

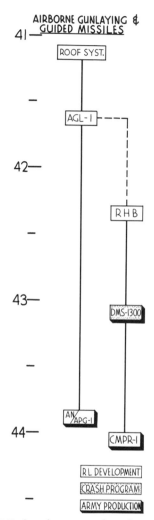

FIG. 12-3. Chart of RL developmental and production systems (airborne gunlaying and guided missiles).

been improved by the use of some form of gunlaying equipment. But lack of effective fighter opposition for bombers, and lack of the need for night-fighters, at the time when gunlaying systems were available for full-scale production, lessened any enthusiasm the Services might have shown.

Lighthouse Tube Systems

The lighthouse tube radar (LHTR) represents an early attempt to design a really lightweight unit, requiring little input power, which could serve as the heart of a number of different types of radar systems. It was chiefly used for fire-control systems, but was also tried in a lightweight ASV and in a radar-guided missile (SRB). In all its applications it was used without internal modification. Its only major competitor, in the field of lightweight, low-power, adaptable radar units, was the low-voltage magnetron transmitter–receiver (XMTR or SMTR), which furnished more output power, and was lighter in weight, but which required more input power.

The lighthouse tube itself, so called from its shape, was developed by the General Electric Company. The lighthouse tube with reentrant cavity, in which form it was used in LHTR systems, was developed in the fall of 1941, though the results were not published until May 1942. Early tubes were unreliable, since a satisfactory method of test had not been worked out. The work of the Radiation Laboratory in Group 52, from the end of 1942 to 1945, was largely concerned with improving the tubes by developing proper mechanical and electrical specifications. The JAN (Joint Army–Navy) tubes finally produced by General Electric and RCA were the 2C43, a large-anode tube for pulsed transmission, and the 2040, a small-anode tube primarily for cw transmission (local oscillators). These tubes were light, simple, and compact, and, by means of an associated circuit, could cover a wide band of microwave frequencies.

The LHTR was a lightweight (25-lb) pressure-tight unit containing a transmitter–receiver and a power supply. It was originally designed for operation at 10.7 cm, but in 1943 the band of frequencies from 11.0 to 12.5 cm was assigned for LHTR operation. The output was low (approximately 1.5 kW) but the input power required was very low also. The LHTR Mark I was developed by H. L. Schultz in 1942 as part of the Airborne Range-Only system. A lighter (15-lb), smaller unit, the LHTR Mark II, was developed primarily for use in a Tail Warning system, but was never put into production.

Airborne Range-Only
(ARO, SCR-726, AN/APG-5, AN/APG-14) Systems

For accurate gunsighting, target range, as well as bearing, is needed. It is impossible to determine range accurately by eye, and the commonly used stadiametric method, in which the gunner adjusts the reticle in his sight to correspond to the wing span of the target plane, involves too much adjustment to be accurate in the stress of combat. It was for this reason that the development of an airborne radar system to provide range only (ARO) was undertaken by the Radiation Laboratory in the spring of 1942, following a request from the Bureau of Ordnance. The Navy had been interested since January 1942 in a system to be combined with the

Ford Instrument Company's fire-control system, but the idea was not then practicable.[8] The Army became interested in June 1942, and most of the subsequent development was undertaken for the Army Air Forces.

ARO was designed to furnish range automatically to the computing sight, with no attention from the gunner. The components were developed during the spring and summer of 1942, with particular emphasis being placed on the oscillator circuits. As it took shape, the system was found to consist essentially of an LHTR unit, a range unit which automatically maintained a voltage output proportional to range, a servomechanism to convert this voltage to shaft rotation, and an 8-in. parabolic antenna. An order for ten LHTR units and six range units was undertaken by RCC. The Philco Radio Corporation and the Galvin Manufacturing Company both received small educational orders.[9]

During the fall of 1942 the Laboratory-built ARO was given flight tests which were followed by firing tests at Eglin Field. These tests resulted in an Army order for 400 sets from Galvin. Meanwhile the Navy was conducting exhaustive tests at the Aircraft Armament Unit at Norfolk, Virginia. The Army laid out a series of future tests which included installation of ARO in a B-25G for use with 75-mm cannon (this resulted in the Falcon system), firing tests of ARO in the Emerson lower ball turret of a B-17, and mockup of a B-29 for ARO.

RCC production began in June 1943. Galvin produced their first six (preproduction) systems in April 1944. Since it was then obvious that there was little chance of quantity production in 1944, the Army agreed to cancel all orders except for 25 preproduction systems, while the Navy requirement was reduced to 80 sets. After various modifications to permit the use of ARO with the Fairchild K-15A gunsight, Galvin resumed production in May 1945 and produced 1230 sets.

These production delays had a number of assignable causes. Lighthouse tube production was at first far from satisfactory, both in quantity and quality. The completely automatic operation meant stringent design requirements that were difficult to reach. There were conflicts with other LHTR systems, more urgently needed or easier to produce. The military requirements for fire control varied sharply with time, so that a few months after an urgent military requirement existed, the production requested at that time would be drastically curtailed, since the requirement no longer existed. There was, finally, some feeling that the equipment was not sufficiently accurate. On investigation it was found that most computers had inherent inaccuracies that did not show up with the inaccurate stadiametric measurements of range; the K-15 sight was the first to be accurate enough to warrant the introduction of inherently accurate radar range.

Combat experience with ARO was small. In January 1944 plans were made to install ARO (under the designation AN/APG-14) in B-29's as part of Project Wasp.[10] Only one such installation was made, and used for tests. Project Figaro, the installation of ARO in five B-17's and five B-24's, was carried out under B. P. Bogart. The B-17's were sent to the 15th Air Force in Italy at the beginning of 1945; the B-24's did not arrive until May. The B-17 installations were successfully flown on many missions, but no tactical experience was gained since no enemy fighter opposition was encountered.

AGS (AN/APG-8, AN/APG-15)

In October 1942 the Radiation Laboratory received a request from the Army Air Force to consider the feasibility of a modification of ARO for gunsighting.[11] The AGS (airborne gunsight)[12] project was set up November 1, 1942, under the guidance of J. V. Holden. It was decided to develop a lightweight system based on ARO, without a computer, for installation in the Emerson tail turret of the B-24 bomber. As a long-range project, the system was to be adapted for use with lead-computing sights.

The necessary modifications to the ARO system were the addition of conical scan, the precise coordination of the radar axis with either the sight or boresight line, and the addition of an indicator unit. The indicator presentation chosen was a spot whose position was an indication of target bearing, with wings whose size was inversely proportional to range. Since there was some doubt of the adaptability of the LHTR to such a system, an alternative development of a low-voltage magnetron transmitter–receiver, the SMTR or AN/APG-10, was initiated. This, which would have given greatly increased rf power, with only a small increase in input power and weight, was canceled at the end of 1943 because of the time required to put it into manufacture, and because sufficient manpower was not available. By April 1943, a complete AGS Mark I, with an endfire array antenna replacing the original paraboloid, was undergoing flight tests in an AT-11 plane. This was followed by tests in the Emerson tail turret of a B-24. An NDRC contract for 25 AGS Mark I systems was given to Galvin, on the possibility that an urgent military need might arise.

Parallel development continued on various AGS systems, and by October 1943 there were several recognized versions. The AN/APG-8, under production at Galvin, had no computer and was designed for installation in an Emerson tail turret. The AN/APG-15 was a modified form of the AN/APG-8, for installation in the tail of the B-29. There was an SMTR version of the AN/APG-15, and an LHTR system for installation in an Emerson tail turret in conjunction with the Fairchild lead-computing sight. The Army, encouraged by preliminary tests, set up a requirement of 3500 sets of AGS in 1944, with contracts to General Electric and to Galvin. In May 1944 when Falcon was given precedence over other LHTR systems the AGS contracts were curtailed. AGS, like other fire-control systems, suffered from changing tactical requirements. The production was later reinstated, with total production of 9376 AN/APG-15 sets.

In August 1944 the Army requested assistance from the Radiation Laboratory on Project Wasp. This was a crash program to install a modified version of the AN/APG-15 in seven B-29 planes; it was originally planned to install AN/APG-14 (ARO) also, but this installation was carried out only in the one plane that was given exhaustive tests at Alamogordo, New Mexico. The principal modification was the development of the AN/APX-15 (Ella), an IFF equipment which detected B-29's by means of their propeller modulation. It was designed for use with the AN/APG-15, though it could have been used with other, similar, systems. The other modification, which was incorporated in all subsequent systems, was the feeding of radar range direct to the computer. All systems produced after February 1, 1945, had this modification, and were adapted for use with the AN/APX-15. Five of the Project Wasp B-29's were sent to the 58th Wing, 20th Air Force, in India, for combat testing. The systems operated successfully, but little tactical information was obtained. This was also the case with the installations in the planes of the 315th and 316th Wings of the 20th Air Force.

The installation of AN/APG-15 represents an example of good planning; had there been serious enemy opposition the equipment would have been available. The set did sacrifice accuracy and range for simplicity and lightness and perhaps if it had been used intensively a more effective system would have been desired.

Falcon (AN/APG-13A), Vulture (AN/APG–13B), and Pterodactyl (AN/APG-21)

After successful tests of the ARO system in the spring of 1943 the Army Air Forces Materiel Command requested tests of the system in a B-25G for control of the 75-mm cannon. Such an aircraft has a fixed gunsight mounted parallel to the cannon boresight axis. The pilot therefore can aim the cannon by aiming the plane at the target. Except in point-blank firing, determination of range by eye is too poor for accurate gunnery; the 75-mm cannon is normally used for ranges up to 6000 yards, which places the plane out of danger from anti-aircraft fire. Performance with the 75-mm cannon had not, in general, been satisfactory, chiefly because of poor ranging.

After tests in September 1943 the Radiation Laboratory began to modify the basic ARO components to produce the Falcon system. This involved elimination of the automatic range unit, range servo, and calibrator, which were replaced by an M scope, and the addition of a ballistic cam to make the rotation of the range-adjusting shaft linear with range. The system was composed of an endfire array antenna, LHTR unit, M scope, pilot's sight, and power supply. The radar operator provided range adjustment of the gunsight by turning a hand crank so that the electronic marker on the cathode-ray tube was always in coincidence with the target signal. The hand crank was connected to the sight through a flexible shaft. The presence of an operator who could retune the set, if necessary, somewhat reduced the stringent engineering requirements which had made production of ARO difficult.

By the end of 1943 the first Falcon system was ready for tests at Eglin Field and a crash program for 30 systems had been initiated by the Laboratory at RCC. These were later assigned to a special experimental squadron. No production equipment was at first contemplated, but when RCC finished the first program in April 1944, the Army asked for 120 more systems. At the same time an order was placed with General Electric for 880 sets to be built on a crash basis: the M scope at General Electric and the LHTR units and antennas at Galvin.

In June 1944 the first experimental squadron, with Falcon equipment, was operating with 5th Bomber Command in New Guinea. The equipment performed in a satisfactory manner, though ship targets were already scarce.[13] By fall the 14th Air Force in China was using Falcon with great success. However, there was one serious fault in the equipment from the operational point of view. The gunsight was designed for an airspeed of 250 mph during the firing run; in operation this speed was often not reached. A request was made to the Laboratory for a method of correcting this, and a mechanical attachment was therefore developed so that the actual airspeed could be set in to the sight just before making the firing run.

During the summer of 1944 the Army requested a unit which would permit mechanical tracking in place of hand tracking; at the same time it became obvious that a modification of Falcon for use over land was needed. This resulted in Vulture (Overland Falcon). The Vulture presentation, in which all targets appear on the indicator, but with the chosen target signal clearly differentiated, was proposed by E. H. B. Bartelink in May 1944. Falcon was easily convertible to Vulture by the addition of a subpanel to the M scope, the replacement of the antenna by an AGS scanner, and the addition of an aided tracking unit to replace the hand crank in feeding range information to the sight. The operator, instead of keeping the target signal in the electronic marker manually, had merely to set a range marker on the chosen target, after which tracking was automatic. The signal from the chosen target (i.e., the target at which the pilot was aiming) was obvious to the operator, since it was the only unmodulated signal on the scope, and appeared as a solid line, whereas all other signals were modulated and appeared as broken lines.

The first Vulture system was successfully flown in November 1944. In February 1945 General Chennault requested 30 modification kits to convert Falcon to Vulture. The 14th Air Force had used Falcon with great success, but they had been pushed back in China and needed an overland system. RCC undertook a crash program which was completed in August 1945. Meanwhile, during the spring and summer a good deal of thought had been given to the problem of adapting Falcon or Vulture for rocket firing. The Army Air Forces Board at Orlando, Florida, had conducted successful tests with Falcon, and recommended the development of a suitable ballistic cam.[14] The Applied Mathematics Group at Columbia University gave a good deal of assistance to the Radiation Laboratory on the design of suitable cams.

By November 1944 Pterodactyl (Automatic Vulture) was ready for flight tests. This was a completely automatic system which searched in range and only locked on a target when both range and directional information were correct. In June 1945 the Laboratory began work on five systems to be made from the improved ARO and AGS components then in production. RCC had an order for 12 systems, which was completed in October 1945.

TW (Tail Warning) Systems

The usefulness of a radar system which would automatically notify the pilot of a fighter or bomber of the approach of enemy aircraft in the rear of the plane is readily apparent. The necessity of tail warning was pressed by the Army. Work on such a device, which would give an audio or visual warning, was begun at the Radiation Laboratory in November 1942, with the idea of making use of the LHTR unit. The problem of developing a satisfactory system was complicated by the varied requirements of the different Services.[15] The Navy wanted a system primarily for bombers, which meant that adequate range (approximately 3000 yds) was required, but weight was not critical. The Army Air Forces wanted a system for day fighters, which meant that it must be very light in weight (approximately 40 lb) but that range could be limited (approximately 500 yds). The British Air Commission wanted an intermediate system, suitable for night-flying planes. All systems had to have a minimum range so that formation flying would be possible; the minimum range varied with the application, from 500 yds to 3000 yds.

The TW systems worked by "floodlighting" the area to the rear of the plane with a conical beam. Since the great advantage of microwave over long waves is the greater ease with which microwaves can be focused, it was soon realized that it would be more efficient, as well as easier, to develop a TW system operating on 60 cm. Such a system would have longer range, simpler circuits, wider beam, and lighter weight. While the Radiation Laboratory, partly as an experimental venture, was working on TW systems based on the 10-cm LHTR, the Westinghouse Manufacturing Company and RCA were developing 60-cm systems, with some advice from the Laboratory. By June 1943 there were six different systems under development in which the Radiation Laboratory had some part. The TW project was, however, terminated at the Laboratory on September 15, 1943. The various systems are shown in Table I.

TABLE I

Set	Frequency in cm	Weight in lb.	Range in yds.	Special characteristics	Manufacturer
TWS-1	10	85	700 – 1000	LHTR; audio warning	
TWS-2	10	65	700 – 1000	LHTR; visual warning	
TWS-3	10	55		LHTR Mark II	Delco (none produced)
TWS-4, ASJ, AN/APS-17	10	Less than 100 lb.	1000– 1500	SMTR	Philco (4 sets)
TWL-1, AN/APS-16	60	Less than 100 lb.	2000	Modified ASB	Westinghouse (1500 sets)
TWL-2, AN/APS-13	60	Less than 40 lb.	500	Modified SCR-518 altimeter	RCA and GE (45 673 sets)

Low-Voltage Magnetron Systems

The low-voltage magnetron systems are closely allied with the light-house-tube systems. Both are lightweight, have low power, and are adaptable for a variety of uses. The low-voltage magnetron systems, the 10-cm SMTR and the 3-cm XMTR, give approximately ten times the power output of the LHTR, with very little increase in input power. By the beginning of 1943 these systems had reached a stage of development which suggested their use as alternatives to the LHTR AGS systems, had these not proved satisfactory. Some work was also done on a low-voltage magnetron tail-warning system, the ASJ. The only low-voltage magnetron system that was pushed beyond the initial stages of development, however, was the AN/APS-10, also known as BFR (blind-flying radar). This lightweight navigational system was derived from the LWASV, a lightweight LHTR system begun in November 1942 and terminated in June 1943 because it did not meet the Army requirements for weight and performance.

At the time the LWASV project was terminated the Army expressed an interest in a very lightweight ASV system, to be called the AN/APS-10, which would be simple and reliable, with relatively short range, suitable for use with beacons and IFF equipment. During the summer of 1943 H. L. Schultz began work on the XMTR, a 3-cm, low-voltage magnetron transmitter–receiver, designed for use as part of a lightweight ASV or lightweight AI system. As development proceeded, it appeared that such a system would be even more useful as a navigational device than as an ASV or AI system.[16] The need for such equipment had been strongly urged by Colonel Stuart P. Wright on his return from the Pacific. The avowed aim of the Airborne Division was to make this system part of the regular aircraft equipment, for all aircraft not otherwise equipped with navigational radar.

By the beginning of 1944, the first model was successfully flight tested. This had an 18-inch csc^2 antenna, and an XMTR unit, but only a makeshift indicator. Various improvements were made so that at a joint Army-Navy meeting on February, 1944, the Army was able to decide that the AN/APS-10, based on the XMTR, would definitely meet its requirements for a lightweight, simple navigational device for use on transport planes.[17] Several engineers from General Electric, the Army's choice for manufacturer, arrived at the Laboratory to work with the system. On March 22 there was a meeting at which the AN/APS-10 program was set up. The Laboratory agreed to act as consultant, with A. Longacre as project engineer for production, while R. L. Sinsheimer remained project engineer for developmental work at the Laboratory. General Electric was

to start delivery in January 1945, and have quantity production established in April 1945.

The General Electric engineers remained at the Radiation Laboratory
until the summer of 1944; by the time they left the engineering drawings
were complete. The Laboratory built three prototype systems based on
these drawings, one of which served as General Electric's prototype model, while the others were used for tests. The AN/APS-10 had become a
rugged and reliable system, simple to operate and easy to service. It con-

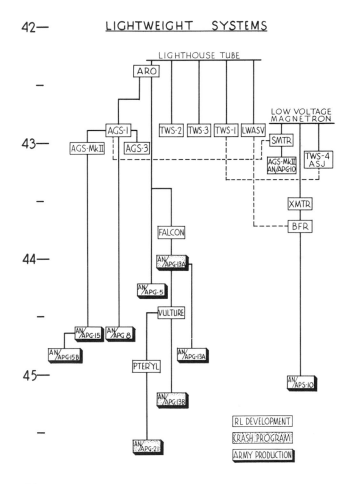

*FIG. 12-4. Chart of RL developmental and production systems
(lightweight).*

sisted of an 18-inch csc^2 antenna, an XMTR unit, a control central, a 5-in. magnetic-synchro PPI, a power supply, and a pressurizer. The number of controls was sharply limited so that reliable automatic tuning was imperative; the AFC circuit was designed by Schultz. For easy servicing the various units were designed to be readily replaceable and completely interchangeable.

Various improvements were incorporated in the system. General Electric made the receiver simpler and easier to manufacture. Since the Laboratory was not satisfied with the General Electric scanner, a new and improved scanner was designed and manufactured at the Houston Corporation; 200 of these were purchased by the Army through General Electric, as extra equipment. Experiments were made on the use of AN/APS-10 in conjunction with Micro-H.[18] With some difficulty, an AN/APS-10 system was installed in an A-26 attack plane together with an AN/APS-5 (LAB), for use as a low-altitude bombing system. At the beginning of 1945 the system was demonstrated to Troop Carrier Command (TCC), which was very enthusiastic. A 30-in. scanner was developed for TCC, to give improved resolution for easier navigation. Plans were made to install the AN/APS-10 in all C-47 transports.

Production at General Electric was slow, due largely to procurement difficulties and necessary design changes. The 50 preproduction systems were completed by April 1, 1945. Since General Electric had subcontracted the manufacture of components, merely assembling and testing the sets, quantity production was delayed until June. By August, 300 sets a month were being produced. Such sets as were available before the end of the war were installed in C-46 transport planes for TCC.

A chart of Radiation Laboratory developmental and production work for all the lightweight systems is given in Fig. 12-4.

Airborne Guided Missiles

In the spring of 1942 work began on a method of bombing in which the bomb would be released from the aircraft at some distance from the target, and would be guided to the target by radar. The RHB (radar-homing bomb) project, under the direction of E. M. Lyman, was regarded as an entirely new system, and, instead of using modifications of components already in use in other systems, as had been largely the practice up to that time, used newly developed components. The bomb itself contained only a radar receiver, antenna, and discriminator device; it homed on the echo obtained by illuminating an isolated target by means of an ASV system contained in the control aircraft.

The Navy (Bureau of Ordnance) was interested from the beginning, and by the summer of 1942 Project Pelican, in which the RHB system was carried by a glider bomb developed by H. Dryden in Division 5 of NDRC, was in full swing. RCC was given a contract for 100 systems, of which 50 were allocated to Project Pelican. In October 1942 work began on SRB, in which the direction-indicating circuits of the RHB were added to the LHTR to form a radar-homing bomb which carried its own transmitter. Low laboratory priorities on the LHTR postponed testing of the SRB until February 1943.[19] A system was built, flown, and tested before the project was terminated in June 1943.

From January to September 1943 a large number of flight tests were made with varying results. At first there was found to be inadequate synchronization by the RHB of the echo with the signal transmitted by the ASV; this was corrected by the addition to the RHB of a "rear view" antenna. There were a considerable number of aerodynamic problems and difficulty was experienced in maintaining the bomb in steady flight. A number of accurate flights were, however, obtained.

Small-scale production began at RCC at the beginning of 1943. In February 1943 both General Electric Company and Zenith Radio Corporation engineers had arrived at the Laboratory in preparation for production orders. The production at RCC was interrupted in the fall of 1943 when RCC experienced a temporary crisis in production. The partially completed systems were sent to the Bureau of Standards, where they were completed by Navy trainees under the supervision of the Radiation Laboratory. Production models were available by January 1944; total production of the missile, now called CMPR-1, was 1220 by Zenith and 220 by General Electric.

When the project was terminated at the Laboratory in the spring of 1944, several staff members were transferred from the Laboratory to Division 5 of NDRC to assist in further development of the project. It was felt that flight tests had proved the feasibility of the radar-homing bomb as a Service weapon. There were, however, many disadvantages. It is obvious that no adjustment is possible once the bomb is launched; this means that maintenance and testing must be rigorous and perfect if the bomb is not to be wasted. It was sometimes felt the RHB was not tactically useful since it required the control plane to remain in the vicinity of the target and to fly a restricted course to keep the target properly illuminated.[20] In this respect SRB was preferable. The difficulty of selecting a target that was not isolated was another disadvantage, though some methods, such as an improved target selector, and the use of the fine discrimination of the Eagle radar, were suggested. In spite of disadvantages it is believed that radar-homing bombs were actually used operationally.

The Radiation Laboratory's contribution to the development and production of airborne guided missiles is included in Fig. 12-3.

NOTES

1. H. M. Jeffers, "Report on the IB-30 Ships at Langley Field," n.d.

2. J. F. Koehler was the initial project engineer, but on his entrance into the Navy he was replaced by Murrell, assisted by D. L. Hagler.

3. U.S. Fleet Anti-Submarine Bulletin, February 1944.

4. D. M. Robinson, "Successful U-Boat Attacks Using ASG," 16 August 1943.

5. E. L. Bowles to L. A. DuBridge, 18 June 1941.

6. D. T. Griggs, "Wright Field Conference on Airborne Gun Laying Systems," 2 October 1941.

7. H. G. Brewer, Trip Report (to ARL, Wright Field), 1 December 1943.

8. L. N. Ridenour, "Meeting at Ford Instrument Company January 6, 1942."

9. See Chap. 34 for an explanation of this term.

10. This was primarily the installation of the AN/APG-15 in B-29's.

11. J. V. Holdam, Jr. and L. M. Jones, "Aircraft Gun Sighting," Radiation Laboratory Report 91-11/21/42.

12. Gunsight systems must be manually adjusted to track the targets, whereas gunlaying systems are completely automatic.

13. J. H. Gregory, Report No. 1, "Falcon (AN/APG-13) Project in the South-West Pacific," 15 June 1944.

14. Report of Army Air Force Center, Orlando, Florida, "Test for Use of AN/APG-19A Radar Ranging Device for Rocket Firing," 6 August 1945.

15. R. L. Sinsheimer, "Conference on Tail Warning Equipment, February 11, 1943," February 17, 1943.

16. L. A. DuBridge to Brig. Gen. H. M. McClelland, 23 November 1943.

17. Airborne Division Notes, 1 March 1944.

18. A beacon bombing device.

19. Minutes of the Coordination Committee, Vol. V., No. 12, p. 262, 9 February 1944.

20. W. M. Brezeale, "Survey of the Navy's Assault Drone Program," 7 June 1943.

CHAPTER 13
Airborne Development on Three Centimeters

From the very earliest days of the Laboratory's program, it was generally understood that a strong effort should be made to devise microwave radar on a wavelength even shorter than 10 cm. To conservatives in military procurement this must have seemed a doubtful enterprise. It was already presumptuous enough to try to develop 10-cm equipment to supplant the longer-wave radar which was only just going into production. To expect equipment on still another unexplored wavelength to be ready in time to be of use in the field was no less than audacious. The problem for I. I. Rabi's Transmitter Group was what wavelength to choose. It was possible either to attempt a radical improvement by going to a wavelength as short as a few millimeters or to stay closer to 10 cm, a band where techniques already being perfected could presumably be adapted, by scaling-down and other modifications, to the new wavelength. A somewhat arbitrary compromise was 3 cm. It was close enough to 10 cm to offer a fair chance of success in a reasonable length of time, yet the gain of about threefold in resolution would be worth a very considerable engineering effort.

The emergence of the first successful 3-cm magnetrons from the Raytheon model shop in the early spring of 1941 led Rabi to form an Advance Development Group, with N. F. Ramsey as chairman, whose task it was to build an experimental 3-cm roof system. This system, the first to use a waveguide,[1] was in operation in the middle of May, giving echoes from ground objects 6 miles away. It included a receiver with a 60-MHz IF strip and klystron local oscillator, a 3.3-cm magnetron "with power comparable to that of the 10-cm tubes,"[2] a Type A indicator, and separate transmitting and receiving paraboloids (12-in. diameter) fed by crude pyramidal horns. This system was followed soon after by a similar one, the XT-2, installed in a truck.

XT-2

During the first week of May 1941, a project was initiated under the direction of E. M. Purcell, of the Advance Development Group, for the assembly of the Laboratory's first mobile 3-cm installation, the XT-2. It was an experimental general search system intended for operation from

341

locations away from the Radiation Laboratory, especially against ships along the coast. Its components were similar to those of the 3-cm roof system. Initially the XT-2 employed a Raytheon Type-D2UY1 10-kW magnetron. The reflector, a 4-ft paraboloid, was mounted on top of the truck.

One month after the Japanese attack on Pearl Harbor, the XT-2 was placed in operation at Deer Island in Boston Harbor. Operating side by side with the XT-3 (10-cm) truck system for almost three months, it demonstrated to visiting Army and Navy officers its superiority over the 10-cm system in obtaining higher resolution of targets.

In February 1942 D. E. Kerr of the High Frequency Group (later the Propagation Group) took over the XT-2 for propagation studies and operated it side by side with the XT-3. In August 1942 the XT-2 was used to prove that submarine periscopes were detectable by radar at distances of 3 to 4 miles. Throughout the next two years it was used in conjunction with the XT-3 at various locations in New England to obtain information about sea return, signal fluctuation, and atmospheric refraction of signals. This propagation work with the XT-2 and XT-3 enabled Kerr's group at an early date to recognize "storm fronts," an observation which led to work which was of considerable importance to the AAF during the war and which is sure to be of great value to shipping and to civil air transportation in the future. The new science of radar meteorology was directed into new channels as the result of these early propagation studies performed with both the XT-2 and the XT-3 truck systems.[3] The Radiation Laboratory's extensive contributions to this new science will be discussed in a later chapter.

AI on Three Centimeters (AIA)

On May 21, 1941 a discussion took place between E. G. Bowen of the British Air Commission and L. A. DuBridge, during the course of which it was agreed that the success of the 3-cm development suggested an immediate application in an AI radar for single-seater fighters.[4] Because of the pressure of previous commitments, the lack of a Service request, and the shortage of manpower, very little could be done immediately about making such an application of a 3-cm radar. The AIA program, as it came to be known, did not take definite form until September 1941, when the Bureau of Aeronautics expressed its interest in a combined interception and gun-aiming set for its carrier-based, night-fighter version of the Vought-Sikorsky F4U-1 aircraft, the F4U-2. The Navy's specifications called for a compact radar having a weight of not more than 250 lb, an antenna system with negligible drag, an accuracy sufficient for blind gun aiming, a useful search range of 2 miles at altitudes of 2000 ft and higher, and a minimum dependable range of 500 ft.[5] Rabi was convinced that a 3-cm AI with a specially designed antenna and indicator could be built to satisfy these requirements.

and a minimum dependable range of 500 ft.[5] Rabi was convinced that a 3-cm AI with a specially designed antenna and indicator could be built to satisfy these requirements.

The antenna was built to include an 18-in. paraboloid mounted in the wing of the airplane and fed from a waveguide through the wing. It was to employ spiral scanning during search and early tracking, and conical scanning during gun aiming. The indicator was to be a 3-in. scope on the pilot's instrument panel, which during spiral scanning would furnish range, azimuth, and angle of elevation of the target and during conical scanning would function as a gun-target indicator.[6] Concurrent with the AIA development the Bureau of Ships was persuaded to sponsor the simultaneous development, by a newly created High Power Group in the Radiation Laboratory, of a height-finding and general-control radar set for aircraft carriers. This set, which became the SM, will be discussed later.

Group 93 in the Airborne Division of the Radiation Laboratory designed and developed three AI systems, one experimental (CXBJ) and two preproduction (both XAIA), while assisting the Sperry Gyroscope Company in the research, development, and assembly of ten preproduction systems. All of these used a 10-kW magnetron, the output of which had to be fed through a long length of waveguide to the antenna assembly in the wing nacelle.

The CXBJ, the first of these three systems, for which J. Halpern served as project engineer, was completed and tested by the Laboratory early in June 1942 and delivered to the Naval Air Station at Quonset, Rhode Island, for bench testing and preliminary pilot training late in the same month.

The Laboratory's first XAIA, with N. H. Moore and L. M. Langer as project engineers, was completed and sent to Quonset for pilot training early in September 1942,[7] and the second XAIA was completed and sent to Quonset late in October.

With the assistance of the Laboratory, Sperry's first preproduction set was assembled and delivered to the Navy in February 1943[8] and underwent tests during the following two months. By the end of April 1943 the company had delivered its ten preproduction sets.

Because Sperry's production program progressed rather slowly (287 production sets were delivered as of May 1, 1944), the company was able to incorporate in all of its AIA production sets a higher-powered magnetron (40 kW), which had been under development at the Bell Telephone Laboratories. Sperry's contract for 604 AIA production sets was not completed until October 1944.

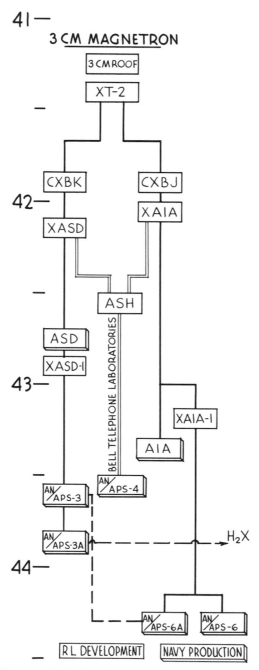

FIG. 13-1. Chart of RL developmental and production systems (3 cm).

In the spring of 1943, having virtually fulfilled its commitments on the AIA, the Laboratory turned its attention to the development of an approved AIA called the AIA-1.

AIA-1 (AN/APS-6; AN/APS-6A)

As early as January 1943 the Navy had conferred with the Laboratory about the development of the "product improvement of AIA" to be known as AIA-1.[9] On March 16, 1943 the Laboratory agreed to develop such a system to operate at an altitude of 30,000 ft or above and to incorporate an important new development, the higher-powered 3-cm magnetron perfected in the interim.

But the real impetus for the AIA-1 program came from the installation difficulties encountered with the AIA.[10] As in AIA the scanner was to be placed in a wing nacelle. In the earlier set this had meant running lengths of waveguide through the airplane wings to carry the rf energy from the magnetron to the paraboloid. The superiority of AIA-1 from the point of view of installation and maintenance was the result of the development of the pulse transformer, an impedance-matching device which made it possible to eliminate the long waveguide. This was done by locating the magnetron immediately in back of the paraboloid in a unit called the rf head, in which were placed, besides the pulse transformer, the TR box and other rf components. This improvement had first been introduced in the parallel 3-cm ASV development, the ASD-1. (See Fig. 13-1.)

On June 14, 1943 the Radiation Laboratory accepted an additional responsibility, that of consultant to the Navy on the AIA-1 production set designated AN/APS-6, to be built by the Westinghouse Electric and Manufacturing Company. With R. M. Alexander as project engineer, the Laboratory completed its own experimental system XAIA-1 in September 1943, while assisting the company in the design and development of a production set. Westinghouse sent its first preproduction set to the Laboratory for testing on December 1, 1943 and its first production set on April 25, 1944. Because of difficulties encountered in the design of an rf head, large-scale production at Westinghouse was considerably delayed. Indeed, for the early production sets it was necessary to use the rf head then being manufactured for ASD-1. These modified early production sets were therefore known as AN/APS-6A. As of April 1, 1945 the Navy had received 791 production sets from the company, 741 of which were AN/APS-6A. (See Fig. 13-2.)

The AIA-1 proved to be a much sturdier and better designed set than its predecessor and was being used in increasing numbers, in conjunction with the SM, aboard carriers of the Fleet as the war came to an end.

FIG. 13-2. AN/APS-6 (AIA-1) installed in Navy fighter F6F.

ASD

In the autumn of 1941, coincident with the inception of AIA, preliminary investigations of 3-cm ASV applications were being made by the
Radiation Laboratory at the request of the Navy. As early as November
1941, components were being assembled at the Laboratory for installation
in a Navy JRB aircraft, a two-motored general utility transport, assigned
for the use of the Laboratory at the East Boston Airport. This system,
wholly experimental and not designed as a prototype of any development,
was completed and flight tested in June 1942. It was designated CXBK,
and H. F. Balmer and E. A. Luebke were its project engineers. Much
flying was done during 1942 and 1943 to explore the behavior of this new
frequency for surface search applications. The PPI photographs made
from the air with this first, airborne, high-resolution radar system showed
at once the great superiority of the new equipment. The greater detail and
fidelity with which the 3-cm image reproduced the natural features below
the airplane caused great excitement among those in the Laboratory and
in the Services who were following this development.[11]

Meanwhile the Navy had been discussing with the Laboratory the ad-
visability of developing a 3-cm ASV production equipment. On February

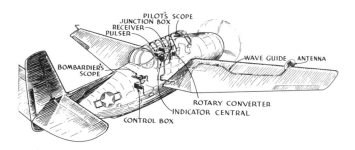

COMPREHENSIVE DRAWING OF XASD EQUIPMENT IN TBF PLANE

FIG. 13-3. XASD installation in TBF.

6, 1942, Rabi formally submitted to the Navy the Laboratory's proposal for a 3-cm system, the ASD, for installation in the Grumman TBF.[12] The ASD would be designed to furnish to pilots of torpedo bombers information as to the position of surface vessels, especially at night or during overcast conditions.

S. J. Simmons was given supervision of the ASD program at the Laboratory. The Navy contracted with the Sperry Gyroscope Company to develop a prototype (XASD) and to manufacture production sets based on that prototype. The Laboratory accepted the somewhat vaguely defined position of advisor to Sperry during this development, and went ahead with the construction of its own experimental system, also called XASD. W. P. Dyke was the project engineer for the latter.

Initially the XASD sets employed a 10-kW magnetron, the output of which, as in the AIA, had to be led through a long length of waveguide to an antenna assembly in a nacelle on the wing of the airplane. After considerable difficulty and many delays, largely inherent in the difficulty of this design, the Laboratory version of XASD was installed and successfully test flown in a Grumman TBF at the end of June 1942. (See Fig. 13-3.) With help from the Laboratory, Sperry completed its experimental preproduction set at about the same time. Ground tests were finished in August and Sperry began production immediately.

The ASD production set weighed 300 lb. Except for the indicator and the antenna mount, the set was identical with AIA. A Type B indicator was at the bombardier's position in the plane and a repeater scope was located in front of the pilot. The set's range was 25 miles on a 5000-ton ship.

Although the ground and flight tests of both the Laboratory and the Sperry XASD sets had been successful, the production sets were not satisfactory.[13] Dissatisfaction with them had two results: first, research was started in the fall of 1942 on a new and improved set to be known as the ASD-1; and second, the production model of ASD was considerably modified and improved. Originally planned for installation in the TBF airplane, the ASD sets were diverted by the Navy to the PV-1 aircraft (Vega Ventura patrol bomber). By June 1943 the Sperry Company had delivered 600 ASD production sets and by April 1944 the company had completed its contract for 3400 sets.

Meanwhile, in a conference at the Radiation Laboratory on November 25, 1942, the Navy and the Laboratory had initiated the ASD-1 program with the Philco Corporation as manufacturer.[14] The essential differences between ASD and ASD-1 were to be an improved antenna, a more stable and accurate indicator, a higher-powered magnetron, and the addition of a specially designed rf head, based on the recently-developed pulse transformer. This eliminated the long waveguide through the wing to the antenna in the wing nacelle. Although designed for the Grumman TBF and the Vought–Sikorsky TBU, it was specified that the ASD-1 should be adaptable to the PV-1 airplane.

Radiation Laboratory was made consultant to the Navy in the Philco contract, instead of advisor, a decision which reflected an important change in the Laboratory's relationship with manufacturers. The responsibilities of the Laboratory during the development and production stages of a project had up to this time not been clearly defined. In certain instances progress had been severely impeded because differences of opinion about engineering matters arose between the Laboratory and the manufacturer which could not be resolved by any final arbiter. This situation became increasingly critical and it became more and more evident that the Laboratory could not terminate its responsibilities on a project at the experimental or "breadboard" stage as had been the tendency at first, but had to follow it through development and production and even into the field.

Experience had shown that a sterilizing deadlock was more liable to ensue when the Laboratory cooperated with the larger concerns having imaginative and forceful research groups. Their long-established engineering departments with traditional ways of doing things did not always take kindly to proposals from a newcomer. It was at about this time that the Laboratory evolved a policy of working to a large extent with smaller electronic concerns, or at least with companies having modest research and development organizations. The Radiation Laboratory became the research and development organization for a group of companies whose production facilities were thereby brought in to relieve the already crowded schedules of the four large electrical concerns which the Army and

Navy had been in the habit of entrusting with its contracts. Late in 1942, the final notion of consultant status was evolved at the Radiation Laboratory to provide a formula by which the Laboratory could continue active participation in the development of equipment as it went into production.

In January 1943, Philco sent three engineers to the Laboratory to assemble a prototype. With the approval of the Philco management, who agreed that two competing research organizations would only impede one another's activity, the Radiation Laboratory served as the development group working directly with the production engineers, almost as part of the Philco organization.[15] Philco's own small but able research group was thereby freed to work on other problems. The Philco engineers completed their prototype at the Radiation Laboratory and shipped it to their factory. Concurrently the Laboratory built an experimental XASD-1 set for comparative tests. It underwent ground tests at the East Boston Airport during March, and flight tests during June 1943. In July the first ASD-1 set built at the Philco factory was installed in a PBN aircraft by the Navy and tested at Anacostia Field. Its performance was comparable with that of the Laboratory system. By the end of August, Philco had delivered nine sets, designated AN/APS-3, to the Navy. Sixteen months later that company's remarkable production line had produced a total of 4924 AN/APS-3 sets.

In October 1943 the Bureau of Aeronautics expressed interest in AN/APS-3 for H_2X (navigation and bombing) applications.[16] Work was begun immediately on an improved ASD-1 with a pressurized rf head. This set was designated AN/APS-3A. Since Sperry had completed some 3400 ASD sets the previous June, the Navy now asked the company to enter into production of AN/APS-3 and AN/APS-3A equipment. Sperry contracted to deliver 1824 sets, of which the last 686 were to be the pressurized set, AN/APS-3A. As of May 1, 1945 Sperry had delivered a total of 1396 sets for the ASD-1 program. The ASD-1 was a distinct improvement over ASD.

Although the ASD was used by the Navy in PV-1 aircraft in bombing in the Aleutians and elsewhere for antisubmarine patrol, it was not wholly satisfactory because of frequent breakdown. Yet in spite of operational failures the ASD set, like the AIA, was a necessary step in the development of 3-cm airborne systems for the armed forces, for out of its faults grew the improvements which appeared in later, more successful, sets such as H_2X.

ASH (AN/APS-4)

Another 3-cm development to which the Laboratory made a significant contribution was a lightweight airborne set intended to perform the combined functions of AI and ASV equipment.[17] The Bureau of Aeronautics hoped this new equipment would eventually supersede the ASB (60-cm) equipment.[18]

In a conference held at the Bureau of Ships, in Washington, on June 30, 1942, representatives of the Western Electric Company, the Bell Telephone Laboratories, the Bureau of Ships, and the Bureau of Aeronautics drew up the general specifications for this equipment, designated ASH.

Specifications called for a high-performance set suitable for small carrier aircraft which would perform radar search operation over large bodies of water for detection of surface vessels, perform mapping and navigational operations, detect and locate other aircraft to permit their interception, pick up coded signals from beacon stations in the 3-cm band to facilitate the return of the searching plane, and operate as the search element in conjunction with identification (IFF) and low-altitude bombing (LAB) equipment.

The ASH radar, as developed by the Bell Telephone Laboratories and manufactured by the Western Electric Company, operated over four ranges: 4, 20, 50, and 100 nautical miles. The indicators were Type B for all ranges of search and Type G for short-range interception. The equipment weighed 180 lb.

Since the ASH program involved essentially engineering problems, the Radiation Laboratory expected to confine its activities to an exchange of technical information with the Bell Telephone Laboratories on 3-cm components. In a conference with Bell representatives on July 7, 1942, Du-Bridge had committed the Laboratory to work with Bell engineers in planning the system and exchanging ideas at the start of development. The Navy, however, felt the need of greater participation by the Laboratory in the developmental stages. Accordingly, the Laboratory appointed R. M. Alexander consultant and project engineer to the Bureau of Aeronautics and to the Bureau of Ships.[19] Although the Laboratory did not undertake the construction of an experimental model nor of a prototype, its representatives gave considerable assistance to Bell in the construction of their experimental model,[20] and through reports from Alexander the Navy was kept informed of the progress of the program.

The first model of ASH was completed in the early part of 1943 and underwent flight tests soon afterwards in the presence of several observers from the Laboratory. Their general impression was that ASH was a well-designed and nicely packaged set which performed the functions of ASV search admirably and of AI moderately well.[21] In July, 1943, at the request of the Bureau of Aeronautics, Laboratory representatives examined a production set, an AN/APS-4.

Production in quantity was under way at the Western Electric Company by September 1943. By April 1944, 403 sets had been delivered. By July 1, 1945, 13,646 sets had been delivered. Of these, Westinghouse produced over 4000 by subcontract. The Army received 466 sets and the British purchased nearly 800 out of the total Navy procurement.

ASH was a very popular set both in the U. S. Navy and in the Royal Navy. Late in October 1944 Bombing Squadron Seven, based on the USS Hancock, made good use of ASH in attacks on Formosa and the Philippines. In December 1944 the British were fitting ASH sets in the following aircraft: Barracuda V, Firefly I, Mosquito (for RAF), and Anson (a training plane). Installations were planned for Barracuda V, Mosquito (for RN), and Goose (a training plane). The British had to reinforce the ASH to withstand catapulting from British carriers since the launching gear on American carriers, for which ASH was designed, exerted much less force.

These airborne radar sets, deriving from the roof system and the XT-2 truck system, were by no means the only 3-cm radar sets. Later developments, particularly in the fields of fire control and ship-search radar, are described in other chapters.

NOTES

1. Interview, I. I. Rabi and H. E. Guerlac.

2. "3.3 cm System, May 20, 1941," Radiation Laboratory Report D-1.

3. Interview, D. E. Kerr and T. A. Farrell, Jr., 22 January 1946.

4. E. G. Bowen, Diary, May 21, 1941.

5. Captain D. C. Ramsey to Cmdr. R. P. Briscoe, 9 September 1941.

6. N. F. Ramsey, "Proposed Night Interception System for F4U-2," 17 November 1941.

7. Minutes of the Coordination Committee, Vol. 1, p. 8, 1 September 1942.

8. *Ibid.*, Vol. II, p. 466, 3 March 1943.

9. I. I. Rabi, "Conference Held at Bureau of Ships, Washington, D. C., January 29, 1943," n.d.

10. J. Halpern, "Proposed Specifications for AIA-1," 20 March 1943.

11. N. F. Ramsey, H. F. Balmer, and E. A. Luebke, "Photographs of the PPI Indicator Tube with 3-cm ASV Over Water and Land," 27 October 1942.

12. I. I. Rabi, "Proposal for 3-cm ASD Equipment for the TBF Airplane," 6 February 1942.

13. Minutes of the Coordination Committee, Vol. II. pp. 129–130, 2 December 1942.

14. S. J. Simmons, Memorandum, "ASD-1 Meeting at Radiation Laboratory," 25 November 1942.

15. M. G. White to S. J. Simmons, Memorandum, "ASD–Philco Conference Between M. G. White and Mr. Grimes," 30 December 1942.

16. LCDR L. V. Berkner to L. A. DuBridge, 13 October 1943.

17. J. T. DeBettencourt, BuShips Memorandum, "Conference on June 30 at the Bureau of Ships concerning plans and specifications for development of lightweight airborne radar equipment of the future," MS-485-31, Serial No. 2547, 2 July 1942.

18. LCDR L. V. Berkner, "Memorandum of Conference at the Bureau of Ships," Aer-E-310l-JTK F42-1/36 (1), 30 June 1942.

19. L. A. DuBridge to Coordinator of Research and Development, Navy Dept., 3 October 1942.

20. R. M. Alexander to Gordon N. Thayer (BTL), 7 October 1942.

21. G. F. Tape to L. A. DuBridge, Memorandum, "ASH Equipment," 3 April 1943.

CHAPTER 14

The Development of Radar Beacons and IFF

BEACONS

Except for their use in identification to be described below, radar beacons at first were only auxiliary ground equipment devised for the essential, if undramatic, role of aiding the short-range navigation of radar-equipped aircraft, principally by enabling them to locate and home directly upon air bases where the beacons were situated. Later, however, after the development of portable and airborne beacons, new and more glamorous assignments were found for these devices. As later developed by the British and ourselves, they became the basis of the most important and effective blind-bombing aids; they were applied experimentally to the spotting of shell fire; and they were used on a limited scale in the radar control of tactical aircraft in support of ground troops; but they rendered perhaps their most glamorous service in the precision control of parachute drops and airborne operations. Beacons were used to identify dropping zones in North Africa, Sicily, southern France, Arnhem, and elsewhere. Before the invasion they were used extensively to drop supplies to the French *maquis* and to members of the Resistance movements in other countries. British and American secret agents parachuted onto a darkened continent by the use of radar beacons.

How Beacons Operate

A radar beacon is basically an echo amplifier. It may be used whenever the echo from a particular target is insufficiently strong or not clearly distinguishable from nearby echoes. When the searching finger of a radar beam—as for example from the radar of a distant plane—strikes the beacon, the latter sends back a signal strong enough, and of such characteristics, that it can be picked up by the distant plane's radar receiver and clearly identified on the indicator despite the presence of other signals. The response is thus a magnified and tagged echo, explicitly located as to range and azimuth.

The essential components of the ground beacon are a receiving antenna, a receiver, a modulator, a transmitter, and a transmitting antenna. Unless special features are incorporated the beacon will respond to all radar beams on its frequency, i.e., to all aircraft having radar operation on the chosen wavelength. It is usually desirable—in order to keep useless replying of the beacon to a minimum—to have the beacon respond only to a special interrogation. If this is desired, certain modifications are introduced into the radar and a "discriminator" is included in the beacon between the receiver and the modulator. To give the signals from the beacon certain distinctive characteristics, they are modified by means of coding circuits inserted between the discriminator and the modulator. Needless to say, the design of these components varies greatly to suit the different types of beacons.

The British Beacon Development

The earliest navigational beacons, as mentioned in an earlier chapter, were developed by the British for the use of the RAF Coastal Command in conjunction with their 200 MHz ASV Mark II equipment. By means of the azimuth and range information supplied by the beacons, crews returning from prolonged antisubmarine patrols over the North Atlantic could home straight to their bases from distances of 70–100 miles. This meant simpler navigation on the final lap of a long mission and sometimes a considerable saving in fuel. Radar beacons were also used during the Battle of Britain by the night-fighters equipped with AI Mark IV. In this case the beacons were used not only to mark the airports, but also to set off the limits of a night-fighter standing patrol. The unprintable RAF song "Orbiting the Beacon" commemorated the long vigils of British pilots patrolling back and forth between two beacons. The earliest production beacons were supplied by modifying available IFF (identification friend or foe) equipment to be described shortly.

The antennas of these long-wave radar sets gave very wide beams and lobe switching was used to obtain the azimuth of the beacon signal as in the case of other targets. The wide beamwidth and relatively long pulse length of these sets led to strong ground clutter. The British found that by having the beacon reply on a frequency slightly different from that of the radar set it was possible to tune out the radar echoes and bring in the beacon signal alone. Since then this procedure has been used for almost all beacons.

Besides ground beacons for AI and ASV, the British had adopted very early a type of IFF involving the installation of replying beacons, called "transponders," in every aircraft. These gave a coded response, indicating the presence of a friendly plane, when interrogated by the radar set of another plane or by a ground station. It turned out to be extremely useful to modify these slightly—by introducing a special frequency of transmission which the pilot could switch on at will—so that when the IFF set was interrogated by a GCI station a continuous presentation would be received. This use is tactically different from the IFF function, and resembles the later use of airborne beacons in close support.

Another outstanding contribution of the British was the development of the Rebecca–Eureka series. These consisted of portable ground beacons (Eureka) operated on 200 MHz, and used companion interrogator–responsors (Rebeccas) which located and homed on the Eureka beacons. The latter were designed to be as simple and light as possible so that they could be carried down by paratroopers in their jumps over enemy territory, set up quickly and turned on. Subsequent waves of troop-carrying planes could then home on the beacon and drop their cargo in well-concentrated groups in predesignated dropping areas.

Early Development of Microwave Beacons

The development and adoption of microwave radar equipment brought with it a need for beacon facilities on these frequencies. Though some preliminary work on microwave beacons was undertaken by the British, they had not progressed very far when a policy decision caused the work to be suspended. America became the source of microwave beacons used in the war.

At the Radiation Laboratory the subject of microwave beacons was first raised in the late summer and early fall of 1941 by Luis W. Alvarez. His proposals were made just as the 10-cm AI development was nearing completion and at a time when the microwave ASV problem, with strong British support, was being given serious attention. British experience had demonstrated the value of beacons, and the enormously improved azimuth resolution of microwave radar led to the expectation that their importance might be enhanced. Accordingly, a group was organized under Alvarez to undertake the development of a 10-cm beacon.

The Problems of Beacon Development

Microwave airborne radar was sufficiently different from radar on longer waves to pose special problems in beacon development. More complicated indicator displays—the B-scan and PPI—were used to present the more accurate information obtained by the narrower beams. Despite efforts to restrict their frequency spread, fixed-tuned magnetrons were used covering an appreciable band. Special duplexing apparatus had been designed using TR and anti-TR cavities of relatively high selectivity. This high selectivity meant loss in sensitivity when the receivers were retuned to a new frequency, i.e., the beacon frequency.

It can be readily seen that a microwave navigational beacon must work with all radar sets of a given type, despite differences in frequency. It must have an onmidirectional antenna system to allow interrogation from any direction. All beacons should use the same reply frequency, so that they can be seen simultaneously. A method of identifying each beacon by its response is required. The equipment should be capable of handling a moderately large amount of traffic. It must be reliable, and should have a range of 100 miles or more.

A radar system must also be specially adapted to work with a beacon. Its receiver must be easily tuned to the beacon transmitter frequency. A special signal for interrogating the beacon is desirable, so that the available traffic capacity of the beacon will not be absorbed by radar sets on search missions which are not, at the moment, using the beacon for navigation. The radar magnetrons must not be too widely scattered in frequency, or the beacon receiver will have to cover an impractically wide frequency band.

Before describing the development of beacon systems it may be helpful to discuss the technical features of the microwave beacons developed at the Radiation Laboratory.

1. *Receivers*. The system of operation adopted was to confine all airborne radar to a given band and to have all beacons reply on a spot frequency just below this band. At 10 cm the "scatter band," as it was called, was chosen to be 66 MHz wide; at 3 cm it was made 90 MHz wide.

Eventually, two satisfactory types of receiver were developed. The first was a simple detector–amplifier receiver, using a crystal detector and a video amplifier. Much of the early work on such crystal–video receivers was done by M. F. Crouch of the Receiver Group and M. D. O'Day of the Beacon Group; later by H. J. Lipkin and J. H. Tinlot of the Receiver Group.

As is well known, the sensitivity of such a receiver is much inferior to that of a superheterodyne, since the crystal detector is a square-law device. Attempts were made to improve the characteristics of crystals for the special purpose of detection rather than for conversion. The BGS, the first production beacon, was eventually equipped with a crystal–video receiver which provided a signal sufficient to trigger the modulator with an input of 5×10^{-8} W from the antenna. With a suitably designed receiving antenna, it turned out that the airborne radar sets then being designed would be able to interrogate the beacon from a range of about 100 miles, which was adequate.

However, it was clear that the crystal–video receiver would not be satisfactory at shorter wavelengths because of the rapid decrease in antenna cross sections at short wavelengths. Accordingly, work was accelerated on a superheterodyne receiver to cover a total band spread of 110 MHz. Much work was done on the design of wide-band IF amplifiers; the solution finally adopted was first used by Philco in its 10-cm production version of BGS, the AN/CPN-3. This was to use a receiver which could cover half the required band at one time, and to tune it electronically back and forth so that each half of the band was covered half the time.

There were also special problems of IF design. Single-tuned or even double-tuned IF stages did not have a sufficient figure of merit to yield, with the tubes available, high-gain IF amplifiers of sufficient bandwidth. Several circuits were developed in the Radiation Laboratory for this purpose. These included the inverse-feedback amplifiers developed by H. J. Lipkin and the adaptation by H. Wallman and J. Tinlot for beacon use of the so-called series-shunt tuned video as an IF amplifier. The latter circuit is noteworthy because it has permitted construction of IF amplifiers which will give a total gain of 80 dB or more with a bandwidth of 100 MHz. It is essentially a video circuit; special means are introduced to cut out the lower video frequencies in order to give it a suitable IF bandpass characteristic.

2. *Antennas*. Beacon antennas were worked on intensively in the Radiation Laboratory Antenna Group by a section under the direction of H. J. Riblet, R. E. Hiatt, G. A. Jarvis, and H. Krutter. They made many contributions to general design. The problem was to obtain a beam which would be omnidirectional in azimuth but which would be sufficiently sharp in elevation. Several approaches were tried. The biconical horn was among the first, but it was bulky for the amount of gain obtainable. Linear arrays appeared to provide a better alternative; and finally, very satisfactory antennas of this type were developed. In these, three dipoles were arranged in a circle to form one element of the array and many such elements were stacked on a coaxial line feed. Such antennas provide patterns of remarkable uniformity in azimuth; gain in elevation is limited only by the length of antenna usable.

Linear array antennas of the above design were extremely satisfactory at 10 cm, but it was considerably more difficult to use them at 3 cm and a new approach had to be made. At 3 cm antennas were developed which consisted of a circular waveguide with slots for radiators. The slots were again arranged in elements around the circular waveguide, in a number large enough to give radiation uniform in azimuth. Elements were stacked in phase at the proper interval so that again a beam as narrow in elevation as desired could be obtained. The use of the circular waveguide was later supplemented by the use of coaxial line. These antennas were rugged and not too difficult to construct, and could be made with adequate bandwidth for the purposes needed and with adequate power-handling capacity for the transmitter.

3. *Discriminators*. The first discriminator was designed by Alvarez and developed by J. H. Buck. In order to keep useless beacon interrogation to a

minimum, it had been decided to design the ground beacons so that they would not respond to ordinary search pulses 1 μsec or shorter in length; however, the radar set would be designed to emit a special pulse when interrogating a beacon. After some preliminary investigation, a pulse length of 2 μsec or more was finally standardized for beacon interrogation for all airborne radars. The later advent of spark and gas-filled tube modulators made this choice of code a particularly fortunate one.

A consequence of this decision, however, was that the beacon designers had to furnish a discriminator circuit which would pass 2-μsec pulses and exclude 1-μsec pulses. This was successfully solved in a variety of ways. The solution which was adopted and has been used in most discriminator circuits is the integrator or limiter type, first suggested by S. E. Golian. When the signal has been brought to a suitable video level, a circuit of narrow video pass band is introduced. The square video pulse is distorted into a triangular pulse with a slowly rising leading edge. If the characteristics are properly chosen, then, no matter how strong a pulse is, the output voltage will be able to rise only at a certain rate determined by the discriminator characteristics. Accordingly, if a trigger is taken from such a circuit at a given voltage level, this level may be so chosen that a 1-μsec pulse can never reach it no matter how strong, and a 2-μsec pulse will reach it even if it is only a few dB above noise. Circuits have been made in which more than 100 dB additional power is needed in a 1-μsec pulse to give an output trigger than for a 2-μsec pulse.

While the discriminator problem was thus successfully solved, other problems arose. It was difficult to get high-gain video amplifiers to give good pulse shapes with widely varying input powers; a certain amount of pulse stretching might occur in the video amplifier. A strong 1-μsec pulse entering the video receiver might, because of saturation, come out of the video as a 2-μsec pulse with which the discriminator is powerless to cope since it can work only on whatever signal it receives. Much work on video amplifier design to avoid pulse stretching was done by Crouch and later by Lipkin and the problem was solved quite successfully. In the BGS, the overall discrimination of the beacon was 60 dB or better between 1- and 2-μsec pulses.

4. *Modulators*. Early modulator circuits were largely the work of W. O. Reed and J. M. Cunningham. The special problems encountered are worth emphasizing. Since radar modulators operate at a constant duty ratio, their design is, in this respect, simpler than beacon modulators where the duty ratio is variable and may be small, large, or zero. The beacon modulator must be able to supply a pulse at any instant. Furthermore, it is almost always desirable to be able to range-code the beacon response. For example, in reply to each interrogating pulse received, the beacon may be required to send out as many as six pulses, with spacing of

10–30 μsec between them. The modulator must, therefore, be able to supply pulses 10 μsec apart and varying duty ratios up to the limit of the transmitter, which ordinarily is between 0.2% and 0.5%. These stringent requirements rule out conventional spark-gap and gas-filled tube modulators.

Triode switch-tube modulators were found to be unsatisfactory. An extremely satisfactory tetrode constant-current circuit suggested by A. E. Whitford was then developed. In this circuit, the switch tube is a tetrode; it has the property that the current through the tetrode is not very sensitive to the plate voltage. Accordingly, if a series of closely spaced pulses comes in and the storage capacitor drops in voltage, the amount of current transmitted to the magnetron (the important factor in determining the magnetron operation) does not change markedly. Such beacon modulators have now been built in all sizes, from modulators supplying a few hundred watts up to modulators supplying 100-kW pulse power. They have proved entirely satisfactory.

5. *Transmitters.* For ground beacons the magnetron was used as the transmitting tube on both the 10-cm and 3-cm bands. The beacon group fell heir to the principal improvements made at various times by the Magnetron Group and its industrial collaborators. For lightweight, low-power beacons the lighthouse tube provided a very satisfactory solution; and it was used in the principal "ultraportable" and airborne sets that were designed.

With the magnetron transmitter the problem of frequency stability—an extremely important factor in beacon operation—was a serious one, especially on 3 cm. For this frequency the absolute variation allowable is the same as on 10 cm, where the conditions are already stringent, and the stability required is correspondingly greater. At a frequency in the neighborhood of 10,000 MHz it is necessary to keep the transmitter frequency constant to within $\pm$ 1–2 MHz.

The problem of stabilizing the frequency was attacked in two principal ways. First, since the thermal expansion of copper is great enough to alter the frequency by 1 MHz for every 6° rise in temperature, it was necessary to run the transmitter at a high temperature and control the temperature to within a few degrees centigrade. In the second place, since controlling the external temperature did not readily control the temperature of the inside cavities, associated rf stabilizing cavities were developed, first by W. M. Preston and later by F. F. Rieke. It was found possible to stabilize the magnetrons quite effectively without undue loss of power. When the war ended the work had been carried to the point at which a 3-cm magnetron adjusted to a given frequency and coupled into a suitable stabilizing cavity will not exhibit a frequency change of more than 1.5 MHz under all operating conditions of current and voltage for which the output is useful, and over a 100 °C range of temperature.

The Development of Beacon Systems

The first attempts made in 1941 to design a 10-cm beacon were not successful. The first system[1] had a klystron transmitter with a pulse power of 10 W and an ordinary narrow-band superheterodyne receiver tuned to the aircraft transmitter. Tests at Langley Field in November 1941 showed at once that neither transmitter nor receiver was satisfactory. The pulse power was adequate only for radar sets whose TR cavities were tuned to the beacon transmitter frequency; it was inadequate if the cavities were detuned. Since the radar magnetrons (and hence the TR cavities) were widely scattered in frequency, the chance that any given one would be on the beacon frequency was remote. Likewise, the narrow-band superheterodyne beacon receiver could not be tuned to all radar sets, but only to one at a time. This difficulty had been foreseen, but the first wide-band beacon receiver, a superheterodyne with an IF pass band of 0–30 MHz, proved inadequate in sensitivity. The tests showed that higher power was needed in the transmitter, mainly to overcome possible losses in the resonant TR cavity off frequency, and that more sensitivity was needed in the receiver.

The next attempt was more successful. The beacon included a magnetron transmitter, a crystal–video receiver, and an omnidirectional biconical horn antenna. For the first time, identification of the beacon response was achieved by a method suitable for a type B indicator or PPI. This method was range coding, in which each response consists of a series of spaced pulses. Two values of spacing, "long" and "short," were used. An example of the resulting display is shown in Fig. 14-1. Discrimination between radar search signals and beacon interrogation was included.

BGS (Beacon : Ground : S Band)

The first beacon of this improved type, called BGS, was installed on top of one of the hangars at Langley Field in August 1942, and tested by R. M. Whitmer and J. M. Cunningham. While serious faults were still evident, it was clear that the general approach was correct.

At this time Service interest in the equipment increased considerably. Colonel Dolan, commanding officer of the First Sea Search Attack Group operating out of Langley Field, gave important support to the program, especially after the experimental installation saved a plane lost at sea in a storm. On this occasion, E. J. Scott, a Laboratory member stationed at Langley, sent the following telegram to Whitmer: "Congratulations from the Colonel and everyone else. Your squawker saved plane lost at sea in ice storm. No visibility. No radio. No navigation. Only you."[2] Even before

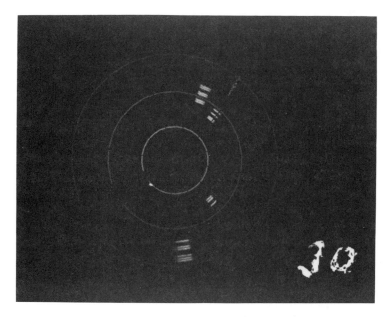

FIG. 14-1. Four 3-cm beacons seen simultaneously on PPI scope of AN/APS-10. Beacons are as follows: Code 2-2-2, Deer Island, Boston; Code 2-1-1, S. Weymouth, Mass.; Code 2-1, Quonset, R. I.; Code 1-2-3, Fisher's Island, N.Y. From range and azimuth data obtained from any one of these beacons, it may be deduced that the plane is 2 mi. north of Woonsocket, R.I., and heading toward Lexington, Mass.

this, General McClelland had written to DuBridge "The beacon program is taking on an ever-increasing importance and every effort should be made to provide reliable beacon facilities."[3] The Navy's Bureau of Aeronautics through Lieutenant Commander L. V. Berkner also lent their encouragement. In addition the British Air Commission asked the Laboratory to build two "crash" beacons to work with the DMS-1000 sets. The first of these was tested with the Laboratory's B-18 plane; and on January 7, 1943, the beacon signal was clearly distinguished at 95 miles.

Partly as a consequence of Army and Navy interest and the success of the new BGS, the beacon effort at the Radiation Laboratory was reorganized and expanded in December 1942. The Beacon Group was shifted to the Receiver Components Division, headed by Bacher, and Arthur Roberts was made Group Leader. The group's two principal tasks were to get 10-cm beacons into the field as soon as possible and to press the development of a suitable 3-cm beacon for airborne radar sets which were expected shortly. Meanwhile the 3-cm beacon development had been languishing for lack of personnel.

Engineers from the Philco Corporation had been associated with the early development of the BGS in the summer of 1942, but had left the laboratory to design their own production equipment, which appeared late in 1943 under the designation AN/CPN-3. Meanwhile, a crash program was started with the Galvin Manufacturing Company to bring out 40 BGS beacons to satisfy immediate Army and Navy requirements. Principally because of the efforts of G. C. Danielson and R. M. Whitmer of the Radiation Laboratory, and of the Galvin project engineer, H. Magnuski, the first of these was ready in February 1943. The last was delivered by June, and beacon navigation was thus made possible for the antisubmarine sea search groups using 10-cm radar.

BGX (Beacon : Ground : X Band)

Under W. M. Preston the development of a 3-cm ground beacon, called BGX, was begun at high speed. It was decided early that a superheterodyne receiver was needed. Even by use of a receiver which covered only half the 3-cm band at one time and whose local oscillator frequency was switched back and forth, they doubted at first the feasibility of covering more than 60 MHz. A conference at the Bureau of Ships, on February 9, 1943, restricted the band to this total width, and centered it at 9375 MHz. The beacon was assigned the frequency of 9335 ± 3 MHz.

The magnetron manufacturers, on hearing the stringent frequency requirements, protested vehemently that too large a number of otherwise satisfactory tubes would have to be rejected as off-frequency, and requested another conference, which was held in the spring. By that time, the Receiver Group at the Radiation Laboratory had succeeded in producing a switched receiver capable of covering 110 MHz. Accordingly, the magnetron frequency tolerance was set at ± 30 MHz for tubes operated into a matched load, and the field tolerance at ± 45 MHz. The beacon was shifted to a frequency assignment of 9310 MHz.

By June 1943 a prototype had been made to work, and Galvin was selected as manufacturer for the AN/CPN-6. Magnuski, having finished his work on the BGS, took over the new project. Lieutenant Commander M. M. Garrison of the Bureau of Aeronautics and L. Carr of the Bureau of Ships both gave their strong support to the new project. With the cooperation of J. R. Watson of the Beacon Group, RCC undertook a crash program for 45 sets, to which the Navy assigned the designation CXEH. The first of the CXEH beacons was finished by August 1943 and four of these were eventually shipped to England along with two Radiation Laboratory

BGX models. All of the Micro-H bombing by the AAF was done with these six beacons, after they had been modified for Micro-H use. AN/CPN-6 production sets were available in September 1944 but they were not put into operational use for bombing because they did not reach the theater in time. By the end of the war, AN/CPN-6 beacons were installed on almost every aircraft carrier in the Navy.

Project Fox (AN/APN-7)

An early attempt to design an airborne microwave beacon was carried out concurrently with the BGS and BGX developments. The first airborne microwave beacon, designed and built by a group under E. R. Gaerttner, was a special purpose beacon with an ordinary superheterodyne receiver only 8 MHz wide. It was manufactured in large quantities for use in a Navy guided missile program known as Project Option.

The Fox beacon was intended to provide range, direction, and identification of a drone aircraft at distances greater than the drone could be followed using reflected signals alone. The radar set of a control aircraft was to trigger the beacon in the drone plane. This beacon, unlike others, did not call for retuning the local oscillator of the radar set and transmitting a different interrogating pulse. The drone received ordinary ASV signals and replied on the same frequency. Identity of the drone was to be accomplished by switching off the beacon by radio control in the plane.

This earliest Laboratory airborne beacon had a number of successors. It used a lighthouse tube transmitter of about 200 W peak power output and was the first microwave beacon to use a triode oscillator and a TR duplexer. The latter had been impossible in the early beacons because no wideband TR cavity had been developed.

Work was begun about January 1942 and at the end of July the first plane installation was made after preliminary testing at Deer Island. Five experimental sets were built at the Laboratory, of which four were turned over to the Navy for testing. S. A. Martin was responsible for operation and maintenance of the beacons during these tests. The Navy entered into a developmental contract with Philco for which the Laboratory did not act as consultant. However, at the request of the Navy, the Laboratory worked on range coding circuits for attachment to Philco's beacon. The Laboratory terminated its work on Project Fox in June 1943. Philco built 1600 AN/APN-7's all of which were delivered by October 1, 1944, although most were not used for the purpose for which they had been designed. When the program was cut back in 1944, many of the beacons were no longer needed. The unused beacons, although they were of a special type, proved to be of great utility later in the war, when they were adapted to at least a half dozen different purposes which required equipment with a minimum of delay.

In June 1943 the two main objectives of the reorganization of the beacon group had been accomplished. 10-cm beacons were in the field, and the 3-cm beacon was ready for production. At this time, Bacher left the Radiation Laboratory for the Los Alamos atomic bomb laboratory and another reorganization of the beacon program took place. Division 7 was established, devoted entirely to beacons, with L. A. Turner as Division Head and Roberts as Associate Division Head. The IFF work, which will be discussed later, was also included in this division.

At this junction, with both BGS and BGX entering the production stage, the future development of beacons was reexamined. It was clear that beacons constituted an additional dimension of the radar art, but the question as to just what direction the limited effort of the Beacon Division should take had to be determined. Two main lines of future activity were agreed upon: (1) the development of beacon bombing and (2) the development of lightweight beacons. These choices reflected the fact that by the early summer of 1943 the radar war was moving into its offensive phase. Both lines of development pointed to the use of beacons as offensive weapons.

BPS (Beacon : Portable : S Band)

The first attempt at a lighter microwave ground beacon was designated BPS and was developed by the Beacon Group under the direction of E. R. Gaerttner. Originally intended to be a lightweight beacon, it became, through the insistence of the Army that no relaxation in specifications or performance was permissible, an improved BGS. The production model, the AN/CPN-8, was made by Galvin and replaced the relatively unsatisfactory AN/CPN-3. Development of this set was completed in the summer of 1943, and two preproduction models were sent with Colonel Stuart P. Wright's low-altitude bombing project to the Pacific, where they served satisfactorily for two years.

BUPS (Beacon : Ultraportable : S Band)

Confident of the value of lightweight beacons, Roberts directed S. E. Golian in June 1943 to start a development aimed at producing a 10-cm "ultraportable" beacon. Two models of this set, called BUPS, ac and dc, were tested at Boca Raton in the first months of 1944. The latter was probably the first microwave radar equipment to be run from batteries. The power supply was a more or less conventional vibrator supply. The transmitter was a grid-pulsed lighthouse triode. The Army's originally cold reception of these sets warmed to a reluctant acceptance through the efforts of Captain N. Caplan, Lieutenant R. Hultgren, Captain L. Kellogg, and Major A. B. Martin.

Early in 1944 Major Martin arranged for a half dozen of the laboratory models to be sent to England for use by the Troop Carrier Command of the 9th Air Force. These were taken to England by Golian in April, and their part in the D-Day operations of the 9th TCC will be described later.

The engineering of BUPS for production was taken over by B. W. Pike early in 1944, and the Hallicrafters Company was eventually selected as the manufacturer.

Rosebud (AN/APN-19)

Early in 1944 the ground radar division of Radiation Laboratory, which had been experimenting with airborne beacons, finally decided to recommend their adoption for aircraft of the Tactical Air Forces in connection with the close support program. The preliminary designs of Rosebud (AN/APN-19), an airborne 10-cm beacon, were made by R. T. McCoy, and the later development was under B. W. Pike. Here again, crash models used by the 8th Air Force were of great utility, and large-scale production by Gilfillan Brothers for the Pacific Theater was under way when the war ended.

The technical features of interest in Rosebud are its light weight and its versatility. Rosebud contains two lightweight, low-drain crystal–video receivers. It can be operated in two ways. It has a wide-band receiver responding to all radar sets in a wide frequency band, and a narrow-band receiver which uses a high-Q cavity in the antenna lead so that only a radar set on one particular frequency can interrogate the system. This was necessary to insure good tracking by the SCR-584 in close-support operation. Rosebud was the first beacon to be tracked by a conical-scanning radar set. Tracking beacons with such a set with a rotating dipole is difficult because of the constantly changing plane of polarization with respect to the receiving antenna of both the interrogation and the response.

Another interesting technical feature of Rosebud was its gas-filled tube modulator for coded response. The light weight and the low-power transmitter (a lighthouse tube) made it impracticable to use a hard tube modulator. A gas-filled tube modulator was developed in which three different tubes were fired in succession in order to provide the successive pips of a three-pip code.

One of the great difficulties with Rosebud was to find locations which were suitable for installing its very small quarter-wave dipole antennas in aircraft and which yet provided good coverage. As might be expected, shielding, reflection, and diffraction effects made the pattern extremely complex and unpredictable. The problem was, however, solved.

BUPX (Beacon : Ultraportable : X Band)

The need for 3-cm lightweight beacons was seen early in 1943. A contract for development of a suitable magnetron and beacon was given to the RCA Laboratory in Princeton, New Jersey. A group at RCA under J. S. Donal developed the 2J41 tube which was eventually used in BUPX. H. H. Bailey of the Radiation Laboratory was assigned to follow the development.

Meanwhile, Gaerttner's group had begun development of high-performance airborne beacons. Late in 1943 it was clear that no great tactical need existed for such beacons, and the group was accordingly shifted to BUPX under the joint direction of Bailey and Gaerttner. Rapid development ensued; by the fall of 1944, when the magnetron design was fixed, designs for production equipment were being completed.

The final production design of BUPX was carried out by groups of engineers from three companies working in the laboratory with the BUPX group. The General Electric Company was selected to manufacture the ac model, AN/UPN-3; Gilfillan Brothers the dc, AN/UPN-4; and RCA Victor the airborne, AN/APN-11. The joint effort was, on the whole, remarkably successful. Production of the sets was scheduled to start in the summer of 1945, but was halted by the end of the war.

Some of the technical features of BUPX are of interest. Bailey's group turned out sets which were remarkable for their technical excellence. BUPX had a very highly stabilized low-power 3-cm magnetron developed especially for it by RCA Laboratories. This tube (2J41) was capable of putting out 500 W peak power and of holding its frequency to within 1.5 MHz except under extremely unfavorable conditions.

Another point of interest was the introduction of a new type of receiver designed by J. H. Tinlot for the ac version of BUPX, the AN/UPN-3. In this receiver, the 110-MHz rf pass band was covered with a single local oscillator by a new method. Instead of switching the local oscillator frequency, the local oscillator was frequency-modulated at a rate of 40 MHz. This results in providing the local oscillator with many sidebands of 40 MHz separation. The first two sidebands on either side of the fundamental are normally the strongest. The IF pass band is now set at one-sixth the total pass band. For a 120-MHz band this corresponds to a pass band from almost 0 to 20 MHz. Now there are three equivalent local oscillators each with two sidebands from 0 to 20 MHz, and spaced 40 MHz apart. The result is a coverage of 120 MHz, with quite adequate sensitivity, with a single local oscillator which requires no switching.

The major drawback of this type of receiver for application to a large ground beacon is that it would make discrimination of pulse lengths quite difficult, if the pulse were in a frequency region near one of the local oscillator frequencies. In this case, an IF of only a few megahertz must be converted to video, and considerable pulse distortion arises if the video frequency is close to the IF. This results in making discrimination rather poor at those frequencies. Since the lightweight beacons had no discriminators, this was not of particular significance in BUPX.

Other applications of lightweight beacons had also been suggested. Among these were two uses of beacons for marking front lines. One of these was to give a reference point for bombing. A beacon offset from a target can designate the target for the benefit of a plane whose bombardier is properly briefed as to the meaning of the beacon. The second use, which takes advantage of the excellent resolving power of microwave radar displays, was simply to mark the contour of the front line more or less accurately, depending upon the number of beacons that could be employed. These uses had aroused considerable interest in the Ground Forces and Air Forces, and were being tested when the war ended. It had been planned to send teams with some of the lightweight 3-cm beacons to the Phillippines so that the idea of front line marking might be tried out in combat.

Another use was also suggested for the airborne version of BUPX. B-29's in the Pacific were having considerable difficulty both in forming up and in making rendezvous with their supporting fighters. The Army Air Forces Board recommended the airborne BUPX (AN/APN-11) for this purpose.

RADAR IDENTIFICATION (IFF)

The work of Division 14 in developing equipment for the identification of friend or foe (IFF) is a somewhat special chapter in the history of the Radiation Laboratory. The responsibility for most of the IFF development and the direction of the overall program was assumed by other agencies both before and during the war. In consequence, only a small group at the Radiation Laboratory was engaged in this work, and the small fraction of the overall program assigned to the group was conducted for a long time under conditions of magnified secrecy and compartmentalization which set it apart from the general radar program of the Laboratory.

Although from the very beginning of the radar program it was recognized that satisfactory identification equipment would be essential, the war ended without this need being met despite an intensive and expensive effort. This failure may have been largely due to the inherent difficulty of the program. Undoubtedly the failure was, to some extent, influenced by the restrictive conditions of secrecy under which the development was obliged to take place. But it was also due in good part to conflicts of policy, the lack of a unified program, and an inadequate solution of the problems of proper training and indoctrination: it was due, in a word, to human rather than technical problems.

Early Work on IFF

The necessity of some sort of device for distinguishing the pips on a radar screen due to friendly targets from those produced by enemy ships or planes was recognized as soon as radar was half a reality. In Great Britain, in April 1935, Watson-Watt discussed the problem at a meeting of the Committee for the Scientific Survey of Air Defense. In America, the need was demonstrated, as has been pointed out previously, during the fleet maneuvers early in 1939 when the XAF was found useless for its principal role of aircraft detection during its actual mock engagement because it could not distinguish friendly from enemy planes. The first tests with IFF schemes were made in America at this time.

Plainly, the ideal solution to the identification problem would be to give the pips from enemy targets a distinctive appearance. This is the old problem of belling the cat faced by the mice in the fable, and as the mice discovered, it requires an unprecedented degree of cooperation on the part of the enemy. The next best solution, to identify friendly targets, was the one adopted, but that turned out to require an almost insuperable degree of cooperation and discipline among friends and allies; agreement, as it were, upon the type of bell to be worn and upon the necessity of wearing it at all times.

All the work done on identification has been based on this principle, so that the designation IFF—implying ability to distinguish friend from foe—is erroneous and misleading. We have never been able to accomplish this and probably never shall. We can only identify friends and assume that the lack of a friendly response indicates the presence of an enemy. This assumption, of course, is not always valid, as in the case of a friendly plane whose identification equipment is not working, perhaps because it was damaged in enemy gun fire. Many disabled planes were shot down by friends during the war because the planes were unable to send out the proper signal indicating their identity. This was the basic flaw of all IFF systems and a principal cause of the lack of confidence in them.

The earliest embodiment of this principle had been suggested by Watson-Watt in 1935, and was later tried out by the British. It consisted of installing on friendly planes a resonant dipole tuned to the frequency of the radar. For identification purposes, this dipole would be keyed slowly in some fashion. Because of its resonance to the frequency of the radar, it would reradiate the received signal, and the echo would usually appear to be several times as large as that of a similar plane without equipment. However, the techniques developed prior to 1939 were inadequate to give sufficient reliability, and the system was dropped.

After the keyed dipole arrangement had failed to give satisfactory results, work was begun by the British on a radar beacon to be carried by aircraft or ships. This beacon, in effect, would provide an enhanced radar echo. When triggered by the radar, it had to send back a coded reply on the same frequency. Since it was expected to supply information to a number of different radars on different frequencies in the same band, it was necessary to devise some mechanical arrangement for sweeping the frequency of the beacon's transmitter and receiver over the band in which the various radars were located. The use of the same tube and tuned circuits for transmitting and receiving made such a scheme practical. When the beacon receiving and transmitting circuits swept through the frequency of the radar which was interrogating the plane, the device would send back a coded reply which on the radar indicator appeared as a periodic variation in the amplitude and shape of the echo. The time taken to sweep through this band of frequencies was fairly large, in the neighborhood of 6 sec, and it is evident that the sweep technique appreciably increased the time necessary to make an identification. However, in the early days of the Battle of Britain the number of planes in a sector at any one time was not great and ample time was available.

The first system of this type, the British Mark I, swept the band 20–30 MHz and responded directly to interrogation by the radar. It was designed to be used with CH stations during the RAF exercises in 1939. Only 30 sets were made; but these worked sufficiently well to demonstrate the principle. This type of equipment reached its maximum usefulness in the British IFF Mark II system. In general, the Mark II swept two bands simultaneously, one in the neighborhood of 50 MHz for CH stations, and the other from $3\frac{1}{2}$–4 times this value, for CHL. The Mark II was disclosed to the United States by the Tizard Mission in September 1940. At that time America had only visual identification systems.

In 1941 the British proposed a system called the IFF Mark III, which was adopted by Great Britain and by the United States in preference to an American system that became known as the Mark IV. The wisdom of their decision has been the subject of prolonged discussion ever since.

In 1941 the British proposed a system called the IFF Mark III, in this III system, the beacon was not triggered by the radar beam; instead, a separate transmitter associated with the radar sent out the challenge and a separate receiver picked up the reply—the challenging pulse from this transmitter being synchronized with the pulse sent out by the radar. At this time, the British introduced a new vocabulary which also was adopted in the United States. The separate transmitter and receiver used to challenge the beacon was called an *interrogator–responsor* and the beacon was called a *transponder.*

Organization of the IFF Group

The Radiation Laboratory IFF Group was organized in the spring of 1942 as a part of Division 7 under L. W. Alvarez. The Laboratory's work in this field was initiated at the request of Alfred Loomis, who was chairman of an IFF Committee of NDRC. Loomis had made a tour of the various Service laboratories and the offices in Washington concerned with identification, and had been told that no provision for an airborne interrogator–responsor had been made for the Mark III system. Since at this time airborne radar was coming into its own, the lack of plans for the development of an interrogator–responsor to go with such equipment seemed especially serious.

As a matter of fact, the information given Loomis had been erroneous. The Army's Aircraft Radio Laboratory had given a contract to the Philco Corporation to engineer such a device. As a consequence the Radiation Laboratory and Philco cooperated in this development—the Radiation Laboratory making the first breadboard models and turning these over to Philco for final engineering and manufacturing. The unit became known as the SCR 729 and proved of considerable value during the war.

An unusually large amount of secrecy was connected with the IFF project and few people at the Radiation Laboratory had access to any material on IFF. Because of secrecy restrictions, later regarded in the Laboratory as excessive, it was necessary to restrict personnel in this group to a number far too small to be effective. The original group consisted of Marcus O'Day, Chairman, and Arthur Roberts, C. E. Stone, George Perkins, and S. Girardet. Special clearance had to be obtained for people working in this group and all of the work had to be carried on behind locked doors. Access to these rooms was forbidden to the rest of the Radiation Laboratory.

It was especially difficult to obtain the necessary clearance for new members. Each addition to the group was opposed by the Navy on the grounds that the more people who knew about the subject the less the security it would have. The procedure for expanding the group was to get a man assigned to the IFF Group and then lend him temporarily to some other group in the Laboratory until completion of his clearance by Naval Intelligence, Army Intelligence, and the Federal Bureau of Investigation.

Following this procedure, a number of other staff members and technicians were added to the group by the middle of the summer of 1942. These included H. M. Brown, John Reed, Jesse Lien, P. A. DePaolo, Wilson Smith, Donald Young, and Richard Rollman. For security reasons, the size of the group was not increased further. In spite of the small size of the group, work on the airborne interrogator–responsor proceeded rapidly. In less than a month a breadboard model was available for flight test.

With the Mark III system a band 30 MHz wide was swept through every 2.5 sec, so that a given transponder would reply for a brief interval of every such sweep. There was no type of interrogation coding, so any pulse on the right frequency could trigger the beacon. Reply codes were provided by regular changes in the duration of the pulses, an especially long pulse serving as a distress signal.

The system had the advantage of being simple and universal, with an element of security being provided by the reply codes. The chief disadvantage was that the lack of interrogation coding not only reduced the amount of traffic which the system could handle, but could give positive identification of our planes to the enemy, if he chose to use the proper interrogator. The long time necessary to read the codes prevented effective use where traffic was heavy. Because of the low frequency the angular discrimination was poor so that the bearing of a plane could not be determined easily. The intermittent replies could not be effectively displayed on the indicators of scanning radars because the radar usually would not be looking at the transponder at the time that it happened to be in tune with the interrogator. The disadvantages were apparent to both British and American engineers in 1941, and plans were even then under way for the development of a different system.

The American Mark IV radar rejected in favor of the British Mark III had been developed before Pearl Harbor. When finally engineered, the transponder of this system was known as the ABA by the Navy, and the SCR 515 by the Army. It had the following characteristics:

1. Upon reception of a cw challenge modulated with 30 kHz and on a carrier frequency of 470 kHz, a relay would operate causing a visual recognition signal to be displayed.

2. Each unit had provision for sending out a challenge of this type to be received by a similar transponder.

3. It would reply with a coded pulse of 493 MHz upon reception of a pulse at a carrier frequency of 470 MHz.

The system had some advantages over the British Mark III. Because of the much higher frequency used, it allowed for directional antennas. The challenge and reply frequencies were different. It was a spot-frequency rather than a swept-frequency system, which made possible the presentation of data on a PPI scope. The disadvantages of the system were that the only air-to-air challenge possible was by means of the recognition feature since no airborne interrogator–responsor had been engineered, and that the bandwidth of the receiver was not sufficient for reliable operation. The receiver was of the superheterodyne type and the local oscillator might vary enough in frequency to prevent reception of the proper challenge. As a consequence, a decision was made by the Combined Communication Board to adapt the British Mark III for universal use, and the pros and cons of this decision have ever since been the subject of argument among the engineers involved.

One argument against the Mark IV system was that the British planes were mocked up for a box of a different shape and they could not carry the American transponder. Accordingly, in the summer of 1942, Alvarez suggested that the American system be reengineered to fit in the same box and with the same cable connections as the British Mark III transponder, and be held in reserve in cases of compromise of the British system. The American system was henceforth known as the Mark IV and it was decided that the still newer system to be designed would be known as the Mark V. A subcommittee of the Radar Committee of JCB, known as the IFF Subcommittee, was organized to handle identification problems. Alvarez served as the NDRC representative on this committee with O'Day as alternate. It was decided to concentrate the Mark V development work at the Naval Research Laboratory. Engineers and scientists were sent over from TRE and other British laboratories, from Canada, from Camp Evans Signal Laboratory, from Aircraft Radio Laboratory, and from the Radiation Laboratory to form what was called the Combined Research Group under the direction of C. E. Cleeton of NRL.

In the fall of 1942 Alvarez authorized O'Day to go to a conference in Washington and offer to take down to NRL the entire Radiation Laboratory IFF Group in order to speed the development of the Mark V System. This offer was refused. It had previously been determined that in order to preserve proper secrecy no more than 20 civilian engineers in the world would be informed about the new system and not more than an equal number of engineers in uniform should have access to the new laboratory. As a consequence only two people, Brown and Girardet, went to Washington to work on the Mark V. Several members of the groups went to the Aircraft Radio Laboratory and the Camp Evans Signal Laboratory to work on Mark III and Mark IV IFF. At the request of the IFF Committee the remaining members of the group worked on a scheme for giving interrogation coding to the Mark III system. This coding project was carried through in cooperation with Signal Corps engineers, and successful tests were held at Camp Evans in the spring of 1944.

In the early summer of 1943; Alvarez left the Radiation Laboratory to work at Los Alamos under the Manhattan project. A new Division 7 containing two groups was formed, with L. A. Turner as head. Group 71 under the leadership of Roberts was concerned principally with the development of microwave beacons. Group 72 under O'Day carried on the work on IFF and related projects. Turner succeeded Alvarez as a member of the JCB Subcommittee on Identification.

In the late summer of 1943 it became evident that the Mark V IFF project was hopelessly understaffed in view of the magnitude of the problem. The group greatly expanded, engineers being drawn from all of the principal radar establishments. Five staff members from the Radiation Laboratory were detailed to the project. Four of these men, H. Brown, B. Cork, L. Mautner, and Z. Wilchinsky, were made group leaders, while Turner served as a consultant to the project.

In the spring of 1944, the Radar Committee of the JCB appointed a special working committee under the chairmanship of W. L. Everitt to study the effectiveness of the Mark III IFF system in use. Information was gathered from war theaters all over the world, and a small party headed by Everitt made a tour to the southwest Pacific. The findings were that the Mark III IFF equipment was well engineered and stood up well in use, but that the system was ineffective because it was not properly handled. This was attributable principally to lack of knowledge at command level of the purpose and potentialities of the IFF system. A program of educating officers was put into effect.

In the early summer of 1944, the Joint Communications Board appointed an *ad hoc* committee to consider the IFF situation and made recommendations in view of expected delays in the completion of Mark V IFF. This committee requested the Racon and Identification Subcommittee to prepare a technical study of all IFF proposals. This was carried through by a working committee, of which Turner was a member. The time required and the probable expense of putting into effect the various proposed modifications of the Mark V system were analyzed carefully. The *ad hoc* committee finally decided to recommend only the completion of the work on the Mark V system at the earliest practical date. Following this decision, work on the Mark III system at the Radiation Laboratory was confined to improving some of the components in various ways that did not imply an actual change in the system, but would merely improve it in operation.

Unfortunately, the decision of the *ad hoc* committee did not satisfy the acute need of the U.S. Navy for improved identification equipment for forthcoming operations in the Pacific. The result was heightened interest in equipment known as the Black Maria. This was a very lightweight

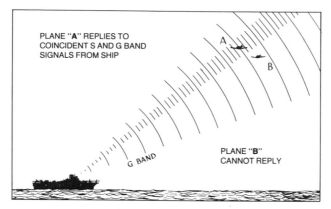

FIG. 14-2. Black Maria coincident signals.

coincidence cross-band beacon designed to work with AEW and other *S*-band radars and was the most important of the IFF projects carried on by Group 72. The almost universal use of 10-cm radar in the fleet made such an IFF system feasible. It was designed to be added to the regular Mark III transponder and did not impair the operation of the Mark III system. However, if a plane carrying this transponder were in the beam of a 10-cm radar it would reply on a 209-MHz frequency if challenged on that frequency. Since a beacon only replied when in the beam of the 10-cm radar, and since it was on a spot frequency, a PPI indication was possible. The traffic-handling capacity of this device was noteworthy, because it would only reply when it was interrogated at the same time by the radar and the 209-MHz challenge. Because of this required coincidence of two interrogating pulses, it had greater traffic handling capacity than a microwave beacon and would reply only to intentional interrogation. (See Fig. 14-2.)

The last IFF equipment developed by the Laboratory was a special lightweight interrogator–responsor for use with AEW. This interrogator–responsor was made to challenge not only regular Mark II transponders but also the Black Maria equipment. The transponder was in production at the Lewyt Corporation and General Electric at the end of the war. It was not only the lightest-weight interrogator–responsor in the Mark III system but also the most powerful.

NOTES

1. R. M. Whitmer, Radiation Laboratory Report 71-7/27/42.

2. Minutes of the Coordination Committee, Vol. II, 9 December 1942. pp. 132–3.

3. Brig. Gen. H. M. McClelland, U.S.A., to Dr. L. A. DuBridge, "ASV Beacons," 29 October 1942.

CHAPTER 15
Blind Bombing

The difficult problem of radar blind bombing (bombing at night or through overcast) is threefold: there must be adequate navigation information to permit blind flying to the target and return; the radar equipment must permit identification of the target; and it must perform the duties of a bombsight by indicating the proper instant at which to release the bomb load, or marker flares. Solutions to the three problems were provided during the war by three types of equipment used separately or in combination: (1) long-wave pulse navigation devices like the British Gee or American Loran; (2) self-contained radar bombsights; and (3) radar beacon bombing systems.

The need for some sort of blind bombing system was clearly expressed by the Army Air Forces as early as June 1940. At that time radar beacon bombing schemes had not been devised and radar techniques were not sufficiently advanced to promise for some time the development of self-contained radar devices. The latter, as it turned out, had to await the development by the British and ourselves of airborne microwave equipment. But techniques were sufficiently advanced when the demand from the Services for such equipment became insistent in 1943 (the year when the strategic bombing offensive was first seriously launched against Germany) to make such devices possible. The earliest bombing devices, such as the British 10-cm H_2S (Home Sweet Home) equipment and the Radiation Laboratory's H_2X, the much publicized "Mickey," were adaptations of ASV equipment. These devices, especially the former, could only be depended upon to identify whole cities or discrete builtup areas; and except where the presence of a river or harbor gave sharp land–water contrast, they could not seem to identify specific targets with great precision.

Bombing isolated targets at sea from low altitudes obviously does not present such difficulties in identification, and this special case was therefore solved earlier and with quite simple equipment. Later radar bombing devices for use over land, in production at the end of the war, had much higher resolution, obtained either by the use of shorter wavelengths or larger antennas; and they would have permitted bombing more nearly comparable in accuracy with the Norden optical bombsight. But in spite of the fact that radar bombsights used during the war mainly permitted only area bombing, they must be included among the Radiation Laboratory's most effective contributions to winning the war.

375

All radar bombsights, as distinct from beacon bombing devices, combine a navigational radar set of the ASV type with some form of computer. In the impact-predicting computer various quantities such as aircraft altitude, slant range to the target, airspeed, wind, and bomb characteristics must be put into the computer. The bomb is released when a bomb-release marker, introduced on the cathode-ray screen by the computer, is seen to coincide with the target signal. When correct values are set into a synchronous computer, a pair of crosshairs automatically follows the target signal as the aircraft approaches the target. This permits corrections to be made continously during the bombing run, and results in greater accuracy. The bombs are released automatically. The quantities which all computers need to solve the bombing problem are the velocity of the aircraft relative to the ground, velocity of the aircraft relative to the air, altitude of the aircraft, and ballistic characteristics of the bomb after its release (see Fig. 15-1). Radar can determine altitude, drift, and slant range, while the aircraft heading and airspeed can be found from the usual navigation instruments. In most cases determination of the bomb release point is made entirely by the radar operator, with the bombardier merely standing by in case visual bombing should become possible; in a few cases, notably with LAB (low altitude bombing) and Visar (visual and radar), the bombardier takes complete control while the radar operator keeps the radar set properly adjusted.

Beacon bombing is extremely precise and permits the bombing of targets that are not easily identified by radar; unlike the self-contained radar bombsights these devices are limited in range. There are two methods of

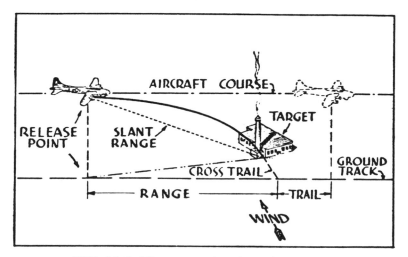

FIG. 15-1. Elementary bombing diagram.

bombing with beacons: *H-bombing*, in which the radar set is located in the aircraft and ranges on two ground beacons, and the *Oboe bombing* system, in which the beacon is carried in the aircraft and range measurements are made by two ground radar stations. In both cases the bomb release point is determined by a process of triangulation, based on the range of the aircraft from the ground stations. While this makes for accuracy, since determining range precisely is radar's specialty, the target must be sufficiently near so that the airplane is always within line-of-sight distance from the ground stations, and the bombing area must be accurately mapped. When it can be used, beacon bombing proves to be the most accurate means. It was used to good effect in the European Theater.

Blind Bombing at Sea

LAB (AN/APQ-5 and AN/APA-5)

Soon after the ASV-equipped B-18 planes started antisubmarine operations in the spring of 1942 it became apparent that while the ASV equipment was extremely satisfactory for detecting targets at sea, it was not so satisfactory for homing on the target at night. Lieutenant Ned B. Estes, one of Colonel Dolan's bombardiers, suggested tying in the radar scope with the Norden bombsight. B. L. Havens, at that time stationed at Langley Field to assist in maintaining the ASV's, took up the idea and suggested including a cathode-ray tube, which would be tied in with the sight so that the bombardier could place the plane on the desired collision course by adjustment of the bombsight controls in the same manner as in visual bombing.[1] This would permit completely blind bombing. An experimental system, which gave azimuth tracking only, was installed in a B-18 and flown successfully from Langley Field in July 1942. Immediately after this flight David T. Griggs, of the Office of the Secretary of War, reported that there was urgent need in Alaska for a blind bombing set with both azimuth and range tracking.

A program to equip two B-24's with BASV (bombing ASV), or LAB (low-altitude bombing), as it was later called, was instituted immediately. B. L. Havens developed the basic principles of the LAB systems and served as project engineer at the Radiation Laboratory during the three-year period when they were being developed and tested.

The set that emerged at the end of the summer permitted automatic release of the bombs. The bombardier had a B scope with a 1-mile sweep; by adjustment of the bombsight controls he "flew" the plane (either directly through the automatic flight control equipment, or indirectly through indication on the pilot's direction indicator) so that the target remained on the range mark. A release mark appeared near the end of the bombing run; when the release mark coincided with the range mark the bombs were released automatically.[2] The ASV used was the SCR-517. Colonel Stuart P. Wright, then Air Forces Liaison Officer at the Laboratory, became extremely interested in the project and assisted in every way possible. In September 1942 the system was flown at Boston, and was then taken to Langley Field. On October 22 a demonstration flight was made, in which several bombs were dropped with good accuracy. The same day the plane was ordered to Africa; a new plane was not received until December, thus setting the project back two months.

As a result of the early tests on LAB, the Army gave a contract to the Western Electric Company for 150 sets of the RC-217 (later AN/APQ-5). This was to be a low-altitude blind bombing set; it was not based, except in philosophy, on the Radiation Laboratory system, and the Laboratory had no status as consultant or advisor. The Laboratory did give help whenever it was requested, but the arrangement was purely informal and the circuits developed by engineers of the Bell Telephone Laboratories were quite different from those in the Radiation Laboratory flight model. Proof tests were run on the Laboratory system during early 1943, and in April 1943 similar tests were run on the first model of the AN/APQ-5. Performance was equally good at low altitudes; at higher altitudes the Laboratory system was better. Following this there was a LAB conference at Wright Field, at which the Radiation Laboratory made several suggestions for improvement of the AN/APQ-5, most of which were adopted.[3] Production started in June, and continued steadily thereafter. Many improvements were incorporated in the system in the course of time. The essential components were the synchronizer; control boxes and tracking unit which contained the controls and circuits necessary to solve the bombing problem; the compensator, which was attached to the bombsight controls and stabilization mechanism to provide yaw stabilization; and the indicator, which had an expanded B scope for search, and a 1-mile sweep for tracking. To track the target correctly, the target was kept in coincidence with the intersection of the central vertical crosshair and the horizontal range mark.

In the spring of 1943 the Army became interested in a modification of LAB which would permit high-altitude bombing. At a joint Army–Navy conference on May 27, 1943, the Radiation Laboratory was requested to develop an improved system, the LAB Mark II or AN/APA-5.[4] This was to operate primarily at low altitudes, but be capable of operating at higher altitudes with somewhat less accuracy. It was for use with 3-cm, as well as 10-cm, radar systems. A LAB committee, with representatives from the Army Air Forces, Bureau of Aeronautics, Bureau of Ships, Aircraft Radio Laboratory, and the Radiation Laboratory, was set up. The Army favored the Western Electric Company as producer, but the Radiation

Laboratory was successful in having the contract given to the Farnsworth Television and Radio Corporation, since it was felt that the Eagle program was absorbing all the facilities of the Bell Laboratories.

During the summer of 1943 various improved circuits were developed at the Laboratory. A system incorporating all desirable improvements was completed for installation in a Navy SNB-1 scout–bomber. The system had an electrical triangle solver, developed by J. J. Lentz and K. E. Schreiner, which gave a voltage proportional to the slant range to the target; extended release and tracking ranges (a maximum slant release range of 30,000 feet, and a maximum tracking range of 30 miles); a coincidence-indication range sweep, developed by R. M. Walker, which improved range and azimuth tracking; sector scan; and an optional automatic tilt control. The use of tables was eliminated during the bombing run. During the early part of 1944 extensive flight tests were conducted; to assist in these a special field station was set up in the spring at the Eastern Point Light off Gloucester, Massachusetts.

Farnsworth's first complete prototype system was delivered in April 1944; in June the second prototype was installed in a B-24 for use with the AN/APQ-13. Acceptance tests were complete by October 1944, at which time procurement was transferred from the Army to the Navy. All subsequent production went to the Navy and the British. Presumably this was a result of the Bell Laboratories' development of the AN/APQ-5B, which also permitted high-altitude bombing, though it did not contain all the refinements developed by the Radiation Laboratory for the AN/APA-5. Delivery of the production systems was completed in November; quantity production, delayed by the transfer from the Army to the Navy,[5] did not start until January 1945.

The AN/APW-5, of which 4634 sets were produced, was widely used by the Army in the Pacific. The Wright Project, a squadron of B-24's equipped with LAB, went out to the Pacific in the summer of 1943, and was soon followed by others. The AN/APA-5 was never used by the Army, but was used by the Navy and the British. The whole LAB program at the Radiation Laboratory was carried out by a relatively small group. The Laboratory priority, while not discouragingly low, was never high—it hovered around twentieth in the list. The support and encouragement came principally from the Air Forces. The project had the enthusiastic support of Colonel Wright and of Colonel Dolan. Dale Corson, who had worked with the B-18 ASV systems at Langley Field, supported the LAB program at the Radiation Laboratory bombing conference on February 16, 1943, as a civilian representative of the AAF; in this David T. Griggs concurred. The operational accomplishments of the LAB equipment will be described in the chapter devoted to radar in the Pacific.

High-Altitude Blind Bombing Over Land

H_2X (AN/APS-15, AN/APQ-13)

H_2X was the first Radiation Laboratory system for high-altitude bombing through overcast to be used operationally. In June 1942 work began under J. W. Miller on a device referred to as NAB (navigation and bombing). It was a 10-cm system, using the lighthouse tube unit recently developed for ARO (since it was considered undesirable to fly magnetrons over enemy territory, and because it was lightweight), a PPI, and a cut-paraboloid antenna with a 360° scan.[6] The project was inspired by a recent observation that cities gave stronger echoes than open country. The first NAB system was installed in a B-18 in October. Flight tests were unsuccessful, since the system did not give the hoped-for discrimination; cities showed up well only when there was land–water contrast. In December, when G. E. Valley became project engineer, it was decided to replace the lighthouse tube with a magnetron. NAB then became a modification of the production ASG set, with a newly developed antenna providing a csc^2 pattern. (See Fig. 15-2.) However, even with this improved system cities were hard to identify, so it was decided in January 1943 to change to 3 cm. This made NAB essentially a 3-cm version of the British bombing system known as H_2S. Work was also begun on an impact-predicting-type computer, which would permit bombing as well as navigation to the target under conditions of poor visibility.[7]

Work on the necessary 3-cm rf components was delayed by the low priority acquired by NAB as a result of the conference on radar bombing held at the Radiation Laboratory on February 13, 1943. Dale R. Corson and N. F. Ramsey, both civilian representatives of the Army Air Forces,

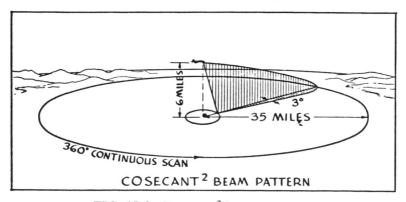

FIG. 15-2. Cosecant ^{2} beam pattern.

and David T. Griggs; of the Office of the Secretary of War, were in agreement that the Army required accuracy equivalent to that of the Norden bombsight for any radar bombing aid. This somewhat visionary accuracy was in no sense claimed for NAB. However, Valley believed in the potentialities of NAB, and work continued after 3-cm rf components were received in April. This new system had improved resolution, partly because of the higher frequency and partly because of the use of the improved ASD receiver, which gave the set better discrimination than had been provided by the older ASG receiver. As a result, H_2X (H_2S on X band) was in existence when Griggs returned from England where he had been investigating British blind bombing techniques at the request of Robert A. Lovett, the Assistant Secretary of War for Air. He brought with him requests from the 8th Air Force[8] to back his demand for the delivery of twenty 3-cm systems by September 1, 1943.

In mid-June a joint Army–Navy cooperative program was set up.[9] The Radiation Laboratory undertook crash production of 20 systems to be delivered to the Army Air Forces by September 1943. Of the 20 sets, 12 were to be installed in B-17's, 6 were to be sent with these planes as spares, 1 was to remain at the Radiation Laboratory for developmental work and as preproduction model, and 1 was to be installed in a Lancaster for the RAF. (See Fig. 15-3.)

The Laboratory was also to act as consultant to the Philco Corporation which was to manufacture a production version, the AN/APS-15, under a

FIG. 15-3. H_2X in B-17; one of the original B-17's equipped with H_2X by the Radiation Laboratory in the summer of 1943.

Navy contract; later the Laboratory agreed to act as advisor to the Western Electric Company for the production, under an Army contract, of a closely similar set, the AN/APQ-13.

G. E. Valley was in overall charge of the H_2X program; he was assisted in the crash program by Lt. (j.g.) R. L. Foote, USNR, T. A. Murrell, Jr., and D. L. Hagler. The H_2X and the AN/APS-15 had the advantage of being built of components already in production. They consisted of the ASD rf assembly, the ASG scanner base, the ASG-3 indicator control, the ASD-1 receiver, and Radiation Laboratory designed 3-cm rf components, an improved csc^2 antenna, and precision-ranging circuits developed by B. Chance and A. H. Frederick. A drum computer was used, consisting of a chart wrapped around a rotating drum, with slant range and altitude lines marked on the drum. Crosshairs were set by means of altitude and range knobs to the proper scale on the chart, causing a bombing circle to appear on the PPI. The bomb release point was indicated by the coincidence of the signal and the bomb release circle, whereupon the bombs were released by the operator. Provision was made for operation with the BGX beacon, to be supplied by RCC. The AN/APQ-13 was similar, but used the components of the SCR-717, of which a supply was on hand at Western Electric because of a canceled Army contract.

Other projects at the Radiation Laboratory slowed down or ground momentarily to a stop; other engineers and physicists stepped aside muttering imprecations, while the "red ticket" H_2X components, commanding top Laboratory priority, were rushed through the shops and assembled under Lieutenant Foote's supervision in a hangar laboratory of the East Boston airport.

By August 1943 one H_2X system had been installed in a B-17 aircraft. By September 15 all 20 systems were complete, and the required 12 were installed in B-17's.

Simultaneously the training of crews, especially the future H_2X operators, was also proceeding on a crash basis. In response to an 8th Air Force request the 2nd Air Force, in July, after fruitlessly combing its list of navigators for men with electronic experience, selected men with no previous exposure to this or any other type of electronic equipment, and sent them to Boca Raton where they received elementary instruction in the principles of radar. Enlisted men who had done sea search duty in the Aleutians taught them tuning, homing practice, and coastline pilotage to the limit of their experience using ASG, SCR-517, and SCR-717 equipment. After four weeks of such preliminary work, the navigators voluntarily undertook a program of inland flying from Boca Raton to Tampa to Orlando and back home. In all, six weeks were spent at Boca Raton during which only 16 hours or so were spent flying, primarily because of equip-

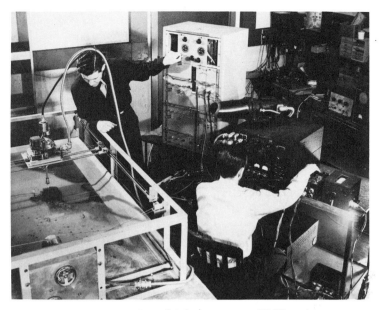

FIG. 15-4. Student (right) operates H_2X trainer.

ment failures. Adequate training by the Air Force was out of the question: no one knew precisely what the training problem was; no one knew precisely what radar operators were supposed to do; no one of course knew what H_2X equipment was, for it was just being built.

Training with H_2X equipment began at Grenier Field, New Hampshire, at the end of August under the direction of David T. Griggs. (See Fig. 15-4.) The group of navigators reached Grenier Field about August 20. Some B-17's had been equipped by the Radiation Laboratory which was trying simultaneously (1) to expedite the installations, (2) to keep the troublesome equipment in an operable condition, (3) to train the crews in H_2X bombing, (4) to train maintenance personnel, and (5) to obtain radar photographs for an evaluation of the accuracy of the equipment.

During the training period it was discovered that rf breakdown occurred at high altitudes. Radiation Laboratory members (principally A. G. Hill, J. B. Wiesner, and F. T. Worrell) went to Grenier Field and installed the pressurized rf system which had been originally designed, but which had not been installed previously because of difficulty experienced in pressurizing the Cutler feed.[10] This difficulty was overcome and later sets were pressurized. In November a crash training program was set up at

Langley Field, using H_2X systems in airplanes and AN/APS-15 sets on the bench.

By September 20 all the installations were completed and the navigators had averaged 25 flying hours. They had learned to home on a target; but they did not yet know how to kill drift. The equipment was still faulty at high altitudes. The 12 planes left the United States for England before the end of the month and arrived at Alconbury—the headquarters of the 482nd Pathfinder Group—early in October. The rest of October was devoted to more training and to a final checkup of the installations under the direction of two members of the H_2X Group, David Halliday and Sims McGrath, who had arrived in England at the end of September in the vanguard of the British branch of the Radiation Laboratory. The first H_2X combat mission took place on November 3, 1943.

Philco production of the AN/APS-15 began in October 1943; Western Electric began production of the AN/APQ-13 in December. After this, quantity production followed smoothly. Meanwhile the Radiation Laboratory was developing improved circuits and components. By the spring of 1944 the design of the AN/APS-15A and the AN/APS-15B had been established.[11] The AN/APS-15A was to have a new dial computer to replace the old drum type, crystal-controlled range markers at 1, 5, and 10 miles, and variable 4-, 30-, 50-, and 100-mile sweeps. In this computer instead of setting in altitude and range separately, the so-called $H + B$ technique was used. That is, slant range was set in directly by using the H-knob to set in altitude, and the B-knob to set in ground speed. The AN/APS-15B had a separate rf head. Plans were later made for an AN/APS-15C, to replace both AN/APS-15A and AN/APS-15B; this was to have a wide-band receiver with antijamming circuits, and a completely repackaged rf assembly. Production of the AN/APS-15A began in August 1944, and of the AN/APS-15B in September; the AN/APS-15C was canceled since it was expected that the AN/APS-30 would take its place.[12] Total production by Philco of all modifications of the AN/APS-15, up to the end of the war, was 7835 sets. Western Electric produced a total of 10,995 sets of the AN/APQ-13.

Nosmo and Visar (AN/APA-46 AND AN/APA-47)

Nosmo (Norden Sight Modification) was designed as a method of converting an impact-predicting computer such as that used with H_2X into a synchronous computer. In order to increase the accuracy of H_2X it became the practice during early operations to have the radar operator call out a number of check points to the bombardier, who adjusted the rate and displacement control of the Norden sight accordingly, so that the Norden sight released the bomb. Various people associated with the 8th

Air Force urged in the summer of 1944 that the Radiation Laboratory develop a device to tie in the H_2X with a synchronous computer, especially since radar permitted adjustment of the bombsight at greater ranges than were possible with visual bombing.[13] Since the Norden rate-end was the most readily available computer it was decided to develop Nosmo.

The use of the Doppler effect to measure drift angle had been suggested by R. Sherr in June 1944, and it was decided to incorporate this in Nosmo. The finite beamwidth of H_2X causes the general ground return to appear modulated at high frequencies, with a large amplitude, when the beam is pointed off the ground track of the airplane. By adjusting the scanner manually, the point of minimum modulation can be determined. This determines the ground track, and the difference between the ground track and the aircraft heading is the drift angle. It was at first thought that the Doppler effect could be observed only on an A scope, but in November 1944 it was found that it could be observed satisfactorily on a PPI provided with a blue filter to eliminate persistence in the sector of the tube being used for drift determination. To obtain synchronization with the Norden sight the radar operator was provided with a Norden rate-end, so that on adjustment of the Norden sight controls the PPI slant range marker tracked the target. Automatic release was possible from either the radar operator's sight or the bombardier's sight, which was kept correctly adjusted.

Initial development work at the Radiation Laboratory was done in Group 63, principally under the direction of R. C. Close. In October 1944 an Army contract for 10 systems was given to the Thomas B. Gibbs Company, and responsibility for flight tests was transferred to Division 9. Extensive flight tests were carried out by the AAF Board at Orlando, and by the beginning of 1945 a large Nosmo order had been given to Gibbs, intended for use with both H_2X and Eagle.

Meanwhile both the 8th and 15th Air Force had developed their own version of Norden tie-in with H_2X. The 8th Air Force version was similar to Nosmo, in that synchronization was carried out by the radar operator, but in the 15th Air Force's Visar (visual and radar) the bombardier was provided with his own scope, and did all synchronization. Modification kits were therefore to be provided to convert Nosmo Mod I (AN/APA-46) to Nosmo Mod II (AN/APA-47 or Visar) where desired. The AN/APA-46 consisted of a control box, servo, antenna attachment, computer, potentiometer unit, and comparator. The AN/APA-47 was the same except that the computer, potentiometer unit, and comparator were replaced by a stabilizer attachment and a different potentiometer unit. Production was held up by a shortage of Norden rate-ends, but the total production of AN/APA-46 at the end of the war was 1823 sets.

The 15th Air Force received two production sets in April 1945; these were flown briefly before operations ceased. Production Nosmo sets did not reach the 8th Air Force until May, too late to be used operationally. Initial tests in this country showed that Nosmo bombing accuracy was twice that obtained when the checkpoint method was used.[14]

GPI (AN/APA-44)

GPI (ground position indicator) was developed as a synchronous computer for various radar sets, first the AN/APS-15, and later primarily for the AN/APS-1. An electronic index was generated which automatically moved across the PPI with target signal, once the index was set on the target signal by means of fix knobs on the control box. Wind was measured by setting the index on any known radar echo; after a few minutes the crosshairs were, if necessary, again fixed on the echo. This automatically entered wind into the computer. Navigation was accomplished by setting the index on a series of points of reference (i.e., taking a series of fixes). For bombing, the necessary bombing data were set in, and the index set on the bombing reference point; one minute before bomb release this fix was checked. Except when the crosshairs were being adjusted the pilot was free to take evasive action.

Work on GPI started, primarily in Division 6, in August 1943. At this time it was planned to use either a Carnegie Odograph or a Bendix API (air position indicator). By the spring of 1944 two models, the Mark I and Mark II, were ready for flight tests, and plans were made for a Mark III version, in which the API was comprehended by the system. It consisted of a control unit—tracking box, servo, and power unit, and computer circuits, and was primarily for use with the AN/APS-1K, though at the same time plans were made for an Eagle GPI. In August the Stratton Committee on Radar Aids to Bombing met to consider a computer for the 1-cm AN/APS-1; it was decided to select GPI, under the designation AN/APA-44 for use with the AN/APQ-34. At first the Fairchild Camera and Instrument Co., which had received a small order from the Laboratory, was considered as manufacturer, but the Army finally decided to give the contract to the Bell Telephone Laboratories. The Bell engineers insisted on redesigning the circuits, which slowed production considerably.[15] The Laboratory then withdrew from the program except for work on a lightweight GPI for navigational use only, which had been requested by the Navy.

Eagle (AN/APQ-7)

Work on Eagle, planned as a precision, high-altitude radar blind-bombing device, began in November 1941. It was desired to build a system with extremely high resolution, obtained by increasing the aperture of the antenna for the frequencies then available rather than to try to develop new techniques for shorter wavelengths. As a result of a conversation with E. G. Bowen, British liaison representative at the Radiation Laboratory, L. W. Alvarez realized that a practicable radar bombsight need have a narrow beam (high directivity) in only one plane, and hence required a reflector large in only one dimension.[16] As Alvarez worked it out, the system was to operate on 3 cm, and would have as its distinguishing feature a linear array antenna, 20 feet long and mounted along the leading edge of the wing of a bomber. The array was originally conceived as a slotted waveguide (leaky pipe). The beam would be scanned by changing—in some fashion not at first determined—the electrical properties of the array. The indicator was to present a true map of the ground, and be stabilized against roll and yaw. It was hoped that the accuracy of the set would approach that of the Norden bombsight, then erroneously thought to be capable of 15-mile bombing under operational conditions.

By February 1942 a nonscanning antenna had been built. Tests on the antenna pattern were disappointing, since three distinct lobes were obtained instead of a single beam.[17] In March, after the antenna had been assigned to R. M. Robertson, various more or less successful methods of eliminating the extra lobes were developed. By the end of April a leaky-pipe antenna had been devised that gave a narrow beam free from large secondary lobes. This, however, did not scan.

In May, Alvarez conceived the idea of the reversed dipole array for use with Eagle and MEW. By reversal of alternate dipoles it is possible to reduce the dipole spacing sufficiently to give a single beam. For use with Eagle such an array could be scanned by varying the cross section of the waveguide.

There was a good deal of skepticism about the practicability of such a long array as had been proposed; it was felt that a 20-foot array would be quite impossible to align and that phase errors would be enormous.[18] A 13-foot fixed antenna was therefore built and tested; it gave a good pattern with a beamwidth of less than half a degree. For some time, however, skepticism remained strong within the Laboratory, except within the Eagle group itself, so much so that the group was forced to work with low priority. In consequence much shop work was done by outside firms.

With the promise of a workable antenna the Eagle Project, then variously known as RBS (radar bombsight), BTO (bombing through overcast) and, more picturesquely, as EHIB (Every House in Berlin, indicating the accuracy expected), was formalized as a Laboratory project with E. A. Luebke as project engineer. He undertook to have a 3-foot scannable antenna constructed, while in July the Indicator Group began work on the

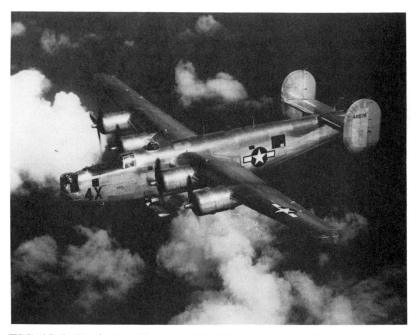

FIG. 15-5. Eagle installed in B-24. The special vane housing the linear array antenna is clearly visible below the port engine.

project under W. A. Higinbotham. The indicator was to have accurate ground range sweeps combined with computer circuits, together with an expanded sweep indicator for bombing.

During the fall of 1942 the idea of a separate computer came into favor; the Norden Mark 15, General Electric, and Librascope computers, the Bell Telephone Laboratories BTO computer, and finally Bell's still mythological UBS (universal bombsight) were all considered. Parallel development on circuits for use with all these computers went on for a year. At the same time it was recognized that the first goal should be a simple straight-line computer with a stabilized indicator giving an accurate map of the ground.[19]

In the summer of 1942 the 3-foot, variable-width-waveguide scanner was tested; the pattern was poor and the losses high but it radiated and scanned.[20] The idea of alternate-end feed with an rf switch was then developed and incorporated in succeeding larger models, together with improved dipoles and chokes. The final production antenna, 16 feet long with a 60° scan, was begun early in 1943, and after some changes was satisfactorily tested in May. In October 1942 Robertson suggested mounting the antenna in a wing-shaped fairing or vane under the airplane, instead of in the leading edge of the wing. A plywood vane was designed at MIT and built by the F. J. Hagerty Company. This vane was attached to a B-24 at Wright Field in May 1943 and the plane was then successfully flown to Westover Field for installation of the 16-foot antenna. The vane did not materially affect the flight characteristics of the airplane.

During the spring of 1943 work was pushed to permit early flight tests under the guidance of J. H. Buck, and at the same time steps were taken to initiate production. In March 1943 the Materiel Command formalized the Eagle program, recommending Western Electric, in the interests of standardization, as the contractor. A contract for five systems, without computers, was given to Western Electric in May with the designation AN/APQ-7. At this time there was an Army–Bell Laboratories–Radiation Laboratory conference, the first of many, to define Eagle. The Army wanted a short-range project with a simple straight-line computer, but with the UBS Eagle as the ultimate goal. As at many other conferences during the next two years, the relationship of Eagle to the overall bombing program was discussed. Many times it was suggested that Eagle be canceled in favor of Bell's HAB (high-altitude bombing, the AN/APQ-10) or, later, in favor of the 1-cm program.[21]

In April 1943 a NDRC contract was given to Douglas Aircraft Company for the development of an antenna vane. Soon after this Eagle became the first Laboratory project to have the full-time services of a Transition Office member. J. W. Eggers contributed much to Eagle, especially in coordinating the antenna and vane production. He also was influential in promoting an interim, simple Eagle, which would permit the use of the high resolution of the Eagle antenna long before the UBS program could become a reality. In June, a group of Bell engineers came to work with the various Eagle groups in the Radiation Laboratory, a beginning of very close cooperation between the two organizations in system design; this was followed in July by the arrival of a Douglas engineer for the same purpose.

On June 16, 1943, Eagle was given its first flight test at Westover Field. The resolution was all that was expected, though there were some troubles with the rf switch and the flaps which shaped the vertical pattern. However, the electrostatic tube required for the completely stabilized, accurate sweeps in the indicator gave very poor contrast, and the complicated circuits were unstable.[22] The tube was replaced by a simple sector PPI magnetic tube, which gave a satisfactory presentation of the high resolution afforded by the antenna. This installation was the one shown to the Army.

Although a complete indicator with stabilized sweeps with an expanded indicator in addition was working on the roof of Building 24, Luebke, Eggers, and Robertson were successful in convincing the Army of the advantages of a simple Eagle with magnetic tube.

In August, Western Electric received an order for 50 AN/APQ-7 sets, and it was necessary to decide just what an AN/APQ-7 set was. Discussions involving the Army, the Radiation Laboratory, and Bell Laboratories culminated in a conference on October 22, 1943, at which the simplified Eagle Mark I was chosen for production, in spite of some opposition.[23] The Bell engineers did not want to give up the tie-in with their UBS, and certain of the Radiation Laboratory indicator-computer people naturally disliked the shelving of all the effort they had put in. In fact, work continued hopefully on the UBS project for some months.

Eagle was the first Laboratory-designed airborne radar system for which complete specifications were set up before production. The design objective was 80-mile bombing accuracy, a figure bettered in production. Douglas agreed to continue work on the B-24 installation, and plans were made for Eagle in the B-29. When Western Electric, arguing that it was not in the airplane business, asked to have the wings and leading edges supplied by the government, the Radiation Laboratory, at the Army's request, gave Douglas a NDRC contract for 50 preproduction wings. The Radiation Laboratory then assisted Bell Laboratories and Western Electric in finding a suitable manufacturer for the antenna; Ex-Cell-O Corporation in Detroit was chosen in December 1943.

Eagle was the first Laboratory-designed airborne radar system for which complete specifications were set up before production. The design objective was 80-mile bombing accuracy, a figure bettered in production. Douglas agreed to continue work on the B-24 installation, and plans were made for Eagle in the B-29. When Western Electric, arguing that it was not in the airplane buisiness, asked to have the wings and leading edges supplied by the government, the Radiation Laboratory, at the Army's request, gave Douglas a NDRC contract for 50 preproduction wings. The Radiation Laboratory then assisted Bell Laboratories and Western Electric in finding a suitable manufacturer for the antenna; Ex-Cell-O Corporation in Detroit was chosen in December 1943.

The B-24 with the Radiation Laboratory Mark I Eagle installed (see Fig. 15-5) went to Boca Raton, Florida where extensive flight tests were carried out during the winter to assess the bombing accuracy and the suitability of the equipment for navigation and for measuring ground speed and drift. The operation of the equipment was quite satisfactory and such bombing as was done was well within the required accuracy.[24] (See Fig. 15-6.)

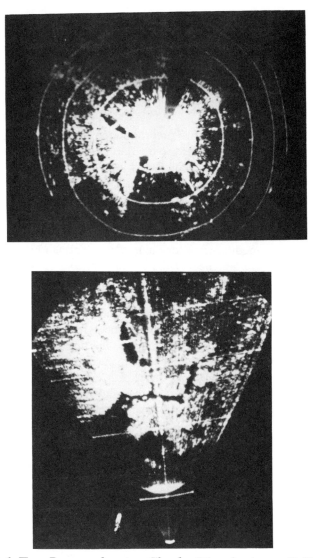

FIG. 15-6. Top: Boston, showing Charles River basin, on H_2X scope, 5-mile range. Bottom: Eagle scope of same area, 10-mile sweep at 14,000-foot altitude. Note bridges across Charles River.

In March there was a temporary crisis in Eagle production. Western Electric declared that no sets would be produced that year unless the Army furnished considerable help.[25] At the March meeting of the Stratton Committee on radar aids to bombing, an attempt was made to have Eagle canceled in favor of the K-band (1-cm) program; K-band bombing systems were claimed to have almost as high resolution as Eagle, and moreover had 360° scan.[26] The limited 60° sector scan of Eagle had always been severely criticized and many in the Army remained opposed on the grounds that 360° scan was necessary for easy navigation, though navigation proved, in practice, to be not especially difficult. Shortly after this, Western Electric decided that it would be possible to meet production schedules after all.[27]

On May 16, 1944 the first Bell preproduction set was successfully flight tested. The flight model was sent to the Aircraft Radio Field Laboratory at Boca Raton for tests; acceptance tests were completed by fall. In September the Bell preproduction order was completed ahead of schedule, and Western Electric, whose order had been increased in June, began production. Total production, up to the end of the war, was 2660 sets.

During 1944 and 1945 the Radiation Laboratory worked on various attachments and modifications for Eagle. Camera attachments for taking scope pictures were designed by the Laboratory and 25 were manufactured by RCC. AN/APQ-16, Eagle with GPI (AN/APA-44), had been under construction for some time at the end of the war. AN/APQ-7A was Eagle with a modified Nosmo attachment. Plans were made for mounting the Eagle antenna in the wings of future bombers, such as the B-42 and B-36, with the possibility of an Eagle antenna in each wing to increase the scan angle. Some inconclusive tests were made in the summer of 1944 on the use of Eagle for tank reconnaissance. Eagle was successfully used for blind approach landings, and there were discussions of the use of Eagle in the control of guided missiles (war-weary aircraft).

In the late summer of 1944 attempts were made to equip planes of the 8th Air Force with Eagle. Considerable difficulty was met in this attempt, partly because the Army, reacting from the H_2X crash program, insisted that the Eagle program be orderly,[28] and partly because the 20th Air Force had priority for its B-17's. Finally one Eagle B-17 was obtained; this arrived in Alconbury in October 1944. Buck went to England to assist in testing and setting up training facilities. The set, when it arrived, was found to have a bad hole in the antenna pattern, which was corrected by careful adjustment of the flaps.[29] An intensive program was initiated at the Radiation Laboratory to correct this situation. It was found that rigid control of tolerance was required, so Western Electric was persuaded to accept the overall supervision of the vane and scanner manufacture and installation. In England flight tests were carried out, and a training program set up for operators and mechanics in the field.

The first Eagle training school for operators and mechanics was set up at Boca Raton in December 1944, with assistance from the Radiation Laboratory. The Laboratory trained the instructors, helped set up proce-

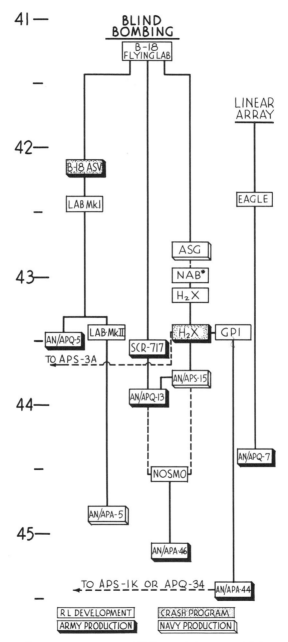

FIG. 15-7. Chart of RL developmental and production systems (blind bombing).

dures, and prepared a training film for the mechanics. When training began in the 2nd Air Force for the 315th and 316th Wings of the 20th Air Force, the Radiation Laboratory cooperated extensively. The operators were given basic training at Boca Raton and at Williams Field, advanced training at Victorville, and crew training at 2nd Air Force bases.[30] Radiation Laboratory men were instrumental in improving crew training. In addition, one man from the Laboratory was stationed at the headquarters of the 316th Wing, and one man at each of the four groups in the Wing. The Laboratory also trained the personnel in the Bowditch Project of photo reconnaissance, an extensive program of scope photography of intended targets, named for its commanding officer.

AAF Board tests of Eagle took place at Orlando in the beginning of 1945.[31] The men conducting the tests were H_2X operators who disliked the limited scan. After somewhat limited tests, using poorly conceived bombing procedures, they turned in a most unfavorable report, though they conceded that Eagle appeared to have a higher bombing accuracy than other self-contained radar bombing devices. This adverse report may have delayed the operational use of Eagle, but did not prevent its use altogether.

Although just too late to become operational in Europe, Eagle was used in several successful B-29 strikes by the 315th Wing in the Pacific. The bombing of the Maruzen oil refinery on July 6–7, with 95% destruction, was the most spectacular of these operations. General LeMay, in a telegram of commendation, called this performance "the most successful radar bombing of this command to date."

A chart of Radiation Laboratory developmental and production work on blind bombing equipment is shown in Fig. 15-7.

NOTES

1. Radiation Laboratory Notebook No. 178, issued to B. L. Havens, 18 May 1942, pp. 46–49.

2. B. L. Havens and D. R. Corson, "Correlation of ASV Equipment with the Bombsight," Radiation Laboratory Report OP-2, 24 July 1942.

3. Minutes of the Coordination Committee, Vol. III, 21 April 1943, p. 149.

4. *Ibid.*, Vol. III, 16 June 1943, p. 345.

5. *Ibid.*, Vol. VII, 22 November 1944, p. 356

6. *Ibid.*, Vol. I, 29 July 1942, p. 11.

7. Airborne Division Notes, 6 January 1943, p. 18.

8. General Eaker to Commanding General, Army Air Forces, Washington, D. C., Spring, 1943.

9. Status of AN/APS-15 (H_2X) Project at Radiation Laboratory, prepared by the Transition Office, 8 July 1943.

10. Minutes of the Coordination Committee, Vol. IV, 6 October 1943, p. 328.

11. *Ibid*; Vol. V, 22 March 1944, p. 440.

12. Airborne Division Notes, 2 August 1944, p. 7.

13. T. A. Murrell to L. A. DuBridge, 31 July 1944.

14. Minutes of the Coordination Committee, Vol. VIII, 3 January 1945, p. 140.

15. *Ibid*., Vol. VII, 6 September 1944, p. 85.

16. Radiation Laboratory Notebook No. 11, issued to L. W. Alvarez, 21 November 1941, p. 134.

17. *Ibid*., 18 February 1942, p. 148.

18. Radiation Laboratory, Notebook No. 297 issued to L. W. Alvarez, p. 3, 9 May 1942; p. 3, R. M. Robertson, "Variable Width Waveguide Scanners for Eagle (AN/APQ-7) and GCA (AN/MPN-1)," Report 840, 30 April 1946.

19. Radiation Laboratory Report 75, 2/12/43, E. A. Luebke, "Project Eagle."

20. Minutes of the Coordination Committee, Vol. I, 26 August 1942, p. 8.

21. E. A. Luebke and J. W. Eggers, Conference Report Project Eagle, 31 May 1943.

22. J. H. Buck, "Report on Eagle Operational Tests May 1943 to May 1944," 1 June 1944.

23. J. W. Eggers to H. R. Gaither, "History of Eagle," 30 April 1945.

24. J. H. Buck, *loc. cit.*

25. J. W. Eggers to L. A. DuBridge, "AN/APQ-7 (Eagle)," 12 April 1944.

26. E. A. Luebke to Eagle File, "Stratton Committee Meeting March 10," 13 March 1944.

27. Minutes of the Coordination Committee, Vol. V, p. 442, 22 March 1944.

28. J. W. Eggers to H. R. Gaither, "Summary of Eagle Project," 30 April 1945.

29. Minutes of the Coordination Committee, Vol. VII, 22 November 1944, p. 363.

30. For a detailed account, see the Minutes of the Coordination Committee, Vol. X, 27 June 1945, p. 212.

31. J. H. Buck, "Trip Report 21 February, to 8 March 1945; Boca Raton and Orlando, 12 March 1945."

CHAPTER 16

Ship Search Radar Systems

In undertaking the development of shipboard radar equipment on microwave frequencies the Radiation Laboratory entered an area in which the Naval Research Laboratory and several of the industrial concerns had already accumulated considerable experience, chiefly with long-wave radar, and were the acknowledged leaders. Moreover, the engineering and installation problems in this area were especially complex compared to ground and even to airborne equipment. For example, a piece of radar gear destined for installation in a capital ship must be designed in terms of the ship as a whole; the weight, size, and location of the radar set are matters of great concern; the radar must be carefully integrated with the other electronic devices, the communication systems, and the other types of detection apparatus of the vessel; it must be carefully designed to withstand the conditions of warfare at sea; and its characteristics must suit the tactical assignment of the type of vessel. The answers to many of these questions were most readily provided by the accumulated backlog of professional naval experience. For these and other reasons, the procurement of new shipboard radar gear by the Navy tended during the war to follow a more formal and traditional scheme than was the case, for example, with the Air Forces; the Radiation Laboratory was less influential in shaping the course of radar development; the unconventional "crash program" seemed less suited to the Navy's needs; and it will surprise no one to learn that the Navy felt it neither necessary nor desirable to encourage free-lance civilian scientific aid in the battle areas to anything like the extent of the Army Air Forces or the RAF. On the matter of the operational use and performance of its equipment in the Fleet, the information reaching the Laboratory was fragmentary.

Several of the Radiation Laboratory divisions worked closely with the Navy; but the bulk of the developmental work described in this chapter was conducted for the Bureau of Ships by Division 10, successively headed by L. C. Marshall, R. G. Herb and E. C. Pollard (serving jointly), R. E. Meagher, and J. C. Street.

The Ship Search Problem

Although ship search serves several purposes in addition to detecting an enemy target at maximum range, this primary function itself has two aspects. When long-range detection of aircraft is involved, the maximum range of detection to the horizon is often sacrificed to obtain all-round coverage or coverage at high altitudes. In the detection of surface targets and low-flying aircraft, the extent of low-angle coverage—i.e., coverage low over the surface of the water—is a rough measure of the effectiveness of the search.

The Navy's earliest long-wave (200 MHz) sets were suitable for installation only on larger vessels because of their heavy stacked array antennas. The accuracy of the bearing determination was only about 3° or 4°, but the range obtained was excellent (over 100 miles) and the first sets were very reliable in operation. Although they provided no low coverage, had poor target discrimination, and would have been easy to jam, the excellence of these long-wave systems for long-range air warning led to their retention until the end of the war, in spite of their limitations.

Microwave sets like the SG and its successors greatly increased the effectiveness of low-coverage surface search, making it possible for ships to be detected at close range. All search radars[1] throw a great deal of their radiated energy on the surface of the sea, and the sea acts as a reflecting surface giving rise to alternating bright and dark lobes as the target moves to higher elevation angles. When the target is in a bright lobe, the range of the set is greater than it would have been if no reflection had occurred; in a dark lobe, the range is much less. Since the lowest lobe is dark, a surface target will be detected by a long-wave set only if the target's superstructure projects into the lower edge of the lowest bright lobe. At microwave frequencies, the lowest bright lobe is closer to the line-of-sight horizon than at lower frequencies, and the microwave sets give better performance against surface targets and low-flying aircraft for this reason.

The effect of "sea return" on the pattern in surface search was not at first understood—although it was realized that it was related to diffuse reflection from optically rough surfaces—and was worked out from experience rather than theory. It was found that clutter is frequently present on the scope to about 5 miles but does not render the search completely ineffective. Turning down the receiver gain would bring great improvement, but this is inadvisable because it reduces maximum range on low-flying aircraft and may result in failure to detect a submarine periscope.

In spite of the fact that microwave search can and should give low-angle coverage, the equipment needs to be stabilized to give best performance. This difficulty enters into long-wave search less markedly because only high angles are covered and there is practically no distortion resulting from the roll and pitch of the ship. When, however, the spinner of a microwave antenna system is fixed on a stable base, the beam may be directed above or below the horizon depending on the ship's motion and the azimuth of the scanning beam. Stabilization for pitch is not a pressing problem, at least on large vessels, but if the period of scan is comparable to that of roll the beam may miss a target for many consecutive scans. The narrower the beam, the more serious this can be. It should also be apparent that without stabilization the roll of a ship limits performance against low-flying aircraft because of their high speed and direction.

In early experimental work, manual spinner control was tried rather than some kind of constant-speed motor drive, since ships do not roll in regular cycles. This was too inefficient and it was soon realized that gyro stabilization would provide the only solution. This is a problem of mechanical engineering but because it could not be done readily elsewhere, the Radiation Laboratory had to undertake an extensive program of research and development on stabilization devices to make its own radar equipment usable and effective.

In the widest sense, search radar is an aid to navigation or to that part of the art called pilotage, the guiding of ships through coastal waters, narrow channels, and encumbered harbors. Here the greatest advantage is that radar visibility in fog and darkness is comparable to that of all other navigational aids under good daylight conditions. In handling of a ship, radar navigation can serve to make landfall and to pick up familiar objects, like marker buoys, and display their positions relative to the ship by a method which does not admit of confusion. This is particularly well done with microwave equipment. The long-wave search sets with which the Navy started the war did not have the advantages of the PPI nor did they have sufficiently good resolution to be well suited to navigation. Although the microwave systems with the PPI and other improved indicators were not able to match optical aids in angular resolution or discrimination between objects of different shape, they did replace the eye as a precise means of determining range and range differences. They were invaluable in convoy station keeping and in amphibious landings, and enabled all sorts of operations to be carried out in the dark.

In general, the requirements of shipboard radar equipment are extremely hard to fulfill. Shipboard sets must be rugged, or they are useless; they must have high precision for fire control and fighter direction, or they are utterly unsatisfactory. The equipment must withstand shocks more severe than those experienced by any airborne or ground equipment and its greater complexity calls for much more careful engineering. The many classes of ships and their multiple functions brought about the design of separate radar sets for each type of vessel and resulted in the installation of a variety of types of equipment on larger ships. The wide variety of sets greatly complicates the training of operators and maintenance men. Some variety is of course inevitable. While a general purpose search set could solve many problems for large ships, on small craft, such as submarines or PT-boats, where space is at a premium, it could not be installed and at any rate would be superfluous.[2] For the small ships, 1-cm sets would have had many advantages had they been available. They are lightweight; their high resolution in azimuth and range and their short minimum range would have aided navigation, especially close to land, or even in large task forces.

Ten-Centimeter SSV (Ship to Surface Vessel)

The SG was the first production 10-cm ship-search set to follow the experimental system on the USS *Semmes*. In the design of this highly successful set the Raytheon Manufacturing Company, the Naval Research Laboratory, and the Radiation Laboratory can all claim a share. The Laboratory's contribution, in addition to the basic research that went into the original *Semmes* equipment and the four later models that were tried out on the ship, took the form of advisory services and of the release of personnel to Raytheon, where the detailed design was largely worked out. The Navy contributed heavily to the sturdiness of the design by recommendations based on severe shock and vibration testing of the prototype equipment conducted at the Naval Research Laboratory.

In anticipation of a Navy contract, Raytheon representatives discussed with the Radiation Laboratory the general layout of a 10-cm search set and the improvements over the *Semmes* equipment that should be included. W. M. Hall, one time Chairman of the Indicator Group, and W. G. Tuller, a member of the Receiver Group of the Radiation Laboratory, entered the employ of Raytheon. G. N. Glasoe of the Modulator Group was given a leave of absence to assist in the design of the SG modulator. The Bureau of Ships gave the contract to Raytheon at the end of June 1941.

It was expected that the Radiation Laboratory would serve in an advisory capacity to Raytheon in this development; but since the precise responsibilities associated with this function had not as yet been worked out, services were rendered on an informal basis only and no Laboratory project engineer was assigned to follow the development.

In December 1941, however, the Navy asked the Laboratory to go over the experimental model and make suggestions to ensure the best performance of the equipment. In February 1942, in response to the need for a more formal arrangement, K. T. Bainbridge was appointed project leader with responsibility for reporting on the progress of work at Raytheon. J. R. Zacharias was made advisor on rf circuits.[3]

The antenna and rf Groups of the Laboratory were the most closely associated with the development at Raytheon. The two groups made a study of fanned beam antennas (narrow in azimuth, wide in elevation) which might be used. The experiments on the USS *Semmes*[4] had convinced the Laboratory of the value of stabilization so their first choice was a stabilized radiator with a full paraboloid giving a pencil beam. However, since no stabilized antenna had yet been developed, a cut paraboloid with a simple waveguide feed was recommended. The design of the 48-in. cut paraboloid with beam width 5° in azimuth and 15° in elevation finally adopted was largely the work of T. J. Keary.[5]

The SG was tested at the Naval Research Laboratory and the first production set was installed on the USS *Augusta* and went to sea on April 5, 1942. The set was still aboard when the *Augusta* served as Admiral Kirk's flagship during the Invasion of Normandy and when it carried President Truman to the Potsdam Conference. Shortly after the first installation, the second and third sets were put in operation on the USS *San Juan* and the USS *Saratoga.* The sturdy construction, compact arrangement, and all-round dependability of the SG won wide acclaim. Six months after the set went into operation, reports showed[6] that the maximum reliable ranges of the SG on surface targets exceeded those of all other types of shipboard radar then in use by the U.S. Navy. According to a report from the *Saratoga*, "... operation has been beyond all expectation.... . The ranges very closely approach the absolute line of sight maximum under all conditions, and even on clear days, detection is often well in advance of that of lookouts equipped with glasses."[7]

When in March 1943 the Bureau of Ships requested Raytheon to develop a higher-powered, stabilized set, which later became the SG-3, Radiation Laboratory was asked to serve as Advisor. R. W. Blue and H. M. Lindgren followed through the research and development. Raytheon worked out its own stabilization method in cooperation with the Servo Mechanisms Laboratory of MIT.

Very much later in the operational life of the SG, in 1944, the Navy asked the Radiation Laboratory to undertake a project to incorporate AJ (anti-jamming) features in the system. This was accepted and N. E. Edlefsen was placed in charge.

The original order for 100 SG sets was increased to 955 and these had all been delivered by November 1943. The high reputation which the SG came to enjoy in the Fleet gave Radiation Laboratory just cause for pride in its collaboration. Although the SG was not installed in the ships which fought the night engagement in the First Battle of Guadalcanal, they soon began to appear in the Pacific and the Navy came to feel that the SG was an important factor in the advantage which our Fleet held over the Japanese. The SG took part also in the invasions of Sicily and Normandy.

SF

SG was suitable only for larger vessels. At the end of 1941, the Navy felt an urgent need for a light, compact, shipboard radar set to install on motor torpedo boats, submarine chasers, and other craft lighter than destroyers. It would be used for the detection of ships and surfaced submarines, and for station keeping in convoys. It was hoped also that it might detect aircraft at short range. The set known as the SF grew out of a request from Lieutenant Commander S. M. Tucker to the Radiation Laboratory on December 26, 1941. L. C. Marshall and members of his group drew up a proposal and recommended the Submarine Signal Company as manufacturer for this equipment. Commander Tucker requested that the Laboratory and the company hold conferences and draw up a formal proposal for the Bureau of Ships.

A series of informal visits was paid by Submarine Signal Company representatives to the Radiation Laboratory and, after conferring with components groups, H. M. Hart and E. F. Smith (of Submarine Signal) and R. G. Herb (of the laboratory) drew up a memorandum to serve as a basis for a conference on specifications.[8] A conference was held at Radiation Laboratory on January 9, 1942, attended by representatives of the Bureau of Ships, the Naval Research Laboratory, and the Submarine Signal Company. Commander A. N. Granum led the conference and described the type of equipment needed.[9] He placed great emphasis on speed in manufacture. The main point of disagreement was over the modulator. M. G. White, of the Laboratory's Modulator Group, recommended the Stromberg-Carlson "vest pocket" modulator which was under development by the Laboratory for airborne installations. The Navy representatives did not think this suitable for shipboard use but agreed to accept the recommendation of the Laboratory. They requested, however, that Submarine Signal Company run tests to compare the "vest pocket" modulator with a thyratron modulator which the Naval Research Laboratory was then developing.[10] The proposal of January 8 called for a simple indicator, with provisions for adding a PPI, and without accurate ranging. Commander Granum, however, requested "the best possible range accuracy without undue complications." It was agreed that ten preproduction sets were to be built, the first by June 1, 1942, and that production would start in August of the same year.

The rough specifications called for a 24-in. paraboloid, in a turret, mounted on the mast, with fixed elevation, unstabilized. Except for the power supplies, the other components were to be contained in two boxes; the "vest pocket" pulser, rf components, and preamplifier in one, and the receiver-indicator and ranging circuits with 5-in. A-scope and 5-in. PPI scope in the other.

No experimental model was made at the Laboratory. R. G. Herb, the Project Engineer, initiated a series of informal conferences at which agreements were reached for Submarine Signal Company engineers to come to work with components specialists at Radiation Laboratory. Herb was assisted by J. Nash and later by J. K. Hilliard and S. F. Johnson. Among the Submarine Signal members, C. W. Barbour worked under R. E. Meagher on the receiver-indicator unit; R. H. Willard under A. E. Whitford on the modulator. L. A. Turner made recommendations on the receiver. R. F. Bacher, who had been consulted on indicator problems, decided that the MM (Magnetic-Mechanical) PPI should be used. When it became evident, some time after the work on the receiver-indicator unit had been started, that accurate ranging would be required, B. Chance, R. E. Meagher, and others conferred with Submarine Signal on suitable circuits. It was decided to use the recently developed Chance circuit with movable gate.[11] The contract for the construction of ten pre-production models was followed by a letter of intent, on March 27, 1942, for the manufacture of 400 production sets. Delivery was not made on the first model until August 12, 1942 because of vicissitudes that illustrate, better perhaps than accounts of more smoothly running programs, the difficult problems that were encountered before the three-way cooperation of interested Service, manufacturer, and civilian laboratory had been satisfactorily worked out. Specifically, complications resulted because of misunderstandings on the part of Laboratory members and the Submarine Signal Company as to the specific requirements of the Bureau of Ships, and because of the undefined responsibilities of the Radiation Laboratory in its status as consultant.

When the project was undertaken, its urgency was stressed and the manufacturer was given to understand that getting the sets out as soon as possible was of paramount importance.[12] Yet later in the developmental period, the Bureau of Ships insisted on keeping to Navy specifications. Changes had to be made which caused delay and confusion. When redesign had to be effected, the Company was hard-pressed and rushed components through without adequate consultations with the Radiation Laboratory.

The controversy over the modulator was the most heated and prolonged. The Modulator Group had strongly recommended the use of the Stromberg-Carlson "vest pocket" modulator. R. H. Willard, of the Submarine Signal Company, working at the Laboratory, built a vest pocket modulator introducing changes which met with the approval of the Modulator Group. Plans were made to put this into production. After a visit to the Laboratory and to Submarine Signal Company, Commander S. M. Tucker indicated that if changes in the layout of the modulator were not made the Bureau of Ships might not accept the set.

Changes were accordingly made that resulted in wiring mistakes. An experimental set was sent to the Naval Research Laboratory for type testing without preliminary checking by the Laboratory. This involved the Laboratory in embarrassment and shook the confidence of the Bureau of Ships in the equipment.

During October and November 1942, a prolonged series of conferences was held to determine the ultimate solution to the modulator problem. For some time, the Bureau of Ships was undecided whether to abandon the hard-tube modulator and substitute a spark-gap modulator as had been recommended by M. G. White at the Radiation Laboratory. It was finally decided to put in hard-tube modulators after the first fifty production sets and abandon any idea of spark-gap modulators.

The receiver-indicator unit development began when Barbour came to the Laboratory to work under the direction of R. E. Meagher. Provisions for a PPI and for accurate ranging necessitated a change in the original plans. The Bureau of Ships requested that the unit be enlarged three times and insisted on the use of special resistors, condensers, and transformers to meet Navy specifications.

This experimental unit was built and tested during April 1942. Although it was not considered quite satisfactory, in the interest of getting the unit into production it was quickly redesigned. Later tests on production sets showed "shadows" on the PPI which were the fault of the circuit. L. A. Turner was anxious to have circuit changes made, and he was much concerned at the haste with which the development was put through.[13]

The first experimental model was in operation at the submarine Signal Company on June 3, 1942, and the Navy put at the disposal of the Laboratory the USS *Gallant*, for conducting tests on the first preproduction model, which was delivered to the Boston Navy Yard on August 13, 1942. The first trial was held on August 20 and the set performed satisfactorily. Signals were seen to 48,000 yds. and a buoy was detected and followed to 6000 yds.

Many changes had to be made in the ship, which was a converted yacht used for training officers, to install the antenna on the mast. The rf line was probably the longest 7/8-in. stub-supported line yet put into service. The distance from the modulator to the spinner on top of the mast was about 55 ft. Although the ship was not a floating laboratory, as was the USS *Semmes,* and the testing must have interfered often with normal operation, full cooperation was received from Captain A. O. Rabideau and the members of his crew during the many months when experimental work was being carried out.

In connection with this test another program was undertaken to experiment with stabilization of the SF antenna. Stabilized spinners from the

Westinghouse Laboratories and the General Electric Company were installed on the ship for comparison with the unstabilized SF. The SF was never stabilized, although L. C. Marshall and G. P. Stout of the Bureau of Ships recommended it, but the operational data collected were of value in future development. Most of the equipment was removed from the USS *Gallant* by the end of March 1943.

A by-product of the SF program was the 3-cm SF modified at Radiation Laboratory by Navy request. The Laboratory was unwilling to make definite commitments but at length undertook the project. The SF (XI), built by F. Niemann, was ready for test by March 1943. The rf components were changed and an unstabilized 24-in. paraboloid, cut at the top and bottom to produce a slight vertical fanning of the beam, was constructed. It had been expected that the system could be installed on the USS *Gallant,* but as this was not possible it was put on the USS *Semmes* for a six months' period. This experiment was valuable for the development of the SU radar which was eventually to follow the SF.

The original order for 400 sets was many times increased—up to 1655 sets—all of which were delivered by August 1, 1944. Some of the later sets were designated SF-1. The latter had different power supplies and provision for remote indicators. Fifty sets were turned over to Britain through Lend-Lease and several SF's were installed on French battleships and destroyers in American ports.

J. M. Wolf of the Radiation Laboratory made a trip to study some SF sets in the field.[14] He interviewed operating crews of five PT-boats and one SC-boat. He found that the SF's were used for the most part only at night, for search and for locating escort vessels and stragglers in convoy; seldom were they used for coastal navigation.

An interesting project, involving the SF, was sponsored jointly by the Bureau of Ships, Submarine Signal Company, and Radiation Laboratory to study icing conditions in northern waters.[15] This was known as Project Indian. The ship USS *Mohawk,* with H. M. Lindgren and M. Kessler aboard, left Boston on March 20 and returned on April 26, 1943. The SF proved to be very dependable, operating 375 hr without failure. Due to mild weather, it was not possible to carry out the purpose of the cruise, but good data on the performance of the equipment were obtained. Airplanes did not show up well; icebergs gave good signals; sea return was noted from 500 to 6000 yds. There was ample opportunity to study the problem of "phantom targets", caused by multiple reflection of more than two real targets. The ship's personnel recommended installation of a remote PPI on the bridge and a true-bearing PPI. Suggestions were also made as to operating technique.

The experiences of Radiation Laboratory in the development of the SF led to conferences with the Bureau of Ships which resulted in a clarification of the Laboratory's position in future Ship Search Radar programs and a definition of the Consultant-Advisor status. It was agreed that the Laboratory should coordinate the microwave program in similar fashion to the Naval Research Laboratory's coordination of the long-wave program; it should not duplicate the work of manufacturers; it should take designs produced by research and act as a liaison group for development, and act as advisor during early stages of development; it should serve as development source of components dependent on microwave technique; work to assemble a "standard set"; and keep in close touch with manufacturers. If the Laboratory's suggestions were acceptable, such suggestions were to be referred to the Bureau of Ships and incorporated in the equipment; if not, the decision would be up to the Bureau. The Director summarized the agreements reached on the role to be played by Radiation Laboratory when acting as consultant or advisor to the Bureau of Ships or to the manufacturer, or both.[16] The Laboratory agreed to make available to manufacturers all necessary information about microwave radar in general, and the special project in particular; to check and criticize proposal designs in all stages of development; to carry out thorough testing of prototype equipment. Recognition of the Laboratory's Project Committee as coordinator not only of the Laboratory's work on a given project but of the relationships of the Laboratory with the Service and with the manufacturer was given. This definition of functions furnished a good working basis for future collaboration on shipboard equipments.

SO Series

The many changes in the course of the SF design had resulted in a system too heavy and bulky for use in the smallest craft, as originally intended. In May 1942, the Bureau of Ships, in accordance with the Navy's program of equipping every type of ship with search radar sets, made a suggestion that a set should be developed specifically for PT-boats.[17] At a conference on June 5, 1942, attended by Commander Tucker, L. A. DuBridge, L. C. Marshall, J. F. Koehler, and others, it was decided to install an SF in a motor-torpedo boat for tests during July 1942. The Laboratory considered it desirable for a formal project to be set up so that they might be sure of having a motor-torpedo boat assigned for the running of tests. Some planning was done along these lines but the project was finally abandoned because of the Navy's decision to have Raytheon Manufacturing Company develop a new set to be known as the "SO" for PT-boats, with the Radiation Laboratory as Consultant. This development turned out to be not one set but a whole series and on two different wavelengths.

As the result of a meeting between Radiation Laboratory and a Raytheon representative on August 8, 1942, in the office of Commander Bernstein, Raytheon was asked to submit bids for the development of eight preliminary models, four 10-cm sets and four 3-cm sets. The first 10-cm model was to be finished by October 1, 1942, and it was planned to start production on January 1, 1943. N. E. Edlefsen was appointed project engineer.

The Radiation Laboratory was not to build the experimental 10-cm model, called CXBX, but the services of the Laboratory were to be used in testing components and advising in all possible ways. The 3-cm experimental system, called CXBY, was to be assembled at the Laboratory using some components supplied by Raytheon. In order to take advantage of the 3-cm experimental work which had been carried out at the Laboratory, a man was sent to the Laboratory to become familiar with 3-cm techniques and to help with the assembly of CXBY. The 3-cm system followed the design of the 10-cm system as closely as possible. At various times during the SO development, the Laboratory was asked by the Navy and Raytheon to carry out several kinds of tests. G. D. Shockels and W. T. Harrold served as assistant project engineers to supervise these tests.

The SO components were divided into four packages: the antenna assembly with paraboloids of varying sizes, depending on the use to which the set is put, with 360° azimuth variation and no vertical variation; the transmitter-receiver to be placed on the mast with the antenna; the indicator-central, containing a 5-in. MM-PPI with 5-, 20-, and 70-mile range scales; and a rotary spark-gap modulator.

CXBX (SO)—The Ten-Centimeter Prototype

The first model (CXBX) was assembled at Raytheon during the week of November 16, 1942, and was installed in a PT-boat at the Charlestown Navy Yard on November 25, 1942. Operating out of the PT-boat base at Melville, Rhode Island, the PT-boat used the set from November 27 to December 7, 1942 in tactical problems between Martha's Vineyard and Block Island and between Brenton Reef Lightship and Point Judith. On November 27, 1942, a tow boat and string of barges were picked up at 5 miles, each barge being distinguished at 3 miles. On December 5, 1942, ranges were: aircraft, 7 miles; buoys, 1–6 miles; net tender, 8 miles; cruisers, 12–14 miles; land, 14 miles or more. In spite of storm and rough seas, the set functioned well and the Navy was pleased with it.

The second model of CXBX was installed in a PT-boat at Melville, Rhode Island on January 14, 1943. A new mount had been designed by

Raytheon, which permitted the antenna to be folded up on deck. The rf package was fastened to the mount near the deck. A preplumbed waveguide, developed by Raytheon, was used in the rf system.

At the request of the Bureau of Ships, Raytheon was asked to turn out 20 model shop SO's, beginning February 1, 1943.[18] The first production-line SO passed its final tests during the week of February 18, 1943.

CXBY (SO-3)—The Three-Centimeter Prototype

It had been agreed at a conference held at Raytheon Manufacturing Company on August 19, 1942 that the Radiation Laboratory could give most effective help to Raytheon by developing a 3-cm model and turning it over when completed and by running tests on components.

The Roof System S was used for the assembly and testing of the CXBY. Tube mounts for the 5-in. MM-PPI tube were redesigned to meet Navy specifications. These were to be used on both CXBX and CXBY. Alternative designs of the rf package were worked on, the RF-Modulator, Receiver, and Antenna Groups all assisting.

The frequency ranges at 3 cm had been developed for airborne systems and it was felt that, if delays in the SO program were to be avoided, these same ranges should be used. The CXBY was first set up on the 3.2-cm band, but Raytheon was developing components on the 3.3-cm band which had been allocated by the Washington Communications Board for Ship-Search radar equipment. At length, the Bureau of Ships called a meeting at the Radiation Laboratory to set up specifications for an rf package on 3.3 cm. This was to be interchangeable with the SU rf system and will be described later. At a conference on February 2, 1943, the division of labor between Radiation Laboratory and Raytheon due to the change in wavelength was agreed upon.

The first SO-3 production model was ready for delivery in March 1944. It was installed on a PT-boat at Melville, Rhode Island (see Fig. 16-1), and in May a group from the laboratory visited the base to see it in operation. Operational crews praised its performance and the resolution was found to be very good. Discussions with officers who had returned from the South Pacific Area and who had seen search radar in action emphasized that high resolution was one of the great needs. They also rated low air coverage as important as surface search.

The Modulator and Power Equipment Groups helped in the development of a procedure for improvement of motor generator sets and modulator components when faulty operation of MG sets caused considerable delays in production. The Laboratory prepared recommendations for test equipment for SO and SG, to aid Raytheon in its test procedure. The

FIG. 16-1. SO on PT-boat; antenna raised.

design of a precision Echo Box for *S*-band cavity dimensions was worked out. The Bureau of Ships placed orders for 1280 units on this band at Johnson Service Company of Milwaukee, Wisconsin.

Most of the sets of the SO series used components that were fundamentally the same, the differences being mostly in the antennas and the power supplies. (See Table I.) Of the two sets that did not serve aboard ship, viz. the SO-7M and the SO-12M, the first is mentioned here because of the Laboratory's contribution to its development, particularly its antenna. The antenna of the SO-7M (a truck and trailer mounted set for the U. S. Marine Corps) was a 3 ft $\times$ 5 ft slotted dish that folded up within the truck. It was designed by the Laboratory's Antenna Group. The final set to be described in this series, though not the final set in the series, is the SO-11, which was proposed by the Radiation Laboratory.

SO-11 ("Zenith Watch")

In April 1943, A. Longacre, while aboard the USS *Lexington* on a shakedown cruise, in conversations with officers, particularly Lieutenant Commander A. F. Fleming, learned that existing search radar sets did not patrol the zenith region from which attacks were likely to come.[19] On his return to the Radiation Laboratory, he put the matter up to the Early

Table I. SO-Series.

Name	For use on	Power supply	Weight (lbs)	Antenna reflector
		10 cm		
SO	MTB	24 V dc	500	24″ uncut
SO-1	SC, PC, LST	115 V dc	1200	48″ × 15″ cut
SO-2	PC (Coast Guard)	115 V ac	1000	30″ × 12″ cut
SO-7M	Truck (USMC)	115 V 60-Hz	1 ton	36″ × 60″ slotted dish
SO-7N	Trailer (USMC)	115 V 60-Hz	1 ton	36″ × 60″ slotted dish
SO-8	PYC, YMS	115 V dc	900	30″ × 12″ cut
SO-9	BYMS, PYC*	32 V dc	500	24″ uncut
SO-11	CV, BB, CA, DD	115 V 60-Hz	1100	24″ × 48″ cut
SO-13	LCC, HTB	24 V dc		24″ uncut. Same as SO except for modulator and a few minor improvements.
		3 cm		
SO-3	MTB, LCC	24 V dc	500	24″ × 8″ cut
SO-4	SP, SC, LST	115 V dc	800	24″ × 6″ cut
SO-12M	Truck (USMC)	110 V ac	1 ton	36″ × 60″ slotted dish
SO-12N	Trailer (USMC)	110 V ac	1 ton	36″ × 60″ slotted dish

*For British Ships

Warning Group, which proposed a low-powered system, and with the help of the Antenna Group constructed an antenna called from its shape the "hog trough" with dipole array. This was installed on an experimental system and PPI photographs were taken. The Laboratory system was examined by officers from the USS *Lexington* and they requested that an SL or SO set be modified for zenith coverage and that it be installed on their carrier as soon as possible. The Laboratory's Steering Committee approved the project and transferred it to the Ship Division.

The first unit was built from SO-1 and SO-2 components. An SO-1 antenna was supplied by Raytheon for conversion. Plans were completed and delivered to Raytheon and to the Bureau of Ships by September 1943. A 24 in. × 48 in. antenna, with special feed to give angular coverage from 20° above the horizontal to 70° and vertical coverage up to 30,000 ft., was proposed.[20]

Table II. SO-Series Orders and Deliveries.

Name	Ordered	Delivered	Date
		10 cm	
SO	525	525	Complete April 1, 1944
SO-1	2047	2047	" November 1, 1944
SO-2	535	535	" August 1, 1944
SO-5*	67	0	
SO-6*	352	0	
SO-7M	100	100	" November 1, 1944
SO-7N	40	40	" July 1, 1944
SO-8	1700	1700	" August 1, 1944
SO-9	110	110	" November 1, 1944
SO-11	204	4	August 1, 1944; 100 cancelled
SO-13	587	587	Complete August 1, 1944
		3 cm	
SO-3	365	300	June 1, 1945
SO-4	700	230	July 1, 1945; 350 cancelled
SO-12M	110	60	July 1, 1945; 30 cancelled
SO-12N	110	33	July 1, 1945; 44 cancelled
		Unspecified†	
SO-10	No orders	0	150 planned; not scheduled
SO-14**	No orders	0	

*Radiation Laboratory was not Consultant on these sets. Raytheon collaborated with the Naval Research Laboratory.

†Radiation Laboratory was not Consultant on these.

**Designation of SO-14 was changed to Mark 33. This was probably the 3-cm, high-resolution part of a fire control system for use by the USMC.

The SO-11 was first tested on January 25, 1944, when it picked up an SNB test plane at a range of 5 miles at 10,000 ft. A B-17 was observed at 7.5 miles when it flew over at 5000 ft.

The first system was sent to the Pacific Coast for installation on the USS *Lexington,* accompanied by Alan Byers, who later supervised the operational trials. The installation was complete by March 1944. With the exception of the antenna, this first unit was considered to be of production design. For trials, the carrier went into Puget Sound. The test was a failure because the antenna was mounted aft of the stack, where the heat destroyed the dielectric material in the antenna. When an attempt was made to substitute mica as dielectric material, a mismatch resulted. Although the SO-11 was not given a fair trial in these tests there were indications that it showed too much clutter to have met Navy requirements. It was felt that with a redesigned antenna the SO-11 would perform its function of "Zenith Watch", although target designation could not be expected.[21] Four preproduction sets had been completed by August 1, 1944, but production sets were not forthcoming in time for combat use.

SO Series Summary

By August 1, 1944, a total of 4704 SO sets had been delivered and 7023 were on order; on July 25, 1945, 6271 out of a total order of 7025 had been delivered. (See Table II.) The SO took part in the landings in Normandy during the invasion of the European continent and in all amphibious landings in the Pacific that were accomplished with radar aids to landing, from the campaign in the Marshalls to Okinawa.

Project Cachalot

The importance of navigational problems and the inadequate methods available to the Amphibious Forces in landing operations led COMINCH to request that the NDRC set up a committee to study the problems, on April 3, 1943. Soon afterwards, NALOC (Navigational Aids to Landing Committee) was formed with J. E. Burchard as Chairman. L. C. Marshall was the first Radiation Laboratory member of this committee. When he left to organize BBRL he was succeeded by R. G. Herb, who in turn appointed M. D. McFarlane to represent him in dealings with NALOC and the Services. NDRC accepted "a project of broad scope to survey and investigate navigational aids for landing craft, to determine additional means which might usefully be adopted or developed for that purpose, and to carry out research and development work on such means as appear promising."[22]

At first, radar aids were not considered, but it was soon realized that no other satisfactory navigational aid for landing operations was available, especially for night operations. It was decided to use the 10-cm SO set which was already in production. O. J. Stephens II, a member of the committee who was a shipbuilder, advised that the LCC, a small boat designed by the Navy for leading group landing craft into a beach, could be modified to take the SO radar set.

The project was surrounded with the greatest secrecy and no mention of the committee's activities was made in divisional reports.[23]

Early tests were run off Fisher's Island, this being the first project assigned to the newly established Mount Prospect Field Laboratory. In these tests the submarine USS *Cachalot* took part and gave its name to the project.

NALOC was presented with the problem of controlling landing craft so as to arrive at a beach within 200 yd. of a target point, within one minute of H-hour, from a position 5 to 10 miles off the coast, in zero visibility. It was soon found that the SO set as normally used was not satisfactory for such precise navigation but when used with the VPR (Virtual PPI Reflectoscope), excellent results were obtained.[24] This simple mechanical device, designed by E. E. Miller and developed by the Laboratory somewhat earlier, superimposed the image of a map on the PPI pattern by optical means.

About the middle of July 1943, the Laboratory set up a program for training LCC crews. In August, the Laboratory instructors, P. S. Clymer, E. G. Nickerson, F. M. Pease, and H. H. Wheaton started a more comprehensive program at Woods Hole, Massachusetts, where the Navy put a boat at their disposal. Thirty-six crews of three men each were trained from August to October. During this early period, the accuracy of the trained crews surpassed all expectations. Landings on beaches were made to within $\pm$ 100 yd. within $\pm$ 1/2 minute of a specified time. The most competent officers were selected as instructors for the continuation of the program which the Navy started at Little Creek, Virginia. Over 800 men were instructed at the Norfolk Training School, and training was continued at the Landing Craft School of the Amphibious Training Base at Coronado, California. Pease assisted the Navy at Norfolk and later at San Diego instructing crews and installing equipment.

As an improvement to the VPR the NMP (Navigational Microfilm Projector) was designed by M. D. McFarlane and the first model was made by Eastman Kodak Company. This device used microfilm charts whose magnified image is matched with the PPI.[25] In a visit to the Hydrographic Office on October 27, 1943, McFarlane was shown charts prepared for landing operations in Sicily. These charts show the shore line in front of each landing barge and inland topographical features using a principle of projection from data onto contour maps. Later he learned that CINCPAC had designated ships for hydrographic survey in the vicinity of landing beaches. McFarlane pointed out that the use of radar would make it possible for these surveys to be carried out at night or under conditions of poor visibility. Medina, of the Hydrographic Office, suggested that a survey be tried under NALOC auspices in Guantánamo Bay and he agreed to prepare charts of the area.[26]

The Navy put USS *PC-576* at the disposal of NALOC to try out the possibility of a "blind survey" with radar aids. In January 1944, several NALOC members, including Burchard and Stephens, set off on a cruise from Norfolk to Jamaica. They used the SF set which had been on the ship for some time, installing an NMP, a VPR, and a plotting table. McFarlane and D. B. McLaughlin from Radiation Laboratory supervised the tests. As a result of the survey, a splendid chart was made of the surface features and shore line of Portland Bight, Jamaica, and McLaughlin prepared "A Manual of Radar Recognition"[27] describing the devices used. At the request of Commander Carmichael, of the Office of COMINCH, the Committee furnished 130 copies for LCC officers and crews.

Non-radar devices such as odographs, fathometers, and communications equipment to be used in amphibious landings were not within the province of the Laboratory and were worked on elsewhere. However, the Laboratory continued to work on methods of improving Radar Mapping Techniques and merged further developments along this line with the original Project Cachalot. Since the NMP was suitable only for SO sets, McLaughlin and C. A. Smith developed an autofocus microfilm projector which would be suitable for navigational purposes with other search systems. This device, built by the Spencer Lens Company under a Radiation Laboratory contract, was tested at San Diego in July 1944. Smith later developed a manual bearing unit to tie in true bearing information with radar sets. Still later, the Radar Chart Projector, a remote indicator with built-in microfilm chart projector, was completed at the Laboratory and the Eastman Kodak Company was given a developmental contract. The program of developing these later projection and chart matching devices was turned over to the Navy before the end of the war. In August 1944, Burchard considered that NALOC had served its purpose in working out a basic method of navigation and should disband,[28] leaving the matter of indoctrination and further refinements up to the Navy. After a final meeting at the Radiation Laboratory on October 9, 1944, the Committee was dissolved.

There is some doubt as to whether the LCC's used the methods of navigation exactly as recommended by NALOC. There is not much doubt that they served an important function in the Pacific and elsewhere in rounding up stragglers during landing operations and that they were also used for communication purposes. Shortly after the training course at Woods Hole, LCC's were sent to England and to the Pacific with trained crews.[29] CINCPAC was anxious to get LCC's as soon as possible in December 1943 because the landing operations in the Gilberts had been so costly[30] because of the inability of landing craft to navigate accurately enough under the conditions they encountered.

The first time the LCC's were used was in the operation in the Marshall Islands.[31] At Kwajalein, six LCC's were used to lead in waves of landing troops. Three waves hit the right spot, the first one minute late, the second two minutes late, and the third three minutes late. Further waves were controlled almost as accurately. The LCC's were especially valuable in passing on correct information. The method of operation[32] consisted in establishing a line of departure, marked by a destroyer, 5000 yds. off shore. The LCC's brought the landing craft from the transports to the line of departure.

In the Normandy invasion, nine LCC's were employed.[33] PC's exercised primary control for the Omaha and Utah beachheads, navigating by Gee[34] to a line 6000 yds. off shore. The beachheads were 800 yds. apart and four LCC's, in secondary control, were employed to lead in waves of amphibious tanks. The remaining five LCC's served as tertiary control in the transport area.

Conditions were very unfavorable to LCC operation because pre-assault bombardment had flattened the coastline. Nevertheless, on the Utah beachhead the LCC's carried out their mission satisfactorily.[35] On the Omaha beachhead, the performance was much more spectacular. There the only available LCC landed her own wave and cut back to pick up a second wave led by another LCC which had been disabled, "reporting with some chagrin that one of these waves had missed the pinpoint by 500–600 yards." [36] Eight days after D-Day, two LCC's made a hydrographic survey in Cherbourg harbor. Shortly following the Normandy operation, Admiral Moon took six LCC's with his Task Force to the Mediterranean.

Three-Centimeter SSV

As the result of a request from Vannevar Bush, in June 1942 a survey was made of all information on shipboard radar equipment designed for detection of surface vessels. L. C. Marshall, head of the Systems Engineering and Production Division, promptly drew up a report[37] covering low frequency gear, as well as the newer 10-cm sets, for detection of ships, aircraft, and submarines. Marshall pointed out as disadvantages of the longwave sets: large, clumsy antenna arrays with excessive weights and mechanical difficulties, exposed to extremes of weather, with pattern effects and a tendency to miss targets at close range. The most reliable 10-cm equipment, the SG, like the SE, SH, and SL, shared the disadvantages common to those that use the "chopped" paraboloid[38] in that the range on medium-sized ships is apt to be less than the normal line-of-sight. The high range accuracy and rugged construction of the SG were definite advantages. However, the weight and size of its cabinets made installation difficult even on a destroyer. No then existing or contemplated 10-cm SSV set, running the gamut from the small SF to the large OXBL, solved all of the problems.

Navy Department regulations came in for sharp criticism. It was believed that more compact sets could meet shock tests, humidity tests, and voltage clearances without adhering to rigid specifications which could only result in production delays. The policy of encouraging a plurality of sets without regard to standardization was leading to a duplication of functions which seemed undesirable. Specific sets for carriers, destroyers, submarine chasers, PT-boats, and auxiliary vessels resulted in creating the situation that by January 1943 about a dozen ship-search sets were in operation or production, requiring different methods for installation, servicing, and supplying of spare parts.

The principal recommendation was that, except for the early warning sets, the newer centimeter sets should supersede the longer-wave gear. The use of standard components was especially urged. At least the various research and development laboratories should standardize on such items as spark-gap modulators, antenna systems, and indicators. It was pointed out that the chief advantage of the 10-cm system is the small size of the antenna, which might be sacrificed by exposure to icing and storm. This could be remedied by housing the antenna in some kind of shelter. To compensate for the roll and pitch of the ships, fanning of the beam was suggested or, as an alternative, the development of stabilization equipment for which very optimistic estimates of weight were given. Attention was called to the recent 3-cm developments which could find ready application in submarines or on small vessels where space and weight must be carefully considered.

At this time the Navy could not have accepted all of these recommendations. Their concern was to see that sets were produced for immediate use and they could not wait for all sorts of refinements. Lieutenant Commander Henry E. Bernstein, Head of the Radar Design and Installation Section of the Bureau of Ships, introduced as much standardization as possible in all the ship-search programs which were to follow.[39] He also introduced the policy of field changes so that the sets already installed could be kept up to date by the addition of components or small parts of the latest design. Many of these changes were made in forward areas. The SG, for instance, had over 150 field changes made during the war.

Standardization of many components was eventually carried through successfully, and in this the Radiation Laboratory played a large part because of the large stock of high-quality component items which were always available there. This trend was especially noticeable in the design of the 3-cm sets, especially SO-3 and SU, in which cooperation between the Bureau of Ships, the manufacturers, and the Laboratory resulted in the standard rf package.

SU

The results of the successful overwater tests on 3 cm made at the end of 1942 were reported to Commander H. E. Bernstein in January 1943, and in March the Submarine Signal Company was advised of orders for a new 3-cm search set to be known as SU. On March 22, a conference on components for the new set was held among representatives of the Bureau of Ships, Raytheon, Submarine Signal, and the Radiation Laboratory. The SU was designed especially for the DE (Destroyer Escort) class of ship which was due to come out late in the fall of 1943. The 2-ft, line-of-sight stabilized antenna (beam width 4°) was to be enclosed in a radome. The most striking innovation was the design of an rf package (containing the transmitter, pulse transformer, plumbing, and receiver) interchangeable with that of the SO-3 system. A PPI indicator stabilized to true north, and an A scope with 2000-yd., 40,000-yd. and 80-mile markers, were to be provided. The indicating system's generating trigger was to be delayed to allow IFF signals to be sent and received synchronously with the target signal.

Radiation Laboratory was appointed Consultant on the SU, and S. F. Johnson became Project Engineer. Edlefsen, Project Engineer for the SO sets, was to have general supervision over the SO-SU rf package. Since one of the most pressing problems concerned the stowing of the rf package, Johnson and J. K. Hilliard made a tour of the Brooklyn Navy Yard in April,[40] visiting PC-boats equipped with SF and PT-boats carrying SO sets, and the cruiser USS *Philadelphia,* to assess troubles which might be avoided in the design of the SU. An out-in-the-open arrangement was not favored by the Navy due to the difficulty of making repairs in rough seas and fear of the unreliability of such components. Johnson felt, though, that since a below-deck arrangement would imply the use of a long rf line little advantage would be gained. It was finally decided to mount the rf system in the open and to carry a spare unit below deck.

Since small vessels do not carry skilled technical personnel, the design of the SU had to be as simple and reliable as possible, designed for replacement with a minimum of matching and adjusting. It had to be packaged in units weighing not over 200 lb and small enough to pass through compartment doors.

In general, modulator and rf components were designed by Radiation Laboratory and Submarine Signal and engineered by Raytheon, although Raytheon designed experimental SO components for comparison purposes and Submarine Signal did a great deal of engineering on the SU components.

In June, Johnson and H. M. Lindgren left for a trip to the Philadelphia and Charleston Navy Yards to study further advantageous positions for the SU rf package. They went aboard one of the new DE's and the USS *New Jersey.* They learned about the new CIC (Combat Information Center), which as Johnson described it "divorces radar from communications." [41] On the return trip, they conferred with A. P. Richards at the Bureau of Ships and learned that 402 SU sets had been ordered for DE's and 170 sets for PC's and SC's. For the present, it was planned to keep the SF's already installed on the submarine chasers.

Westinghouse delivered the first model of the gyro stabilizer for the antenna system to the Submarine Signal Company on June 17 and a conference was held in Springfield, Massachusetts on the next day to draw up plans for testing it.

During the whole summer, the Antenna Group conducted tests on radomes for the SU, not only on the turret shape but on materials. By September, E. M. McMillan and J. White had worked out the details and made recommendations. Submarine Signal placed orders with the Virginia-Lincoln company for the radomes.

Another problem of importance to all ship-search radar equipment came to a head in the course of the SU development. E. Pietz, of the Radiation Laboratory Test Laboratory, called a meeting to discuss shock mounts. The Bureau of Ships, as well as the General Electric Company and Raytheon, sent representatives. G. S. Bean, of the Bureau of Ships, opened the discussion, outlining failures caused by improper shock mounts: certain ships' speeds cause jitter on scopes, tubes short out, swaying during gunfire breaks cable. On shipboard, the deck is shocked and the unit has to be protected. Vibrations set up by propellers and gun shock, not only from bombs, mines, and shell hits, have to be considered. Near misses cause more shock than others because of the force of the water wave front which sometimes moves the whole ship when it strikes.

Mounts permitting metal-to-metal contact had been found unsatisfactory; rubber in compression seemed best. The "Pietz mount" (also called Radiation Laboratory compression mount), manufactured by Lawrence N. Barry of Cambridge, Massachusetts, was recommended by the Laboratory and the Bureau of Ships. [42]

In October, Submarine Signal sent its antenna to the Laboratory for tests. Difficulty with training the antenna was encountered. In December, Bureau of Ships representatives decided to put out the SU set with an interim antenna carrying a balance weight, in the interest of early production. Later it would be replaced by a model of new design for which Submarine Signal engineers would have the aid of M. B. Karelitz of the Laboratory's Mechanical Engineering Group.

By January 7, 1944, SU Serial No. 1 was being tested. The first five production sets were distributed as follows: one to the Radiation Laboratory and two to the Naval Research Laboratory for tests, one to the Navy's Radar School in the Harbor Building in Boston, and one for installation on the USS *Semmes* for trial runs. By March, 12 SU's had been produced. By the end of May, tests on the *Semmes* had been completed with encouraging results,[43] although the antenna was only 45 ft. above water and within 3 ft. of a 12-in. mast. Most ships were picked up at 35,000 yd., land from 70 to 80 miles, and buoys at 12,000–15,000 yd., and 3-in. shells had been followed to 7000 yd.

At about the time the first production SU's were coming off the line, Commander Bernstein was making plans for a new SU, which would give higher resolution, designed to operate with two antennas, one line-of-sight and the other stable-base stabilized.[44]

M. B. Karelitz directed the design of the two new stabilized antennas. The line-of-sight antenna (for the SU-2) was the more simple, requiring no computer, and weighing only 175 lb.[45] In this method, the beam pattern was kept directed toward the horizon by tilting the paraboloid, the feed following the roll and pitch of the ship. Elevation information was supplied to the antenna servo system by means of a stable element designed and developed by the Radiation Laboratory and the Eclipse-Pioneer Company. This system stabilized the beam through a roll of $\pm 30°$ with an expected accuracy of $\pm \frac{1}{2}°$ in a period of 6 sec and with an accuracy of $\pm 7°$ in pitch in a period of 5 sec.

The stable-base mount consisted of a pair of servo-driven gimbals whose axes were aligned with the roll and pitch of the ship. The same stable vertical as for the line-of-sight antenna was used. G. L. Stancliff made major contributions to the development of the stable element and servo mechanisms for both antenna mounts. No computer was required since the gimbals compensated directly for the roll and pitch of the ship.[46] The stable-base antenna system weighed 440 lb. Expected accuracies in the roll and pitch axes were within $\pm 0.°5$.

Two prototype models of each of these antennas were constructed by the Radiation Laboratory. One model of the SU-2 was installed on the USS *Kestrel*, the Laboratory's experimental ship, and a model of the SU-4 was placed in operation on the USS *Catoctin* before the end of the war.

A total of 692 SU sets were ordered by the Navy, of which 426 had been delivered by December 1944; 691 by July 1, 1945. About 50 of these were designated SU-1, differing from the SU only in the power supply used. An additional number, about 170, SU-1's were delivered before the end of the war, a good many being installed on corvettes of the Royal Canadian Navy.

In the fall of 1944, the Commander of DD-DE Shakedown Group, Atlantic Fleet, staged some tests near Bermuda to determine performance and target discrimination of the SU.[47] Three ships were used as targets and the fourth—a DE equipped with an SU set—acted as observing ship. All sorts of maneuvers were tried, including even the laying of a chemical smoke screen which had no effect on the radar set. The SU performed very well. Its angular discrimination was found to be about 4° and range discrimination about 300 yd. Distances of the target ships from the radar ship were 2000 to 15,000 yd. Piloting through channels was extremely easy with the SU; the course to pass buoys was readily determined.

Project Henry (CXGQ)

Naval units in the Southwest Pacific at the end of 1943 were reporting the immediate necessity for increased resolution in radar sets installed on small vessels such as PT-boats and landing craft to combat Japanese successes in moving troops and supplies in barges close to shore, where they could not be detected by the search radar sets then in use.[48] Commander Bernstein discussed the problem informally with Radiation Laboratory members N. E. Edlefsen, R. G. Herb, E. C. Pollard, and I. I. Rabi in January 1944, and a month later a formal request for the construction of 20 "crash" sets was accepted. The Navy did not assign the designation CXGQ immediately and the program was briefly referred to in the Laboratory as the Project Hydra crash program for the South Pacific.[49] This was soon dropped and it was decided to call the set Henry, in honor of Commander Bernstein.[50]

The design of Henry was initiated under the leadership of R. M. Emberson. The 3-cm system was planned to use an unstablized 10 in. × 48 in. aluminum grid reflector fed by a horn feed. Beam widths, at half-power, were 2° in the horizontal plane and 7°.1 in the vertical plane, with no prominent secondary lobes. The hydrogen thyratron modulator, recently designed by the Laboratory, was utilized, one of its advantages being a saving of about 100 lb in weight over the hard-tube modulator used on the SU. Two pulse-length–recurrence-rate combinations were possible with this modulator: a 0.15-μsec pulse and a 0.65-μsec pulse making it possible to use an 80-mile sweep and a faster sweep for close-in searching. The SO-SU rf package with suitable modifications was adopted. The two 5-in. indicators were a PPI with five sweeps from 2000 to 160,000 yards in both range and azimuth, and a B-scope. The center of the B-scan could be set at any point by hand wheels. An electronic selector ring, appearing on both PPI and B-scope, makes it possible for the operator to center the B-scan on any target appearing on the PPI.[51]

Production responsibility was divided between the Laboratory and RCC. The antenna, designed by the Laboratory, was built by the American Type Founders Company. The SO-SU rf units were obtained from Raytheon, and RCC built the modulator, indicator, and power supply units. The Laboratory designed and procured a total of 35 system components, since the Navy expected that the sets would operate in regions remote from supply bases and spares would therefore have to be carried.

The first CXGQ (the prototype units) was assembled in the old MEW turret on the roof of Building 6 on May 12, 1944 to demonstrate its performance to Navy visitors.[52] Performance was highly satisfactory, the B-scope especially proving its value on the long sweeps. Targets which appeared as one blip on the PPI were often resolved on the *B*-scope into several targets. In June, a higher precedence rating had been received but by this time it was apparent that the changing tactical situation in the Southwest Pacific had nullified the requirement for the sets.

The prototype CXGQ was placed aboard the USS *Kestrel* for tests. A second set, the first of the production systems, was shipped to the Amphibious Training Command, U. S. Atlantic Fleet, Norfolk, Virginia, in November 1944.[53] It was installed on an LCS, which was also equipped with SO-2 and SU sets, for tests in Chesapeake Bay during November and December. Radiation Laboratory members acted merely as advisors during the installation, all of which was carried out by regular Navy crews. No Laboratory representatives witnessed the tests, as the Navy was especially anxious to have the set operate without expert assistance. Henry's performance more than came up to expectations.[54] Meanwhile, the second production set was shipped to BBRL and in February 1945 the British Admiralty requested all the Henry sets, through Lend-lease, for installation on small vessels of the Royal Navy for hunting Schnorkel-equipped submarines.[55] They also asked for trained personnel from the Radiation Laboratory. It was decided to send one Naval officer and five enlisted men, who had been trained at the Laboratory and had followed the Henry almost from the beginning. Thirty-one Henry sets were shipped to Britain. The sets were intended for Captain-class Frigates (American-built DE's), but as the Schnorkel threat subsided no installations were made before the end of the war.

One-Centimeter SSV—Project Cindy (CXJG)

As a result of the experimental work on the H_2K systems built by the Fundamental Developments Group and tested at Deer Island and Fishers Island,[56] and of informal discussion between Commander Bernstein, R. G. Herb, M. G. White and J. R. Zacharias,[57] the Bureau of Ships requested that the Radiation Laboratory develop a set for small ships, such as PT-boats, to explore the possibilities of the 1-cm wavelength for ship-search work. It was stipulated that the set should use as many SO components as possible; for this reason the set was first referrred to as SO-K. The result-ing program was called Project Cindy and the Navy later gave the designation CXJG to the high-resolution set which evolved. By the time Cindy

was under way, Project Henry had been started, and this 3-cm high-resolution set was to influence greatly the ultimate design.

The chief advantage of 1-cm equipment lies in its high azimuth resolution with a small antenna, a great advantage in navigation close to shore-
lines and in station keeping, since the shape, size, and heading of ships can
be determined with great accuracy. Targets close to the surface of the
water, such as buoys, floating mines, and devices such as Schnorkel can be
detected much more easily by 1-cm sets because of favorable over-water
propagation, when the antenna is close to the water, than with the longer-
wave sets. However, sea return for a given antenna height is worse at 1 cm
than at 3 cm, so it is necessary to operate the 1-cm set at a low antenna
height. The greatest disadvantage of the 1-cm band is the adverse effect of
atmospheric attenuation at ranges beyond 10 miles and attenuation
through rain. This new development could be justified for ship-search
work only by emphasis on the advantages of higher resolution, light
weight, simplicity, and small power consumption.[58]

A program for the development of a suitable rf head and antenna was
begun in January 1944 by a group composed of W. M. Fairbank, the
project engineer, Edlefsen, Scheckels, and Jane D. Fairbank. The Cindy
equipment consists of three packages weighing about 500 lb. The first
package contains the rf head with some parts copied from H_2K, completely repackaged according to the SO-SU arrangement with some change in
circuits. The hydrogen thyratron modulator in this package was designed
by C. A. Carlson, while the preliminary design of the receiver was largely
the result of work by H. Wallman. This package contains also built-in test
equipment. The second package, which is waterproof, contains the power
supply for the rf head and junction box for the system. The third package,
containing the indicating system and its built-in power supply, was designed by R. E. Meagher so that it could be used, with a few cabling
changes, with other ship systems. The indicator was a 5-in. rotating coil
PPI tube, a remodeled SO type, with 1-, 2-, 4-, and 10-mile sweeps and a
delayable 0–10 mile sweep with variable range markers. Performance
specifications called for a range resolution of 30 yd., bearing resolution of
$0°.7$, and a minimum range of 75 yd. In this system, for the first time, focus
adjustment was eliminated by using a combination of an electromagnet
and permanent magnet in the focusing coils.[59]

The antenna, a 58-in. parabolic cylinder fed by a horn feed, was placed
on top of the rf head. It was designed by J. I. Bohnert and was similar to
the Henry antenna. The narrow beam width in azimuth of $0°.7$, at half
power, made high resolution possible.[60] Stabilization was omitted to reduce weight, but it was really not essential because of the width of the
beam in elevation, 11° at half power.

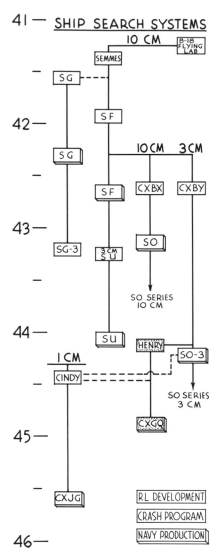

FIG. 16-2. Chart of RL developmental and production systems (ship search).

The experimental Cindy was tested on the roof of Building 6 in September 1944 and installed on the USS *Kestrel*, where it operated from October 1944 until April 1945, when it was removed and sent to the General Electric Company for experimental purposes in connection with Torpedo Director Mark 35. During this time, when SO and Henry sets were also aboard, Cindy was used to test components for the production CXJG set, to demonstrate 1-cm equipment to the Navy, and to draw up specifications for the final set.[61] During March 1945, a special series of tests was run south of Block Island to determine the ability of the SO (10-cm), Henry (3-cm), and Cindy (1-cm) sets to detect Schnorkel in rough water. A dummy Schnorkel installed on a U. S. submarine was used as a target. The Cindy set was shown to have the greatest possibilities for such detection.

In September 1944, the Laboratory proposed to the Bureau of Ships that the Navy consider Cindy for production. The Bureau proceeded very cautiously, preferring to have more tests made on the *Kestrel* to ensure reliability of the system and to wait until more was known about atmospheric attenuation at 1-cm. However, the Navy asked the Laboratory to build two sets for experimental work on 1-cm propagation and later the U. S. Army and British Admiralty each requested two sets.

The Laboratory decided in March 1945 that it would have to stop work on CXJG because of the pressure of more urgent projects, and they asked the Sylvania Products Company of Salem, Massachusetts if they would take the job of finishing the prototype set which the Laboratory had been developing and produce a few additional sets. Sylvania engineers had had previous experience along this line when they built the H_2K rf head for the Laboratory. Sylvania was willing to undertake the project, and received a contract to build 15 additional sets for the Navy. Sylvania had to do some redesigning in the rf package but introduced only minor changes in the power supply and indicator.

Sylvania finished the CXJG and demonstrated it to Radiation Laboratory members on August 30, 1945. Cindy, like Henry, did not see action in the war, but they both served an important experimental purpose in making it possible to assess the advantages and disadvantages of 3-cm and 1-cm ship-search radar sets.

A chart of Radiation Laboratory developmental and production work on ship search systems is given in Fig. 16-2.

NOTES

1. With the exception of "Zenith Watch" sets, which never went into production.

2. It is important for small craft, since they are limited to one set, to have a radar that provides facilities other than that of surface search. The submarine set SJ (produced by Western Electric Company) was both a search and fire control set and, though not designed for the purpose, gave some warning against low-flying aircraft.

3. L. A. DuBridge to Bureau of Ships, Attn.: Comdr. A. M. Granum. February 27, 1942.

4. See Chapter 10.

5. T. J. Keary, "A Study of Fanned Beam Radiators," Radiation Laboratory Report III-10, February 10, 1942. This report is based on the work not only of the writer but also of L. C. Van Atta, G. I. Ragan, G. A. Jarvis, and R. W. Wright of the Antenna Group and W. W. Salisbury of the RF Group. See also G. E. M. Bertram, Raytheon Mfg. Co. Memorandum No. S-51, February 4, 1942, to Messrs. Marshall, Hall, Brown, and Tuller,"Nos.-88522-WX-3873 Radar–R. F. System, Conference Notes." Present from the Radiation Laboratory were J. Lawson, W. W. Salisbury, T. J. Keary, and J. R. Zacharias.

6. R. N. DeHart, "Report on Operational Analysis of Results on Model SG Radar," Radiation Laboratory Report 108-12/3/42. Navy Liaison Office/RL (Comdr. Ferrier). Memorandum to L. A. DuBridge. A7-3/F Serial 099. December 5, 1942, "Extract from Radar Information." Gives extracts from a recent report of SG performance aboard the USS *Saratoga*.

7. Navy Liaison Office/RT, Memorandum Comdr. D. T. Ferrier to L. A. DuBridge, December 5, 1942. A7-3F.

8. Memorandum submitted by H. M. Hart, E. W. Smith of Submarine Signal Company, R. G. Herb of Radiation Laboratory, January 8, 1942, "Simple 3000 megacycle radar equipment proposed for use on submarine chasers and motor torpedo boats."

9. Radiation Laboratory members present were: L. A. DuBridge, L. C. Marshall, F. G. Dunnington, G. N. Glasoe, A. Herb, R. G. Herb, and I. I. Rabi.

10. R. G. Herb to L. A. DuBridge, p. 3, November 19, 1942.

11. This circuit is described in Britton Chance, M. H. Johnson, R. S. Phillips, "A Precision Delay Multivibrator for Range Measurement," Radiation Laboratory Report 63-2, June 1, 1942. See also Britton Chance, "Tests on SF Range Circuits," Radiation Laboratory Report 63-3/5/43. In this report Dr. Chance recommended that at some time in SF production the delay multivibrator circuit be further improved.

12. R. G. Herb to L. A. DuBridge, p. 3, November 19, 1942. "We understood from him (Cdr. Granum) that Navy specifications wouild be relaxed in order to speed up production, and throughout the development period Cdr. Granum continued his insistence on this policy of the earliest possible production dates, even at a sacrifice in performance and dependability."

13. L. A. Turner to J. P. Nash. "Report on Work of Group 61 on the SF Receiver," October 31, 1942. J. P. Nash, to L. A. Turner, "SF Receiver," November 7, 1942. L. A. Turner to J. P. Nash, "SF Receiver," November 11, 1942.

14. J. M. Wolf, "Report of Inspection Trip to Staten Island Operating Base September 4-10, 1943."

15. H. L. Lindgren and M. Kessler, "Report on Radar Behavior during Cruise of USS *Mohawk* in Northern Waters (Indian Project)," Radiation Laboratory Report 103—5/25/43.

16. L. A. DuBridge to the Chief of the Bureau of Ships, Attn: Comdr. H. E. Bernstein, No. 915, "Radiation Laboratory Consultation on Shipborne Radar Projects."

17. Navy Department, Bureau of Ships, Cdr. A. M. Granum, by Direction, to L. A. Du-Bridge, June 11, 1942. S-NXs-2297 (485-1A) Serial S-485-2523.

18. Minutes of the Coordination Meeting, Vol. II, p. 305, January 20, 1943, "Activity Report of SO," January 20, 1943.

19. Felix B. Stump, CO CV-16 to COMINCH, U. S. Fleet 0 V16/S67 Serial 007, n.d., "Zenith Watch Radar–Request for."

20. Minutes of the Coordination Committee, Vol. IV, p. 283, September 22, 1943.

21. Alan C. Byers, "The Operation of the First SO-11 in Fleet Tests," May 14, 1944.

22. D. B. McLaughlin to J. C. Street, "Termination of Project AN-2 (Cachalot), "September 27, 1945.

23. Edward L. Moreland (NDRC) to Dr. Alfred L. Loomis, June 19, 1943.

24. E. E. Miller and D. B. McLaughlin, "Use of the VPR for Matching a Map to the PPI," Radiation Laboratory Report 102—4/13/44.

25. NALOC Progress Report No. 4, February 15, 1944.

26. M. D. McFarlane to Dr. John E. Burchard, December 3, 1943.

27. Seventh Progress Report NALOC, submitted by John E. Burchard, Chairman, n.d.

28. John E. Burchard to Olin J. Stephens, August 26, 1944.

29. Interview, R. G. Herb, P. S. Clymer, and D. B. McLaughlin with H. L. Thomas, March 11, 1944.

30. M. D. McFarlane to Dr. John E. Burchard, December 3, 1943.

31. John E. Burchard to Members, Com. NALOC Log, April 4, 1944.

32. M. D. McFarlane to Dr. John E. Burchard, March 9, 1944.

33. BBRL Reference No. 39, July 25, 1944, E. E. Miller, "Cachalot. Part played by LCC's in Normandy Operation."

34. Gee is described in Chapter 38.

35. John E. Burchard to Olin J. Stephens, August 26, 1944.

36. *Ibid.*

37. L. C. Marshall, "Survey of SSV Radar Sets in Production and Design as of July 1, 1942." This report included all existing Navy ship-search equipment regardless of origin.

38. Since these sets were unstablized, the "chopped paraboloid" was necessary. L. C. Marshall was a strong advocate of stabilization.

39. The standardization that was probably most successful was the almost universal use of the SG, SO, and SJ sets in forward areas.

40. Radiation Laboratory Notebook No. 790, issued to S. F. Johnson, pp. 40 ff, April 22, 1943.

41. *Ibid.*, pp. 72-73, June 7, 1943.

42. *Ibid.*, pp. 117–123, August 20, 1943.

43. Division Ten Notes, Vol. I, No. 5, p. 76, May 31, 1944.

44. S. F. Johnson, "Problems Concerning SU-2 Radar," June 1, 1944.

45. T. J. Keary and J. I. Bohnert, "SU-2 Antenna," Radiation Laboratory Report 659, March 7, 1945.

46. G. L. Stancliff, Jr., Lt. F. E. Swain, USNR, and R. J. Grenzeback, "SU-4 Stabilized Base Antenna System," Radiation Laboratory Report 101—5-18-45.

47. CIC, pp. 44-45, October 25, 1944.

48. Cdr. H. E. Bernstein to Coordinator of Research and Development, January 27, 1944, "Request for NDRC project on high resolution X-Band set for small craft at Radiation Laboratory." (915) S-67-5 Serial No. 07310.

49. Minutes of Ship Committee Meeting, January 21, 1944.

50. Division Ten Notes, Vol. I, No. 2, p. 23, February 24, 1944.

51. George Hite, "Medium Precision Range System for CXGQ (Project Henry)," Radiation Laboratory Report 579, September 9, 1944.

52. Division Ten Notes, Vol. I, No. 6, p. 96, June-July-August, 1944.

53. Minutes of the Coordination Committee, Vol. VII, pp. 376–7, November 29, 1944.

54. Results were reported by the Commander Amphibious Training Command, in a memorandum dated January 6, 1945, File No. FE 25S67/RD163) Serial 0014; cited in R. H. Emberson to John Exter, September 15, 1945, "Summary of Work on Project NS-250."

55. Minutes of the Coordination Committee, Vol. VIII, p. 371, February 28, 1945.

56. See Chapter 20.

57. W. M. Fairbank, "Termination Report of Project CXJG (Cindy)," October 9, 1945; Memorandum, BuShips to Coordinator of Research and Development, November 12, 1943. S-S67-5(915).

58. R. G. Herb to Members of the Ship Committee, December 28, 1943, "Tentative Proposal for a K-Band Search Set"; W.M. Fairbank, op. cit.

59. Fairbank, *op. cit.* Note 57.

60. J. I. Bohnert, "Cindy Antenna. A High Resolution K-Band Radar Antenna for Sea Search," Radiation Laboratory Report 849, November 1, 1945.

61. W. M. Fairbank, W. T. Harrold, G. D. Scheckels, and J. D. Fairbank, "A High Resolution K-Band Ship Search Set," Radiation Laboratory Report 576, December 7, 1944.

CHAPTER 17

Ground Systems and Their Shipboard Counterparts

In this chapter we are to consider only those ground-based radar sets developed by the Radiation Laboratory which come under the headings of aircraft warning, harbor control or coast defense, and GCI (Ground Control of Interception) or height finding. Other ground-based systems, like anti-aircraft fire control radar or blind approach systems, have chapters to themselves. Nevertheless, some of the general considerations we shall offer at this point apply equally well to them.

The ground radar program at the Radiation Laboratory, important and profitable though it finally proved to be, was one that had to sell itself by sheer force of accomplishment to the Armed Services and to some extent within the Laboratory itself. Until toward the end, there was never the ardent interest or the sustaining pressure of Service demand that characterized the airborne program. There were various reasons for this, the principal one being that the Army was in possession of excellent long-range warning equipment on long waves and the advantages of microwaves—low coverage and fine resolution—were thought to be outweighed by the time and effort needed to develop adequate power on microwave frequencies; moreover, the Signal Corps Laboratories, who were felt to be the acknowledged leaders in the field, were given the principal responsibility for devising the lighter-weight equipment needed to supplement the SCR-270 and SCR-271. At no time did the Radiation Laboratory have the direct contact with the Army ground forces in the field, or the full understanding of its tactical problems and radar needs, that it enjoyed with the Air Forces.

Neither was British interest as strong in our ground radar developments as in our airborne work, though the initial stimulus for both did, in fact, come from England. The British, to a greater extent than the U.S. Army, were in possession of an elaborate and diverse accoutrement of ground radar. Their interest was confined to solving the problem of the proper type of centimeter-wave height finder to be used against low-flying planes; and, as it turned out, their own solution was the more successful.

Within the Laboratory itself the ground program was for some time almost a marginal enterprise, carried on because of the conviction and tenacity of the group leaders and division heads involved. For the first two or even three years, if not longer, the airborne projects were the center of attention. Not until 1944 did a ground radar project at any time command a top Laboratory priority. Belated enthusiasm for ground equipment was associated with the demonstration of the tremendous capabilities of the MEW (Microwave Early Warning); with the successful debut of the SCR-584 on the Anzio Beachhead and its dramatic success against the V-1 flying bomb; and in particular with the discovery of the uses to which ground sets of that sort, originally conceived as defensive weapons, could be put in offensive operations. The overall success of the ground radar program at the Radiation Laboratory testifies to the wisdom of the policy which permitted new ideas and new projects to flourish in the early years of the Laboratory's history even when they were not always officially popular, and which had set up an organization that allowed a large amount of local autonomy to the groups and divisions.

In several respects the physical requirements of ground radar are less stringent than those for radar installed in aircraft or on shipboard. Ground radar has neither stabilization nor pressurization problems which complicate the design, respectively, of ship and airborne radar. It does not have to withstand the shock of main battery fire or the effect of guns firing from the same movable platform. Space and weight considerations are less important, except in the case of equipment designed especially to be lightweight and portable. In certain Radiation Laboratory equipment, this freedom in the matter of size and weight permitted the use of extremely large antennas and of high-power components, modulators, etc. The large antennas made possible extremely fine angular resolution even when operating at the lowest microwave frequency (10 cm); while the use of this wavelength, for which microwave techniques were most fully developed, made it easiest to develop high power.

The air-warning effort in the Radiation Laboratory was chiefly in the direction of high power. A group formed late in 1941 put into the field the following year a high-power GCI radar which did not perform as creditably as had been expected, but which did serve to introduce new techniques in system design incorporated into later and more successful systems. During 1942–1943 an extremely successful set, the MEW radar, was under development; a number of crash units were built during 1944 and before the end of that year production equipment had begun to appear. The outstanding feature of this system was the novel antenna design, in which extremely high gain was obtained by means of a long dipole array. Like the systems which followed, it included in addition such important features as preplumbing of the rf section between the magnetron and the TR box; back-of-dish mounting of components; and the use of a complex and flexible system of indicators.

From 1943 on, the Radiation Laboratory had to be reckoned with as a force in the Army's ground radar program. In November 1942, a conference was called by General Saville, Director of air defense, Headquarters, AAF, to discuss the problems of radar for air defense. The meeting was attended by representatives of interested agencies, including the Radiation Laboratory. The stated objective was to establish a closer liaison between the operating and using branch of the Army Air Forces and the research and development laboratories of the country. At this meeting two facts were brought out: that urgent warning from the British had indicated the vulnerability to jamming of all fixed, low-frequency radar systems; that the Air Forces wished to see the radar program adapted to a highly mobile phase of offensive warfare.

An important consequence of this meeting was the establishment, under the chairmanship of J. A. Stratton of the Office of the Secretary of War, of a Technical Committee on Air Warning on which were represented the principal civilian laboratories developing radar. K. T. Bainbridge and J. C. Slater were the Radiation Laboratory representatives on this committee, the main function of which was to prune away the overluxuriant growth of ground radar projects and to recommend a clearly defined program for the Army. The Committee reported its final conclusions in June 1943. The report gave its approval to the Radiation Laboratory ground program and urged, in addition, that the Laboratory undertake a 10-cm height finder. The support and interest of the Army Air Forces and the shift to thinking in terms of offensive war had an important effect upon the standing of the ground radar program.

Some of the functions of ground equipment have their shipboard application. Shipboard warning equipment and fire-control radar are discussed in other chapters; but adaptations of GCI equipment to use on shipboard was an important aspect of the high-power development at the Radiation Laboratory.

Fighter direction from carriers implies the control of the ship's own planes and the interception of enemy bombers and fighters as well as "snooper" aircraft trailing a ship or task force. A radar system suitable for this purpose must have wide angle coverage; it must be effective at close range and at long range; it must give height data on aircraft; it is also expected to keep track of surface vessels and must have large traffic-handling capacity. Aside from long range, all the requirements point to microwave equipment of high output power. Height finding on shipboard calls for elaborate stabilization. The simple ship-search system can work effectively if its stabilization is accurate to within a few degrees; but a fighter direction radar set must be stabilized to permit target designation in azimuth and elevation within $\frac{1}{4}°$.

Not all of these requirements could be met in the first radar system of this type designed by the Radiation Laboratory. In fact it was intended to work in conjunction with a long-wave search set and its height finding turned out to be comparatively slow and unreliable at low elevations. It is also something of a paradox that the higher resolution of this equipment at first brought complaints because clouds and echoes from rain squalls gave what were considered spurious signals. In the Central Pacific, however, this was eventually turned to good account by making it possible for a carrier to spot squalls miles ahead and then change course to avoid them while taking its aircraft aboard. Further development of high-power components and experience gained with ground radar interception and height-finding techniques made it possible for a highly satisfactory system to be in production at the end of the war.

Coast Defense Radar

SCR-582

So successful was the XT-3 equipment in aiding the work of the HECP by supplying accurate range and bearing on all ships entering or leaving Boston harbor (see Chapter 10), that after December 7, 1941 this experimental Laboratory equipment remained on a 24-hour duty for several weeks.

Although the XT-3 equipment was badly needed to serve as a prototype for production equipment, the commanding officer at the HECP refused to relinquish it until the Laboratory provided a suitable substitute. In January 1942 a duplicate system, called the HECP system, was assembled in three weeks time under the direction of E. L. Hudspeth. The HECP radar was not truck mounted but it was otherwise very similar to the XT-3 equipment. Principal differences were in the use of a 30-in. Sperry spinner, a Raytheon modulator, and the important changes to crystal mixer and Sutton Type TR-box. The two major changes marked a shift in Laboratory practice resulting, as previously indicated, from reports on British developments.

Colonel R. V. D. Corput, Director of the Camp Evans Signal Laboratory, one of the many visitors from the Army and Navy to whom XT-3 had been demonstrated, decided that the Army needed a permanent radar installation based on the XT-3 design. Samuel Seely, the Radiation Laboratory project engineer for XT-3, visited Ft. Hancock in December to confer with the Signal Corps on specifications and after a further conference on December 30, 1941 the Signal Corps ordered a crash production of 50 sets.

The production of these sets, to which the Signal Corps assigned the designation SCR-582, was undertaken by the Research Construction Corporation, NDRC's factory-sized model shop, which went into operation late in 1941. Dale Bagley and Curtis E. Abbott were assigned by the Laboratory to assist RCC, and R. M. Emberson was appointed Chief Engineer to expedite production.

Both the XT-3 and the HECP radar system influenced the design of SCR-582, for the final design made use of the crystal mixer and the Sutton TR-box. Among the new features were a 4-ft spinner, a Radiation Laboratory 60-Hz (modified SG) modulator, a Stromberg–Carlson receiver, and a selsyn-amplidyne spinner-indicator linkage. Although the SCR-582 was to be built for fixed installations, provisions were made so that it could be truck mounted if the Army needed a mobile version. Another important innovation was the use of "preplumbing", that is, the preselection of the length of line from the magnetron to the TR box so that the received signals would go through the TR box with no appreciable leakage of energy back to the magnetron.

The first production SCR-582 replaced the HECP radar at Deer Island on June 29, 1942. A permanent structure had been built there to house it. The second was installed at Ft. Story, Virginia, on the Cape Henry side of the entrance to Chesapeake Bay. Second Lieutenant C. E. Abbott, who had meanwhile been commissioned in the Army and reassigned to the Radiation Laboratory, was sent there by the Laboratory to assist in the initial operation of the set.

In the summer of 1942, E. L. Bowles visited the Panama Canal Zone and found General F. M. Andrews anxious to have the Caribbean Defense Command used as a field laboratory for testing new microwave gear. Soon afterwards L. A. DuBridge and L. C. Marshall visited Panama and recommended that a SCR-615 and two modified SCR-582 sets be sent to the Canal Zone to supplement the longer wave early warning systems already in operation there.

The Laboratory project to provide these systems for Panama—which constituted the Laboratory's first venture into overseas field service—was given the code name Yellow Jack. In addition to supplying radar sets the Laboratory agreed to send Laboratory members to supervise installation, early operation and training of military personnel. Two SCR-582X sets were sent to Panama to be incorporated into the Air Warning System set up by the 26th Fighter Command, AAF. The "X" sets were equipped with a 6-ft spinner and were the first radar systems to use a spark-gap modulator. R. M. Emberson was project engineer in charge of the two SCR-582X stations which were located at Ft. Sherman (Station 61) and at a site five miles east of Santiago (Station 63), Panama. Emberson arrived in December and returned to the United States in February 1943. He was accompanied by C. J. Black, C. Hopkins, D. J. Shekels, and C. A. Smith. Later he was joined by E. L. Hudspeth, C. E. Moore, and S. F. West. The sets which were shipped from Cambridge in November did not arrive in Panama until January 1943. On June 7, 1943 the NDRC transferred these two sets and the SCR-615, which was also in the Caribbean Defense Area, to the 26th Fighter Command, AAF.

In addition to providing the only effective radar warning against low-flying aircraft in the Caribbean Defense Area, these two sets performed another important function. From September 1943 storm detection by means of radar was a regular part of the weather service in the Caribbean (Sixth Weather Region). Up to September 1944 over 7000 rain areas were reported by weather observers and Air Warning Station crews. The SCR-615 and SCR-582 stations all contributed to this program.

One more SCR-582X set, known as the "super-X", because it had a 10-ft dish, was built for installation at the Lakehurst Naval Air Station. Ten other sets were built for the Signal Corps destined for mobile use. These were called SCR-582, Mark III, and later received the Army–Navy designation AN/MPS-2. Cleveland Hopkins served as project engineer, relying heavily on the advice of Captain J. D. Coultrop who had been sent back from North Africa to obtain mobile radar sets. Special features of the Mark III are a push-button control to replace the amplidyne control mechanism and an expanded A scope. Three of these were requested for overseas shipment by January 1, 1944. Modification of the vehicles for nine units was done by CESL and the sets were shipped from Cambridge to Camp Evans for installation by February 1944. The tenth SCR-582 Mark III was sent to the AAF Proving Ground at Eglin Field, Florida, to be used in radar countermeasures work being carried on there by the Radio Research Laboratory of Harvard University. Nine units saw service in both the Mediterranean and European Theaters; four were shipped to England in the spring of 1944.

Five of the original crash units accompanied the American Forces in the invasion of North Africa and were the first microwave ground equipment to see action. The five SCR-582 sets with operating teams arrived at Oran about January 27, 1943. One set was installed on the top of a grain elevator in the harbor of Casablanca, and became operational on March 6, being used principally to direct shipping through the mine field. The second was installed near Fort Limon (Oran) and became operational on February 15. This set reported convoys to the port authorities so they could ready the harbor.

Three sets were loaded in 2½-ton trucks and moved to Phillipeville where they were made mobile in the shops of the Royal Navy (see Fig. 17-

FIG. 17-1. Mobile SCR-582 being set up on Algerian seacoast, early in 1943.

1). The first unit was sent to Tabarqa by February 22; the second was sited at Phillipeville on March 1; the third at Cap Degarde to cover the harbor entrance at Bône, by March 11. The principal use of these units was defense against mine-laying by German E-boats and aircraft, but they also served as a navigational aid for convoys.

Experience with the SCR-582 led the Army to request a further modification for use by mobile seacoast artillery battalions. Such organizations are semimobile and often occupy the same position for weeks or months. Radar for these battalions should be provided in transportable rather than mobile form.

Toward the end of 1942, Crosley Radio Corporation was given an Educational Order by the Radiation Laboratory for the development of five SCR-682 sets. The Signal Corps followed this with a further order for 45 sets. R. P. Scott was designated project engineer by the Laboratory to follow the development and production. Crosley completed its prototype model on April 10, 1943. This higher-powered set differed from all other modifications of SCR-582 by having a back-of-the-dish mounting of the rf components. This set was carefully engineered to adhere to rigid Army specifications. Its total weight was about 1500 lb. and no single unit weighed more than 400 lb.

The SCR-682 saw service in the European Theater and in the Pacific. An instance of its use in ETO shows the potentialities of this set. During the first two weeks of July 1944 a SCR-682 was located northeast of Catz at the time when the front line was but a few miles away, the town of St. Lô (15 miles south) still being in enemy hands. The radar set observed a stream of objects moving towards St. Lo from a point 75 miles to the southeast. A plot covering a 4-hr period was made and it was concluded that a large German convoy was moving toward St. Lo. This was reported to G-2, U.S. First Army. It was later confirmed that the SCR-682 had observed reinforcements against the American sector. This report had served as a valuable check on tactical air reconnaissance.

After trials at Oahu in 1944, the Coast Artillery Board pronounced the SCR-682 to be the set best suiting the requirements for general seacoast and harbor surveillance purposes. This type of set played only a modest supporting role during the war—it was after all a navigational and a defensive weapon in a war that soon shifted to the offensive.

HR (SCR-802)

The HR lightweight hand radar was the last of a series of range-only radars, development of which was instigated by a Navy request in the spring of 1942 for a radar to supply range (50-yard accuracy at 5000 yards) for use by the Sperry Draper sight. This set, the RO-1, was to be self-contained and operable on all types of ships. Although the Navy made the first request, it was the Army that was mainly interested in the HR.

The RO-1 was tested by the AA Artillery Board at Camp Davis, North Carolina for two weeks in June 1942, and then demonstrated by the project engineer, G. E. Valley, to the British at ADRDE in the autumn of 1942. When Valley returned to the United States, the set was left with the British.

After tests of the RO-1 had been completed, work began on the RO-1-Z, a portable set, assembled at the Radiation Laboratory (certain parts being built at Zenith Radio Corporation) under the supervision of the Laboratory with the assistance of Zenith representatives. This was the first instance of this type of cooperation between the Laboratory and an industry. The RO-1-Z was to be the prototype for five GPR sets (General Purpose Radars), an NDRC contract having been given to Zenith the previous May. The set used medium power at 10 cm, a PPI scope for indication, and a 30-in. paraboloid which gave a beamwidth of 6°. It used the first vest-pocket-type pulser ever made. Scanning in azimuth and elevation was by hand.

The RO-1-Z was to be used to supply range for fire-control purposes on beachheads, harbors, and airfields. The set was delivered to the Laboratory in August 1942 and installed on a T-13 experimental director, which fed range to 37- and 75-mm guns. This director, manufactured by the Frankford Arsenal, had been proved to have insufficient range accuracy by itself and the Army hoped that the addition of the RO-1-Z to give range would improve its accuracy. The unit was first tested at Camp Davis and then at AAFSAT, Orlando, with poor results.

Following the tests of the RO-1-Z, the Air Corps proposed an HR (Hand Radar) set similar to the RO, but able to furnish 20-mile ranges and serve as an all-purpose radar wherever such range was useful. Conferences at the Laboratory in mid-October 1942, attended by representatives from Zenith, Radiation Laboratory, the Air and Signal Corps, the Ordnance Department, and NDRC, served for discussion of the characteristics of such a set. G. A. Fowler was appointed the Laboratory project engineer; the contract for five sets was extended to include ten units originally on a Radiation Laboratory purchase order, making a total of 15 HR units. Another conference later in October, attended by Mr. E. S. Shire, ADRDE, was held for discussion of possible applications of the HR to a British director.

The first production HR was received at the Laboratory on December 29, 1942. This was a portable, one-box set, mounted on a tripod, and weighing about 200 lbs. The minimum operational and maintenance crew was two men. With a crew of three men, the HR could be set up and operated in eight minutes, provided that a power plant was already available.

The 15 HR's produced by Zenith were fundamentally alike, differing chiefly in details. They used the GPR modulator and receiver. In particular, the indicator system was a main point of difference. Some HR's used 5-in. PPI scopes but the majority used a double scope system consisting of a B scope and a J, or circular sweep, scope for fine positioning in range. These were 3-in. tubes, with the J scope giving range accuracy of ± 20 yards. Elevation positioning was by means of a pointer attached to the unit box.

Of the HR's produced between the end of 1942 and the fall of 1943, three were used in theaters of operation. One of these sets, HR No. 2, was shipped to Panama in February 1943, and there made operational 30 minutes after arrival. The set was carried about on a jeep and used for tactical and siting purposes. One of the sets for which it helped determine the site was the SCR-615. In Panama, the HR traveled 2100 miles by land and 1500 miles by water, at times reporting land signals at ranges of

65,000 yards (looking over water). The set was shipped back to the Laboratory in May 1943.

Another HR set was put on board the USS *Lexington* in the spring of 1943 to serve as a standby set. Radiation Laboratory personnel accompanied it during the first week of operation. Title to the set was transferred to the Navy Department in March 1943.

In October 1943, the 11th HR set to be produced went to England to ADRDE, as the result of Shire's visit to America in October. In England, the set was used to test direction. The British originally intended to use a modified HR set to furnish range to anti-aircraft gun directors, provided they could buy X-band HR's to go with them. The Signal Corps had frowned on this idea.

With the exception of the above-mentioned sets, HR radar was used in the United States chiefly by a wide variety of groups for experimental purposes. HR sets were tried out at Fort Ord, California for use in amphibious operations; by the Laboratory at Deer Island for use as a stand-by set for the SCR-582; under the auspices of Johns Hopkins University for tests to obtain coincidence between plane and shell signals (5-in. shells were followed to aircraft targets at ranges of 6000-8000 yards); for training; at Deer Island for following shipping in and out of Boston Harbor; at Frankford Arsenal for use with various trackers and directors (made by Frankford Arsenal and by Eastman Kodak Co.); at Fishers Island for tests by the Marine Corps in locating men on a beach and men in rubber boats (this was an HR converted to X-band—with a resultant reduction in beamwidth from 6° to 2°—signals were obtained on men in rubber boats at 1000 yards); for meteorological tests (tracking weather balloons to 35,000 yards) (another HR conversion to X-band); and at the Radiation Laboratory for testing the possibility of detecting moving objects by the pulse Doppler principle.

An important experimental use of HR was the gathering of comparative data on the propagation of microwaves over water. For this purpose, one HR at 10 cm, and two others, converted to 3 and 1 cm, were used in tests. The conversion to 3 cm was done by the HR group, while conversion to 1 cm was accomplished by R. S. Bender and H. R. Northington of group 41. (The set converted to K-band was one that had been previously used by Zenith for type-testing.) In the late spring of 1943, the units were tested at Fishers Island for over-water effects.

In June 1943, an order for 25 HR sets (10 at 10 cm, and 15 at 3 cm) was given to the RCC. These were to be similar to the Zenith model except for seat-type mounts in place of tripod mounts. In October 1943, when termination of the HR project was requested by Division 14, as the sets "have apparently not fulfilled the Service requirements for tactical use," the order to RCC was canceled. "Henceforth, the HR group will be classified under general research and will be primarily concerned with the use of radar in detecting ground objects from the ground."

High-Power Microwave Radar

When K. T. Bainbridge returned from England in June 1941 after a three-month study of the British radar installations and immediate radar needs, he was deeply impressed with the importance of long-range early warning systems, and especially with the lack of suitable height-finding provisions. Bainbridge had observed the British long-wave (Type 598) set which had plenty of range for high altitudes but could not give accurate height determination which was specified by the British to be ~500 ft. J. L. Lawson, who was also in England during July–August of the same year, stated the need for wide vertical coverage.

In mid-September the High-Power Group (Group G) was formed under Bainbridge's leadership to explore the possibilities of high-power radar (which then meant radar handling any peak power well in excess of 100 kW) and to design an experimental system with height-finding features to serve as a GCI and under certain conditions as a low-coverage general warning set. In this endeavor E. C. Pollard, who was later to become leader of the High-Power Group, acted as coordinator of the project, which by the end of the year had increased its staff to nine.

Bainbridge later laid down the requirements of a 35-mile range and 1% accuracy in height finding. This immediately raised problems concerning the high-power magnetron to be used, the kind of modulator, and the means of handling high power. The equipment was to be large, with the biggest reflector considered up to that time.

The group had been hoping for 500-kW peak power output but were willing to get along with 350 kW. At that time, Fisk, of BTL, was working on magnetrons which they hoped might give the desired power. These were not necessary, for just at this time the British announced that an ordinary magnetron could be run at 500 kW, if the pulse recurrent frequency were reduced from 2000 (AI frequency) to about 800 (search frequency). The magnetron then available (the Collins right-angle strapped K7RF4) used in this manner turned out to be very good. Ultimately, Raytheon manufactured this magnetron from designs furnished by the Radiation Laboratory.

Bainbridge worked with M. H. Kanner, of the Modulator Group, in designing a high-power modulator, which was needed within three months. M. G. White, head of the Modulator Group, announced that they would have to use their hard-tube modulator circuit in order to meet this date. The Model 9, made by the Link Company, was developed to go with this circuit. Later, this was replaced by a rotary spark-gap modulator, which was smaller and lighter.

In view of the high power contemplated, waveguide, instead of coaxial line, was chosen to carry the rf energy because of its smaller attenuation and its greater ability to handle power without sparking. High power also introduced the question of crystal protection. From England, favorable reports were received on the CV-58 diode mixer which was attractive because there was reason to fear inadequate crystal protection at 500 kW. Andrew Longacre developed the resonant slot TR box, called the "beetle," and J. S. Hall the waveguide attachments for the diode, which were used in the original model. Bainbridge drew up the original plans for the rotating joints to work at high power, by scaling them up from drawings of successful joints at 3 cm. E. M. Purcell and M. H. Preston of the High-Frequency Group also worked on these. Longacre finished the joints, which were satisfactory except for their narrow bandpass, and afterward, with H. K. Farr, worked out the magnetron coupling into the waveguide. Later, around April 1942, Farr joined the rf Group and J. R. Zacharias, the Group Leader, adopted these techniques.

Receiver sensitivity, in the early days around 22 dB (including the TR), was afterwards improved to about 12 dB, using crystal information from L. D. Smullin of the rf Group and M. Chaffee who was working on the MEW system. Much of the crystal information, which was adapted to the grown-up HPG system, came, in fact, from the MEW system. MEW, in turn, both borrowed from and contributed to the HPG system. The incorporation of the MEW group in the High-Power Group in 1943 enhanced the interchange of techniques between the two systems.

The experimental system built on the roof, called HPG (High-Power Ground) used a 6-ft reflector (later 10 ft, and finally 8 ft) carried on a Sperry 60-in. searchlight mount. A Sperry thyratron control was modified for automatic following. PPI echoes on the first HPG system were obtained in the first week in February 1942, on airplane targets at about 25 miles. By March 1942, the HPG system was performing quite well.

In the next few months it was resolved to incorporate two important features: (1) conical scanning for precise positioning in both elevation and azimuth and (2) instantaneous presentation of height. Since the height of the target is given by multiplying the slant range to a target by the sine of the elevation angle of the antenna, electrical circuits were designed which performed this computation automatically, so that height, corrected for the effect of the earth's curvature, could be read directly and continuously on a meter.

The roof system was successfully tested as a GCI system against the laboratory B-18 plane on May 15, 1942. Service interest was suddenly focused on the HPG development in the early part of 1942, soon after the United States entered the war. The Navy was the first to show interest. I. I. Rabi, of the Laboratory, who was directing work on the 3-cm AIA set, invited Commander Stedman Teller (OpNav) to visit the Laboratory for an inspection of the HPG system. Since the AIA had a range of only one or two miles, it was believed that an accurate SCI (Ship Control of Interception) System with a 35-mile range would be useful to control it. Commander Teller was greatly impressed by the work, and as a result of this visit the Laboratory was directed to develop a height-finding and general control set for aircraft carriers. Eventually after passing through the stage of a prototype system, called the CXBL, this resulted in the Navy's SM radar.

SM

In the spring of 1942, preliminary specifications for the shipboard SM were completed in conference with Navy representatives headed by Lieutenant Commander S. M. Tucker and in the summer a letter of intent was given to General Electric Company for the manufacture of 25 sets. Specifications included dependable detection in range of three fighters or one medium bomber at 35 nautical miles, though if possible detection of these aircraft at 50 nautical miles was desired. Work was begun on the prototype CXBL at the Laboratory. Andrew Longacre and M. B. Karelitz were the project engineers for the prototype system. J. S. Hall was project engineer for SM. The CXBL was used not only for testing components and design features of the SM, but also, installed on the USS *Lexington* (CV-16), contributed to the development of tactics under actual battle conditions.

An experimental HPG system was set up in July 1942 at a field station, Spraycliff Observatory, on Beavertail Point, Jamestown, Rhode Island, near the Naval Air Station at Quonset Point.

The field station antenna height was adjusted to the equivalent antenna height of the future SM on board a CV-class carrier. HF and VHF equipment, and a borrowed SC (150-cm) search set, were also supplied in order to provide a complete operational unit for test with SNJ, F4U, and JRB naval aircraft. The Spraycliff HPG was never a complete CXBL. As finally assembled, it was a mixture of three systems; the mount was from SCR-615, the modulator from CXBL, and the console from the SM production model. This HPG System was not used primarily for components testing but it did serve an important function in working out SCI techniques. At one time Spraycliff provided information which directed a rescue vessel to two boats which foundered in a storm. The installation was used in addition for training fighter director officers.

The SM was to operate at 10 cm, with a 500-kW peak power output. It was eventually made for scanning circularly in azimuth at a rate of 6 rpm, with elevation angle adjustable from $-3°$ to $\sim 75°$. The set could also scan helically through a total elevation angle of 12', this angle being chosen anywhere within the $-3°$ to $\sim 75°$ range. For accurate positioning, the SM used conical scanning.

On August 6, 1942, at a conference with representatives of the Bureau of Ships, it was determined that the Laboratory was to furnish General Electric with circuit drawings, pilot models of components (chiefly rf and receiver components), and parts lists of circuit components. The company was to furnish monthly reports to the Radiation Laboratory and to the Navy. In order to shorten the time customarily spent at the Naval Research Laboratory in testing new equipment, it was proposed by G. P. Stout of the Bureau of Ships that two SM units be completed ahead of the others, so that one of these could undergo mechanical tests at NRL, while the other was being given an overall performance checkup at some other Naval installation. Actually, because of the pressure of time, the first few SM's were tested at the factory and then installed directly on ships.

At the conference, it was also pointed out that many of the SM's would be put on new carriers which would be equipped with a special mount to anticipate the SM radar base. In effect, this mount would be a hollow, raised sub-base, which would allow removal of internal assemblies for replacement, since in the case of the new carriers it would not be possible to have a deck opening under the radar set for this purpose. Installation time on the new carriers would also be shortened, since much of the necessary wiring could be done while the ship was on the ways.

As the method of quickest production of the radar antenna mount, it was to be manufactured on subcontract by the American Machine and Foundry Company, who had also made the CXBL mount. A conference was held at the company on September 18, 1942 to acquaint Navy representatives with details of the mount as used on the CXBL. This mount was to weigh about 5000 lbs. The installation on a rolling ship of narrow-beam, microwave SCI equipment brought with it the acute problem of stabilization. With earlier wide-beamed ship sets, errors caused by rolling and pitching of the ship were not sufficient to lose a target, or to leave regions of sky unscanned. With narrow-beam radar sets, however, there was need to keep the beam in a closely defined direction, if the full accuracy of the set was to be realized. The importance of this problem was recognized as early in the history of the SM as March 9, 1942, when a conference was held at the Ford Instrument Company, to discuss the stabilization. General features of the problem as presented by the CXBL were described.

There were three possibilities for radar stabilization systems. The two-axis system corrects for roll and pitch of the ship, but does not correct for deck tilt. This system gives poor azimuth data. The stable-base system supports the radar set as if the ship were perfectly stationary; is accurate to a few minutes of arc and can be used with high rates of scan, but has the disadvantages of being heavy, complicated, and expensive to build. The three-axis system is lighter, cheaper, and less complicated than the stable-base type, and although not so accurate, is accurate enough (10 to 20 min of arc) for the comparatively low rate of scan (6 rpm) of the SM.

On August 20, 1942, a conference was called between representatives of the Bureau of Ships, the Ford Instrument Company, and the Arma Company, to discuss the stabilization—particularly as regarded the RASD (Radar Antenna Stabilization Director) project which had been undertaken by the laboratory. The stabilization system used on the first few SM's consisted of parts from Ford, Arma, and the General Electric companies. It was accurate to a few minutes of arc but was complicated and expensive. Another disadvantage was the short lifetime of some of its parts, especially under the conditions of 24-hr use which would be required of them. All of these difficulties were eliminated by the development in the Radiation Laboratory of the RASD system.

The original RASD project was in the design stage at the Laboratory in the fall of 1942, with W. B. Ewing as project engineer. The advantages of this system, which combined stable element and computer (widely separated components in the General Electric system), were compactness, cheapness, and long life. The accuracy was not so great (about 20 min of arc) as the General Electric system, but was sufficient to meet requirements. This first RASD was not designed as a production prototype, but as an experimental unit which would indicate the order of accuracy to be expected from the method. A contract was placed by the Laboratory with the International Business Machines Corporation early in 1943 for the construction of one model, but the instrument was not delivered until the spring of 1944.

After the Laboratory had started work on the RASD, Westinghouse Electric and Manufacturing Company began to develop a production version of the system, which proceeded so quickly that units were installed on some of the earliest SM's, as well as on future sets. In fact, a Westinghouse RASD system was installed on an SM equipment before the Laboratory's RASD model had been delivered by International Business Machines.

Frequent and belated changes proposed by the Laboratory and by the Navy during the spring of 1943, however, caused the delivery date of the SM to be shifted ahead, until in June 1943, the first SM was promised for delivery some time in October. These last-minute changes included such important features as substitution of an 8-ft for a 6-ft reflector, and of a crystal for a diode mixer. Although the Laboratory made no conversion kits, it did furnish the company with several crystal mixers for the first few SM's.

Before the end of March 1943, the CXBL equipment was completed and installed by Navy and Radiation Laboratory personnel on the USS *Lexington* in the Charlestown Navy Yard. Radiation Laboratory members who went along for the sea tests in April were Lt. Comm. L. P. Tabor, A. C. Byers, and H. W. Royce. This experimental unit saw extensive use in the Pacific before being displaced by SM production equipment.

The first SM set was scheduled to go on the USS *Bunker Hill,* a new carrier which had finished her shakedown cruise and, in the summer of 1943, was waiting in Boston Harbor for the radar set before proceeding to the Pacific. The scheduled sailing time was the first week in September, although the Captain told the laboratory's project engineer that he was so eager to have the SM on board that he would hold up the sailing date a week, if necessary. Hall was informed by the General Electric Company that the only way to speed up SM production was to have the present precedence rating changed, to obtain the services of a Navy expediter, and, finally, to step up the morale of the company by furnishing a letter from "an admiral" telling of the extreme need of the equipment. The expediter and the precedence rating came through, and a letter written to General Electric from Admiral Horne cited favorable operational reports received from the CXBL installation on the USS *Lexington* and asked for the "first SM" in September and two or three models as soon thereafter as possible."

By August 24, 1943, the first SM was dockside, and by September 4, installed. It was "debugged" en route to the Pacific. The second SM, delivered early in October, was put on board the USS *Enterprise,* and thereafter production SM's were delivered at a regular rate. By the middle of November, four SM's had been delivered to the Navy; by the middle of February 1944, 13 had been delivered. By October 1944, production was nearly complete. Out of the 25 sets requested, 23 or 24 had been supplied.

SP

From the SM equipment came the SP and the SM-1 sets, both lightweight versions. The SP was thought of as early as the fall of 1942, when Bainbridge conferred with the Bell Telephone Laboratories about the possibility of building a lightweight SM. Bell refused the project, however, not only because they thought the Navy specifications could not be met, but also because they were occupied with production of fire-control sets. General Electric finally contracted for the production of 50 SP sets.

Although the SP was thought of as a lightweight SM, the two sets had in reality few electrical components in common. An important difference was that the SP used two pulse lengths, 1 and 5 μs, and was able to switch from one to the other in about 3 sec. The 1-μs pulse was used for better discrimination, the 5-μs pulse for greater range. Use of two different pulse lengths required a new receiver in which it would be possible to switch bandwidths, and also a new modulator which would furnish greater range and provide the two different pulse repetition frequencies (600 and 120, respectively) for the 1- and 5-μs pulse lengths.

Two versions of the SP were built, using a 6- and 8-ft reflector; delivery of the former was expected to begin about the middle of 1944, delivery of the latter in 1945. P. J. Rice and J. S. Hall of the Radiation Laboratory followed the sets through production. More than 150 SP's were finally produced. The 6- and 8-ft sizes weighed, respectively, 1700 and 2300 lbs, an important saving over the SM's 5000 lbs. The 8-ft SP had 40% more range than the 8-ft SM, while the 6-ft SM had about the same range. The smaller size SM was put on destroyers and light carriers, while the larger SM's were put on large carriers and battleships. The antenna and mount for the 6-ft SM was subcontracted with the Grinding Tool Company of Vermont, while the 8-ft size was made by the General Electric Company.

SM-1

SM-1 was the designation given to an SM set adapted to the requirements of the Royal Navy. The British Admiralty ordered 24 sets through Lend-Lease. The SM-1 used a 6-ft reflector on a lightweight mount weighting abut 1500 lbs. The sets had to be built for dc operation instead of the ac common to U. S. carriers. They were manufactured by General Electric Company. One SM-1 was delivered in 1943, and the rest in 1944.

SP-2

An oscillating feed was applied to the SM radar set to make the new SP-2. On January 27, 1945, L. A. DuBridge suggested to the Chief of the Bureau of Ships that a formal request be initiated, taking the form of an extension of another project, for converting the SP to provide height-finding facilities using a vertically vibrating feed, and an RHI type of indicator. In the SP-2, the vibrating feed replaced the conical scanning feed and gave the beam a rapid vertical scan of 10°. The Project Committee at Radiation Laboratory concurred on February 1, 1945, with plans for the conversion of the SP set to a range-height-finder. Conversion work at the Laboratory, however, had been going on since December 1944.

The application of a vertically oscillating feed enabled the SP-2 to find height on a number of different targets, without having to interrupt the azimuth scan. The new type of feed resulted, however, in a decrease of about 25% in range. Changeover of the SP's, as envisaged by the Navy, would be accomplished through modification kits, obtained from General Electric, and installed at advance service bases.

SCR-615

After the SM program had been started a parallel development was initiated to produce the SCR-615, a GCI set for the Army. Although SCR-615 was actually completed first, since the special plumbing and stabilization for CXBL were time consuming, the SCR-615 should be considered a by-product of CXBL.

In April 1942, the Roof Group HPG system was inspected by Army representatives, and a discussion was held on height finding and general GCI problems. Colonel Gordon Savile, Director of Air Defense, outlined the problems from the Army's point of view. It was appreciated that the HPG system was not perfectly suited for ground applications because of siting difficulties and because the angular measurements necessary for height finding really called for a large antenna, while the HPG system had been committed, by Navy requirements, to a comparatively small dish. The Army had high hopes for a Canadian set (which afterwards became the SCR-588) which combined the functions of early warning and height finding, but they feared that it might not get into production soon enough. A GCI was needed immediately and would also be useful in the interim for training purposes.

Accordingly, the Army decided to order a transportable version of the HPG system; and a letter of intent was given to the Westinghouse Electric and Manufacturing Company on August 13, 1942, to manufacture 40 sets under the designation SCR-615A. Radiation Laboratory delivered one of the HPG experimental sets to Westinghouse, in September 1942, to be used as a prototype. By October 1943, ten sets designated SCR-615T1 (since they were to be used for training purposes) had been delivered to the Army by Westinghouse. The remaining 30 were delivered by February 1944.

The Western Electric Company received a letter of intent for 160 additional SCR-615's. These were designated SCR-615B since several changes were introduced. They used a mount engineered by Westinghouse, simplified circuits, a different console, and a spot indicator. An automatic helical scan from 0°–12° was also introduced with provision for shifting this sector to a maximum elevation angle of 90°. The original letter of intent was modified so that only 70 were actually manufactured.

Two HPG installations were made by the Laboratory, one for AAF-SAT (Army Air Forces School of Applied Tactics) at Orlando, Florida, and the other in Panama, as a part of Project Yellow Jack, which had opposite effects on the Army's SCR-615 program.

The second HPG installation (if we consider Spraycliff to have been the first) was a fixed ground system set up in September 1942 at Kent Station of the Fighter Command School, about 20 miles west of Orlando. The site had been previously chosen by members of the Microwave Committee and Radiation Laboratory but was not ready so the HPG was located temporarily at Kent. E. G. Schneider was placed in charge of this project which was called "Orange Crush."

After a few weeks of operation it was moved to the permanent site on a hill about 30 miles west of Orlando. The permanent site was not nearly as good as the temporary site because of ground echoes. This site, to which the set was moved late in November, was referred to in Laboratory and Army parlance as "Betsie." Although generally the 615 worked well in its new location, there were numerous complications. The Army filter system in Florida received information from long-wave radar sets having little or no low coverage, with the result that only plots above a few thousand feet were usually recorded on the filter board. The SCR-615 at Betsie furnished range, chiefly, for low plots close in, and so cluttered up the filter system by picking up what were considered to be spurious plots, although actually they were often targets below about 5000 ft. The set also had very elaborate circuits and mechanisms, and was difficult to keep in repair. The Army did not like the height finding (by pip matching), because it was too slow for their purposes. The overall result was a scathing report by the AAFB, which slowed up the further development of a good SCR-615.

The Yellow Jack set, SCR-615X, was set up on Taboga Island (Station 62), in the Caribbean Sea, during January 1943. E. C. Pollard, the project engineer, arrived in Panama in November 1942 and returned to Cambridge in January 1943. Although the set, and its associated prefabricated housing and air-conditioning units packed in over 100 boxes, was ready for shipment in November, the boxes did not begin to arrive in Panama until December. Pollard's Radiation Laboratory crew who were responsible for installation and early operation consisted of B. R. Curtis, C. Rider, A. R. Tobey, and Lieutenant W. J. McBride, Jr. who had been trained at the laboratory and who had assisted in the development of the equipment. The site had been formerly occupied by a long-wave SCR-271 set which was useless in that place because of ground clutter on the side lobes.

Ground echoes were somewhat bothersome to the SCR-615 also but it was soon evident that the set had good low-flying coverage. Ranges of 80–

90 miles on aircraft were reported, and by June 1943, when the equipment was turned over to the 26th FC, AAF, some 700 targets per day were being turned in. In September 1943, along with the two SCR-582 sets, the SCR-615 was incorporated into the Sixth Weather Region Storm Warning Service in the Caribbean. It fitted in well with this program because of its ability to see rain clouds.

The attitude of the Caribbean Defense Command toward SCR-615 was extremely favorable. To a large extent this was the result of a flight which the Commanding General, Lieutenant General Brett (successor to General Andrews), made in an attempt to show up the whole canal warning system by coming in low, below the beams of the long-wave sets, part of the time at only 10 ft above the water. The SCR-615 was the only set which was able to track him in all the way. When he landed, fully confident that he had not been picked up, he was shown his track carefully reported for the whole distance by the 615.

The Army reports to Washington from both the Panama and Orlando field stations came in at about the same time. The Panama SCR-615 was reported to be about 85% percent efficient for daylight interceptions (measured in terms of number of interceptions completed versus number of planes known to be in the area.) Washington did not know what to do. A conference was held in the Office of the Chief Signal Officer in the Pentagon on December 9, 1942, to consider the Florida results and to determine further Service interest. The Radiation Laboratory was represented by Bainbridge, Pollard, Curtis, and Longacre. The Orlando set was criticized as being precise but too slow at height finding, and as not having enough range; it had been compared in operation and performance with a SCR-588 set, located about ten miles away. It was pointed out by the Radiation Laboratory representatives that it was hardly fair to compare an experimental and hurriedly-built system with one elaborately engineered. The unfortunate siting in Florida was also pointed out. The meeting ended rather inconclusively as far as the Florida model was concerned, except for the decision to continue the operational trials. On April 26, 1943, the performance of the SCR-615 at Betsie was improved, when three laboratory members installed an 8-in. dish and a crystal mixer. In August 1943, the Florida equipment was returned to Cambridge where it was installed in trucks. The mobile set remained at the Laboratory until the summer of 1944 when it was sent to Bar Harbor, Maine, to assist in tests for Project Cadillac.

Discussions between Army and Laboratory representatives continued during 1943 on the operational characteristics of the production set SCR-615B. On June 7, 1943, Major General C. C. Williams, Hq. AAF, requested that the Laboratory be retained as consultant. It was decided to adapt the Chrysler mount, used on the SCR-584, to the improved SCR-615.

Simplified range circuits and improved data and plotting presentation were planned, as well as an 8-in. dish and provision for IFF. The new range was expected to be 75 miles. Sixty-eight of these sets were delivered to the Army by July 1944.

Of the two other experimental HPG systems built by the Laboratory, one went to Camp Evans at Belmar, New Jersey, where it was used for training purposes. The last HPG was sent to England in answer to a request from Air Vice-Marshall V. H. Tait, seconded by J. D. Cockcroft, at the end of 1942, as a possible solution to the problem of a low-coverage GCI. It was felt at first that one of the early production sets might be sent, but delayed delivery caused the HPG Group to build a crash model. This system, given the code name of Pea Soup, arrived in England on April 4, 1943, and was set up at Worth Matravers. Two Laboratory members, C. H. Rider and B. G. Farley, followed, and put the set in operation by about the first of July. The British were not satisfied with its performance as a GCI and subsequently used it chiefly for low-coverage early warning and for surface search. Pea Soup was used against sneak enemy raids, as well as for mine detection and ship rescue. Eventually, the RAF took possession of the set and asked for ten more through Lend-Lease. With production sets available by the summer of 1943, the HPG set at Belmar was no longer needed for training purposes, and so was released to the laboratory. This release was made known to the British, through BBRL, and the Air Ministry promptly asked for the set. It was shipped to England, where it was dismantled and used for spare parts for the Pea Soup set.

Of the production SCR-615's on foreign duty, one was used for warning by the Admiralty in a fixed location in Dover Castle. It reported on surface vessels, and was used during air attacks on enemy shipping. Two other sets in mobile form were sent to France. Two went to Corsica, one serving as height finder to the MEW on Cap Corse which covered the invasion of southern France, the other working as a GCI station at Ajaccio, where it was credited with bringing down a Ju-88. A 615 was used successfully for surface watch covering the entrance to Antwerp Harbor. Several more SCR-615's went to the Pacific. One was installed on Guadalcanal. Two members of the Laboratory, R. T. McCoy and B. R. Curtis, went with the sets to Corsica, and two others, I. H. MacLaren and G. F. Harpell, went to the Pacific to help with training and maintenance there. The SCR-615 was not widely used or thoroughly satisfactory as a GCI set; in fact, it did not see extensive service even as a warning device. The set was extremely complicated and difficult to maintain, though this was somewhat less true of production models. Its chief value was as auxiliary equipment.

Microwave Early Warning (MEW)

As a high-power set for early warning, a function for which it had not been primarily designed, the SCR-615 was soon outclassed by a new illustrious set called MEW, standing for Microwave Early Warning, which received the joint Army–Navy designation of AN/CPS-1. MEW was an independent outgrowth of the same interest in high-power sets which had given birth to the SCR-615. In particular, the set was conceived during the months immediately following the Pearl Harbor attack when it appeared likely that a powerful early warning system might well be needed to protect the West Coast against Japanese air assault or even invasion.

The chief limitation of ground early warning radars in 1942 was lack of range at sufficiently low coverage. This limitation often prevented warning of enemy bombers from being given in time to permit fighters to "scramble" and climb to the necessary altitude to intercept them. There were also the related limitations of site dependence and spotty tracking. M. H. Kanner, at Bainbridge's suggestion, went to England in the early winter of 1942 to study high-power technique and overall radar warning problems. Upon his return he drew "radar horizon" charts on which he plotted the minimum radar coverage needed for interception. These charts showed better than anything else the shortcomings of existing early warning sets. To produce the desired coverage it turned out that a fan beam 3° wide in the vertical plane and extremely narrow in the horizontal plane was desirable; while the beam was narrow enough in the vertical plane to give low coverage its breadth enabled it to sweep the sky without recourse to complicated scanning. The high gain required in the horizontal plane in order to attain the desired range gave the set the incidental benefit of extremely high resolution.

Among the men with whom Kanner discussed his findings when he returned to the United States in February 1942 was Luis W. Alvarez, then head of the Laboratory Special Systems Group. Alvarez was in charge of several projects, among them the 3-cm blind bombing system, EHIB, later known as Eagle, for which he had been trying to find an antenna that would give a fan-shaped beam and provide electrical scanning. It was his early work on this antenna which eventually led to his idea for the MEW.

Alverez's first antenna for the blind bombing system was a "leaky pipe" linear array, in which a succession of holes cut in waveguide, a wavelength apart, served as the sources of radiation. The first model leaky pipe was tested on February 18, 1942, and gave a pattern of one lobe and two secondary (or side) lobes about 45° away. Much time was spent trying to eliminate the two side lobes. The offending dimension was known from diffraction theory: the radiating holes, although one guide-wavelength apart, were more than one air-wavelength apart, allowing constructive interference in more than one direction. Finally, W. W. Hansen, on a visit to the Laboratory, suggested to Alvarez the idea of filling the guide with polystyrene, a dielectric material which would reduce the wavelength inside the guide and thereby allow the holes to be spaced closer together. By April 17th, Alvarez, with help from Randal M. Robertson, had built a leaky-pipe antenna that gave a narrow beam free from large secondary lobes. The difficulty remained, however, that it was not easy to vary the amount of radiation from each hole.

The development of the leaky pipe antenna for the Eagle project gave Alvarez the idea for MEW on May 9, 1942. In June, the plan to build a Microwave Early Warning radar received the approval of DuBridge and a special Early Warning Group was set up in the Special Systems Division then presided over by Alvarez. Morton H. Kanner was appointed project engineer. Alvarez indicated that an early warning radar for protection of the West Coast against Japanese attack should have the following characteristics: range comparable to the SCR-270, around 200 miles (high power, waveguide instead of coaxial line); low and gap-free coverage, site independence (microwaves); accuracy (microwaves, large antenna, short pulse length); large traffic-handling capacity (several flexible indicators, probably of the PPI type); close-in heights (wide vertical beam); ease of calibration (adequate test equipment, simple tuning procedures); electrical stability (preplumbing, AFC). MEW was to represent the first application of the leaky pipe on 10 cm. Ultimately, all of these features were achieved.

In early discussions Mt. Palomar, California, was suggested as a possible location for the first of a series of MEW's to be erected on the West Coast against possible Japanese bombing attack or even invasion. The discussions centered around a cylindrical parabolic reflector, 100 ft long by 10 ft high. It was planned to test the first 100-ft pipe, with holes a wavelength apart, on the roof of the National Guard Building at the East Boston Airport. The pipe was to be double-stub tuned, with an end-match-to-air previously designed by A. Longacre for the SCR-615. Eventually, the pipe was to be mounted on railroad type cars, one end on each of two cars, the entire assemblage scanning by revolving around its center.

The immediate problem was to eliminate the side lobes produced at 45° on either side of the main lobe. The first plan, to fill the guide with polystyrene, was not tried because it would have resulted in too much attenuation of the signal over the 100 ft length of the MEW array. On May 19, 1942, Alvarez announced what proved to be a practical solution to the problem, reversing the dipoles alternately along the array and spacing them one-half the guide-wavelength apart. By making the dipoles probe fed, the radiating power of each dipole could be adjusted within very wide limits.

On June 3, 1942, the first MEW array was made, using six wire dipoles, reversed alternately and spaced a half-wavelength apart. The dipole dimensions were chosen in the absence of data simply because they seemed logical. In general, development of the MEW antenna started from scratch, helped, of course, by the simultaneous work on linear arrays being done by Robertson.

The first wire dipole had important advantages over the hole radiator; it also had the disadvantages of low power capacity (small dimensions, plus sharp wing tips tending at high power to corona discharge), mechanical weakness, and asymmetry of pattern due to the unequal radiating lengths of the two wings. A second wire design was tried to overcome asymmetry by employing a reversed bond in the grounded wing to make the radiating lengths of these wings more nearly equal. An array of ten of these improved dipoles was used on June 12, 1942, to take the first antenna pattern in MEW history.

By this time consideration had to be given to assembling a system on the roof to provide data for component improvement. With this in mind, a 15 ft array using 64 wire-type dipoles was bench tested on June 29th. This was the first MEW array for which a definite illumination pattern (a modified sine function) had been worked out. The Theory Group later reported that uniform illumination was best for maximum gain but would give large side lobes as well. The eventual compromise in MEW's case was a gable-type illumination, determined by the maximum side lobes which the set could stand.

The final roof model of the linear array was 16 ft long; the reflector 16 ft long and 10 ft wide. The choice was made for two reasons. The early-warning radar coverage charts had shown the requirements. Although cylindrical parabolas were new to the Laboratory, L. C. Van Atta, Chairman of the Antenna Group, on the basis of his own research, was able to suggest that a 10-ft dimension would give about the required beam. The 16-ft dimension of the reflector was limited by strength considerations on the roof.

During August, finishing touches were applied to the Roof System. A gigantic plywood radome, raised above the second level of the penthouse of MIT's Building 6, housed the antenna and all the rf components. The indicators, power supply, and high-power Link modulator were in a room underneath. On August 26th, Alvarez announced that "all system components are on hand, and it now appears that signals can be looked for in the first week of September." Actually, this prediction was missed by one day. Alvarez reported later that "land signals were observed September 8, 1942, with Dr. Kanner's new MEW system."

The antenna problem had now resolved itself into the design of an efficient, high-power-capacity dipole. L. L. Blackmer, a new member of the MEW group, was assigned to work on the problem. Most of the work was done at 3 cm, instead of at 10, for two reasons: dipoles and dishes were easier to handle at the shorter wavelength; Robertson was developing dipoles for a 3-cm scanning array at the same time, and his data could thus be utilized.

The work on dipoles yielded five separate designs, the final one of which was used on the roof and in later arrays. This design, called the cut-off dipole, had the advantages over preceding types of symmetry of radiation, maximum coupling coefficient (consistent with MEW's operating frequency range), high power capacity, mechanical strength, and ease of manufacture.

Many of the characteristic features of MEW resulted from the necessity of handling large amounts of power, for it was hoped that a system could be designed to give a megawatt (10^6 W) of pulse power. The "back-of-dish" design is a case in point, for it made it unnecessary to carry rf power over long distances. Waveguide, which the earlier High Power Group had introduced for 10 cm, was even more important at the high power levels envisaged for MEW, for though coaxial line might have carried a few hundred kilowatts without arcing, waveguide could carry ten times that amount.

On the first MEW model, the back-of-dish mounting was a wooden shelf, holding all the rf components and fastened to the back of the reflector. In later models, the smaller spark-gap modulator, which replaced the Link modulator, was also carried back-of-the-dish, and the shelf was replaced by cabinets.

During the first days of operation on the roof, the mount was rotated by hand since a servo drive unit which was on order did not arrive for several weeks. A shift was about half an hour long; a group member sat on a box, and pulled the dish back and forth with a rope, while the others downstairs eagerly watched the indicators for signals. The maximum sector scanned was limited to 120° by trailing cables.

The weeks following September 8th were devoted to cleaning up and testing the new system. Beamwidth was measured on a standard ground signal by rotating the dish slowly in azimuth while the signal varied on the X scope downstairs. Although these measurements were rough, they were sufficiently accurate for Kanner to report on October 7th that antenna measurements were in accord with values predicted from the dimensions and illumination distribution.

Preliminary flight tests with the system showed occasional aircraft signals at 80 miles and fairly consistent good results at 40 to 50 miles. With improvement in tune-up of the system, however, a new series of flight tests was begun, the results of which indicated that the range of MEW was apparently limited by the radar horizon. On November 25th it was reported that "the most distant tracking has been made on a PBY at 17,000 ft...out to a distance of 177 miles."

This brief statement scarcely conveys the interest which accompanied this first flight. The MEW room was filled with roof personnel, all trying to watch a pip on a single 200-mile PPI. As the pip moved away on the scope, its range and azimuth were called out and pin-pointed on a map on the wall outside. The PBY was flying in a southeasterly direction which took it over the ocean beyond Nantucket. The pins followed, ran off the edge of the map, and started across the plywood wall. Over an hour after the flight started the last out-bound pin was put in, at 177 miles, 6 in. from the map border. Pointed out to visitors for months afterwards, this line of pins on the wall was testimony to the first real effort of the MEW.

On November 13th, an era of precison for MEW began with the installation of a test signal generator. Although the signal generator had been designed by B. E. Watt, especially for the MEW system, its characteristics and versatility were such that it was later adapted to other systems, among them V-beam and AEW. The signal generator was valuable not only in tuning the system but in flight testing, for it gave quantitative results that could be compared from day to day. The signal generator pulse, appearing on the A-scope as an artificial signal, could be used to measure the strength of target signals directly.

The flight test results also showed that an upper beam was needed for coverage at high altitudes out to the 100-mile range. The first such beam was an ordinary lobe shape, supplied by a 2-ft by 15-ft cylindrical parabolic reflector, which normally would be fitted on top of the main reflector but which, for clearance reasons, was mounted in front of the reflector on the roof. The beam lacked sufficient gain because the 3-in. waveguide array shielded too large a portion of the 2-ft reflecting aperture. Early in March, tests on a model of a csc^2 reflector indicated that this new beam shape (see Fig. 17-2) would give proper coverage above the main beam. The model consisted of a half-section parabola, with a horizontally polarized array at the focus, and a lateral reflecting strip just below the array to direct energy toward the dish. Horizontal polarization was required for reflection from the lateral strip.

By the end of March, a full size (5×15 ft) reflector was ready for high-level flight testing using the Beaufighter from East Boston Airport. When this shape proved to have undesirable vertical side lobes, Blackmer worked out an improved shape by bending the bottom edge of the reflector toward the feed an effective quarter of a wavelength. The gain on this dish

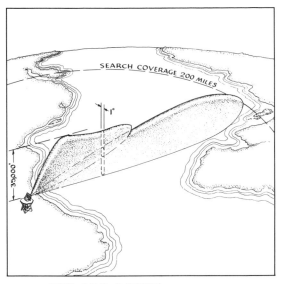

FIG. 17-2. MEW beam pattern.

was measured and found to check the calculated gain, and the pattern of a fixed signal was found to check the calculated pattern.

The antenna provided only the first of three main difficulties to be solved. The rf components and the indicators each furnished problems, which did not come to a head, however, until after the first MEW was in operation. These were of different natures, the rf problem being concerned with improving waveguide techniques demanded by a new order of transmitted power and the indicator problem with devising or choosing the types of scope best suited to the capacity of a new system.

The extremely high power levels gave rise to problems that had not been encountered, or at least not in such critical form, with low-power systems. Arcing of the magnetron input feed and serious crystal burnout were the most persistent problems. The latter was due to the unsuspected strength of one harmonic of the radiation frequency which leaked through the TR box and injured the crystal. The development of a special crystal mixer with a choke to eliminate this harmonic was a necessary and important step. The mixer, designed by M. A. Chaffee, was the first in the laboratory to work successfully at MEW output powers. The other rf difficulty arising from the magnetron feed, a simple probe-type antenna protruding into the waveguide, was more difficult to solve.

In the spring of 1943 F. F. Rieke, of the Magnetron Group, suggested that by use of a "doorknob" type input the magnetron could be coupled into the waveguide in such a fashion as to increase its output power. In effect, the magnetron was made to "pull", i.e., to operate at other than its stable frequency. This was possible in MEW's case because of the electrical stability of the large antenna and the protection of the receiver against small frequency changes (AFC); such coupling allowed an output power (on the first field set) of between 750 and 800 kW. Cold air had to be blown on the magnetrons to keep them from melting away. The higher power available with the field set revived the problem of crystal burnout. This time the trouble came from a large initial "spike" on the transmitted pulse which passed the TR box before it had fully fired. The trouble was cured by "prepulsing" the TR box, causing it to fire before the spike had reached it.

The indicator problems were concerned with selecting scope presentations best suited to the combination of narrow beam, slow scanning speed, and high power. The indicators used on the roof were designed and built before the system was actually put into operation and turned out to be inadequate to handle the unexpectedly large amount of information the MEW provided. They were a 7-in. PPI, with a maximum sweep of 100 miles, and a 5-in. A scope. The most necessary addition to these was a PPI with a 200-mile sweep, to cover the full range of the system. Shortly afterward, a "micro-B" circuit was added to the PPI chassis which enabled the sweep to be delayed in range, thus permitting an expanded annular view of the sky. This 7-in. PPI carried most of the burden of indication in the first months of flight testing. As early as July 1942, Kanner predicted that an expanded scale PPI would be the best solution. In later months, after a variety of indicator types had been tried, an expanded type of PPI was finally chosen for the production set.

It was evident in December 1942 that the experimental MEW had already outgrown its indicating system. Numerous flight tests had shown that a highly controllable expanded scope was necessary; the 100-mile PPI on hand could be delayed in range but was limited in azimuth expansion; also, a need was soon to arise for a scope that could be assigned permanently to the upper beam. By the end of the year, the Indicator Group was building a new system to fill these deficiencies and to obtain information on the kind of indicators to be built for the preproduction and production sets.

The indicating system, finished late in the winter, had four scopes designed to meet the new requirements. There was a 7-in. B scope with 50- and 20-mile sweeps which could be positioned in any 90° sector and delayed in range in steps of 10 miles. A 5-in. A scope had a fixed 200-mile sweep, as well as 50- and 20-mile sweeps which were controlled from the B scope. There was also a pair of 7-in. PPI's, one with 200-mile sweep for the lower beam and the other with a 100-mile sweep for the upper beam. Certain improvements in presentation and chassis arrangement were destined to be carried over into later indicators. Electronic angle and range marks were put on both the PPI's and the B scope. The scopes were mounted in console-type cabinets with tilted front faces for easy reading, and shelves were provided for the operators.

By the end of the winter, plans were underway at the laboratory for a field station MEW, to be erected at a still undetermined site. Florida was finally chosen because of its location within the AAF ground radar testing grounds, its flat country, and its climate. Developmental work on indicators became only part of the work of components for the station. The indictors, rf components, and the receiver either were being built or had been contracted for by different groups in the laboratory. The Link (hard tube) modulator represented the biggest change in components, being replaced by a rotary spark gap modulator, smaller, lighter, and delivering three times as much power (3 mW). The mount and reflector were being built by the Morgan Construction Company of Worcester. This size of the reflector, 8 ft by 15 ft, was determined by Army requirements for coverage up to 40,000 ft at the full range of the set. The reflecting surface was made solid instead of screened, despite the severe military requirements that MEW stand up in a 125-mph gale and operate in a 90-mph wind. Although a screened reflector seemed to be called for, measurements showed that a spacing which would appreciably reduce wind resistance would cause back lobes (serious because of the high power of the system and its probable location in mountainous country) and would also weaken the dish. In view of Blackmer's declaration that he could not tolerate torsional deflection of more than a quarter inch, a solid surface seemed the best solution.

Late in March 1943, B. E. Watt left for Florida to investigate possible sites for a field station. With components being finally assembled, it appeared that the shipping date would be somewhere near the middle of April. On April 19th, 1943, at a conference in the office of Colonel T. J. Cody at AAFSAT, Orlando, Watt reported on his three weeks investigation of the Gulf Coast region within the range of the Orlando filter area.

Since this was to be a testing site, the requirements were somewhat stringent: as high a location as possible (even though MEW was to be mounted on a 100-ft tower); a view in all directions, over water and over land. Then there was the desirability of being near a town. The site finally chosen was a hill about three miles west of Tarpon Springs on the Gulf of Mexico. The hill was 47 ft above sea level (high for Florida) and, since the set was to be placed on a 100 ft tower, provided an excellent opportunity for tests over water and land.

In April, Signal Corps specifications for an MEW production model were presented to the General Electric Company, the Western Electric Company, and Westinghouse Electric and Manufacturing Company. A further conference was held at Lieutenant Colonel Winter's office in the Pentagon on May 20, 1943, when it was decided to give the contract for 102 sets to GE. The Radiation Laboratory was to serve as consultant. Procurement was made contingent on the Tarpon Springs field model giving range and general performance acceptable to the Army Air Forces Board. On June 11, 1943, General G. P. Savile outlined a ground radar development and procurement program for AAF which followed closely the recommendations of the Stratton Committee. This letter recommended that procurement of MEW be delayed until the field tests had been run and necessary modifications made. Later, after preliminary tests at Tarpon Springs, this recommendation was withdrawn, especially in view of the fact that a letter of intent had been issued to GE.

The experimental period and the beginning of the field testing of the MEW coincided with the death, on June 10, 1943, of the project engineer Morton Kanner. Kanner lived to see the high-power techniques to which he had made significant contributions incorporated into the MEW system but the group was saddened by the fact that he was never to see the Tarpon Springs installation to which he had looked forward. E. G. Schneider carried on as Project Engineer through the Tarpon Springs period. M. A. Chaffee and A. G. Bagg accompanied the first MEW to be sent to England. L. C. Mansur became Project Engineer after Schneider went to BBRL.

By the middle of April, MEW No. 2, with the exception of the indicators, had been shipped to Florida. Chaffee, Bagg, and Blackmer went along to set it up. By May 1, a concrete foundation had been poured, the MEW had been set up on the ground while the tower was still being erected nearby, and signals had been received on a 5-in. A scope and a 7-in. PPI borrowed from the SCR-615 group at the Laboratory.

The name given by AAFSAT to the Tarpon Springs MEW was "Rosie," chosen in advance, perhaps because of service skepticism about Laboratory claims as to the set's performance. From her location on the ground, Rosie was confined to an eastern land view. Nevertheless, plane signals dotted the PPI when it was first turned on.

A week later, the tower was up. It consisted of the bottom half of a 200-ft SCR-271 tower, with the upper 15 ft modified to carry an octagonal wooden platform 35 ft across. The antenna was mounted in the center of the platform leaving approximately 5 ft of walking space around the outside. Preparations began immediately for putting MEW on the tower. The task took less than a week on May 16, from Cambridge, Pollard announced that the Florida MEW was in operation on the tower.

The six consoles which arrived in Tarpon Springs on June 26 were a step toward solving the indicator problem. These indicators, including two PPI's, three B scopes, and a micro-B scope, were more flexible than previous designs, both in numbers and in types of presentation. With the exception of the micro-B, all used 12-in. tubes. The MEW group preferred 7-in. tubes because of their better focus. The Army, however, wanted 12-inch tubes because of their operational advantages, the larger screen being easier to watch and to plot upon. Nevertheless, their poor focus (later improved) resulted in a decision to return to the 7-inch size in the indicators which went abroad with the preproduction MEW's. Eventually, these were replaced by 12-in., off-center PPI's.

The 7-in. micro-B scope was the newest thing in long-range indicators. Essentially a highly expanded B scope, with ranges of 10 and 25 miles, and with azimuth sectors adjustable from 20° to 90° degrees, the micro-B was the magnifying glass of the system. It could be used for ordinary tracking, for counting planes in formation, or for detecting targets through window jamming. Extra advantages over the B scope were its full delay and its ability to choose azimuth sectors anywhere through 360°. Thus it could follow the movement of the target easily. Because of its versatility, the micro-B was made the model for later MEW B scopes.

In turn, these gave way to off-center PPI's, which combined the controlling advantages of both B scopes and regular PPI's. Off-center PPI's, which permitted the center of a large PPI-type sweep to be moved at will, were developed during the winter of 1943–1944 at BBRL by John Hexem, of the indicator group of the Laboratory. A total of 24 standard PPI's were modified and 8 new indicators made, to include the off-center feature.

MEW at Tarpon Springs was used at first for acceptance tests by the Army and, later, for actual operation during which time the set reported plots directly to the AAFSAT filter center at Orlando. Carl Hundstad, an engineer from GE, was stationed at Tarpon Springs to observe the workings of the MEW equipment, which was now designated AN/CPS-1. For purposes of the GE contract the Army specified that tests should be over by September 1, 1943.

The acceptance tests were intended to outline the effective, or "tracking," lobe shape of the set. Flights conducted over a period of about three weeks by Army representative Lieutenant R. Bennett, of the Army Air Forces Board at Orlando, were run using B-17's and other aircraft flying at varying altitudes over water and over land. On August 4th, Schneider announced that results indicated a free-space beam which should give 50% tracking to a maximum of 200 to 225 miles over water at 40,000 ft.

Two points were brought out during the testing period. First, the MEW indicator system, as then set up, was still wholly inadequate to meet the needs of the set. The six indicators, which had been intended for both upper and lower beams, were inadequate for just the lower beam. This led to the decision, made jointly by Radiation Laboratory representatives and the Army, to include five B scopes and five PPI-scopes with each of the preproduction models. Second, the extraordinary traffic-handling capacity of the MEW made a great impression upon the Army officers at AAF-SAT. With existing radar equipment it had been customary to report all isolated information to the Orlando filter center; but with the advent of the MEW, which on certain days could have reported as many as 12,000 plots, the filter center would have been jammed with information telephoned in from the MEW site. A system of "pre-filtering" at the MEW site, i.e., sending on only clearly defined tracks to the filter center, was evolved to avoid clogging the center. Prefiltering techniques were improved with each succeeding MEW system, growing in efficiency in later models until, in England and on the Continent, the MEW became control headquarters for its own information.

The filtering system in Florida centered around a semicircular table, 15 ft in diameter, with the Gulf coast outline painted on it. MEW was a point at the center of the table. Around this point, concentric circles and radial lines represented range and azimuth, respectively. Since these were the coordinates of the scopes, these were made the coordinates for plotting scope information. For "telling in" tracks to Orlando, the table was further divided into grid squares, in rectangular coordinates. Thus on the MEW prefiltering board, plane positions were plotted in terms of range and azimuth and read off in terms of grid squares. In later sets, the filter board had similar markings but was mounted vertically so it could be seen by everyone in the control room. The board was of glass; tracks were marked on from behind, while the controllers read their information directly off the front.

The officers at AAFSAT were well pleased with the general design and performance of the MEW. Colonel T. J. Cody, head of the Air Warning Department of AAFSAT, under whose supervision the tests had been carried out, found that his own first impression that the Air Forces had "hit the jackpot" was satisfactorily confirmed. The new and extremely powerful radar was readily adapted to Air Forces thinking which had already decisively shifted to the offensive; at AAFSAT considerable thought was paid to the possible uses of MEW as a control instrument in aerial offensive warfare, and the result was a campaign, in which Colonel Cody was a leader, to have the MEW designed so that it could be made mobile.

At a conference held in General McClelland's office in the Pentagon in August 1943, it was agreed that the production sets would probably not be available before early 1945 and that operational experience with the equipment was urgently needed. The Laboratory agreed to build a total of five MEW sets on a crash basis (this was later increased to seven) with plans at the time to send one to England for the use of the Eighth Air Force, the rest to go to the Southwest Pacific, the Central Pacific, and the Aleutians for use in training. This crash program was carried out under the supervision of E. C. Pollard, with E. G. Schneider as Project Engineer.

Height-Finder and Mountain Sets

It may seem strange that since height finding was so necessary to the solution of interception problems, the laboratory did not make a second attempt to design such equipment until rather late. The main reason was lack of manpower. The High-Power Group was committed to the SM and MEW programs and they felt that the Army should use SCR-615 although they realized its slowness in finding height and its general cumbersomeness. Also, the British had some good height finders, the Type 16 and the greatly improved Type 24, which could be and were largely used with MEW on the Continent.

Early in 1943 the Army ground radar program was in a chaotic state. About a dozen early warning and height finding equipments were being worked on by different development laboratories. There was tremendous duplication of effort and no immediate solution was in sight. Early in 1943, E. L. Bowles, Expert Consultant to the Secretary of War, set up an *ad hoc* committee, headed by J. A. Stratton of his office, to review and make recommendations regarding the general Army ground radar program. The committee, on which K. T. Bainbridge served, submitted its report on July 11, 1943; it was accepted by the Army and resulted in a sharp curtailment of its program.

Beavertail (AN/CPS-4)

The report encouraged the development at the Radiation Laboratory of a 10-cm height finding set known as Beavertail, because of the shape of the beam it produced. It was officially designated AN/CPS-4. Designed as an auxiliary to a search set, it was capable of giving heights of all aircraft within a given sector simultaneously. Work on Beavertail was started by A. Longacre, working with a small group of men including J. Millman, S. B. Wells, and H. P. Stabler. The first problem was to investigate the best type of scan, mechanical or electrical, and the best way of producing it. There were questions as to the electrical characteristics suited to optimum

GCI tracking. Longacre conferred with members of the Army Air Forces Board at Orlando on some of these points. In general, the development of Beavertail was looked upon as meeting certain definite needs of the AAF. It was a set that could be more easily sited than others then in use; it would have increased accuracy and low coverage, plus the antijamming properties of microwaves. It was thought desirable by the Army that the set give 90% shows on a bomber to 60 miles, and height readings of $\pm$ 1000 feet at 45 miles. The antenna was to be narrow enough in the vertical plane to permit direct reading of height from measurement of the angle of the antenna. Ultimately, the Beavertail antenna was horn fed, with a reflector 20 ft by 5 ft shaped as an elliptical section of a paraboloid. The antenna gave a beam 1.2° wide in the vertical plane, and to pick up targets in altitude, this was raised and lowered 25 cycles per minute by nodding the antenna structure. This was the height finding principle finally favored by the British. Signals were displayed on a special indicator (RHI, range height indicator) which plotted elevation angle against range (out to 90 miles) and cut out ground clutter except for the base line of the scope. By September 1943, Longacre, who had been investigating a second height finding idea on the side, transferred to another project and A. R. Tobey took over as project engineer of the AN/CPS-4. In September also, there was a reorganization and the High-Power Group became part of a new division (Division 12) devoted to air warning problems.

On October 30, 1943, the Steering Committee, investigating the program of air defense for Division 12, recommended that the AN/CPS-4 be actively pursued as the best thing that could be done in height finding at the time. Shortly thereafter, it was decided by the group to construct an operational Beavertail, consisting of a modified "Mickey Mouse" (SCR-547) mount, obtained from Camp Evans, with a Beavertail antenna put on it. The function of this "breadboard" model was to determine the final design of the production set. By this time Beavertail was far enough along to raise the question of procurement. In October, representatives of the Federal Radio and Telephone Corporation, who were interested in building the set, met with members of the Radiation Laboratory and representatives from the Office of the Secretary of War. By the middle of November, negotiations between the Signal Corps and the Company were proceeding slowly, and Federal finally decided not to undertake the project because of the plant expansion that would be required.

In December 1943, the General Electric Company, the manufacturers of the AN/CPS-5 set, with which Beavertail might serve as an auxiliary equipment, received a contract for 100 models of the AN/CPS-4, with the Radiation Laboratory appointed Consultant. A conference at the laboratory on December 17, 1943, served to clarify the Laboratory's status and to bring General Electric up to date on the latest developments in components and the overall system design. By the beginning of 1944, General Electric had begun engineering work on the production set and the Morgan Construction Company of Worcester, Massachusetts, had been offered the subcontract for the mount.

In April 1944, the Laboratory's experimental Beavertail was set up at Bedford Airport to check general performance, height accuracy, and range. Height tests showed an accuracy better than $\pm$ 1000 ft at 60 miles. By May 1944, a final draft of specifications had been given to General Electric, and a production schedule set up which called for first deliveries in May 1945.

In June 1944, Tobey left for BBRL and R. H. Muller was appointed project engineer. In August, the Bedford model moved to Leesburg, Florida, for operational tests in conjunction with the V-beam equipment, under the direction of the Orlando Army Air Forces Board. These tests indicated height errors of the order of $\pm$ 300 ft. An average of 3.6 sec was required to fix the height of a target, once its range and azimuth had been obtained from the associated search set. This Beavertail system was eventually transferred to the Watson Laboratories.

The first production model AN/CPS-4 came out in June 1945. Several sets were shipped to the Pacific during July and August, but none arrived in time for combat.

Portable GCI (V-Beam)

The Army, faced with the possibility of fighting in mountainous territory, needed a PGCI (Portable Ground Control of Interception) set to give wide coverage and rapid handling capacity as well as rapid and accurate height finding. And it had to be free from siting difficulties. After several months of canvassing the possibilities of separate search and height-finding systems, the laboratory began the design of a single system operating on 10 cm to answer both requirements. The V-beam system, AN/CPS-6, whose laboratory name was derived from its double beam, used an antenna system larger than any other American radar set, consisting of one antenna giving a vertical fan beam for search and a second "slant" antenna giving a beam at an angle of 45° from the other (see Fig. 17-3). The range of the search beam was about 200 miles; the auxiliary antenna used with the other in height finding was useful out to about 140 miles. Height finding depended upon the fact that as the two beams scanned simultaneously, a target was picked up first by the vertical beam and then by the slant beam, and the angle difference depended only upon the slant range to the target and its altitude.

Radiation Laboratory thought on a V-beam system began in the spring of 1943, when D. B. Langmuir of the OSRD London Mission, in a letter to

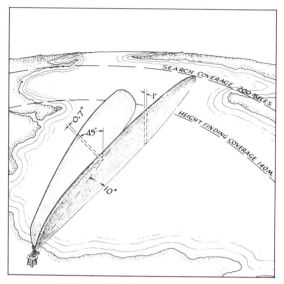

FIG. 17-3. V-beam pattern.

K. T. Bainbridge, mentioned a British scheme for a double-beam radar (both beams slanted) to obtain heights. The two beams, maintaining their relative positions at all times, swept in azimuth, and the height of the target was determined by the angular separation of the two pips. The beams met near the ground, however, making height readings on low-flying planes difficult if not impossible. Longacre proposed that this difficulty be cleared up (at the same time maintaining the V-beam principle) by separating the beams at the bottom by an angle whose optimum value he determined to be 10°.

During May 1943, Longacre, then working on both the Beavertail and the V-beam ideas, visited the GCI control division of AAFSAT to study control problems, and in July wrote a brief sketch of the proposed system. Bainbridge approved the project verbally, but the Steering Committee would not. In September 1943, Project SC-74, originating in the Office of the Chief Signal Officer, was accepted by the Steering Committee as covering development in the laboratory of the V-beam system, as well as Beavertail.

The antennas for V-beam had multiple horn feeds, providing energy from a total of five magnetrons, and produced fan beams 1° wide in azimuth for fine resolution and 30° in elevation for all-altitude coverage. In addition to PPI's and B scopes, the V-beam used a special height indicator which displayed returning signals from both beams so that height could be read directly.

Work in the laboratory on the V-beam system was first carried out by Longacre and several members of the group that had worked on Beavertail, including J. Millman, who later became Project Engineer. In June, members of this group, together with a representative of the Laboratory's purchasing department, visited a carnival in the Boston area, with the result that an old carousel frame was purchased to serve as mount for the new system. With the advice and assistance of the Antenna Group, dimensions of the two antennas (32 ft by 10 ft and 25 ft by 10 ft) were worked out, and by November patterns had been taken on the smaller dish and the larger dish assembled on the ground. In February 1944, the first laboratory V-beam (lower beam only) was in operation and receiving signals at the Bedford Airport. The system had all its components, except for the indicators, mounted on the merry-go-round frame, the whole having the aspect of a giant back-of-dish system. Thus the V-beam gained the advantage previously enjoyed by the HEW, of carrying most of its components with it, thereby eliminating rotating joints.

By March 1944, two planes, a B-24 and a B-47, had been assigned to the system and the upper beam was on the air. These first tests showed a weakness in the antenna structure which caused the middle section to be deformed with resulting increase in beamwidth. Since later tests at Bedford were satisfactory, E. C. Pollard went to Orlando, Florida, in April 1944, to confer with members of the AAFB about more detailed tests of the V-beam system and also to try to get a commitment on the military need for the V-beam. Plans were made to locate the equipment for test at Leesburg, Florida.

The set arrived in Florida on April 16, 1944, and was located on "Tomato Hill" about three miles from Leesburg. Five days later, the set was on the air. It was agreed that during these tests, the equipment would remain under the technical control of the Radiation Laboratory, and also it was clearly understood that the equipment would be considered experimental and that principles and operation, not engineering, would be tested.

The first real calibration flight, run on May 10, 1944, using a B-19 airplane, gave more accurate results than had been indicated by preliminary tests at Bedford. Although the results could not be considered completely certain, because the planes were not equipped with absolute altimeters, about 50% of the altitude readings were within 500 ft of the pilot's altimeter readings. During the period from May 15 to May 20, about 70 successful missions were flown with a B-25 airplane, and in 95% of the runs, contact was made. A variety of tests were made to determine possible uses of the V-beam. The set was tried out for control of day-fighter interceptions, and of night-fighter interceptions with and without searchlight assistance; for the control of fighter sweeps and escort missions; and for blind bombing. The V-beam, in common with other high-power microwave sets, showed that it was useful for determining the position of storm centers. Tests to explore the effect of "window" showed that although the set could follow targets through it, the slant beam was jammed so much worse than the vertical beam that height readings were possible only about 10% of the time. The reasons were incorrect polarization, which was later changed to vertical, slightly wider beamwidth than the vertical beam, and bad side lobes due to distortion of the dish. The Army suggested that the V- beam try to get MEW range by breaking the lower beam in two and using two magnetrons. This suggestion was later followed.

On the basis of the Florida tests, the Signal Corps requested that the Laboratory build six preproduction models of the V-beam, or AN/CPS-6, and this order was later increased to 8. The Stone and Webster Company was given a contract to design a suitable mount, the mounts being manufactured according to their design by the Walsh Construction Company.

The plan of manufacture of the preproduction V-beams called for testing the components at the Radiation Laboratory and then assembling them in a location where the considerable space needed was available. After a fruitless search of the Boston area for a satisfactory location, it was decided to establish a field station near Orlando, Florida. The advantages of such a location would be close cooperation between the Laboratory group and the AAFB, and the opportunity to obtain assistance in building the systems. The plan also included giving theoretical instruction at the Laboratory to a cadre of two officers and eight enlisted men for each set, and following this by field experience in Florida on the actual system.

By March 1945, the first preproduction V-beam was in operation from a 25-ft tower at the station. By October, five of the eight sets had been completed. Of these, three stayed in this country for training purposes, one was shipped to the Pacific area, and one went to the Panama Canal Zone.

In December 1944, a contract was let to the General Electric Company by the Army for 60 production AN/CPS-6 sets. In anticipation of this contract, two engineers had been stationed at the Laboratory since September 1944. The Radiation Laboratory was to act as consultant. To save time in production, the V-beam set, operating on the same wavelength as MEW, had been designed to use as many of the MEW components as possible, so that the production V-beams were able to use components diverted from the MEW production line.

All property at the Orlando Field Station was formally transferred to and accepted by the Property Officer of the Florida Field Station of the Watson Laboratories, Leesburg, Florida, on October 23, 1945.

Light Mountain Set (Li'l Abner)

In the winter of 1943–44 an interim system, Li'l Abner (AN/TPS-10), was started to fill the Army's need for a lightweight set which could be transported into mountainous country and used to detect low-flying planes. In the Laboratory, the AN/TPS-10 was thought of more as a warning device than as a pure height finder; and, in this light, was expected, as far as low-flying planes were concerned, to be a stopgap for the slower-moving MTI program. The Moving Target Indicator (MTI), a special attachment which was designed to eliminate ground clutter and present only the signals from moving targets, will be discussed in a later chapter. Tactically, the AN/TPS-10 was thought of in terms of present experience (late 1943) in the Italian Theater and in terms of future experience in the China–Burma–India Theater.

The AN/TPS-10 set was designed so that it could be broken down for hand transport (with a few exceptions, no piece weighed more than 40 lbs) and to produce (at 3 cm) a flattened pencil beam 0.°7 in elevation for height discrimination and 2° in azimuth (see Fig. 17-4). The beam nodded

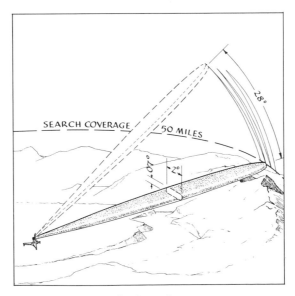

FIG. 17-4. Li'l Abner beam pattern.

rapidly in elevation while scanning slowly in azimuth. The AN/TPS-10 overcame the difficulty of ground clutter on the scope by using a single Range-Height Indicator (adapted for Beavertail) which plotted elevation angle against range, thereby restricting clutter to the base line of the scope.

In November 1943, Pollard discussed with General McClelland the possible use of the SCR-615 to aid the ferry service operating between Burma and China, countries whose mountainous terrain made it difficult for existing sets to provide early warning against low-flying aircraft. In order to get experience for further discussion of the problem, an HR set was taken in January 1944 to the mountains of North Carolina by Belmont Farby to gather data for further study. These tests showed that when the beam was elevated two beamwidths above the ground, the clutter practically disappeared from the scope. From these tests it was concluded that a set designed with a very narrow beam vertically, which could be elevated, and a short pulse length, would enable planes to be seen better than fixed ground echoes without recourse to special Doppler or comparison techniques. In conferences with Schneider and Longacre, Pollard proposed that a set be built which had a narrow beam capable of scanning in the vertical direction. Accordingly, Longacre produced the basic design for a 3-cm equipment, which used the Beavertail (nodding) type of scan and which also used the Beavertail (RHI) type of indicator. Such a system could be used where another equipment employing ordinary PPI presentation could not.

On December 1, 1943, an informal meeting attended by Pollard, Schneider, Captain Fogle, and Lieutenant Colonel Sheppard was held at the Radiation Laboratory to draw up military characteristics and to lay plans for a pack-type equipment which could be carried on men's backs and be set up quickly. To the end of 1943, the Li'l Abner project (as it was known in the Laboratory) had been regarded simply as a research project in Division 10. At a meeting on February 25, the Ground Committee approved Li'l Abner as a formal project and recommended its acceptance by the Steering Committee. B. G. Farley was appointed Project Engineer, but was later succeeded by C. E. Moore. It was stated that, in Division 10, Li'l Abner would partially replace MTI as a formal project and would use some of the personnel working on MTI. In March 1944, the Steering Committee gave its approval, and in May, a project request came from the Army to cover the development of Li'l Abner, with the stipulation that only one prototype equipment should be constructed.

By the first of the year, the Mechanical Engineering Group, under M. B. Karelitz, was studying designs for the antenna mount; and by February 1944, a 10 ft × 3 ft reflector (a section of a paraboloid) which could nod vertically was being built of plywood and screening. Using available components, a 3-cm system was set up on the roof of the Laboratory to see if the desired 50-mile range on a medium bomber was possible. This system also provided for testing the components to be used in the first field system. By the middle of 1944, a Li'l Abner system had been installed at the Bedford Army Airport. Although it used a heavy modulator designed for a 10-cm system, plans called for a 3-cm hydrogen thyratron modulator whose peak power was expected to be about 200 kW. Radio frequency and modulator components rotated with the mount, the whole setup, including mount, weighing about 550 lbs.

By April 1944, the system at Bedford was undergoing tests. Ranges were short, of the order of 30 miles with occasional pickups at 40 miles, partly because of the low power of the modulator and partly because of low receiver sensitivity. The tests indicated, however, that with certain improvements, ranges would be satisfactory. Between April and June, these chief improvements were made: electrically, a higher-power modulator (borrowed from the H_2K) and a more sensitive receiver (SO-SU type); mechanically, a new reflector and new elevation and azimuth movements.

Meanwhile, Service interest was varied and expressed only informally. The Marine Corps, wanting a height finder for beachhead use, weight no object, were willing to wait for production equipment. The British, on the other hand, wanted ten preproduction sets with the possibility of taking 100 production sets, if they were built. In May 1944, the British Air Commission requested and received from the Laboratory three Li'l Abner's with stabilized platforms, for use for fighter direction aboard carriers to replace the SM-1 systems, which were not completely satisfactory.

In July 1944, the improved Li'l Abner was truck mounted and driven to North Carolina for tests in hilly country. It was set up on Rich Hill Mountain, near Asheville. These tests, which were witnessed by representatives of the Army Air Forces and the Signal Corps, indicated that a medium bomber could be tracked at ranges of 40 to 50 miles, with approximately 50 % success, and that target discrimination was possible where the target was separated from the ground by at least one beamwidth. On the basis of these tests, the Signal Corps requested the Laboratory and RCC to initiate a crash program for 40 sets in addition to two prototype models.

The North Carolina model was moved to Leesburg, Florida, for operational trials during November by the AAFB, and also for use as a training set. Li'l Abner was tested both by itself and (for GCI) in conjunction with the AN/TPS-1, a portable Army search set. The second prototype AN/TPS-10 was tested mechanically at East Boston airport and then sent to Warner–Robins Field, Georgia, where it was used for training Army personnel. Production of the 40 sets covered the period from January to June 1945, with the manufacturing being done in the laboratory by officers and enlisted personnel to whom the sets were later assigned.

These 40 sets proved to be reliable, and had been built with an eye to immediate use in theaters of operation. Besides being used for training in the United States, AN/TPS-10's went to France, Italy, Saipan, Iwo Jima, and the CBI theater. By January 1945, the second production model was shipped to Saipan by air, and later was sent to Iwo Jima. A set was shipped in February 1945 to England; and one was sent to Italy in April. Put in front-line operation under a former SCR-270 crew, it turned in 250 plots during the first 24 hr of operation.

In February 1945, a letter of intent was issued for 100 sets (increased in April to 150) to the Zenith Radio Corporation, delivery to start in June 1945. Meanwhile, in the spring of 1945, plans were being made to design a new 3-cm rf head for a more sensitive, higher powered Li'l Abner, which was to have about 70 % more range. As of August 15, 1945, however, further production of AN/TPS-10 was terminated. All equipment on hand was tranferred to the Watson Laboratories.

Ship Control of Interception (SCI)

CXHR-SX

During the early part of 1943, the Navy believed that the SM and SP radar sets, with IFF equipment, could protect our carriers against all forms of air attack. When it became evident later that the Japanese would eventually attack our ships with ground-based planes, thereby greatly increasing the number and frequency of such attacks, the Navy realized the limitations of these sets. Their traffic-handling capacity was small because they had to stop azimuthal scan and point directly at the target in order to find its height; the maximum range of 50 miles was insufficient; the presentation system caused delay because data had to be plotted; and height finding was reliable only at elevation angles exceeding 2°.

Informal conversations between Navy representatives and Radiation Laboratory members were held for several months previous to a conference called at the Laboratory on August 18, 1943, at which a project was initiated which was first called SCI. At this conference, Commander H. E. Bernstein, of the Radar Design Section of the Bureau of Ships, outlined rough specifications for a set which he had long envisaged to combine the functions of a MEW and a height finder. It was asked that the height-finding antenna give vertical coverage of 10° and range on a medium bomber of 40 to 50 nautical miles; and that the search antenna furnish range on a medium bomber of 70 to 80 nautical miles. The wavelength was to be between 10 and 3 cm. Weight was limited to 2500 lbs topside. The set was to scan continuously at about 4 rpm, and to give the height of all aircraft within 50 miles every 15 sec and at the same time provide high and low microwave coverage. Accuracy comparable to the SP in height and azimuth was expected.

A formal request for the development of the SCI at Radiation Laboratory was made on August 30, 1943, and accepted by NDRC on September 13, 1943. By this date, research work on the system had already begun in the Laboratory with J. S. Hall as project engineer.

A scanner, designed by C. V. Robinson of the Antenna Group, was selected as the feed for the height finder. At 8.5 cm the beavertail beam of the height finder is 3.5° wide and 1.1° high. This beam scans from the horizon to an elevation of 11°, 600 times per minute. As the mount rotates in azimuth the height data are continuously presented on RHI's (range-height indicators) at the several consoles. At 4 rpm the height beam hits the target $3\frac{3}{4}$ sec later than does the search beam. The search beam is 1.7° wide in the horizontal direction and is fanned up to 18° vertically.

Since the weight of the antenna system was an important design limitation, it was planned to place both antennas on the same stabilized mount. It was thought that stabilization could probably be provided by RASD.[1] All the information presented by the height finder and the search system (on the same mount) is presented at any one of five mutually independent consoles. Each console has a PPI, an off-centered PPI (for greater target discrimination in early warning), and an RHI[2] (to give height vertically and range horizontally). The coordination of three scopes on each console, the use of five independent consoles for each set, and the ability of the system to provide height data under conditions of continuous scan are reasons for the greatly superior traffic handling capacity of the SX when compared to the SM or SP sets.

The most difficult problem was that of designing the Robinson feed. An experimental system involving only the height finder was put into operation on March 1, 1944 on the roof of Building 6 at MIT. It was flight tested on March 8, 9, and 11. An SNB airplane served as target for the tests and a reliable range of 45 nautical miles was obtained. This was the first height finder to indicate accurate heights using a beavertail beam during continuous scan. It differed from the SM and SP, which used conical scanning. It also differed from the Beavertail (AN/CPS-4) set which used a fanned beam, because (1) the feed moved instead of the dish and (2) height data were furnished as the mount rotates.

A building with the largest radome ever built (25 ft high × 26 ft in diameter at the top; 19 ft in diameter at the bottom) was constructed at the Laboratory's Spraycliff Field Station, on Beavertail Point at Jamestown,

Rhode Island, to house the equipment and the mount. During April the height finder was moved to Spraycliff and combined with an early warning dish placed on a United Shoe Machinery Mount. This experimental system was designated SCI and served a purpose other than its primary one of components testing. It was used to direct night interceptions in the night fighter training program at Beavertail.[3] Many pilots trained on information from SCI later saw action in the Pacific.

Earlier in the year, in January 1944, W. O. Gordy had taken over the responsibility of carrying out experimental flight tests with an experimental search system at the Mt. Prospect Field Laboratory at Fishers Island. Flight tests were run throughout the year and Gordy worked closely with the Antenna Group at Ipswich. A 7-ft $\times$ 11-ft grid antenna with a nine-dipole feed giving a csc^2 pattern coverage was first tried out. The early experiments showed a considerable amount of reflection from the water and the resulting peaks and nulls gave spotty coverage. Later a 5-ft $\times$ 14-ft. dish with three-horn feed was constructed and Gordy obtained results which seemed adequate. He found that a two-motored plane could be followed to 80 nautical miles up to altitudes of 20,000 ft and in to elevation angles of 15° to 18°.

In April 1944, the General Electric Company was brought into the program, and plans were made for the construction of two units, later designated CXHR by the Navy, to be modeled after the Radiation Laboratory's final design. General Electric wanted to design its own console, however, and it was decided that it should proceed independently in that direction. The Navy was anxious to have both consoles tried out and it was decided that both would be demonstrated to the Navy at Spraycliff.

Under an OSRD contract, the Laboratory and the company agreed to have two preproduction (CXHR) models ready for delivery by March 15, 1945. G. W. Fyler, who had supervised the production SM sets, was to direct the General Electric work at Bridgeport. The General Electric Company agreed to build most of the components for the CXHR. The Radiation Laboratory was to design and build the two main control panels and the two rf heads. The latter were to be placed between the reflectors on the stabilized mount.

The complete SCI system was demonstrated to Navy representatives at Spraycliff in July 1944. The General Electric console, designed by G. W. Fyler, was called the "soda fountain." With this type of console all the fighter director officers would sit together and obtain their data from a single skiatron (see Chap. 26 for a description of this device). In the Radiation Laboratory system each fighter director officer was at a different console separated from the others. The Navy felt that the latter was the

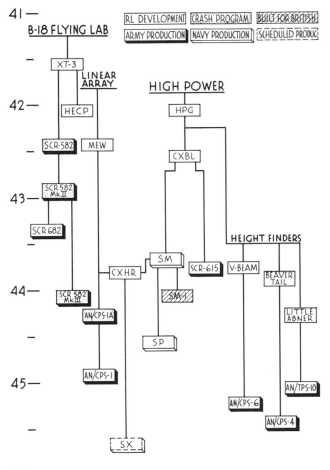

FIG. 17-5. Chart of RL developmental and production systems (ground systems and ship counterparts).

better solution and General Electric agreed to build eight consoles according to Radiation Laboratory specifications.

In the fall of 1944, the Bureau of Ships ordered four production sets from General Electric to which the designation SX was finally given. The first SX was to be delivered in August 1945. It was decided that the SX sets were to be delivered without a stabilization system which would be procured by the Navy direct from Westinghouse. Early in September 1945 a group of General Electric engineers headed by G. L. Hollingsworth came to the Laboratory and worked for about four months on the design of components of the SX system. During this same time the Laboratory built control panels and rf heads for the two CXHR's. The CXHR's used the Mark 8, Model 4, Westinghouse stable element and computer, the Westinghouse equivalent of the Radiation Laboratory's RASD.

The expected delivery of the first CXHR was delayed, however, for a number of reasons. The hurricane of September 1944 created damage to the CXHR installation at Spraycliff, which took a month to repair. The first of General Electric's eight consoles was delivered to Spraycliff on January 10, 1945. The first prototype CXHR mount was not delivered to the Laboratory until April 10, 1945, when it was installed on the roof of Building 20 for tests. The first CXHR system was flight tested in May. Subsequent flight tests showed that the height finder had a range approximately ten miles greater than that which had been obtained previously on the roof of Building 6. The early warning system had the same coverage as that which Gordy had earlier reported.

In June the second CXHR system was installed at the Naval Air Station, St. Simons Island, Georgia. The first CXHR (the one on Building 20) was installed on the carrier USS *Midway* during July 1945. The first production (SX) set was installed on the carrier USS *Franklin D. Roosevelt* in August. The Bureau of Ships had increased its order, on February 24, 1945, to 41 SX sets, with delivery to start in May 1946.

In the SX the Navy finally obtained its general purpose set for search, warning, and fighter direction, but too late to show its mettle in combat. All the resources of high power and stabilization research, the microwave early warning and height-finding techniques, were combined in what might have proved an ideal shipboard set.

A chart of Radiation Laboratory developmental and production work on ground systems and their shipboard counterparts is shown in Fig. 17-5.

NOTES

1. See above, p. 441.

2. The RHI was similar to the one being developed for AN/TRS-10 and AN/CPS-4.

3. See below, Chap. 41.

CHAPTER 18
Fire-Control Radar

The problems of fire-control radar[1] will be considered from the point of view of the two military agencies which were responsible for them, the Army Ground Forces and the Navy Bureau of Ordnance. The problems are different in each case and were not solved to complete satisfaction in either Service, although some remarkable results were achieved. In general, radar equipment for the ground forces had to be mobile, although in coast defense and harbor surveillance fixed installations could be used; air-transportable equipment would have been extremely useful, but it was only coming into production at the end of the war. As for the Navy, the same requirement of careful engineering that applied to ship-search radar had to be met in fire-control equipment.

In the Army Ground Forces, the Coast Artillery Board and the Anti-Aircraft Command were the principal agencies familiar with radar before the war. Their problem was location of targets for the purpose of destroying them by heavy gunfire. Ships and aircraft were suitable targets. The possible extension of radar to other uses such as road watching, spotting enemy gun positions, mortar location, surveillance of river crossings, and detection of moving tanks was recognized by AGF Headquarters but fulfillment was achieved only by combat troops using anti-aircraft radar sets rather than equipment designed and procured for these uses. These anomalous circumstances arose from the difficulty of the problems involved and from the magnitude of training required.

For use by the infantry and the field artillery, radar equipment must be rugged, durable, and simple. On the other hand, high resolution and the elimination of ground clutter are harder to attain in ground radar than in any other applications. The large number of troops involved, and especially the large number of field commanders to be trained in the tactical application of radar equipment, in itself, presented an enormous difficulty. Finally, radar for the infantry and artillery was useful, in general, only under special conditions of terrain and battle. This made it difficult to set up tables of organization and allowances. It was greatly to the credit of the combat troops that they were able to meet such trying conditions using large anti-aircraft radar sets.

The Navy started the war with only an embryonic fire-control research and development section in the Bureau of Ordnance and with the Radar Sub-section barely started. Nevertheless, two fire-control systems (both

473

operating in the 40-cm region) with associated gun directors for main and secondary batteries were in production. These sets had excellent range performance and beams sufficiently wide for resolution and easy target acquisition. Both were used with few replacements until the end of the war.

In the field of automatic weapons and small directors for naval 5-in. guns, the radar program was confused by a large variety of director designs, some designed before the advent of radar and few of which saw any practical use. A family of blind-firing radar sets was developed but none went into combat.

Among range-only radar sets, two were produced but one was never used. An excellent director in this class was ready only at the close of the war. It was not until 1943 that an attempt was made to design a fully automatic gun-director system around the radar equipment with the result that this system (Mark 56) was not ready until the end of the war.

Naval gun directors are roughly of three kinds: main battery, heavy anti-aircraft (secondary) battery, and machine gun fire-control systems. Main battery fire control handles only surface targets at long ranges. Long-wave equipment of this type was available before Pearl Harbor and was gradually being replaced by microwave sets at the end of the war. These long-wave systems should have been useful in shore bombardment but their most serious limitations were poor angular discrimination in close waters and inability to do deflection spotting at maximum range. They were heavy, complicated to operate, vulnerable to direct hits because of their size, and liable to jamming. They could be installed only on the large ships for which they were designed.

With naval heavy anti-aircraft and machine gun fire-control systems the most stringent requirement is target designation. All fire-control equipment depended on the radar search systems for target acquisition and designation. They were thus limited by the slow radar search speeds, hampered by the difficulties of exchanging information, and unable to handle the wide variety of aircraft targets presenting themselves. Operator fatigue contributed greatly to the inefficient use of these systems and automatic range tracking, at least, was almost a necessity. At no time did the Navy have in operation automatic tracking or a radar set comparable in performance to the Army's SCR-545 or SCR-584.

Ninety percent of the equipment used by the Army Ground Forces stemmed from Radiation Laboratory research and development (including coast surveillance as well as fire-control equipment). On the other hand, the Bureau of Ordnance relied mainly on the Bell Telephone Laboratories and the Western Electric Company for their gun-director radar systems. Fire-control radar systems were developed in the Radiation Laboratory for a short time in the Project II Group headed by L. N. Ridenour, and later in Division S, which was headed for a while by K. T. Bainbridge but for the greater period of time by Ivan A. Getting. Getting served also as a member of Division D-1.5 and as Chief of D-7.6 of NDRC and as Expert Consultant to the Secretary of War. One member of Division 8, Lieutenant Colonel A. H. Warner, a physicist who joined the Laboratory as a civilian[2] and was soon ordered to active duty in the Army Ground Forces, was extremely useful to the Army Ground Forces' program because of his clear understanding of both Laboratory problems and military needs. Another Laboratory member, Robert A. Patterson, a scientist who was first assigned to the Navy Liaison Office, where he gained a knowledge of the Laboratory as well as an appreciation of Navy requirements, was appointed by L. A. DuBridge liaison member in the office of Commander D. P. Tucker in the Bureau of Ordnance. He remained in Washington from November 1942 to the end of the war and a great deal of the successful cooperation between the Laboratory and the Bureau was due to his efforts.

RO Radar for Director Mark 51

After the Radar Mark 9, which has been described in Chapter 10, the next problem presented by the Bureau of Ordnance to the Radiation Laboratory was to design a blind-firing radar set for Gun Director Mark 51, which turned out to be a very successful director. The Laboratory did not advise putting a blind-firing radar set on such a small director but agreed to design a range-only system.

The heart of Director Mark 51 was the lead computing sight, designed by C. S. Draper of MIT, developed and produced by the Sperry Gyroscope Company and Crosley Radio Corporation. When Commander Murphy and Lieutenant Horatio Rivero (who had worked at MIT as a graduate student under Professor Draper) first inspected the sight on May 28, 1941, they learned that Sperry was designing a version of it for the British Admiralty. Commander Murphy realized that this sight, which was later known as Mark 14, would be most valuable mounted in a small gun director and a contract was entered into with Sperry which resulted in the delivery of the first few production gun Directors Mark 51 in June 1942. The director was used with 40-mm guns.

With this director, however, the tracker could follow a target only under conditions of good visibility and the Bureau of Ordnance questioned the Radiation Laboratory on the practicability of providing radar range for the gunsight Mark 14 as early as October 1941. After conferences at MIT and with Sperry, the Laboratory accepted a project in December, but the radar system was not clearly defined until April 1942 after further discussions with the Navy.

H. A. Kirkpatrick and E. H. Krohn served as project engineers. The RO system consisted of an 18-in. paraboloid mounted on the director, from which a 7/8-in. coaxial cable ran through the base of the mount to a cabinet (15 in. × 15 in. × 30 in.), weighing about 300 lbs, containing all the rest of the radar components. The range scope was mounted flush with the top of the cabinet. Through selsyns the radar range was supplied to the gunsight Mark 14.

One of the Mark 51's arrived at the Radiation Laboratory on July 18, 1942, and by August the RO radar was ready for tests. Tests were made at the Bureau of Ordnance Test Unit, Dam Neck, Virginia. The Mark 51 was connected to a set of 40-mm guns firing at sleeve targets. The holes were counted and compared with records of visual firing already available at Dam Neck.

The RO installation on the Mark 51 Director was completed on September 8, 1942, and the next day it left for Virginia. The tests during November and December did not run too smoothly. The gunsight Mark 14 had to be sent back to MIT for readjustment and the guns had to be realigned and boresighted; to obtain more reliable data, synchronized cameras were provided by the Laboratory. Another series of tests was run from January 22 to February 12, 1943. The averages from a Bureau of Ordnance report, dated February 16, 1943, were: percentage of hits with radar 1.11 %; percentage of hits without radar 0.151 %.[3]

Although the radar system was considered adequate for the ranges used by the Mark 51 (400–3400 yards), the Navy felt that the performance of the director would not be improved sufficiently to warrant production of this equipment, but that the experience gained in these tests would be valuable in the design of a new director-radar system. At the Radiation Laboratory the RO system became the ancestor of the Hand Radar (SCR-802) sets which are discussed in Chapter 17.

Director System Mark 151

In a conference with L. A. DuBridge on December 30, 1942, Commander Murphy requested that the old Mark 51 project be continued in order to use the information which might be obtained in designing a radar set for the new intermediate-range gun Director Mark 52, designed by C. S. Draper and under development at the Sperry Gyroscope Company. The Mark 52 was essentially the same as Mark 51 but heavier and more accurate. It was intended for use with a 5/38 caliber gun. An improved Mark 14 sight was being worked on. The new sight was eventually known as Gunsight Mark 15. Its incorporation into a pedestal made it the Director Mark 52.

The new project was approved by the director of the Radiation Laboratory because conversion of the Mark 52 for blind firing required a certain amount of fundamental development, although it involved the Laboratory in a cooperative project with Section T of OSRD.

Several conferences were held, the most important of which was the one on February 13, 1943, attended by Commander Murphy, C. S. Draper, and I. A. Getting. Getting made two general proposals which were adopted. First, he recommended that the Radar Mark 11 (range only) already in production by GE and RCA be used for the Mark 52 Director. This would not permit blind firing but "should be considered as a requisite for operating the equipment at all."[4] The outcome of this suggestion (Mark 26) will be considered later. He next suggested a new project to be carried out in two steps: the experimental conversion of Director Mark 51 for blind firing by using a 10-cm AGL system, and the eventual incorporation of automatic tracking into the design of the Mark 52 by using elements similar to AGL, operating at 3 cm. The director system as finally evolved was known as the Director System Mark 151.

The operation of the system may be briefly sketched. The gun and director point to a future target position. To the director is added a computer (consisting of an axis converter similar to the RASD, designed by W. B. Ewing). The angle converter receives present target position from the radar (which switches to automatic tracking as soon as it picks up the target) and the angular position of the director, and computes by synchro outputs which are transmitted to synchro motors geared to mirrors in the sight. The mirrors will thus be oriented to project the image of a reticle representing the target.

No definite specifications were set up by the Navy but the group at the Laboratory adopted some general characteristics: a weight limit of 5000 lbs.; range out to 3000 yards with an overall blind firing error of not more than 4 mils (1 mil = 1/6400 of a circle = $0°05625$).

Division 9 contributed an AGL-1 with outdated 3-cm waveguide and a production model from GE was promised for April 1943. H. A. Kirkpatrick, the project engineer, assigned responsibilities on the 10-cm system to E. H. Krohn, Ensign F. E. Huggin, G. J. Plain, and H. S. Sommers. R. P. Scott was made responsible for the 3-cm system. An optical laboratory was set up in Building 22 to test the auxiliary mirrors for the system. A mirror system suggested by Radiation Laboratory was built and engineered at the Applied Physics Laboratory at Johns Hopkins University by N. P. Heydenburg (Section T).

By the middle of July, a receiver which suited the system was being developed. The angle converter received in August had to be rebuilt because of faulty workmanship. In September the modified Gunsight Mark

14, received from Johns Hopkins, gave a double circle image and had to be corrected at MIT.

The system was finally ready for test by this time, as an oil servo mount and 6-ft dish had replaced the old AGL-1 antenna mount. J. A. Joseph, of Section T, arrived in Cambridge about the middle of November to participate in tracking tests which were run until November 24. C. S. Draper did much of the tracking himself and expressed approval of the system. The probable tracking error turned out to be 1.5–2.0 mils for overall blind-firing performance. The RO worked well throughout the tests and during the visit of the British Radar Mission.

Following these tests the Naval Research Laboratory requested that the angle converter and the modified sight be shipped to the NRL, Chesapeake Annex, to be tried out with the Radar Mark 7, a system designed by the Bell Telephone Laboratories. H. S. Sommers and R. V. Harris accompanied it to Washington and witnessed tests from December 7–16. The equipment performed better than expected, analysis of four runs giving an average tracking error of 1 mill. It was apparent, however, that the Bureau of Ordnance considered the automatic tracking too complicated to put into production in this form and the laboratory considered the project terminated on December 15, 1943, and turned over the equipment to NRL. Tests of the Director System Mark 151 furnished the first proof that such a system was practicable with radar.

Radar Mark 26

The second line of attack on the problem of a radar for the Director Mark 52, was the modification of the 10-cm Mark 11 called the Mark 26. The Mark 11, a range-only radar, was originally designed for the gun Director Mark 49 and was produced under a joint contract by General Electric and RCA. Plans to make the Mark 49 blind firing led to the decision to adapt Mark 11 for mounting on the Mark 52 Director.

In a letter dated May 31, 1943, Rear Admiral W. H. P. Blandy, USN, Chief of the Bureau of Ordnance, requested that the Laboratory examine the Mark 11 and make recommendations for its improvement. This request followed a conference held at the Confidential Instruments Development Laboratory of MIT, on March 30, to which the Bureau of Ordnance and the General Electric Company had sent representatives. As a result of this meeting and futher conferences, it was decided that the radar transmitter and receiver unit should be located apart from the director, and that only the antenna and dish would be mounted above deck in front of the director. The mounting would be supplied by CIDL. General Electric agreed to furnish a 24-in. dish for experimental purposes and as soon as possible furnish parts necessary to convert it to a 36-in. reflector. Radiation Laboratory was to serve as advisor on changes in radar components.

Radiation Laboratory's project engineer, H. A. Kirkpatrick, asked the following members of the components group to write recommendations: L. C. Van Atta, antenna; G. N. Glasoe, modulator; L. M. Hollingsworth, rf. Britton Chance was assigned to confer with H. Mayer, of General Electric, on the range circuit.

The Laboratory's report was submitted to DuBridge for transmission to the Bureau of Ordnance on August 4, 1943. Although various suggestions were made, the most important contributions of the Laboratory to the development of the Mark 26 were broadbanded rf plumbing and the slotted antenna reflector which it furnished. Kirkpatrick found by inquiry that no experiments had been made on the effect of a steady wind. Wind tunnel tests of the grating type paraboloid had shown that the effect of wind is about 40% of the effect on the solid dish.[5] For this reason a slotted type of construction was recommended to reduce the effects of wind and gun blast.

The Laboratory believed that with a 30-in. dish, the recommended rf components, and an improved receiver, the range of the Mark 26 would be somewhat near 18,000 yards. It was considered to be a satisfactory range finder for gun Director Mark 52 and it was the simplest and smallest fire-control radar in the fleet.

With this project the Radiation Laboratory's direct participation in Bureau of Ordnance gun director systems ceased until the request for the design and development of the Director System Mark 56. In the meantime, however, a great deal of effort was put into cooperation with Section T, OSRD. On completion of its work on the Proximity Fuze, Section T (headed by Merle A. Tuve) became involved in naval fire-control problems. One of these was assisting Radiation Laboratory on the Director System Mark 151; another was assisting CIDL in completing Gunsight Mark 15 and gun Director Mark 52.

Their main effort, however, went into the design of gun Director Mark 57 and the subsequent versions Mark 58, Mark 60, and Mark 63. In the course of design Section T used an indicator from Radar Mark 19, designed by Bell Telephone Laboratories and related to the Radiation Laboratory's Mark 9. They also used a modification of the SO–SU rf head, developed by the Radiation Laboratory. The Laboratory assisted Section T as advisor on all its radar problems and furnished a variety of parts and test equipment.

SCR-584

Soon after the first automatic tracking experiments, plans were made to take the XT-1 to Fort Hancock, New Jersey for demonstration to the Signal Corps. On November 30, 1941, the truck left Cambridge accompanied by Davenport, Getting, and A. H. Warner, and arrived at Fort Hancock on December 1. The XT-1 remained there for a month, although, since war had broken out in the meantime, planes were not available for tests. Blimps, ships, and occasional aircraft were tracked. The XT-1 returned to Radiation Laboratory and after some improvements was taken to Fort Monroe, Virginia, for service tests by the Coast Artillery Board. Before the truck left Ft. Hancock, General Electric Company representatives arrived to make drawings of the system, for NDRC had placed a contract with the General Electric Company for two copies of the XT-1 in anticipation of future production. The General Electric Company made its deliveries of XT-1A's, as they were called, in May 1943.

Tentative military characteristics of the XT-1A were drawn up at Fort Hancock, on January 7, 1942,[6] by Major Paul E. Watson, Captain L. R. Moses, and Edwin A. Goodwin (all of the Signal Corps Laboratories) and Getting and Ridenour. It was planned that SCR-268 would serve as a long-range radar search set for the XT-1. The elevation coverage expected was from $-15°$ to $+87°$. Probable error in tracking aircraft targets should be not greater than 5 mils on straight courses, involving elevation angles not greater than 85°, and angular rates not greater than 15°/sec, while the probable error in range measurement could not be greater than 50 yards. Angular tracking was to be fully automatic or aided manually. There was no provision for automatic range tracking. The operating range on aircraft required was not less than 14,000 yards; it must be capable of tracking targets to 32,000 yards with a minimum range not greater than 500 yards. Weight of the entire unit was limited to 17 tons. The Signal Corps asked for two additional sets and eventually 4 XT-1A's were delivered.

After the return to Cambridge, a 72-in. dish was substituted for the 48-in. dish and power plant was put into a separate truck. On February 3, 1942, the XT-1, with L. L. Davenport and Lieutenant Colonel Warner (who had been ordered to active duty at the Radiation Laboratory) in charge, set out for Fort Monroe, stopping off at Fort Hancock on the way, arriving on February 6. Two weeks later the new power truck, accompanied by F. B. Abajian, who had been in charge of building it, and G. B. Harris, arrived.

The Coast Artillery Board tested the XT-1 as Project 1218 to determine the accuracy of tracking in azimuth and elevation, the accuracy of range, as well as maximum and minimum ranges. Tests were conducted for several weeks during February after a week of final adjustment and training of service crews in the operation of the equipment. The general program for the tests included kinetheodolite and camera recording as well as sleeve and flag target firing. The new NDRC-sponsored T-10 director, designed by BTL, was coupled to XT-1 for the firing. T-10 was later standardized as Gun Director M9. Over 60 courses on an O-47 target plane were recorded.

Although the results showed that the probable error in angular tracking was somewhat less than 1 mil and the probable error in range measurement was about 20 yards, certain objectionable features in tracking and range measurement were uncovered. A velocity lag in angular tracking was found. Range measurement errors were due to a large extent to the operators, indicating that automatic range tracking might be desirable. While the XT-1 was at Ft. Monroe, a PPI recently developed at the Radiation Laboratory was added. Its use was so successful that a recommendation was made by the Coast Artillery Board for its inclusion in every set.[7] Colonel McGraw was very enthusiastic about the performance of the PPI which in early experiments indicated the possibility of target acquisition at 98,000 yards and locking on the target at 32,000 yards. Two sweep ranges were later provided, the maximum 70,000-yard sweep on the PPI.

The Coast Artillery Board recommended that, with the modifications which were indicated, the XT-1 be considered a standard radar gun-laying set and that procurement be authorized. The CAB concluded that[8] "The Radio Set XT-1 is superior to any radio direction finding equipment yet tested by the Coast Artillery or Anti-aircraft Artillery Boards for the purpose of furnishing present position data to an anti-aircraft director." After the conclusion of the formal tests, informal tests were carried involving some firing with 90-mm AA guns and plans were made for firing tests at Camp Davis, North Carolina, later in the summer.

In April, L. W. Alvarez requested the XT-1 be used for blind-landing experiments. Experiments showed however, that the "Lloyd's mirror effect" made this set unreliable for measurement of elevation angles below 2.5°, a region of particular interest for blind-landing systems. As a result, Alvarez was forced to consider other lines of approach.[9]

At a meeting held on April 2, 1942, at the Office of the Chief Signal Officer in Washington, it was decided to order 1256 units, functional copies of XT-1, to which the designation SCR-584 was given, from the General Electric Company and the Westinghouse Electric and Manufacturing Company. This order was increased several times until nearly 1700 SCR-584's were finally delivered. The first production SCR-584 deliveries were made by the middle of July 1943, 15 months after the first meeting to draw up specifications for production. In this contract (for the first time) the Radiation Laboratory was named consultant to one of the Services to follow its own experimental radar system through production. This was done at the instigation of E. L. Bowles, Expert Consultant to the Secretary of War, with the approval of Colonel Tom C. Rives, who was in charge of Signal Corps procurement.

One of the greatest problems in the production of SCR-584 was to find a manufacturer who could design and build a pedestal mount since General Electric and Westinghouse were unable to undertake the job. A mount had been designed by the United Shoe Machinery Corporation early in 1942 but USMC was not in a position to manufacture a large quantity. The Laboratory then attempted to interest the American Machine and Foundry Company but they, too, could not undertake quantity production.

In April 1942, Getting, as consultant to the Signal Corps, undertook to interest the Chrysler Corporation in manufacturing the mount, upon the advice of Robert McMath, president of the Motors Metal Manufacturing Company. With the assistance of K. T. Compton, K. T. Keller, President of Chrysler Corporation, was persuaded to accept a developmental contract from the Signal Corps for the design of the A-69 mount. Chrysler served as subcontractor to General Electric and Westinghouse, furnishing the trailer, elevation mechanism, stabilizing jacks, and, above all, the precision antenna mount for all the production SCR-584's.

XT-1, rather than XT-1A, served as the prototype of the SCR-584. During the year May 1942 to May 1943, when the factory prototype was delivered, XT-1 was completely rebuilt; greatly improved rf and indicating systems and receivers were furnished by the Laboratory. Nevertheless, all these improvements were incorporated into the production set so that when the first production SCR-584 appeared it was an up-to-date design. This is in contrast to the usual practice wherein a production design is usually about one year or so behind the latest Laboratory design. So sound was the basic design of XT-1 that changes were principally of an engineering nature to fit the set for combat use. SCR-584 was designed to feed data to the M-4, M-7, and M-9 gun directors, although its best results were obtained with Western Electric's M-9 director.

While production was under way, in August 1942, XT-1 was driven to Camp Davis, North Carolina, for firing tests by the Anti-aircraft Artillery Board. Here it was possible to compare its performance with other gun-laying sets. At this time the only American gun-laying ground radar set, besides the XT-1, which existed in prototype form and had undergone Service test was the Canadian GL-III-C, manufactured by Research Enterprises Limited, in Toronto. Four handmade models had undergone tests in Canada, England, and the United States. This was designed as an accurate position finder for large-caliber AA guns. It had separate transmitting and receiving 48-in. antennas mounted on a cabin. A separate unit, known as ZPI, contained a long-wave early warning unit, with PPI indicator. All tracking was done manually.

Contracts had been let in 1941 for two other sets, the SCR-541, to be manufactured by Westinghouse, and the SCR-545, by Western Electric. SCR-541 was intended for searchlight control and was to have been a close copy of the GL-III-C. By the omission of range measurement facilities and because of its type of construction, the SCR-541 was limited to searchlight control work and could not be used as a position finder for fire control.

The SCR-545 was a very different sort of equipment and in it SCR-584 found its nearest rival. It was expected that a preliminary model would be tested late in the summer of 1942. SCR-545 had associated with it a long-wave (50-cm) early warning unit, and its tracking was completely automatic. A single antenna mount carried the 10-cm fire-control antenna, the long-wave early warning antenna array, and a smaller IFF array. Radiation Laboratory representatives felt that of the existing sets, XT-1 represented the more satisfactory design, and of the proposed equipments the SCR-541 was both too elaborate in design and too limited in function, while the SCR-545 was perhaps too ambitious a design for prompt manufacture.[10] Actually, the first production models of the SCR-545 were produced somewhat sooner than SCR-584. The contract for SCR-541 was finally cancelled when Westinghouse received a part of the SCR-584 contract.

While the XT-1 was at Camp Davis, A. M. Grass of the Radiation Laboratory installed a Photoelectric Automatic Range Unit (PEAR) and later William Hahn and Chester Rice of the General Electric Company installed a unit which they called the General Electric Automatic Range Unit (GEAR). Both units achieved successful automatic tracking in range but the Army reversed its decision to include this feature and the manually aided tracking was kept in the production set. Radiation Laboratory members concurred, feeling that automatic range tracking was not needed and would introduce production delays.

Trials during August and September were for the purpose of testing XT-1's performance with the T-17, M9, and M7 gun directors but there were tests which showed it was possible to follow shells to 7000 yards and to observe burst, and there were many demonstrations. On August 12, 1942, Secretary Stimson visited Camp Davis and two XT-1's performed perfectly for him, tracking smoothly and accurately. A drone (PQ8) was used as target for a test at 6000 yards range and at 500-ft altitude. It was hit on the fourth shot and the visitors were disappointed because the show was over so soon. In later XT-1 tests with drones, the fuzes were cut short so as to save the PQ8's.

By this time, XT-1 was already a venerable set and had operated longer than experimental equipment could be expected to. Nevertheless, it was used at Aberdeen Proving Ground in tests to determine the trajectories of bombs and shells; and at the Marine Corps Field at Cherry Point, North Carolina, it was used, at the request of the Bureau of Aeronautics, in dive tests on the Model F6F-3 airplane which were held under the auspices of the Langley Memorial Aeronautical Laboratory. The Navy made arrangements to procure production equipment for use in future research. The XT-1A and early production SCR-584's were also used in other research projects to track aircraft during experimental and training flights to measure the performance of new planes, especially in test dives.

Before the invasion of Normandy quite a number of SCR-584's were in England and in the Mediterranean. Davenport made two trips to the ETO. The first trip, in July 1943, was made with Colonel Warner (who represented AGF HQ.) to bring information of new radar developments to American military personnel in England. A part of the duties of this mission was to train American and British operators and maintenance personnel for a testing program of SCR-584 to be held under British direction. Davenport made a second trip to BBRL in the summer of 1944 to develop a satisfactory close-support modification of SCR-584 for the Tactical Air Commands of the U.S. Army Air Forces and the Royal Air Force.

Leo Sullivan was sent first to North Africa to check equipment and later to BBRL. He spent five days on the Anzio Beachhead with the SCR-584's and then went on to BBRL to assist Davenport in the Close Support development. Abajian went to BBRL in February 1944 to assist in anti-buzz-bomb defense.[11] Colonel Warner and a staff of Army Anti-Aircraft personnel were assigned to SHAEF for the same program.

The SCR-584 first went into action on the Anzio Beachhead, in February 1944, where the long-wave SCR-268's had been put out of action by German ground jammers and "Window" attacks. SCR-545 went in with SCR-584. Its long-wave search system was jammed but it could still track and its automatic range tracking was a great advantage. Eight SCR-584's were soon set up, operated by crews who had been trained in North Africa on textbooks only and who had never seen a set until they arrived in Italy. Nevertheless, in the first night after landing the one SCR-584-controlled gun battery in operation brought down 5 of 12 enemy aircraft. After this, raids stopped for a week, making it possible for supplies to be landed and the beachhead expanded. When the raids were resumed, the SCR-584-controlled guns built up some remarkable scores; in one month they destroyed over 100 aircraft.

In England there was a pool of SCR-584's ready for the Invasion. Most of them were next used against the German V-1 bomb. As the result of a personal appeal from Prime Minister Churchill to President Roosevelt, General Eisenhower lent the British 200 sets from the American pool. Working in conjunction with the British No. 10 Predictor (the M9 director with British 3.7-in. gun ballistics), the proximity fuze, and both British and American guns, SCR-584 destroyed more than 1700 V-1 bombs in the defense of London. A great part of the operation was in overcast weather.

The speed with which the Radiation Laboratory could meet a military need was clearly shown in the planning for the defense against the Buzz Bomb (V-1).[12] Abajian, at BBRL, sent a teletype request in March 1944 for a modification kit for use with the SCR-584 which would provide automatic scanning over an adjustable sector. The tactical plans provided that each SCR-584 in the Buzz-Bomb belt scan over a specific sector 24 hr each day. Abajian's request was received in Cambridge three days before L. A. DuBridge was to leave for England. Within this time N. B. Nichols designed and built (from stock on hand) a "sector scan" device which enabled a sector, say 20° to 120° in azimuth, to be searched with no elevation movement. It was flown to New York where DuBridge was awaiting transportation for England. In less than five days from the time of request the unit had been designed and built on this side of the Altantic, flown across the ocean, installed in London, and operated. A total of 215 were built by RCC and put to use in V-1 defense as well as close support and mortar location functions in Europe, the Mediterranean, and the Pacific. Another request from the field met with an equally prompt response. On February 8, 1945, General McNarney, from MAAF Headquarters at Caserta, Italy, requested ten sector scan units for detection of mortars in counter-mortar operation. He asked that they be shipped to Pisa by March 8. The Radiation Laboratory was notified of the request at 3:00 P.M. on March 3. By working 36 hr, RCC delivered the units to Air Transport Command on the morning of March 5.

Up to this time the versatility of the SCR-584 had been little appreciated. It had been designed and used as a defensive weapon but from the time the set reached the Continent it became an offensive weapon. Instead of confining their activities to shooting down attacking aircraft, SCR-584's began giving support to ground troops by controlling friendly aircraft covering advancing columns. And for the first time in warfare the Ground Forces used radar to locate the enemy's moving vehicles behind his own lines and destroy them by radar-directed artillery fire.

The British, after their experiences in the Egyptian Campaign, had made some attempts to start a program of ground support but because of difficulties of organization it had been given up except for the adoption of the Rebecca–Eureka beacons by the RAF.[13] Work had begun at BBRL to modify the SCR-584 for close support bombing in April 1944 under Davenport's direction. The use of the set in this application is very similar to the use in artillery fire in that it makes it possible for the troops to fire at the proper place at the proper time. It differs from bombing with navigational aids such as H_2X and Shoran in that with SCR-584 control the assignment of the target is the responsibility of the ground control set. This eliminates the difficulty of target recognition and makes it possible for target assignment to be changed. It prevents bombing of one's own troops in mobile warfare. The target can be approached from any direction and the pilot can use any type of evasive tactics since he is being accurately tracked by the control set. If the plane is lost or misses the target on the first try it can be quickly oriented. No extra equipment needs to be carried in the plane.

In order to fit the SCR-584 for GCI modification the range was extended to 96,000 yards under the direction of E. G. Schneider, by B. E. Watt and H. J. Hall. Davenport, assisted by G. Huff, E. Brazell, and Lieutenant C. Schedlbauer, built a 90° automatic plotting table for Close Support application. They used the 96,000-yard modification already developed. Ralph McCreary later joined the group and developed a technique for the use of the Norden Bombsight with the plotting board to compute bomb release time. In August 1944 it was decided to build a 180° plotting table for close support to increase the coverage of the equipment. The Bell Telephone Laboratories were given a contract to produce this new plotting table which was much more successful than the original 90° table. Indication of the position of the aircraft was given by a moving pen which gave a visible trace. This table was mounted in a separate van which was, in effect, an operations room annex to the SCR-584 trailer. The weakest point in the system was the VHF communications link which could have been jammed by the enemy but never was, probably because the enemy preferred to keep the advantage of "listening in" on the instructions given to the pilots.

The first Close Support station was set up three miles north of the St. Lô breakthrough. The SCR-584's attached to the third Army played a very small role, except in the "Falcise Pocket," south of the Seine, where they performed four night missions. Although the set could be set up and in operating condition in 30 min, General Patton was not giving them this extra time. Consequently, morale was very low in the SCR-584 crews. The usefulness of the modified SCR-584 equipment to the Tactical Air Commands had not yet been realized. This situation was changed partly as a result of a visit of L. N. Ridenour and David T. Griggs, of the Advisory Specialist Group, to the IX Tactical Air Command and their recommendations to the Commanding General and to E. L. Bowles, in Washington,[14] and partly to the strong recommendations of BBRL. General E. R. Quesada made an inspection tour and took immediate action.[15] The six SCR-584's in IX TAC then became a real instrument for close cooperation of the Air Forces with the Ground Forces. In Eastern Belgium and Western Germany, with non-operational weather a good part of the time, horizontal bombing above the overcast became normal operational procedure. In the Ardennes breakthrough, ME7 and SCR-584 working together, when the ground was covered with snow and it was impossible to tell visually which tanks were German and which were ours, provided the aircraft with necessary navigational control.

Several of the Close Support SCR-584's were in France when it was decided to make another version, which was really a modified Oboe ground station. The fixed Oboe station was being out-distanced by the rapid advance of the Allied armies. The first SCR-584 station of this type was set up near Rheims with the Ninth Bombardment Division, 9th Air Force, on December 15, 1944. The BBRL plotting table was used until February 7 when the first BTL plotting board arrived, accompanied by a Bell engineer, A. J. Borer, who assisted in setting it up. A new flight plan was worked out, using coded signals so that only in emergencies would voice communication be necessary.[16] The operational site finally chosen for this equipment was 12 miles southeast of Verviers, giving good coverage into Germany.

Another problem for which the modified SCR-584 was pressed into service was in tracking the German V-2 ("Big Ben") rockets which began to be launched with London as the objective in October 1944 and toward the end of the month against Antwerp and Brussels. Colonel Warner at the Air Defense Division, SHAEF, obtained the services of Leo Sullivan at BBRL to put an SCR-584 into operation at Steenbergen to locate launching sites in The Hague area. The set went into operation on February 7, 1945, using a plotting board hastily put together at BBRL from whatever materials could be salvaged. It was possible to lock on the rocket in the ascending portion of the trajectory and to obtain data from which the origin as well as the terminal point could be established.

Hardly less spectacular was the use of the unmodified SCR-584's with the AA 90-mm gun battalions. The anti-aircraft operations at the Remagen Bridgehead, after its capture on the afternoon of March 7, 1945 by the Ninth Armored division, made this area the "most heavily defended vulnerable area since the Normandy beaches."[17] AA defense besides the 90-mm guns were barrage balloons, manned by 974th Squadron RAF, searchlights, and a SCR-584 which was used in river surveillance upstream of the bridge to detect floating mines, swimmers, etc. During the period of the Rhine crossing, March 7 to 21, 441 enemy aircraft were active. Of these, 142 were destroyed and 59 probably destroyed by AA fire. The German Air Forces made "pinpoint" attacks since one well-placed bomb could do irreparable damage. AA commanders decided on protecting the Ludendorff Bridge and the bridges under construction by "umbrella" type barrages fired over the bridges by SCR-584 control. An Automatic Weapons Officer stationed in the SCR-584 Surveillance Post gave warning and commands to fire. He broadcast to all gun units by means of a SCR-534. When an enemy aircraft was tracked to within 10,000 yards of the bridge he ordered a standby to high or low barrage depending on the height of the aircraft. He gave the command to fire when the aircraft approached to within 6000 yards. By March 16, 1945 when the Ludendorff Bridge collapsed, other bridges had been completed and the AA defenses had successfully completed their mission.

In the Seventh Army, XV Corps, SCR-584's demonstrated the ability of ground radar to detect moving ground targets during the drive to establish the bridgehead over the Saar River in February 1945. At Sarreguemines the set was dug in 1500 yards from the front line. A German spotter could have seen the camouflaged antenna easily; therefore, it was raised only at night. Enemy gun emplacements were located by the echo of shells as they left the gun muzzle; infantry regiments were furnished with the locations of enemy troops when they lost contact with them.[18] The set at Sarreguemines was a veteran of the Anzio Beachhead and still had a hole in the front of the truck from an 88-mm direct hit. An SCR-584 operating behind Maginot Fort (W658512) located enemy and friendly personnel successfully for six weeks and was instrumental in the destruction of German tanks. Two vehicles were located at a range of 28,500 yards, the greatest range, up to February 15, 1945, that the set had attained on moving ground targets.

In the Mediterranean four of the sets from the Anzio Beachhead were used by the 64th Fighter Wing of 12th TAC as GCI sets. A Radiation Laboratory member, R. T. McCoy, helped convert the PPI's so that the linear 50-mile sweep was obtained. The available material for this purpose consisted of equipment salvaged from German radio sets. Their GCI method was to use one SCR-584 to track the enemy and another the friendly fighter. Each set would searchlight and the output of both was presented on a common PPI.

The program of putting SCR-584's in use against mortars began too late in the American Army for extensive combat use of the equipment in this role. In the latter part of the Italian Campaign the ground troops had discovered that SCR-584 could be used to locate enemy mortars. At about the same time the British School of Artillery at Larkhill conducted tests to determine the effectiveness of the set in locating enemy mortar positions; they were also testing their own GL-III (Radar AA No. 3 Mark II).[19]

The British concluded that SCR-584 could locate the mortar shell, plot the trajectory, and locate the firing position with an accuracy of 25 yards in range and 3 mils in azimuth and that with a few modifications it would be applicable to the counter-mortar role.

The tests conducted by the counter-mortar school of the 15th U.S. Army were not completed until April 15, 1945. They proved that satisfactory mortar location was possible with SCR-584. The Army, however, felt that a smaller set, requiring fewer operators[20] and more easily transportable should be developed. The Radiation Laboratory began to develop a 3-cm Mortar Location Set, under the direction of E. H. Krohn and L. J. Sullivan, following a conference at Field Artillery Headquarters, Fort Bragg, North Carolina, on January 21, 1944, but it was not ready for use at the end of the war. A 1-cm rapid scan system was also built by J. S. Foster and H. R. Worthington, Jr. This set, with indicators built by the Camp Evans Signal Laboratory, was tested by the Signal Corps and used in preparing specifications for the mortar location set, AN/TPQ-2.

Another use to which the SCR-584 was put was assisting in the development of an American radar-controlled guided missile. When the German V-1 bomb was first announced, Getting suggested to DuBridge and to representatives of the Army Air Forces and the Army Ground Forces that the Radiation Laboratory adapt the SCR-584, by use of beacons, to control a guided missile. Getting considered this as long-range artillery. At the request of E. L. Bowles, General Arnold set up a project to study the possibility of using war-weary B-17's, filled with 15 tons of explosives, to crash V-1 sites, operating under some kind of remote control.

During the AAF experiments which were called "Project Willy, Mother and Daughter" two aircraft were used. The "daughter" plane was flown by television remote control in the "mother" plane. The failure of these experiments led General Arnold to start another project at Eglin Field to determine the possibility of remote control of a missile by a ground radar set. Getting suggested that SCR-584 (with variation in the frequency repetition rate added) and the BTL Automatic Plotting Board be used to send commands to a controlled aircraft. In the first tests it was planned to mount on the plane a beacon, AN/NPN-2 (BUPS),[21] together with a selector unit developed for the AAF by Laurens Hammond, of the Hammond Instrument Company. Because of the shortage of BUPS beacons, the final choice was the AN/APN-21 beacon with another selector.

The new project was called "Willy Orphan." Since Willy Orphan was almost identical in concept to close support bombing, it was a small step to remove the pilot from the controlled aircraft. This made the project come close to the definition of a guided missile. The Radiation Laboratory modified 200 AN/APN-7 (Fox) beacons to type AN/APN-21 and furnished conversion kits for SCR-584 Ground Stations.

Meanwhile, stimulated by Assistant Secretary of War for Air Robert Loveti, the Army Air Forces had started a project at Wright Field to copy the German Buzz Bomb; the U. S. version was called JB-2. The experience of Radiation Laboratory in close support bombing and in the Willy Orphan Project indicated that it should participate in providing radar control for a guided missile. A remote control system, AN/APW-1, was started at Wright Field and the Laboratory's Beacon Division assumed the sole of advisor, to provide a more modern beacon, AN/APN-19 (Rosebud), on which to base the design. The ground station was to be SCR-584 with the addition of a keying device supplied by the laboratory.

Fears expressed by Radiation Laboratory that insufficient information was available in the Air Forces to overcome the technical difficulties which might arise in the design of AN/APW-1 were finally borne out. On the advice of the Laboratory, coding was added to APN-1 to make the beacon part identical with the AN/APN-19A. This version of APW-1, with double-pulse coding, was called AN/APW-1A.

A program was also started at Radiation Laboratory with L. L. Davenport as project engineer, to develop AN/APW-3. This device differed from APW-1 in providing interrogation at 10 cm and reply at 3 cm, with an elaborate system of interrogation coding and a greatly improved high-sensitivity superheterodyne receiver. Model AN/APW-3 was built but flight testing was not possible before the war ended.

Although SCR-584 was introduced more slowly into the Pacific Theater[22] it was used in a number of actions in the Phillippines and elsewhere and was defending supply bases at the time war ended. There was little opportunity in the latest actions of the war, when attacking Japanese planes did not appear over land, for SCR-584 to display the high resolution, long range, and high accuracy which had given the Ground and Air Forces not only a new weapon but a new tool of measurement and control.

Gun Fire Control System Mark 56

The Gun Fire Control System Mark 56, of which the Radar Mark 35 is an integral part, is an outgrowth of the automatic tracking developed in XT-1 and V-1. It is an attempt to arrive at a balanced integration of radar and computer into a director system functioning as a unit and represents a break with the policy of trying to adapt radar sets to gun directors designed before the advent of radar or planned without radar as the primary source of data. The Mark 56 is an intermediate-range director to control Navy 5-in./38 AA guns.

The Bureau of Ordnance request for the Mark 56 grew out of conversations between I. A. Getting and H. L. Hazen with Captain Murphy in the summer of 1942.[23] Captain Murphy was pleased with Getting's suggestion that the Radiation Laboratory would like to undertake an integrated job for the Navy. Further conversations with Lieutenant Commander Irven Travis, in charge of the AA Direction Subsection (Re4c), brought out the information that the Bureau of Ordnance would like NDRC to carry the project through the prototype design. The project was requested and endorsed by the Navy's Coordinator of Research and Development on May 19, 1943. It was requested that Division 7 of OSRD cooperate on the computer schematic.

The project was coordinated at the Radiation Laboratory by I. A. Getting, and H. A. Kirkpatrick was appointed project engineer, assisted by R. P. Scott, for Gun Director Mark 56, and H. S. Sommers for the Radar Mark 35.

Functionally, the Gun Fire Control System Mark 56 was divided into three systems[24]: Radar Mark 35, which had the functions of locating and tracking the target and of supplying position and rate to the computer; the Director proper whose function was to furnish present position data to the computer plus target angular rates in a stabilized system; and the computer which supplied accurate gun orders and fuze time.

The Mark 35, operating on 3 cm, was the first radar to combine fixed polarization with conical scanning and automatic tracking. Fixed polarization should give freedom from "window" and other kinds of jamming. Spiral scan (wide beam) was used for target acquisition. When the target was picked up, the operator switched to automatic tracking and conical scan (narrow beam). A nutating antenna kept the plane of polarization always vertical. The range system, with its very narrow gate, permitted tracking of an aircraft when it was separated from another by as little as 25 yards in range (the improved range gate on SCR-584 was 60 yards long).

The director made use of a precessing line-of-sight gyro and computer, designed by the Laboratory,[25] which together with a vertical gyro and stable vertical (adapted from a vertical designed by the General Electric Company) kept the system on an even keel so that tracking information was always given in stable coordinates.

Getting and others from the Laboratory attended a meeting at the General Elecric Company on December 30, 1943, to discuss details of the OSRD contract which General Electric had agreed to accept for the development of components and the eventual construction of two prototype models of the Mark 56 system. From this time on there was an exchange of information between GE and RL and a division of responsibilities for the development of the model, Mark 56X, set up by the Laboratory at its Heathfield Station at Fort Heath, and also an experimental model at Schenectady. The Laboratory built two other experimental models, Mark 56A and Mark 56B.

The first radar signals on Mark 56X were obtained on December 28, 1944. From this time until the end of the war, tests were run at Heathfield. The complete system Mark 56 Model A was set up at Heathfield while Model B without the ballistic computer was installed aboard the destroyer USS *Winslow* in December 1945. In the fall of 1945, the Navy placed a production order with the General Electric Company.

SCR-598 (AN/MPG-1)

In 1941 the Coast Artillery Board—having found after several years of study that the coast defenses of the United States against motor torpedo boat attack were inadequate—initiated the development of the 90-mm Gun M1, of Gun Data Computer T-13 for the 90-mm gun, and of Gun Data Computer M8 for 6-in. and 8-in. harbor defense batteries. In May 1942, motor torpedo boat attacks on our harbors were considered a possibility and while there was an excellent general surveillance set, the SCR-682, there was no radar equipment to detect and track small and highly maneuverable craft.

On August 3, 1942, after conversations with Laboratory representatives, Colonel W. S. Bowen, President of the Coast Artillery Board, proposed military characteristics for a radar to replace SCR-296, and designed for installation in the battery commander's station in emplacements for 90-mm guns.[26] He recommended that Radiation Laboratory be requested to design and develop the radar unit. On September 2, 1942, the Laboratory received the formal request for a pilot model designated SCR-598. The resulting system, for which Henry A. Straus served as project engineer, was the most effective fire-control radar for seacoast artillery in operation at the end of the war and one of the most remarkable 3-cm systems to emerge from Radiation Laboratory research and development.

The great accuracy of the guns to be controlled and the merciless requirement of handling 70-knot boats at 500 yards range determined certain extraordinary design features. A beam extremely narrow in azimuth and with low side lobes and a 0.25-μsec pulse were necessary to give angular discrimination and range resolutions required.

The most original features of SCR-598 are the antenna and the rf system. Optical mirror theory solved the antenna problem. James G. Baker of Harvard University was asked, by I. A. Getting, to design an unspecified system with curved focal surface, concave, around a horn. The fundamental equations which Carl Schwarzschild had derived in 1905, giving the effect of mirror separation and the position of the focal point in a two-mirror system, under rather general conditions, were used. For this reason the SCR-598 antenna is known as "the Schwarzschild." The astigmatism which has to be guarded against in an optical system of this kind did not enter into the electrical problem since the antenna focuses only in one plane. The dipole source moves on a circular arc, and Baker's problem was to devise a system whose focal curve would have the desired radius. The radiator of the SCR-598 is a folded sectoral horn. The folded sections of the horn are equivalent to the diametral sections of mirrors in the optical system which was used as a guide in design. In operation rf energy is fed into the folded horn from a flared waveguide which scans in the horizontal plane. A plane wave 0.6° wide, swept through 10° in azimuth and sweeping 16 times per second, was produced. For search, the scanning system was stopped and the entire antenna was rotated through 360°. The antenna assembly was housed in a plywood shell, called the "bathtub," supported on a pedestal similar to the lower half of the one used for SCR-584.

Several indicators were provided. For search there was a 7-in. PPI with two scales, 30,000 and 80,000 yd; either continuous or sector scan could be used. The 7-in. B-scope presented an expanded version of a section of the PPI and covered an area 2000 yards deep and 10° wide and could be centered anywhere within the tracking range of the set. A second B scope was provided for shell spotting which made it possible to read range and azimuth deviations from the target to the center of impact, so that after correction future rounds should fall directly on the target.

The Radiation Laboratory prototype was assembled in mobile form and after five days of testing around Boston Harbor, it was driven to Fort Story, Virginia, in November 1943. Straus and other members of his group (R. H. Caston, R. W. Illman, and W. P. Manger) supervised installation and tests. Spectacularly successful firing tests were carried out in the presence of U. S. Army and British observers.[27] The mean tracking error was found to be five yards in range and 0.03° in azimuth. In convoys, large ships such as carriers could be distinguished from smaller craft such as destroyers and even PT-boats.

The Bendix Radio Division of Towson, Maryland, had been called in to prepare for production and had sent engineers to the Laboratory to follow the development in February 1943. At the end of the year, however, the military situation had changed to the offensive and it was no longer thought necessary to produce the fixed installation. Therefore, the SCR-598 became the transportable AN/TPG-1 for use with mobile seacoast artillery and at the request of the Marine Corps the mobile version AN/MPG-1 was ordered.

The fixed version AN/FPG-1 was at first abandoned and later revived when it was planned to install the equipment at fixed harbor defense batteries in Hawaii, Panama, and Alaska.

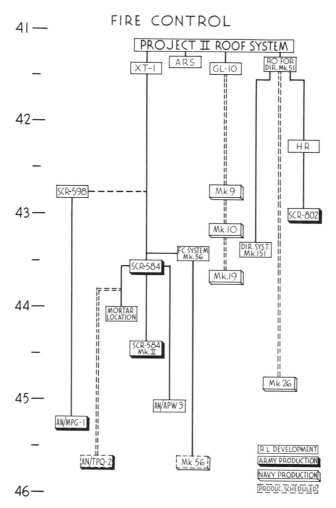

FIG. 18-1. Chart of RL developmental and production systems (fire control).

In the spring of 1944, the Radiation Laboratory agreed to send the prototye SCR-598 to the Pacific for tests at Oahu and to furnish personnel to install and operate it there. R. H. Caston and W. B. Sheriff went to Hawaii with the set. Extensive demonstrations were carried out with 90-mm, 155-mm, 12-in., and 16-in. batteries. With a locally trained crew, average tracking and spotting accuracies were found to be about 0.05° in azimuth and ten yards in range. Ships could be distinguished as separate targets when separated by only 50 yards. Splashes from 155-mm shells could be spotted at ranges out to 28,000 yards.

After these tests SCR-598 was transported to Iwo Jima where it arrived on April 25, 1945. Some time was taken in arranging for a semipermanent site and so the set was not turned on until May 2, 1945. Although the AN/MPG-1 was in production and several sets had been shipped to the Pacific by V-J day the Radiation Laboratory prototype was the first and only one to see service in a combat area.

A chart of Radiation Laboratory developmental and production work on fire control systems is shown in Fig. 18-1.

NOTES

1. Surface and anti-aircraft ground and shipboard fire-control radar sets are dealt with in this chapter. See Chapter 15 for airborne gunlaying sets.

2. Holding a reserve officer's commission from World War I.

3. Minutes of the Coordination Committee, Vol. III, p. 437, February 24, 1943; pp. 450–451, March 3, 1943.

4. I. A. Getting, New Project, Mk 51–Mk 52 Navy Directors, February 26, 1943.

5. H. A. Kirkpatrick to Dr. M. A. Tuve, November 1, 1943.

6. L. N. Ridenour, "Memorandum of Meeting Regarding XT-1 Equipment Held at Fort Hancock, January 7, 1942."

7. Louis N. Ridenour to Dr. A. L. Loomis, March 10, 1942. "In discussions with representatives of the Radiation Laboratory, members of the Coast Artillery Board, notably Col. W. S. Bowen and Lt. Col. J. McGraw, have proposed certain modifications in the design of the XT-1 which would make the unit useful tactically. The most important of these modifications is the provision for continual searching by the antenna to give early warning. You will recall that in our early discussions of the use of the XT-1, it was anticipated that it would receive notice of the approach of planes and their approximate range, bearing, and elevation from some long-range radar equipment located elsewhere... . In particular, in discussing the XT-1 with Signal Corps, we proposed that an SCR-268 would be a suitable type of early warning equipment... . The idea of the Coast Artillery Board is that the 268 is too large and unwieldy a unit to serve this purpose and too expensive besides. According to their ideas, it would be best to arrange each set so that it could do its own searching."

8. Report of Anti-aircraft Artillery Board on Project No. 1218, March 20, 1942, pp. 17-18; quoted in Col. A. H. Warner, "Report of the A. A. B. Test on XT-1 at Ft. Monroe, Virginia, February-March 1942," Radiation Laboratory Report 81-2, July 30, 1942.

9. See Chap. 19.

10. Col. A. H. Warner, "Memorandum on Centimeter Radar Position-Finding Equipment," March 13, 1942.

11. Operation Crossbow.

12. The first V-1 bombs arrived on June 13, 1944.

13. I. A. Getting, "Some Considerations on the Use of the SCR-584 in Close Support Bombing," Radiation Laboratory Report 8-6/7/44,

14. Louis N. Ridenour, to Commanding General, IX Tactical Air Command, 19 September 1944.

15. David Griggs to E. L. Bowles, 17 October 1944.

16. H. J. Hall, "Report on Long Range SCR-584 Ground Control Station," February 26, 1945.

17. SHAEF Air Defense Review, No. 8, p. 14, 30 April 1945.

18. Col. Milo G. Cary, to Hq. ETO, USA, War Dept., Observers Bd. APO 887, 17 March 1945, "AGF Report No. 737–Detection of Ground Targets by Radar."

19. Report No. 3034-44, from Military Attache, London, 30 December 1944.

20. Capt. John E. Anderson, Lt. George E. Reynolds, Tech. Obs. Carl C. Bath, Tech. Obs. Ozro M. Covington, to Director, Technical Liaison Div., OCSigO, Hq. ComZ, ETO, APO 887, U.S. Army, 26 May, 1945, "Report on the Use and Operation of Radio Set SCR-584 by AAA Gun Battalions in the European Theater of Operations."

21. The beacons mentioned in this chapter are discusssed in more detail in Chapter 14.

22. H. B. Abajian spent over six months in the Pacific on field service work with the SCR-584's. He arrived at Guadalcanal in March 1944, moved to other island bases later, and eventually landed on Leyte with the invasion forces. See Chapter 42.

23. I. A. Getting to K. T. Compton, December 29, 1943.

24. Radiation Laboratory Report 39.1, July–August 1945.

25. I. A. Getting was primarily responsible for the design of the Mark 56 System; he suggested using the line-of-sight gyro and vertical gyro. R. S. Phillips, Leader of Division 8's Theory Group was responsible for the design of the computer.

26. I. A. Getting to J. G. Trump, September 4, 1942, "Antimotor Torpedo Boat Fire Control Set."

27. "Report of the Coast Artillery Board on Project No. 1254 Service Test of Radio Set AN/TPG-1 (Experimental Model)," 20 January 1944.

CHAPTER 19
Blind Landing

Long before the war, civil aeronautics had been wrestling with the problem of blind landing of aircraft. Many agencies, civilian and military, had devised systems, none of which was entirely satisfactory for military purposes. The CAA–MIT experiments from 1937 to 1941 have already been described. With the advent of radar, which made it possible for missions to be made despite bad weather and darkness, the Army and Navy were faced with the prospect of not being able to carry out radar missions because of inability to land their planes under adverse conditions. The systems already devised were difficult for the pilot to fly and special equipment had to be installed in the plane which was too bulky to go into any fighter aircraft. The physiological condition of the pilot had not been sufficiently considered. All these systems involved interpretation of data by a pilot returning fatigued from a battle mission. It was necessary to provide a system which required only the use of military communications equipment to give the pilot instructions which could be easily followed.

Reports from England of work on jamming, beacons, and blind landing resulted in the creation of a group at the Radiation Laboratory, under the direction of Luis W. Alvarez, to study these problems. The group began to lay out a blind-landing program in September 1941. The accident that the XT-1 experimental prototype of the SCR-584 had twice been observed to follow airplanes all the way in to a landing suggested the use of radar for blind landing.[1] Alvarez had been following the preliminary experiments of the Precision Gun-Laying Group and conceived the idea that with the GL set an operator on the ground could determine the position of an aircraft with respect to a predetermined glide path. When this information was given to the pilot he could then correct his position and bring the plane in on the glide path to a landing. If the gun-laying equipment could be used, the only remaining problem was to devise a scheme by which the pilot could be given information on the magnitude and direction of his deviation from the predetermined glide path.

Two contrasting systems were ultimately devised. One resembled in general outlines the early MIT continuous wave systems but, using a pulsed glide path, required receiving apparatus in the aircraft. The other, a "talk-down" system requiring no special gear in the aircraft except its normal communications system, was subsequently used in the GCA (Ground Control of Approach) radar. If, it was argued, a ground opera-

497

tor could tell so easily by radar the precise location of an incoming plane, why would it not be possible to convey this information to the pilot by radio telephone?

PGP (Pulse Glide Path)

The first scheme explored was the Pulse Glide Path or PGP system; and it was decided to test a rather simple version while waiting for the XT-1 radar (a piece of equipment then much in demand) to be available for blind landing tests.[2]

This straight-line glide-path system was devised by David T. Griggs (who was himself a pilot) and used comparatively simple equipment. The ground transmitting system used a horizontal 10-cm dipole which was nutated through a 3-in. circle about the focus of a 43-in. paraboloid. The conical beam thus produced was divided into four distinguishable quadrants by switching the repetition rate every 90° by means of a mechanical commutator. The operation of the system was based on the fact that, in the receiver aboard the aircraft, pulse repetition rates could be discriminated from one another by the use of suitable audio filters. The amplitude of the signal strength in the various quadrants was compared by means of a cross-pointer meter. The line of equal signal strength corresponded to the desired glide path for the aircraft. After preliminary experiments during November and December 1941, J. H. Buck took over the duty of project engineer and it was decided to use a 3-cm system and to install it in a truck for further tests.

Meanwhile the Director of OSRD appointed Alfred L. Loomis chairman of a committee known as the *Ad Hoc* Committee on Instrument Landing to study the military aspects of blind landing, to consider the future needs of the Services, and to recommend programs of research and development. The committee was also directed to consider the British requirements. The committee, consisting of Army and Navy representatives as well as NDRC members, held its first meeting on December 4, 1941, and issued its final report on February 16, 1942.[3] Luis W. Alvarez and Donald E. Kerr of the Radiation Laboratory studied all existing blind landing systems, British as well as American, and submitted a report which was edited by E. L. Bowles, Secretary of the Microwave Committee. The *Ad Hoc* Committee reported that the Army had found no solution for the instrument landing of fighter-type aircraft and that the Navy had no system in prospect for carrier landings. The Navy placed emphasis on the need for a unified system for all Services. The Committee recommended that the Radiation Laboratory explore as intensively as possible the application of some GL "talk down" method of instrument landing since most aircraft could not be burdened with receivers. Also the Laboratory was advised to continue the pulse glide-path work to fulfill immediate Army requirements.

By Army request the PGP system was demonstrated,[4] in conjunction with two other systems, at Indianapolis, Dayton, and Pittsburgh, in the presence of representatives of the Army, Navy, and the British Air Commission from September 15 through November 26, 1942.[5] The outcome of these tests was a recommendation by the Army that the Sperry Gyroscope Company undertake production of a glide-path system based upon the Radiation Laboratory equipment, for it seemed that this might be produced sooner than any other. Sperry agreed to undertake the development.

The Laboratory had previously ordered some models of the PGP equipment from the Delco Radio Division of General Motors to serve as prototypes. The first Delco model was tested in East Boston from January 12 to February 5, 1943. The Laboratory's participation in this project ceased with these tests. Because Sperry did not display marked interest in the system, and because the Army did not push the project with any enthusiasm, this equipment never reached production.[6] In the meantime the possibilities of a "talk-down" system were being explored and the blind-landing program eventually proceeded along these lines.

In April 1942 the XT-1 truck was made available for blind-landing experiments. For about two months (April and May 1942) tests were carried out at Olympia Field, near Norfolk, and at NAS, Quonset Point, Rhode Island. The results were, on the whole, rather poor. The GL antenna, except under conditions of anomalous propagation, would not give low enough coverage to land airplanes. It was evident that a special radar must be designed if a "talk-down" system were to be feasible. The solution to the problem came with the decision to use the linear-array antenna, recently invented for MEW and radar bombsight Eagle. The landing project then started off in another guise as GCL (Ground Control of Landing).

GCA (Ground Control of Approach)

By July 1942 an experimental system, for which L. H. Johnston served as project engineer, was set up at the East Boston Airport and designs were drawn up for two trucks, an antenna truck and a control truck, to house the Mark I system. The antenna truck housed two 3-cm antennas of the "leaky pipe" variety with cylindrical paraboloid reflectors. A vertical antenna scanned in azimuth and fed a PPI. A horizontal antenna of the same variety scanned in azimuth and fed another PPI. These scopes were later replaced by B scopes. The power supply and rf system took up the rest of the space. The control truck housed the indicators and operators,

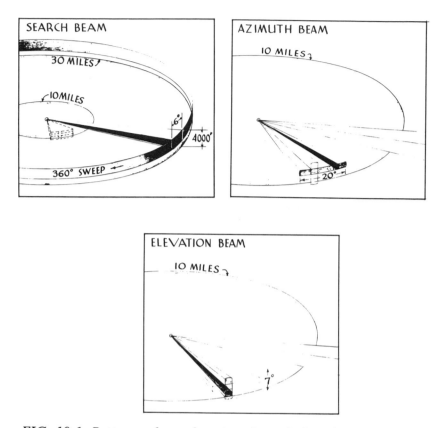

FIG. 19-1. Patterns of search, azimuth, and altitude beam move-ments for GCA.

shows the pattern of motion of these two beams plus the search beam, and Fig. 19-2 shows photographs of the corresponding scope pictures. These scopes were later replaced by B scopes. The power supply and rf system took up the rest of the space. The control truck housed the indicators and operators, the communications equipment, and a 10-cm PPI radar search system for traffic control. A director mechanism for indicating deviation of a plane from an ideal flight path was added later.

The equipment was designed to perform two functions: first, when one or more aircraft are to be landed, the search radar can "stack" all but the one plane to be landed and keep them circling in a traffic pattern about the field; second, the radar's precision units provide the operators with con-tinuous information as to the position of the plane and precise instructions are given to the pilot over the air–ground communications system so that

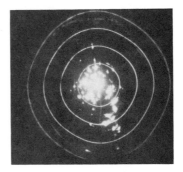

PPI

AZIMUTH   ELEVATION

FIG. 19-2. Photographs of search, azimuth, and elevation scopes of AN/MPN-1 (GCA) showing a plane making radar-controlled approach to a runway. PPI on 8-mile sweep (range works at 2-mile intervals) shows tracks of two planes, one coming in for landing, the other circling. Azimuth screen on 10-mile sweep shows track of approaching plane from about 7 miles out. Note corrections in azimuth at 4- and 2-mile range. Elevation scope at 10-mile sweep shows track of plane starting glide at 6 miles.

he is "talked down" to a point which is in line with the runway. No extra equipment has to be carried in the airplane.

While improvements in the Mark I system, to make it compact enough for one truck, were in progress, the two-truck system was taken to the Naval Air Station at Quonset Point. On December 22, 1942 the first completely blind landing under control of GCL was flown by a Navy SNJ aircraft piloted by Ensign Bruce Griffin, USN.[7] Several hundred successful landings were made here and the results led the High Command to request a demonstration at the National Airport in Washington in February 1943. The AAF immediately afterwards decided to initiate quantity production of this system. The British Air Commission requested that the trucks be sent to England for demonstration.

Before any Army or Navy contracts had been considered, the OSRD entered into a developmental contract with Gilfillan Brothers Company for the manufacture of ten systems which the Laboratory felt might be allotted to the Services for experiment and training.[8] Gilfillan engineers, H. G. Tasker, C. L. Johnson, C. W. Hargens, and F. J. Reinsch, came to the Laboratory soon after July 1942 when research was in the initial stages and they participated in early tests. This was the second time in the Laboratory's history that a manufacturing company was brought into a developmental project to work with the Laboratory system in its initial stages. In the early fall of 1942 steps were taken toward the development of a second version, called Mark II. The principal improvement was the introduction of the scannable array designed by Alvarez and Randal Robertson for the Eagle program. This was an important step but, as it required a long period of development, it could not be considered in the Mark I version. Two prototype arrays, an $8\frac{1}{2}$- and a 14-ft, were tested in the early spring of 1943. A newly designed precision indicator was also added. The Mark II system was planned for one truck and a trailer.

Several conferences were held at Camp Evans Signal Laboratory and specifications were drawn up with the help of Radiation Laboratory members. L. H. Johnston of the Radiation Laboratory was detailed to work at Gilfillan as NDRC representative and he was replaced as project engineer by G. C. Comstock. The Army decided to make Gilfillan its prime contractor and the joint Army–Navy designation AN/MPN-1 was given to the set because the Navy also ordered equipment from the Bendix Radio Division in Baltimore. Radiation Laboratory was named consultant on the Mark II GCA sets. At the Radiation Laboratory the short title was changed to GCA (Ground Control of Approach) since the military characteristics merely required the placing of an incoming airplane over the runway at an altitude of about 30 ft from which point a visual landing could usually be made.

The old Mark I system, accompanied by five Radiation Laboratory members, Alvarez, G. C. Comstock, C. A. Fowler, B. F. Greene, and E. H. Wallin, and six Army men who had been trained at Radiation Laboratory led by Lieutenant Neal A. Jolley, was sent to England for trials by the RAF during July and August 1943. On August 23 the GCA landed 21 Lancasters returning from a raid in 1 hr. and 38 min. Only four failed to make satisfactory approaches and land at the first attempt.

After the final RAF operation, Group Captain Saker, officer in charge of trials, decided to recommend that all contracts for other approach systems be stopped and the American GCA adopted instead.[9]

The first production unit of Mark II GCA was undergoing preliminary field tests during the month of January 1944. A total of 236 GCA's were delivered before the end of the war: 112 to the Army by Gilfillan Brothers Company, 49 to the Navy by Bendix Radio Division, and 75 to the Army by Federal Telephone and Radio Corporation, but the sets were slow in getting into combat use.

The landing technique with GCA involved a standard left circle of the field. The plane was picked up, as it neared the 15–20-mile range of the 10-cm search unit, by the First PPI Operator. He identified the plane by maneuver and moved him to the proper region, stacking and orbiting him if necessary. This operator was in HF and VHF communication with the pilot and the airport tower and could handle 3 to 6 planes at once. The Second PPI Operator took the plane out of the orbit and brought him on the course at 6 to 8 miles from the base. On the course the plane was taken over by the Approach Controller who used the 3-cm high-precision system, giving continuous information on the plane's position, to guide the pilot down the glide path. The rate of handling on the final approach was one plane every 2 min.

By the end of the war Air Transport Command had GCA sets operating in Iceland; in the Azores; at Gander Field, Newfoundland; at Presque Isle, Maine; and at other of its bases. Similarly, Naval Air Stations in the United States at Quonset Point, Rhode Island; San Diego, California; Alameda, California; and Whidbey Island, Washington (Seattle), were equipped with GCA and about 20 more installations were planned.

In Europe,[10] theater requests did not keep pace with the available supply of sets. The first requirement for GCA sets, issued by USSTAF in September 1944, was for only 7 units, but this was later enlarged. The Technical Requirements Section of USSTAF assigned the sets to the different Air Forces, but administrative control was under 5th Wing Army Airways Communications System (AACS). Since the equipment was completely new in the ETO, BBRL members A. P. Albrecht, C. A. Fowler, W. M. Porter, and C. S. Thompson, were assigned the duty of

FIG. 19-3. GCA (AN/MPN-1) on landing field at Etain, France, near Verdun, spring 1945.

introducing the equipment to field personnel and assisting the Signal Corps team, consisting of Captain Joseph Bryan and Sergeants Cope, Bowkamp, and Sproul, who had been sent over to supervise installation and maintenance problems.

Two GCA's were in use in the Mediterranean Theater: one at Fano, the other at Pisa. In the European Theater of Operations General P. O. Weyland was especially anxious, in January 1945, to have each group of the Tactical Air Forces and the Air Transport Wings equipped with a GCA. At that time an allotment plan for 12 sets was drawn up but, as it turned out, only eight sets were operational by the time of German surrender. Three of these were operated by the 8th Air Force in England. The first set to go into combat on the continent was located at A-82 (the night-fighter field) near Verdun, under the 19th TAC (see Fig. 19-3). Early in February 1945, before the regular crew had arrived and before the set was really considered operational, a skeleton crew of two BBRL men and a special installation crew of four Signal Corps men landed a C-47, two P-61's and a flight of P-47's under emergency conditions. This equipment was later moved into Germany. Another GCA was under control of 9th TAC, at

Florennes (field A-78), near Charleroi, Belgium; this was also moved to Germany. The 9th Bombardment Division of the 9th Air Force operated a GCA first at Peronne, France, later at Venlo, Holland (field Y-55). A seventh unit, under the 12th TAC, was used first at Lunéville, France, then in Germany, while the eighth unit was put in at the end of April 1945 near Munster, Germany, with the 29th TAC. Over 40 emergency landings were safely carried out by these first few GCA sets to go into operation.

There were no GCA's in the China–Burma–India Theater but quite a number were installed at or on their way to other locations in the Pacific Ocean areas by the time war ended. The Army had sets operating on Iwo Jima, Leyte, Okinawa, Tinian, and Saipan. The Iwo Jima GCA saved several lost or damaged aircraft, including some B-29's returning from raids on Japan. Sixteen or more sets and 16 trained crews were en route to planned installations in the Aleutians, Guam, and elsewhere in the Pacific at the conclusion of hostilities.

The main criticism of GCA from the military point of view is its inability to handle large flights of aircraft. At the Kunming Airport, for instance, where 600 planes were taking off and landing each day (about one a minute), the GCA would have been able to land comparatively few of these planes. Both the Army and Navy, near the end of the war, tended to separate the approach function from the traffic control function.[11] The AN/MPN-1 combined an effective approach set with inadequate traffic control features. The Radiation Laboratory was working on a microwave radar airport traffic control set (AN/CPN-18) at the close of hostilities.[12] In the future, radar warning systems and beacons with GCA and possibly some as yet undevised traffic control set will make possible the true all-weather airway on which flights will operate in spite of bad weather, making the whole trip without sight of the ground.

NOTES

1. Radiation Laboratory Notebook No. 11, issued to L. W. Alvarez, pp. 111–112, August 27, 1941.

2. B. Chance, D. Griggs and R. C. Raymond, "Special Report on Bolometer Blind Landing System," Radiation Laboratory Report B-2S, December 15, 1941.

3. Report of Ad Hoc Committee on Instrument Landing, February 16, 1942.

4. Col. Hobart R. Yeager (ARL) to Chief Signal Officer, Washington, D.C., July 23, 1942, subject, "Demonstration of 3 cm Glide Path and Talk Down Equipment."

5. J. H. Buck, "Report on PGP Demonstrations at Indianapolis, Dayton, Cincinnati, and Pittsburgh," December 2, 1942.

6. R. Davies to Dr. J. H. Buck, April 13, 1943, "Sperry Visits."

7. Radiation Laboratory Notebook No. 1126, issued to L. H. Johnston, p. 31, December 22, 1942; Minutes of the Coordination Committee, Vol. II, pp. 237–239, January 6, 1943.

8. Minutes of the Coordination Committee, Vol. I, August 26, 1942.

9. Capt. C. A. Ridings and Capt. J. C. Conly, Memorandum to: Brig. Gen. H. M. McClelland, 3 September 1943, "Report of GCA Trials."

10. A. P. Albrecht, C. A. Fowler, W. M. Porter, and W. S. Thompson, "Survey of GCA (AN/MPN-1) in the ETO," ASB Report No. 13, 24 May 1945.

11. G. C. Comstock to J. R. Loofbourow, March 22, 1945, "Navy Request for Advisor Service on New Simplified GCA."

12. Radiation Laboratory Notebook No. 4642, issued to G. A. Fowler, pp. 1–47; 51–85, May 2–November 28, 1945.

CHAPTER 20

K-Band Radar at One Centimeter

The third and last of the microwave regions explored by Division 14 during the war was the so-called K band in the neighborhood of 1 cm (30,000 MHz). Work was carried out in this region between 1942 and 1946. Although no production sets were built on any wavelength in this band, and although no American K-band radar saw military action, a number of experimental systems were built and a concentrated effort was carried on to perfect components for such systems. The development of the fundamental art for the band was the main accomplishment of the K-band research effort.

The 1-cm region constituted the radar frontier being explored at the end of the war, and might have provided one of the principal reservoirs from which new radar applications would have been drawn had the war lasted longer than it did. The greatly increased resolution provided by the K-band equipment—three times that of X-band radar for a given antenna aperture—gave rise to proposals for high-resolution ship equipment, for improved bombing equipment, and for radar to locate enemy mortars by tracing shell trajectories. In a prolonged war other uses of this high resolution would surely have suggested themselves.

Thus as insurance against a long war, unexpected military reverses, or successful enemy countermeasures, the K-band development program needs no apologetics, even though the equipment was not actually employed in the field. Moreover, from the long-run scientific standpoint the K-band development will probably prove as important as any other phase of Division 14's activities. The work is of great practical and theoretical significance for the advance of pure and applied science because it effectively closes the gap between the shortest radio waves and the longest waves of the infrared. The wartime K-band investigation itself gave rise to interesting experiments that provided experimental contribution to fundamental physics made by Division 14. There is every expectation, moreover, that readily available radio energy in the 1-cm band will provide a research tool of great value to post-war physics, a tool perhaps second only to the precision methods evolved during the radar program for the accurate measurements of extremely short time intervals.

Although as a military weapon it was only being held in reserve—except by the RAF which used K-band radar on a small scale to bomb Berlin in the closing days of the European war—it is worth devoting a full chapter to the K-band development because of its scientific interest and for two additional reasons: (1) because numerous improvements were introduced into the radar art on 10 and 3 cm as a result of discoveries made in the course of the K-band development; and (2) because the story of the K-band program throws a spotlight on the hazards encountered in applied science, particularly in the high-speed development work of a war laboratory.

Beginning of K-Band Research

An account of the initiation of K-band research in the spring of 1942 should deal principally with the steps taken to develop a satisfactory K-band magnetron. This is more completely described in the chapter devoted to the work of the Columbia Radiation Laboratory where the magnetron development took place.

The Laboratory at Columbia was created for this express purpose early in 1942 at the instigation of I. I. Rabi, at a time when it was apparent the 3-cm radar was emerging to everyone's satisfaction from the fundamental development stage, and that its successful introduction into use hinged principally upon engineering improvement and system design.

While the magnetron investigation was getting under way at Columbia, work on other K-band components was started at MIT. Most of this activity centered in E. M. Purcell's Fundamental Development Group (Group 41) which simultaneously began to shift to other groups the work on 3-cm components which they had carried on so successfully.[1] The first concern of Group 41 was the development of rf circuit technique on the new wavelength; some members of the group also participated in developing certain types of rapid-scanning antennas; while a third section cooperated with the Theory Group (Group 43) in fundamental waveguide studies, the results of these measurements and calculations proving of general value in designing waveguide components for all three microwave bands.

Development of a suitable local oscillator was the most important product of the early K-band research at the MIT Radiation Laboratory. This source of cw energy was of course required for two purposes: primarily for a K-band receiver; but also for all sorts of bench measurement work in connection with the development of other components on the new wavelength. This work was undertaken in the Receiver Group (Group 61) by H. V. Neher, assisted by G. A. Hobart and C. Z. Nawrocki. By August 1942 Neher had built about a dozen models of this local oscillator, and work was going forward at the Laboratory on assembling the first K-band experimental system in anticipation of the delivery of the first 1-cm magnetrons. Quite by accident—and it was a significant accident, as it turned out—some of the best models of the local oscillator operated at a wavelength of 1.25 cm.

Meanwhile, the British had a K-band magnetron and local oscillator operating at 1.6 cm. Their choice of wavelength seems to have depended solely on the fact that this wavelength was the second harmonic of the established X-band frequency, and that consequently an experimental receiver could be quickly developed.[2]

To coordinate the 1-cm work of the various American groups, and to prepare for ultimate cooperation with the British, a conference on wavelength and waveguide sizes in the 1-cm region was held at Columbia on August 7, 1942. The meeting was somewhat informal, but because its most notable result was the selection of 1.25 cm as the established K-band frequency, it subsequently became known as the Standardization Conference. Present were J. Kellogg and Arnold Norasieck of the Columbia Laboratory; G. C. Southworth, Sloan Robertson, and T. M. Odarenko of the Bell Laboratories; Purcell, Ramsey, Neher, C. G. Montgomery, and his physicist wife, Dorothy Montgomery, of the Radiation Laboratory; and E. L. Ginzton of the Sperry Gyroscope Company.

At this meeting it was decided that a wavelength of 1.25 cm would be the most suitable choice, and that the rectangular waveguide should have certain specified dimensions. The reasons officially cited for the decisions are as follows, listed in the order of their importance.

1. It was felt that any wavelength appreciably longer than 1.25 cm would not be a sufficiently great change from 3.2 cm.

2. Present evidence on atmospheric absorption at this wavelength does not lead one to anticipate any greater difficulty at this wavelength than at any other wavelengths discussed.

3. This is a good wavelength for the available low-power oscillator.

4. The $1/4 \times 1/2$ in. guide has a reasonably low attenuation at 1.25 cm.

5. A guide of this size can also be used at 1.6 cm, which is the wavelength of the existing British oscillator and magnetron.

6. Choke couplings with a circular groove designed for 1.25 cm can be built around this guide.[3]

The problem before the conferees was to suggest a standard wavelength so that a coordinated attack could be made by all the research groups

concerned with opening up the new band. Some number had to be selected. In addition to the value 1.25, the discussion centered around the British choice of 1.6 and the contrasting possibility of going all the way to 1.0
cm.

The Division 14 view was that since the large task of developing a new
band was being undertaken mainly for the sake of getting more resolution,
1.6 was not a big enough step forward. As had been done in going from S
to X band, it seemed desirable in going from X to K to get at least a
threefold increase in resolution. From this viewpoint 1.25 had all the
attraction of a compromise candidate, especially since a satisfactory oscillator was already available.

Attention was paid at this conference to the possibility of unfavorable
atmospheric effects, notably the absorption of energy of this frequency by
atmospheric water vapor. Theory led physicists to suspect that the water
molecule would show some absorption in this region of the spectrum, but
the exact absorption behavior could not be determined. The available
information afforded no reason to avoid or to select any particular frequency in the band.

An important consequence of this conference was that negotiations
were opened with the British to persuade them to accept the American
frequency of 1.25 in place of their own value of 1.6. This they had decided
to do by the time the members of the U.S. Special Radar Mission (the
Compton Mission) met with their British opposites in London late in
April 1943.[4] At these meetings the British tentatively agreed that to
further the cooperative radar research program it was desirable to concentrate the development of K-band components in the United States. By the
end of the year this had become the established Anglo-American policy.
Throughout the remainder of the K-band development, close liaison was
maintained with the British, especially through the British Air Commission office at the Radiation Laboratory. As many as a dozen British scientists worked in the Laboratory for extended periods.

First Experimental K-Band Systems

After the wavelength had been selected, work on assembling the first
K-band system proceeded rapidly at the MIT Radiation Laboratory. In
October 1942, this equipment began operating on the roof of Building 6.
The system was a crude one, and is perhaps chiefly interesting as an example of the primitive state of the K-band art. Ground signals could be seen
at only $3\frac{1}{2}$ miles. The magnetron was one of several 14-slot vane and sector
tubes which had been sent to Group 41 by the Columbia group. These
were not sealed off, but operated on the vacuum pump. Two antennas
were used because a satisfactory TR tube for K-band was lacking. These

were 8-in. paraboloids fed from the rear with scaled-down X-band wave-guides. The crystal—a standard cartridge type of the kind used for 3 cm—performed poorly. The local oscillator was Neher's early high-voltage tube, his low-voltage oscillator which marked an important advance in K-band development being still two years away.

In December 1942, a second version of the roof system was completed, with a single 8-in. paraboloid for transmitting and receiving, and a sealed-off Columbia magnetron giving 3 to 4 kW of peak power. Range was still poor; in a typical performance on January 21 an aircraft was tracked only to $1\frac{1}{2}$ miles, although this small range was partly due to difficulty in training the narrow beam. To lessen this difficulty an optical sight was mounted on the dish.

The system, however, contained two important improvements. One was an experimental model of the integral-cavity-type TR tube which was developed in cooperation with Westinghouse. This tube—the first noteworthy component contributed by K-band research to the overall microwave art—became the standard type for X band as well as K band because of its lower transmission loss and longer life, and because it was easier to replace and simpler to manufacture.

The other improvement was the stubless rotary joint, copied from British 3-cm equipment to serve the needs of K-band. Because it proved simple to make and would stand high power, it was later introduced into American X-band equipment. Both the TR tube and the stubless joint were developed early enough to be widely used in X-band radar sets that went to war.

The general purpose of the research up to this point had been to get an operating K-band radar system so that the properties and possibilities of the new frequency could be explored. The development had not been focused upon any specific military application.

The chief anxiety about K band concerned its propagation characteristics, not so much the possible deleterious effects on system performance of energy absorption by the molecules of atmospheric water vapor, as attenuation resulting from absorption and scattering of the energy by rain, fog droplets, or snow. It was expected that the second effect would be pronounced, for it was well-known that even with 3-cm equipment heavy rain could be picked up on the scope and often obscured targets. With the decrease in wavelength, the scattering effect obviously would be greater; and considerable attention was devoted to observing it. Late in January 1943, for example, Purcell reported to the Coordination Committee that ground signals from objects at a distance of $4\frac{1}{2}$ miles appeared to be reduced to about half strength by a heavy snow storm. At this time he invited discussion of possible systems for which K band might be applicable.

Early in 1943, a K-band ship project (SWK) was started to test rapid scan antennas with which to capitalize on K band's extraordinary resolution in bearing and to investigate the use of very short pulses (pulses as short as 1/20 of a microsecond) for greater range resolution.[5] Out of this project, somewhat indirectly, there later developed Project Cindy—a light-weight, high-resolution radar system for ship navigation. This radar was designed particularly for PT boats, and was accepted by the Navy for production near the end of the war.

Paralleling the SWK development, a 3-cm Hand Radar (HRX) was adapted to make it a portable K-band system. The chief purpose of the K-band version of HR was to obtain propagation data conveniently, although the possibility for developing a lighter-weight Hand Radar through use of a shorter wavelength was also a factor. In the summer of 1943 this system was taken to the field station at Fishers Island and used to study the behavior of 1.25-cm radiation over water in comparison with S and X band radiation.

H₂K Systems

The K-band application upon which the Laboratory soon resolved to concentrate was a high-resolution bombing system, a sort of super-H_2X with approximately a threefold increase in resolving power. This decision was made in June 1943, at a time when bombing problems commanded top priority, and when the importance of such equipment was being dramatized by the feverish activity of the H_2X crash program. Interest in a K-band bombing device had been expressed by the U.S. Army and Navy and by the British during the London meetings with the Compton Mission. DuBridge reported on his return from London that the most urgent radar problem in the war just at that moment and "for the foreseeable future" was radar blind bombing; and that the problem in which the British were most interested was the improvement and multiplying of the equipment used and the extension of its range. "The British," the Director reported, "take an interest in the U.S. Eagle project, which they hope will be pushed. However, they feel that the H_2S type of set can be pushed to still higher accuracy and are studying the possibility of using the K-band wavelength. This possibility should also be investigated in the U.S. as something which might be simpler and quicker than the eventual Eagle set."[6]

At this time, also, the Army and Navy expressed an interest in adding the K-band frequency to the proposed universal airborne radar, AN/APS-1. When the AN/APS-1 Technical Planning Committee convened at the Radiation Laboratory in June 1943, it was believed that by developing standardized S, X, and K-band components, with the appropriate antenna and indicator equipment and attachments, it would be feasible to design a radar to operate on the several frequencies, and serve for air search, navigation, or bombing, so that a single interchangeable set of components could replace a number of different airborne radar sets.

Responsibility for the development of the AN/APS-1 was assumed by the Bell Telephone Laboratories, and the Radiation Laboratory agreed to participate in the program as a consultant. The Laboratory's first contributions were to include the building and testing of an H_2K system, and a study of K band for bombing.

The H_2K project began in Group 95 with H. Fahnestock as project engineer. The work had a low priority, and like the rest of the Laboratory projects was completely eclipsed by the H_2X crash program which, nevertheless, could be thought of as paving the way. Work was able to proceed quite rapidly on assembling the components then available into an experimental system in an aircraft.

In August 1943, when H. Bartelink, a physicist with engineering training and experience who had resigned from one of the commercial laboratories, joined the Radiation Laboratory, he found the work, of course, still in the experimental stages. The chief difficulties were that 1-cm band components, especially the local oscillator and the magnetron, were not sufficiently perfected; but even in its crude state this equipment was flown for the first time in the summer of 1943. Because of growing Laboratory and military interest—the latter chiefly on the part of the RAF—in the prospect of getting a more accurate bombing device, in August the H_2K project was raised to 9th place on the Laboratory priority list.

During the fall, sparked by the excitement over the operational debut of the H_2X equipment, and the progress of magnetron development at Columbia, there was a sudden burst of official enthusiasm for the H_2K project. Development of the 3-cm Eagle bombing set had been somewhat sidetracked by the H_2X crash program. Now, since 1-cm radar promised as good resolution as Eagle or even better, yet with a greater field of view (Eagle saw only a 60° sector of the area ahead) and without the cumbersome Eagle array, influential opinion in the councils of the Laboratory became convinced that the AN/APS-1 (H_2K) would be the bombing equipment of the future.

At this time the range obtained with 3-cm equipment was far superior to that obtained with K band; but it was reasonable to expect that with improved magnetrons and better engineering this shortcoming of K band would largely be remedied.[7] The guiding principle was that the greater resolution obtainable with 1-cm radiation would more than compensate for slight decrease in range. Strong support was soon brought to this view when information on the performance of H_2X with the 8th Air Force began to reach the Laboratory. Radar navigators, following the results of 482nd Bomb Group use of H_2X, pointed out that at least 75% of the operator's task was to navigate to and to identify the correct target. It seemed that the greatest single need for improving blind bombing was to increase the resolution of the equipment for target identification purposes.

In October 1943, I. I. Rabi, who had been largely instrumental in pushing through the 3-cm radar development so that equipment was already going into action, well ahead of schedule, assumed the overall direction of the H_2K project; T. W. Bonner, a staff member experienced in the ways of airborne radar, was made project engineer; and an intensive effort was begun to turn the K-band equipment into a reliable system. In January 1944, the Laboratory raised H_2K to fourth place on the project list. At this moment the equipment looked so promising that there was some disposition to recommend K band for consideration in place of any high-precision version of Eagle for the B-29 program. By April 1944, after successful tests during the winter, H_2K received top billing, which it held until the end of the summer. The Stratton Blind Bombing Committee in its report on April 28 recommended that the development of a K-band blind bombing device be pushed, as the third step in a long-range program following the AN/APS-155 and AN/APS-13 version of H_2X and the Eagle Mark I.

During this period a number of important improvements were incorporated in the equipment and the difficult K-band design problems yielded one after another to engineering solutions. By the fall of 1943 the Laboratory had developed a satisfactory rf head; and with the encouragement of RAF interest expressed through the British Air Commission, a $300,000 production contract for 50 units was given to the Sylvania Company. In February 1944, the first production models of the rf head were delivered to the Laboratory.

Only a few units of this first model, known as Type I, were produced—the idea being simply to get production under way as soon as possible—but they incorporated two important K-band contributions to the microwave art. One was the shield cartridge crystal, which had been developed largely by the Bell Laboratories and which was mechanically much superior to the crystal design previously used in all microwave radar. The other, and more important, improvement was the adoption of waveguide outlets for magnetrons, in place of coaxial line; for the Type I head included the 3J30 magnetron with waveguide output developed at Columbia. Abandonment of the coaxial line technique was a major improvement for X and S as well as K band, for it made it possible to take more power from the magnetron.[8]

About the first of April, Sylvania Type 2 rf head was received at the Laboratory. It was notable for containing the higher-powered 3J31 (Rising Sun) Magnetron, as well as the first anti-TR tube to be developed for 1-cm radar. This tube, produced by the General Electric Company under an NDRC contract, with C. G. Montgomery a consultant, was broadbanded and hence required no tuning. The new head also included the Oxford tube in place of the high-voltage local oscillator that had been developed by Neher and his associates. The British tube actually required 300 V more to operate it than the Neher tube but was indispensable at this stage, for the Neher tube had proved virtually impossible to manufacture.[9]

The installation of the greatly improved Type 2 head on the H_2K experimental system in the spring of 1944 paved the way for the discovery that there was strong absorption of 1.25 cm radiation by atmospheric water vapor. During the preceding winter the H_2K had performed well. An operable system was flown in December, using an experimental rf head. Flight tests in the course of that month gave range of 45 miles on Boston. These were detection ranges, indicated by the first weak signals, on the PPI, but it was a promising performance. In January, Boston was picked up at a range of nearly 60 miles. The arrival in April of the Type 2 rf head made the early H_2K system seem primitive indeed; not only did it now have a better transmitter and receiver but better engineering had eliminated a few meters of waveguide, and with it a corresponding amount of waveguide attenuation. Automatic frequency control had been incorporated for the first time in the rf head, but it worked extremely well. In short, there was reason to believe that a successful high-definition bombing radar of the H_2S type was a virtual certainty in the near future.

But as the equipment got better and better the range obtained with it got worse and worse. The men who were flying daily with the H_2K began to suspect that the decreased performance was connected with the increased humidity that accompanied warmer weather.

As the spring advanced and the humidity rose, the performance continued to drop, until in May the ranges were about half of what had been reported during the winter—a 20-mile detection range on Boston being good performance. By this time it was nearly certain that no equipment defect could account for the loss in range, and since this loss appeared to parallel the increase in humidity, it became highly probable that absorption by water vapor was the only explanation.

The Measurement of the Absorption Effect

That there would be some absorption of K-band radiation by the atmosphere was known before K-band research had been undertaken because of knowledge of water vapor absorption lines which have their peaks in the infrared region. As has been pointed out previously, theory had also led physicists to suspect that there would be an absorption line which had its peak in the 1 cm region. The exact position of this line and its width were not known, but its intensity had been calculated in April 1942 by J. H. Van Vleck of the Harvard Physics Department who was serving as a consulting member of the Radiation Laboratory Theory Group. His confidential laboratory report on atmospheric absorption of microwaves[10] predicted strong absorption by oxygen due to certain lines in the neighborhood of 0.5 cm resulting from the fact that the oxygen molecule in its normal state has a magnetic moment. Van Vleck's prediction of the intensity of the oxygen absorption was subsequently confirmed by experiments, as were his calculations of the intensity of the water vapor absorption. At the time the report was written, however, sufficiently accurate infrared measurements had not been made to determine whether the water vapor line was broad or narrow. Van Vleck stressed the fact that the available infrared data were admittedly inaccurate,[11] and pointed out that if the relatively weak water vapor absorption line were narrow it could cause trouble for radar operating over a long distance.

The drop in H_2K ranges in the spring of 1944 indicated that the line might be narrow. As soon as the range difficulties became apparent, G. E. Uhlenbeck, head of the Theory Group at the Radiation Laboratory, on leave from the University of Michigan, was consulted. He suggested that it might be possible to improve the previous infrared measurements and that the University of Michigan, an infrared research center, might undertake to measure accurately the width of water vapor absorption lines in the far-infrared region, since on theoretical grounds it could be expected that the linewidths in the infrared would be the same as the width in the 1 cm region. After Uhlenbeck had talked over the problem with his colleague, Professor David Dennison, it was learned that a sufficiently accurate infrared measurement was probably obtainable.

Consequently, the NDRC contract was arranged with the University of Michigan, and the work was completed in about three months principally by Professor Arthur Adel. The results confirmed the surmise that the line was narrow enough to have been responsible for the trouble.[12]

While the Michigan work was getting under way, a Radiation Laboratory team went ahead with plans for controlled flight tests using a carefully calibrated H_2K, to ascertain under controlled conditions whether or not water vapor was the cause of the great attenuation of K-band radiation. A H_2K set in a B-17 was provided with fixed attenuators which could be switched in or out of the system. Both rf and IF attenuators were used, and care was taken to calibrate the attenuators under the conditions of the experiment, which consisted of flying a straight level course away from a target and determining the maximum range with and without attenuation in the system. A 4-ft corner reflector was used as a target to give an easily recognizable signal of constant strength—and meteorological data on the amount of humidity were obtained. The first flight tests with the system were made over New England during the summer of 1944. The suspicions were confirmed.

To get additional direct measurements of atmospheric absorption, and to obtain data on other matters, a series of flights in the tropics, where the humidity is high even during the dry season, were carried out from January 2 to February 7, 1945. With the cooperation of Brigadier General H. M. McCelland, the Air Communications Officer, the H_2K-equipped B-17 was taken to Recife, Brazil and Borinquen Field, Puerto Rico. R. S. Bender, A. E. Bent, and J. W. Miller conducted the tests for the Laboratory; these involved extended flights over the Caribbean and parts of Central America and of northern South America. These experiments showed the attenuation to be 0.043 dB per nautical mile for 1 g of water vapor per cubic meter of the atmosphere. Actual maximum ranges obtained on a 4-ft corner reflector were 27 miles with 11.1 g/m^3 of water vapor at Borinquen; 17 miles at Recife with 16.6 g/m^3; and 38 miles at Bedford, Mass., with 2.9 g/m^3.

In addition to the flight tests which gave a direct measurement of the absorption coefficient over a long path, and the infrared measurements which gave the value of the linewidth, what was needed was the measurement of several points in the neighborhood of 1.25 cm to permit the plotting of a water vapor absorption curve, such as is shown in Fig. 20-1. Two methods were employed: one which used the ingenious radiometer invented earlier by R. H. Dicke, and a second and more accurate method which was devised and carried out at the Columbia Radiation Laboratory.

The Radiometer Experiments

Dicke's radiometer may be described as a microwave receiver that serves as a thermometer. Radio waves can, of course, be considered as infrared radiation of long wavelength, and the basis of the radiometer is that any object having an appreciable temperature (a person, the sun, a cigar, a house, even water vapor in the atmosphere) will radiate microwave frequencies of thermal origin. The temperature of the emitting object can be measured by the amount of super-high-frequency energy received by the antenna of the radiometer. This received radiation is the cause of what is called "antenna noise", as distinguished from the noise generated in the receiver itself. In other words, a measurement of this

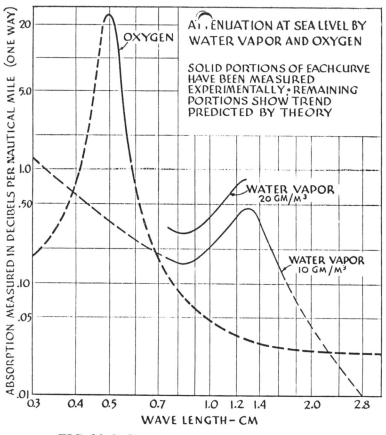

FIG. 20-1. One-centimeter absorption chart.

antenna noise will measure the temperature of the source of the radiation received by the antenna.[13]

The idea of using a microwave receiver to measure temperature had occurred to Dicke before the K-band absorption problem had raised its head, and he had worked out a design for which a patent application was filed by the Laboratory Patent Office. The chief problem was how to detect small changes in antenna noise in the presence of the much larger amount of noise originating in the receiver. This was solved by means of a switch that gave a rapid automatic comparison of the noise output of the receiver when connected to the antenna with the noise output when connected to a dummy resistor consisting of a graphite-Bakelite wheel in the waveguide. The wheel was eccentrically mounted on a shaft so that the wheel dropped in and out of the waveguide 30 times a second. As a result of this and other design features, the radiometer was made extremely sensitive to thermal radiation in a small band (0.06%) in the neighborhood of 1.25 cm. Temperature changes of less than 1°C in an object could be registered on a needle-type indicator.

When the absorption problem arose, Dicke suggested to Purcell that a radiometer be built and used to measure the absorption of K-band radiation by water vapor. The general notion was that by measuring the thermal radiation generated by atmospheric water vapor, it would be possible to determine the total absorption of the K-band radiation along a vertical path through the atmosphere at a given time. More precisely, the idea was to determine the absorption coefficient of water vapor for various frequencies by measuring the radiation coefficient for these frequencies, which thermodynamics requires should be equal.

In general, the method used was rapidly to make two types of measurements: one, with the antenna pointed straight up; the other, with the antenna pointed at a slant angle through the atmosphere. The difference between the first and second temperatures, which depended upon the difference between the lengths of the paths through the water vapor of the atmosphere, was thus observed, and the radiometer was sufficiently sensitive to measure this small difference. The experiments were carried out in Florida at an Army weather station by Dicke, Robert Beringer, Arthur Vane, and R. H. Kyhl. After an extensive series of radiometer measurements, combined with measurements of the humidity that had been obtained by aircraft soundings and from radiosondes, a one-way attenuation value of 0.04 dB per nautical mile resulted. This value, which agreed with the direct measurement, was subsequently confirmed by the more extensive and accurate measurements made at the Columbia Radiation Laboratory.

The Columbia Experiment

The Columbia method, like the radiometer technique and the method of direct measurements from an aircraft, made use of a long path—an essential requirement since the absorption was so weak. In this case, the path was obtained by multiple reflections within a large copper-lined box in which the atmosphere traversed was controllable. Thus, the element of uncertainty over the accuracy of the meterological data was eliminated.

The original idea of the Columbia experiment was to use a series of magnetrons to send pulses of energy into the box, a cube approximately 3 m in size, with varying but known amounts of water vapor present, and then to measure the rate of decay of the energy. This general idea was an old one. For example, early in 1942, before the K-band development had been started, S. A. Goudsmit had suggested measuring the "ringing" time of pulses in an echo box as a means of measuring absorption in the K-band region. Early in 1943, G. A. Hobart, III, had, in fact, attempted this measurement with the primitive K-band components then available but had discovered that the observed decay did not yield a simple exponential curve, but oscillated violently and erratically because of the interference of the different resonance modes. It was suggested, however, that perhaps if the box were made large enough, sufficiently smooth decay curves could be obtained for a measurement of the curves to be possible.

The Columbia scientists had begun to work on this assumption. However, after a few experiments late in the fall of 1944, the suggestion was found to be impractical. Late in December it was decided at Columbia that these difficulties could be avoided by abandoning attempts to measure decay curves and measuring instead a quantity proportional to the time average of the energy density at a point in the box, averaged in turn over many such points; in other words, to measure the area under the curves. This was done by distributing through the box chains of thermocouples, the "hot" junction of which had been coated with material absorbent at microwave frequencies. Voltages generated in the thermocouples were measured on a galvanometer. Since the galvanometer would measure only the change over several seconds, and since the thermocouples did not respond to beats happening in a fraction of a second, the beat-frequency difficulty was thus eliminated. And by choice of a random distribution of chains of thermocouples through the box, space variations were eliminated. Finally, the energy piped into the box by waveguides was directed against a copper fan on the ceiling of the chamber which served to prevent the building up of fixed wave patterns.

The new method made it possible to determine with great accuracy relative changes in energy absorption within the box. In order to reduce the results to absolute terms, some standard of absorption had to be provided. This was done by opening a window of known area in one side of the box. The amount of radiation passing through the window, and thus lost or "absorbed", could easily be calculated. The window device, in short, served to calibrate the box.

By the spring of 1945 water vapor absorption was measured at seven wavelengths between 0.96 and 1.69 cm. Later the box was used to check the possibility of water vapor absorption in the 3-cm band, and to extend the range investigated in the 1-cm band. The Columbia measurements determined that the peak of the absorption curve was at 1.32 cm, for which the attenuation was 0.044 dB per nautical mile for 1 g of water vapor per cubic meter. The accuracy of the experiment was extremely high, the measurements being trustworthy to a thousandth of a decibel.

The experiment itself was, of course, far more difficult and complicated than is indicated by this brief and inadequate summary. All measurements were made with the temperature of the box at 45 °C. To keep the box hot, it was necessary to heat to 115 °F the laboratory room in which the box was located. Measurements were made from an adjoining room; but during the 8 or 12 h required for a single run, prolonged visits into the overheated room were necessary. The experimental work continued through two summers.

Nearly all the Columbia staff were, in some manner or other, involved in the project, over which Kellogg had general supervision. S. Millman and M. Bernstein built satisfactory magnetrons for the several wavelengths, a requirement that would have been difficult if not impossible to meet before 1944. P. Kusch built the box. W. Lamb and A. Nordsieck, the theoretical physicists at the Columbia Radiation Laboratory, made the calculations on the errors involved in the experiment and on the proper location of the 700 thermocouples to obtain a random distribution. The experimental measurements were made by S. Autler and G. Becker.

After the discovery of the K-band absorption, emphasis in the bombing radar program was put on the Eagle set and the 3-cm band bombing radar, AN/APQ-24. However, experimental or preproduction models of H_2K were completed. These models were the AN/APS-32 and -34 and the AN/APQ-34, and were completed shortly before the end of the war. The AN/APS-32 and -34 were made by Philco for the Navy, with Longacre and Halliday as project engineers. It was planned to use the sets for low-altitude bombing, and particularly for strafing operations in enemy harbor areas. The AN/APQ-34 was made by Western Electric for the Army. It consisted of the Ground Position Indicator AN-APA-44 and the AN/APS-22, a K-band radar that superseded the K-band version of the AN/APS-1. The AN/APQ-34 was designated to use the ground position indicator part for the navigation, and the high-resolution K-band part for the bombing.

The AN/APQ-34 contained Neher's low-voltage tube and a valuable 1-cm contribution to the microwave technique: the balanced mixer, including the "magic tee", which will be described in Chap. 25 dealing with the developments of rf components. This mixer canceled out the local oscillator noise which was a special 1-cm band problem because the intermediate frequency was a smaller percentage of the carrier frequency. Both the type 1 and type 2 H_2K heads, in fact, had 70-MHz IF receivers, because of the noise problem.

The RAF, which continued to push the K-band developmental program largely because it would be the best available radar for detecting Schnorkel-equipped U-boats, also decided to use K-band radar for bombing German cities beyond the range of the Oboe-type bombing system. Since K band gave remarkable resolution below 10,000 ft—even railroad tracks could be noticed on the PPI—the RAF decided late in 1944 that the proper way to use K band was to equip pathfinder planes with it to locate the precise target for other bombers, or else have the pathfinders come in low and drop bombs on special targets, rather than attempt to use K band for high-level strategic bombing. Accordingly, a squadron of Mosquito aircraft and a squadron of Lancaster bombers were equipped with K band radar sets. A bombing raid was made on Berlin with K band, but the war ended before any widespread use of the equipment could be made by the RAF.

Only one other K-band system was developed. This was an anti-mortar system, and had its origin in a complicated chain of circumstances. In June 1944, at the request of the National Research Council of Canada, Foster, C. G. Montgomery, and Dorothy Montgomery went to Canada to assist Canadian K-band work and test a Laboratory-built system containing a Foster rapid-scan antenna. Upon returning, Foster, in cooperation with Worthington, began to build a special double-beam rapid scan antenna for a 1-cm-band anti-aircraft fire-control equipment planned by the Laboratory's Fire Control Division. This equipment was nearly completed when U.S. Army observers who had been impressed by the high-resolution possibilities of the equipment they had seen at the Canadian test approached the Laboratory for aid on mortar fire detection work. Since enemy mortars were the outstanding cause of casualties in various theaters, and since the anti-aircraft equipment was suitable for anti-mortar work, it was decided to drop the anti-aircraft project and turn the system over to Camp Evans.

Experiments with this equipment at Camp Evans in the spring of 1945, incidentally, resulted in the development of a simple method for overcoming the scattering of heavy rain on a 1-cm radar indicator. After a rain storm had blotted out mortar shell trajectories on the Camp Evans equipment he was observing in the spring of 1945, Purcell decided to try out circular polarization to reduce the intensity of the rain echoes. (A related suggestion, cross polarization, had been made in a British report.) The general notion was that since raindrops were approximately spherical, circular polarization would so reduce the echo strength from rain that echoes from a noncircular object such as a mortar shell, while reduced in strength, would still be visible. This notion proved practical. A quarter-wave plate was installed in front of a Foster antenna at the Laboratory, and reduced the rain echoes by a sufficiently large factor that it would have been possible to track a mortar shell in a rain storm as well as in clear weather.

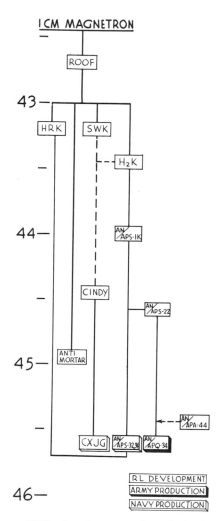

FIG. 20-2. Chart of RL developmental and production systems (1 cm).

A chart of Radiation Laboratory developmental and production work on 1-cm systems is shown in Fig. 20-2.

NOTES

1. Members of Group 41 in the spring of 1942 included, beside the Group Leader, Purcell, the Canadian scientist, J. S. Foster; N. Marcuvitz, who did the waveguide measurement work; C. G. Montgomery, Dorothy Montgomery, W. M. Preston, and F. T. Worrell. Five staff members were detailed to other groups at this period: R. H. Dicke, R. Beringer, W. P. Strandberg, T. S. Saas, and J. B. Wiesner.

2. The method of establishing a new frequency by choosing the harmonic of a familiar frequency was already a standard technique used even in the earliest days of long-wave radar. Indeed a wavelength of 3.2 was chosen for X band because it was the third harmonic of 9.6 cm. Power considerations may have dictated the choice of the second, rather than the third, harmonic of 3.2 cm.

3. "Report on Inter-Laboratory Conference on Wave Length and Wave Guide Sizes in the 1-cm Region, August 7, 1942."

4. See below, Chap. 37.

5. The SWK system, developed jointly by Group 41 and Group 102, was well into a preliminary assembly stage late in April 1943. In June an experimental 6-ft rapid-scan antenna was installed, and during the summer the system was moved to Deer Island. Here the resolution was found to be excellent, although the range proved troublesome.

6. L. A. DuBridge, "Report on Visit to England, April 28 to June 11, 1943," page 3, WA-696-14.

7. It was of course expected that some decrease in range performance would accompany the shift to the shorter wavelenth.

8. With a coaxial line, only about 50 kW can be drawn from the X-band magnetron, for above this figure the line would arc over. The waveguide outlet was one of the improvements which made it possible at once to design 3-cm magnetrons to give 300 kW.

9. Neher's low-voltage tube, operating on 450 V instead of the 1800 V required by the earlier tube, and providing a statisfactory solution of the K-band local oscillator problem, was not developed until the fall of 1944.

10. J. H. Van Vleck, "The Atmospheric Absorption of Microwaves," RL Report April 1942.

11. Especially with regard to line widths, the latest publication on water vapor [D. A. Dennison, Rev. Mod. Phys. **12**, 175 (1940)], indicated that there should be an absorption at 1.25 cm, which was close to the true value of 1.32 subsequently determined.

12. OENsr-1360, University of Michigan.

13. R. H. Dicke, "The Measurement of Thermal Radiation at Microwave Frequencies," RL Report 787, 8/22/45.

CHAPTER 21
Loran

Between the spring of 1941 and the end of January 1942 Project III was changed from an ultrahigh-frequency system into what became known as Loran. The original plan of Project III contained the basic Loran idea for hyperbolic navigation. Had the original plan been followed without modification, an American version of the British hyperbolic navigation system, known as Gee, would have been developed if equally good techniques had been worked out.

The change in direction of Project III away from a 30-MHz system to one working on approximately 2 MHz occurred during the summer of 1941. A number of factors were responsible for the change. First there was the question of what frequency would be most suitable for a long-range ground-wave system. This was a central consideration since the hyperbolic navigation system depended upon a synchronized ground wave between the master and the slave station. Because there would be little bending over the horizon of the 30-MHz radiation, there was an obvious range limitation on synchronizing a ground wave between distant transmitters. Upon joining the group, Stratton emphasized the ground-wave range advantages in going to a lower frequency, and while awaiting the delivery of the equipment on order, the group proceeded to investigate lower frequency possibilities.

During the same period the group also explored the possibilities of using waves reflected from the ionosphere. Pierce particularly influenced the ionospheric aspects of the work, and during investigation of the reflection of low-frequency radio waves by the ionosphere, it was discovered that the lower or E layer was stable enough at night to reflect radiation reliably in the 2-MHz region. This meant that it would be possible to use waves reflected from the ionosphere (sky waves) to get greater range from navigation. In other words, the 2-MHz frequency not only gave a longer-range ground wave, but at night gave a sky wave yielding still greater range.

Another factor that contributed to the reorientation of Project III was the realization that a system should be developed that would not only serve for aircraft navigation, but would have range enough for use by ships. Although small navigational errors may not be serious in peacetime, accurate navigation in foul as well as fair weather was required by wartime convoys and escort vessels. There was a particular need for a navigational service in North Atlantic convoy operations—the Cape Sable to Ireland lend-lease route having some of the world's worst weather. Finally, satisfactory equipment for the ultrahigh-frequency system had not been developed by September of 1941.

Early in the summer of 1941 the Loran Group began building equipment to test the propagation of low-frequency radiation. The first tests were made with a Bell Laboratories transmitter station set up in Deal, New Jersey. A tunable receiver, made by D. Davidson and R. B. Lawrence at MIT and incorporating a new trigger-sweep indicator built by R. H. Woodward, was installed at Weston, Massachusetts, where the facilities of a field station run by Harvard's Cruft Laboratory had been made available. Another receiver was set up in Lawrence's room at the MIT Graduate House. Sky wave reflections of 150-μsec pulses at a medium frequency of about 4 MHz were used in these tests.

Two abandoned lifeboat stations had been obtained from the Coast Guard—one at Montauk Point, on the edge of Long Island, and the other at Fenwick Island, just off the coast of Delaware. The original Project III transmitters were set up at the stations, which provided a base line entirely over water of 209 nautical miles. After the Deal tests it was decided to continue the investigation of sky-wave possibilities by running a test between these coastal stations and receiver stations in the Middle West. The general idea was simply to get data on the behavior of sky waves over land. The transmitters were modified, and made tunable between 2 and 9 MHz. Pulse modulations and other items of equipment needed for the tests were designed and by August the group was ready for the tests.

The main receiving station was set up by Donald E. Kerr in Ann Arbor, Michigan, in the home of a University of Michigan scientist, S. A. Goudsmit. The transmissions were made with a power of 5 kW at frequencies of 2.9, 4.1, 5.4, 6.9, and 8.5 MHz. Kerr received the signals at Ann Arbor, while J. A. Pierce and Mason Garfield, a former research worker at the Cruft Laboratory, made control observations with a receiver mounted in a station wagon (at Springfield, Missouri, and Frankfort, Kentucky.) All the tests were made between September 3 and 22.

These tests had the important result of indicating strongly the possibilities of stable sky-wave transmission. Even with the low power used for the Ann Arbor tests, ground-wave ranges proved greater than those anticipated at the higher frequency, and sky-wave reflection at night from the E layer of the ionosphere produced good signals at ranges of more than one thousand miles. The tests, in fact, were so promising that the group decided to abandon the ultrahigh-frequency system, even before delivery of much of the equipment on order, and to concentrate on the development of a low-frequency system.

Because in the course of these tests the circular-sweep indicator proved wholly unsatisfactory, it occurred to Pierce and Kerr that the problem of accurately measuring the time differences at the navigator's position might be solved by presenting each of the transmitted pulses on a separate linear trace on the face of a cathode-ray tube. This conception of measuring the time delay between the two traces to get a direct measurement of the difference in the two pulses was the fundamental idea of what became the Loran indicator-receiver. The traces were displayed physically to allow inspection of each pulse separately, and were separated in time. The approximate time displacement was measured by counting the number of time-marker pips that appeared on the base line between the two pulses. With this technique, sections of the traces on which the received pulses appear could be selected and compared separately on a greatly expanded time scale, thereby permitting precise matching of the pulses. After the matching, the time interval by which one trace had to be displaced from the other gave the difference in transmission time between the two pulses. The points at which the time interval is the same lie on a hyperbola; thus the measurement of the interval fixes the position of the receiver as somewhere on the corresponding hyperbola. Measurement of the interval from a second pair of stations establishes a second hyperbola, and the receiver is then located at the intersection of the two. A Loran chart consists of sets of such hyperbolas (see Fig. 21-1).

Kerr built the first Loran indicator upon returning from Ann Arbor, with Pierce and J. C. Street contributing a number of essential ideas. Simultaneously, Street and Woodward began to develop an indicator to achieve a double-line display also, but using a counter-circuit method in place of the multivibrators which were used in Kerr's indicator to divide timing frequency, provide signals to initiate sweeps, and furnish time-marker pips. Woodward developed a method of feeding back an output pulse into the counting circuit. This technique not only turned out to be an improvement over the multivibrator method but it permitted obtaining pulse rates which were not necessarily in equal submultiples of the master time frequency. The result was that a greater variety of synchronized pulse recurrence rates were obtainable, and hence more separate Loran channels, since a navigator could select station pairs on the basis of pulse rates assigned to the stations. Thus, more channels were achieved not through having more Loran radio frequencies available, but through having more pulse rates on a single frequency.

The Loran indicator provides another example of parallel invention. The key idea for the double-line indicator that measured difference in the transmission time of pulses from two stations had been used by the British in the Gee system, and had occurred independently to the mid-western field party in September 1941. With Gee the two readings are made simul-

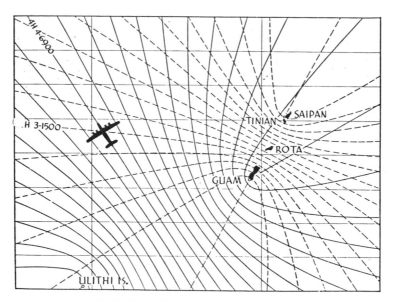

FIG. 21-1. Simplified Loran chart.

taneously, whereas with Loran they are taken successively; basically, however, the two indicators operate on the same principle. The Gee techniques had not been available to the American scientists who built the Loran indicator. A. G. Touch of the British Air Commission disclosed the general multiple-trace idea of the Gee indicator in a visit to the Laboratory in September 1941, and the field party upon returning from the Middle West proposed the same idea to the Project II group.

While the indicator and other equipment for the new system were being developed, plans were made for trying it out. It was decided to use Montauk, Long Island, and Fenwick, Delaware, as transmission stations, and Bermuda as a "ship." In January 1942, the month after Pearl Harbor, J. A. Pierce and E. J. Stephens, a former radio operator, took the receiving equipment to Bermuda.

The two transmitters, operated synchronously at frequencies of 7.7, 4.8, and 2.9 MHz and with a peak power of 5 kW, did not send ground waves as far as Bermuda; but the Bermuda test was decisive for the development of Loran in that it established the practicability of a system using night-time sky waves from the E layer. The problem of accurately measuring the time difference in the arrival of pulses was found to be solved by the new indicator, and the reliability of the sky waves was definitely established. The average of all the readings agreed with the calculated figure within a microsecond; the average error of the readings was only ± 2.8 miles.

The Bermuda tests revealed that the wide-band Loran receivers which had been developed were unsatisfactory for receiving short pulses because of interference in the 5- to 10-MHz region. It was also found that the night-time reflection of sky waves by the E layer was best at the 2.9-MHz frequency. Consequently, the Loran group decided to concentrate on developing a system on 2 MHz and to develop higher-powered transmitters for another test.

During the spring of 1942 procurement of 100-kW transmitters was expedited, along with the development of new and simplified transmitter timers. The frequency selected for the high-power test was 2.2 MHz, but when the first tests were made it was discovered that this channel was being used on the Great Lakes for a ship telephone system. As a result, the 2.2-MHz pulses began to ring telephones all over the Great Lakes, and the Laboratory was promptly instructed by the Navy to abandon that channel. The radio channel of 1.95 MHz, which had been used by radio amateurs before Pearl Harbor, was open and Loran quickly claimed it.

On June 13, 1942 the first "operational" test of Loran was made when a Navy blimp equipped with a Loran receiver–indicator took off from Lakehurst, N.J. Shortly thereafter the first trials were conducted, again with the cooperation of the Navy. The trials were so successful in demonstrating the possibilities of a more accurate navigation system that high-level interest in the Navy was immediately aroused. An early meeting was held of representatives of the Navy, Army, and NDRC, to consider the most effective way of applying the new navigation system to the war effort.

The Navy requested NDRC to procure equipment for, and to install at the earliest possible date, a chain of Loran stations to be located in Canada, Newfoundland, Labrador, and Greenland. Sufficient funds were promised to underwrite this undertaking and also to procure a number of suitable shipboard receiver–indicators for key vessels of the U.S. and Canadian Atlantic Fleets.

The Army Signal Corps was given the responsibility of procuring airborne receivers in large quantities for all services. Subsequent meetings were held during the summer of 1942, at which time NDRC agreed to increase their equipment orders to cover contemplated installations in the Aleutian area and the Northeastern Atlantic. In the meantime the Bureau of Ships and the Coast Guard were requested to set up suitable engineering, installation, and maintenance departments to assume the full responsibility for the continued procurement, installation, operation, and maintenance of equipment and for the training of operators and technicians for both ground and shipboard equipment.

During the summer of 1942, while further operational tests were being conducted by Radiation Laboratory observers on weather ships in the North Atlantic, work was begun on the two stations in Nova Scotia which had been sited during the spring with the cooperation of the Canadian Navy. At the same time a flying survey party of Laboratory and Navy personnel selected sites in Newfoundland, Labrador, and Greenland. Equipment for these stations was rushed to completion and assembled into station packages; shipments started in the late summer and continued throughout the fall. The two Canadian stations were for the most part completed during September 1942. The southern one of this pair was to be synchronized with the Montauk Point station and with its northern mate at the other end of Nova Scotia. In this way the first Loran chain was set up, wherein four stations provided three families of lines of position through double use of the intermediate stations.

Regular 16-hr daily operation of these four stations was inaugurated on October 1, 1942. This is the first time that regular Loran navigational fixes were possible and therefore may be said to have inaugurated Loran service. The initial operation of the system was accomplished with U.S. Coast Guard and Royal Canadian Navy trainees standing all watches under the supervision of NDRC engineers.

Manufacturing delays, shipping difficulties, and bad weather made it impossible to get the northern three stations in Newfoundland, Labrador, and Greenland into regular operation before the early Spring of 1943, although the Radiation Laboratory engineers[1] who supervised these installations during that critical winter ran unusual risks, such as traveling through U-boat–infested waters without waiting for escort. As soon as reliable operation had been attained in the spring of 1943, the entire seven-station system was turned over to the U.S. Coast Guard and the Royal Canadian Navy.

The second and third priorities for Loran coverage after the North-western Atlantic were the fog-bound Aleutian area and the Northeastern Atlantic. During the first months of 1943, the equipment for the Aleutians was built and tested. Early in March, this equipment was shipped; that for the Northeastern Atlantic was dispatched in May. Both these sets of installations were handled entirely by the Services—by the Coast Guard in the Aleutians, and by the Royal Navy, under Coast Guard auspices, in the Iceland–Faroes–Hebrides chain. Regular service was established in the Aleutians before the winter of 1943. The Iceland–Faroes–Hebrides installation service was inaugurated early in 1944, thereby completing the coverage of the vital North Atlantic convoy route, with reliable night-time service as far south as the Azores.

For some time after the Coast Guard assumed full responsibility for all new installations under a directive of the Joint Chiefs of Staff, the Radiation Laboratory continued to supply the necessary equipment, much of which was built at the Laboratory to avoid the delay incident to training an outside manufacturer. However, this procurement function was taken over item by item, during 1943 and early 1944, by the newly created Loran section in the Bureau of Ships. The first designs procured by the Bureau were exact copies of the equipment supplied by NDRC; modifications were made gradually as succeeding models were brought out. This was not easy, for to avoid prohibitive delays in procurement, much of the original equipment had been built around components which were not Navy approved. In many cases there simply was no such thing as an approved component to fit some of the unorthodox circuits. Many waivers had to be granted in order not to interrrupt the smooth transition to standard Navy procurement. But by the introduction of such things as unusually soft shock mounts, the shipboard equipment was able to stand up in spite of constructional shortcomings while the transition was made to entirely Navy-approved designs.

One of the most critical aspects of the rapid expansion of Loran coverage to include a large fraction of the navigable world was the tedious computation, drafting, and reproduction of the necessary navigators' charts and tables. Although the initial charts and tables were produced by the Radiation Laboratory group, the Navy Hydrographic Office took over this responsibility early in the program. By V-J day the Office had produced charts covering over 50,000,000 square miles of the earth's surface. During April 1945 alone, 230 482 Loran charts were distributed to all services. At the end of the war about 125 000 were being produced each month.

SS Loran

One of the recognized shortcomings of standard Loran was its relatively short range over land—150 to 200 nautical miles for the ground wave as opposed to 700 to 800 nautical miles over sea water. After sunset, however, the sky-wave signals travel as well over land as water, with a minimum usable range of about 200 nautical miles and a maximum of about 1300 to 1400 nautical miles. Analysis of the sky-wave readings made during the early days of the system tests indicated unusual stability with respect to timing, a stability which increased with distance. Accordingly an experiment was arranged in the spring of 1943 between two of the regular east coast stations located at Bonavista, Newfoundland, and Fenwick Island, Delaware, to test sky-wave synchronization.

Fenwick was instructed to attempt to synchronize with the sky-wave signal from Bonavista during one of the night-time periods when regular operation was not scheduled. The tests were observed in Cambridge, with all readings being made on sky waves only. Not only was excellent sky-wave synchronization maintained by the Fenwick station, but the readings made in Cambridge, after the application of simple sky-wave corrections, revealed that the probable line-of-position error was very small, about 0.5 miles. This experiment marked the birth of a subspecies of Loran known as SS (Sky-wave Synchronized) Loran.

The first obvious application of the SS Loran principle was for the navigation of bombing planes over Central Europe at night. Stations could be set up in Scotland and North Africa (see Fig. 21-2) in such a way that the service area would fall largely between the ground stations, thereby providing unusually high geometrical accuracy and uniformly good fixes over most of Central Europe. After a study of this proposal by both the AAF and RAF high commands, it was felt that the system would be best suited to the night-time operations of the RAF Bomber Command.

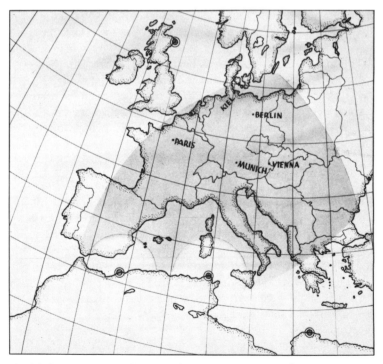

FIG. 21-2. SS Loran chart night-time coverage of Europe. Stations are shown by double circles.

The AAF did not contemplate any large-scale night-time operations and therefore was not in a position to make effective use of such a system. Before the proposal could be definitely approved, however, a complete field test of the entire system was requested. Hence, while the Navy was assuming the responsibility for what is now called "standard Loran", the efforts of the Radiation Laboratory group were concentrated on demonstrating a full-scale SS Loran system in this country.

With the help of a group of RAF trainees, stations were set up near Duluth, Minnesota and on Cape Cod, Massachusetts, for an east–west base line. Key West, Florida and Montauk Point, Long Island, were used for a north–south base line. This test system was ready for full operation in the early fall of 1943. Night after night, Army, Navy, and RAF observation planes flew throughout the east central United States navigating entirely by SS Loran. The observers included high-ranking officers of all three Services. At the conclusion of the tests late in October 1943, a complete report covering ground station performance, as well as the detailed results of the navigational tests, was prepared by the RAF delegation in Washington. The probable error of hundreds of navigational fixes proved to be within one to two miles over the entire service area.

At the conclusion of the SS Loran tests, a joint committee consisting of representatives of the Chief of Naval Operations, the Air Ministry, and the AAF decided that the system was of such operational value as to justify the diversion of much-needed U.S. Navy ground station equipment to the European Theater for RAF use. The test system was immediately dismantled and the equipment returned to Radiation Laboratory for reconditioning. Seven Laboratory engineers and a computer were sent with the equipment to the European Theater to help get the equipment into operational use.

Early in September 1944, the system went into regular operation for bombing German cities. One of the first groups to use this new system was a U.S. Eighth AF Photo-reconnaissance Squadron. RAF Pathfinder planes made the initial use by the British. Throughout the winter, as confidence in the method became established and over 500 planes became Loran-equipped, the number of SS Loran-navigated raids steadily increased. Results showed the probable errors in the vicinity of Berlin to be about 1-1/4 miles, or slightly better than those obtained during the system tests in the U.S. Because this compared favorably with the results being obtained in radar blind bombing, it was finally decided to conduct area bombing operations entirely by SS Loran. For many nights large targets like Berlin—accuracy was insufficient, of course, for small targets—were raided regularly by SS Loran-guided aircraft.

Shortly after the SS Loran system went into operation in the European Theater, the Laboratory engineers began the development and production of an air-transportable ground station for tactical use. The first experimental units, which were built for a field trial of the idea of using Loran over mountainous country, performed sufficiently well, and were so much in demand, that they were shipped to India for use in operations over the Hump to China. Although this equipment was not tropicalized and had not been carefully engineered mechanically, the Laboratory engineer, J. A. Waldschmitt, who accompanied it, was able to get the equipment into successful operation on both sides of the Hump during the fall of 1943 and the spring of 1944. These systems continued to function up to the end of the war. As a result of this success, the Radiation Laboratory was requested by the AAF to design and produce five introductory three-station chains of a completely air-transportable Loran system. This equipment (AN/CPN-11 and 12) was completed early in 1945. The war, however, moved along so rapidly that important operational use of it could not be made.

Another and more fundamental problem was also studied—that of using a much lower radio frequency for greatly extended ranges both by day and by night. Extensive studies were made during the early spring and summer of 1944 on atmospheric noise in the regions between 150 and 250 kHz. The results were sufficiently encouraging to warrant making some experimental transmissions. As the Navy and Army had both expressed an interest in the possibility of obtaining greater ranges by using lower frequencies they arranged an experimental frequency allocation at 170 kHz. In addition to use of the new low frequency, entirely new transmitting and synchronizing techniques were developed which showed much promise for increasing the signal stability and timing accuracy of all Loran systems.

In April 1945, a full-scale low-frequency Loran system test was set up along the East Coast by the Laboratory under a joint Army/Navy directive, and was manned by Army Air Forces and Coast Guard trainees. Because of the temporary nature of these test installations, located at Key Largo, Florida, Cape Fear, North Carolina, and Brewster, Massachusetts, it was not considered practical to erect the larger towers required to radiate the low-frequency Loran pulses efficiently, but barrage balloons were used to support 1300-ft vertical single-wire transmitting antennas.

The daytime signals were reported at distances as great as 2000 nautical miles from the stations, and night-time ranges were more than double this figure. Because of the long pulse lengths and uncertainties created by multipath (sky-waves and ground waves mixed) transmission, the actual probable error of reading a time difference on the low-frequency system was considerably greater that that at standard Loran frequencies. However, the geometrical advantage of longer base lines nearly compensated for this, so that navigational errors were only slightly greater than those experienced with the standard system. At the end of the war it was expected that continued experiments with sharper pulses would solve this difficulty, so that the greatly extended ranges might be used without any sacrifice of accuracy.

With the end of the war, work was continued on the Loran system to explore thoroughly its various potential uses. For example, methods were under investigation for developing a Loran set that would continuously and automatically give the navigator his position. Also under consideration was the possibility of making a connection between the automatic presentation of position on a map and the rudder of a ship so that a predetermined track could automatically be followed. A variant of this equipment could provide a simple right-left indicator to show an airplane pilot whether he is to the right or left of a Loran line he wishes to follow.

Relayed fixes—that is, devices for retransmitting the Loran indications from the receiving point to a remote indicator—were likewise the subject of research at the end of the war. One suggested use was for the study of ocean currents, with Loran providing precise and continuous data in any weather and over long periods of time. It was believed by leaders of the Loran project that an important peacetime use of such a system would be the standardized installation of relay equipment in lifeboats. The information received could be more useful for rescue work than directional data, because it would permit the searching vessels to determine immediately not only the direction but the distance to those in need of assistance. Such a program, of course, would require general use of Loran on shipboard, and a frequency channel exclusively devoted to such operations.

Since Loran navigation does not require transmission of any data from the vehicle under control, it is a system with obvious potentialities for the two-dimensional guidance of automatic projectiles. If flying bombs should become the all-weather air force of the future, no other system affords such immediate possibilities for the mass control of large numbers of such bombs. Without elaboration on such possibilities it appears plain that Loran may be termed one-world-or-none equipment.

The total cost of the research and development from December 1940 to the close of the Loran project was approximately $1,500,000. At least $75,000,000 worth of Loran equipment had been delivered before V-J Day, so that a conservative estimate would indicate that more than $100,000,000 was invested in Loran if shipping, installation, and operational costs are included. Hence, the cost of the research and development was no more than 2% of the investment in the equipment, even under the conditions of war research when money is sacrificed to buy time.

NOTES

1. J. A. Waldschmitt, W. Vissers, Jr., C. E. Henson, A. A. McKenzie, M. H. Feldings, H. Whipple, N. Taylor, A. W. Gunlon.

CHAPTER 22
Project Cadillac

The project which absorbed the largest portion of the Laboratory's attention and effort in the last year of the war was the development of a highly complex mass of equipment first referred to as Airborne Early Warning (AEW) and later for security reasons called Project Cadillac, after Maine's Mount Cadillac on the peak of which an experimental version of the equipment was tested for several months. Cadillac was America's most urgent radar project in the several months before the end of the war; it can fairly be described also as the most complex electronic undertaking of the war from an administrative as well as a technical standpoint. The organization and conduct of this program, involving the close cooperation of many separate groups, with the Radiation Laboratory serving as the principal coordinating agency, is a wartime research and development story well worth careful study.

The purpose of Project Cadillac was to overcome the chief weakness of shipboard search radar, namely, its inability to see beyond the horizon. The Japanese fully exploited this horizon limitation, coming in on our ships as low over the water as possible. In 1944 this technique was used in their effective Kamikaze attacks; a variety of tactics was used in making the attacks, but in particular the Japanese found they could circumvent the radar defenses of the fleet by making their final attacks from the zenith where there was no radar coverage, or close to the surface below the beams of the long-wave search radar of the ships. The best defense for all tactics was to pick up the planes at as great a distance as possible as they approach the task force. The Kamikaze attacks made the extension of the fleet's radar warning range a top-priority Navy problem just when the emphasis in the American war effort shifted from Europe to the Pacific.

The Cadillac equipment could also serve an important role—increasingly so as our forces in the Pacific grew in numbers and complexity—in coordinating Task Force and amphibious operations. For example, in such an operation as the invasion of Japan, when vast armadas of ships and planes would have to be used, the airborne relay radar system could provide all the Combat Information Centers simultaneously with comprehensive data on the disposition of both friendly and enemy ships or aircraft over a wide area.

The airborne relay radar, which was the key device for realizing the Project Cadillac idea (see Fig. 22-1), grew out of an earlier Laboratory

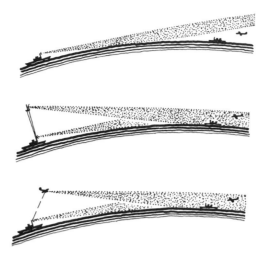

FIG. 22-1. Extending range of radar coverage.

development. In June 1942 the Committee on Joint New Weapons and Equipment had suggested developing a relay link to extend the range of a radar set; and soon afterward, the Navy requested the Laboratory to investigate the possibilities of such equipment and to develop a unit for a Service test. A television transmitter–receiver, loaned to the Laboratory for two weeks by RCA, was set up in Building 24. On August 14, 1942 transmission on a radio link between Building 24 and an experimental radar system on the roof of Building 6 was successfully demonstrated.

Difficulties encountered in these tests led to the decision to substitute frequency modulation for amplitude modulation. In September an experimental FM television transmitter was lent by Zenith Radio Corporation, and in the following months successful FM transmission between the two buildings on the campus was obtained. Plans were then made for transmission from aircraft. A 100-MHz receiver was designed and built. By May 1943, satisfactory PPI reproduction had been received at the East Boston Airport through relay radar in an airplane flying over the island of Nantucket at 10,000 ft. At this time the radio link was reliable for about 50 miles. Shortly thereafter, in July 1943, the relay radar (AN/APS-14) was demonstrated to Naval officers at the East Boston Airport; a short film illustrating the system was prepared for COMINCH, which resulted in a request that the reliable range be extended to 100 miles or more. This was done.

Since immediate production had not been decided upon at the end of December 1943, Division 14 asked that the relay radar project be terminated. A month later, however, the Navy proposed a project to develop an

Airborne Early Warning (AEW) system, incorporating a high-powered relay radar device. The AEW project was then developed under the Laboratory code name of Project Cadillac.

The basic idea for Cadillac—extending the ship search antenna into the air—was simple enough. Achieving a workable system, however, was one of the most complex and difficult developmental problems attacked by the Laboratory. The project was not only the largest in the history of the Laboratory, but was of a new order of magnitude. In addition to engineering of the complicated airborne system, an equally complicated shipboard system was required. So too was identification (IFF) equipment, test equipment, and a suitable voice communication system. In short, several types of electronic equipment, integrated into a system, were needed in the shortest possible time.

The Cadillac project was first proposed in February 1944. After a series of conferences between representatives of the Bureau of Aeronautics and the Laboratory, the Navy in April 1944 formally requested the NDRC to establish the project. In March 1945, just 13 months after the first request, the first production system was delivered to the Navy.

The Cadillac production achievement was made possible, of course, by the scale of the cooperative effort. A large proportion of the members of nine of the Laboratory's eleven divisions worked on the problem. The Bureau of Aeronautics, the Bureau of Ships, the Naval Aircraft Modification Center at Philadelphia, the Naval Research Laboratory, several Navy contractors, and a number of Radiation Laboratory subcontractors all contributed substantially to the equipment as it finally evolved. The scope of Project Cadillac at the Laboratory is indicated by the fact that direct outside purchases for the project (not including large stockroom withdrawals) constituted 12% of the total expenditures of the Laboratory for materials and services for the entire five years of its existence. At the peak of the program, during the summer of 1945, approximately 20% of the time of all the Laboratory's technical staff members working on specific projects was spent on Cadillac. Navy personnel at the Laboratory who were an integral part of the Laboratory program reached a total of 160 officers and men.

The AEW Cadillac I System

The AEW Cadillac I system as it reached the fleet was in two basically separate sections: an airborne section, carried in a modified torpedo bomber, and a shipboard section for the presentation in visual form of the information relayed from the airplane. The complexity of both these sections was an outstanding aspect of the project.

The airplane, a TBM-3, was redesigned to carry the estimated 2300 lbs of equipment. The Naval Aircraft Modification Center (NAMC) at

Philadelphia undertook to adapt the plane and to conduct the necessary wind tunnel and flight tests. An 8-ft-diameter bulbous radome was mounted between the aircraft's wheels to house the radar antenna. The ball turret, armor, and armament, including the torpedo bay, were removed. Two additional tail stabilizers, a high-power supply operated by the engine, substantial modifications to the interior of the plane, and the mounting of nine different antennas at various locations on the wings, tail, and fuselage, completed the plane modifications.

The radar equipment, AN/APS-20, operated at 10 cm with a peak power output of 1 mW and a 2-μsec pulse. The heart of the airborne section was a complex synchronizer and a radar receiver with many new features. The airborne section also included the IFF interrogator–responsor, AN/APX-13. This was designed with the highest available peak power (2 kW) and most sensitive receiver in any airborne IFF development, to make possible the identification of targets on both of the standard Navy A and G bands at ranges comparable to the detection ranges of the radar equipment. The relay-radar transmitter, AN/ART-22, which was based upon the earlier Laboratory project, broadcast on any of several channels around 300 MHz as selected, both radar and IFF information for reception by the shipboard section. Both the IFF and relay equipment were synchronized by the radar synchronizer, which also coded their outputs prior to relaying so as to "get through" interference or enemy jamming. A modified fluxgate compass used to orient the radar information, a new-type radio-control receiver AN/ARW-35 making possible control of functions of the airborne equipment from the shipboard section, a relay system AN/ARC-18 for relaying voice communication between ships and planes or other ships over the horizon, and the standard IFF transponder, voice communication equipment, radio altimeter, and homing receiver completed the airborne electronic gear.

Any ship equipped with the shipboard section of AEW could, if within relay range, receive and display the information relayed from a plane. The shipboard section of AEW was also composed of several different devices, depending upon the requirements of the particular installation. The relay receiving equipment used either an omnidirectional or a horizontal diversity receiving system to pick up the information broadcast by the relay transmitter in the plane. Adjustable bandpass tuning cavities in the antenna line, and line filters for all other shipboard systems which transmitted side band energy on the relay frequency, were developed to minimize interference with the relay reception. The relay information, after it had been decoded by the complex and precision-adjusted decoder, was piped to two, three, or more PPI's located usually in the ship's Combat Information Center (CIC).

The picture on each indicating scope could be expanded in various ways. Facilities were devised so that the AEW airplane's motion could be eliminated and the picture centered on the receiving ship. Another innovation of the indicating system was the delayed PPI (also available in the plane) by which any 20-mile region of the main PPI picture could be expanded for detailed examination over the complete face of the tube.

A transponder beacon (YQ) of the "Black Maria" type responded in code on the IFF G band to interrogation from the plane, making possible identification of the receiving ship in the midst of other shipping. Interrogation could be radio controlled from the ship and was accomplished by the coincident reception of 10-cm radiation from the airborne radar and G-band reception from the airborne IFF transmitter; or, if so desired, by coincident reception of the 10-cm radiation and a trigger signal transmitted over the radar relay link. A radio control transmitter AN/ARW-34 and standard voice communication equipment completed the shipboard electronic equipment.

Development of Cadillac I

To speed the development of a coordinated system, a ground radar set which simulated at lower power the projected AEW radar performance was established on Mount Cadillac, near Bar Harbor, Maine, where it operated for several months. Five complete air and ship AEW experimental systems were planned and constructed. The first airborne section was completed and flown in August 1944; the other four airborne sections, each improved somewhat over the one preceding it, followed at intervals of about a month. Shipboard sections were completed at the same rate, though started approximately one month later.

The AEW organization began in rather modest terms. W. P. Dyke, earlier the project engineer for ASG and ASD-1, was put in charge. Somewhat independent investigations of various aspects of the system were initiated in the Transmitter Division and Receiver Division groups concerned. However by May 1944, when acceptance of the project was recommended to NDRC by Division 14, a complete organization was established as shown in the chart (Fig. 22-2) on which only the key individuals are indicated.

In October 1944 the first full scale demonstration of the AEW system was given to a large group of Navy and Army leaders. For two weeks preceding the demonstration, two 8-hr shifts of personnel operated at the Bedford Airport to get two planes and one shipboard set into good operating condition. Although briefly plagued by aircraft engine trouble, the demonstration was successful; indeed, too successful, many Laboratory

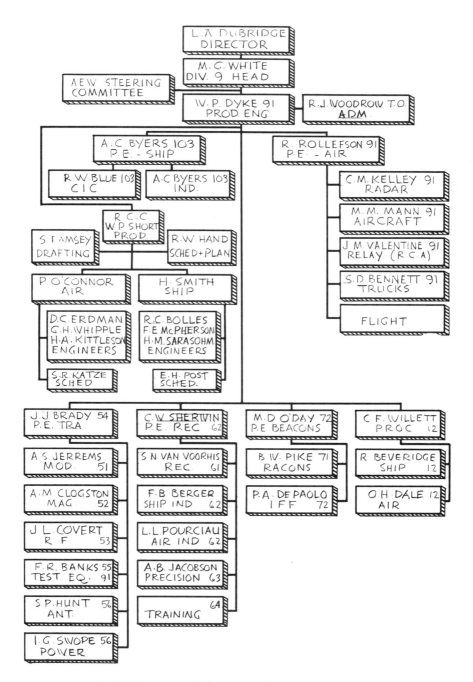

FIG. 22-2. Block diagram of AEW project T.O.

members felt, since temporarily thereafter the urgent requests of the leaders of other projects for more personnel were not met.

Flight Testing

Starting with the first AEW-equipped plane, whose radar operated reasonably well on its first flight in August 1944, continuous flight testing of AEW was carried on from the Bedford airfield until the end of the war. Three experimental planes were eventually fitted out at Bedford. A fourth set was kept operating on the bench to try out new ideas on the ground, and also to serve as a spare whose components could be substituted at short notice for defective components in the planes. The first two experimental shipboard systems were also set up at Bedford and operated from there for many months. The third shipboard set, scheduled for Navy trials at Brigantine, New Jersey, was put into operation during December 1944 in Building 20 of the Laboratory, in order to simulate more closely the heavy interference conditions expected in actual operation.

To the dismay of the research workers, the complex system jammed itself; that is, interference was so bad that rotational data, transmitted by a double-pulsed code over the relay link, was almost completely jammed. Under the threat of possible failure for the program, however, a triple-pulsed coding system was devised and, by around-the-clock work, incorporated into all experimental synchronizers, relay receivers, and decoders. With many misgivings on the part of the engineers because of inadequate testing of this change, shipboard experimental set number three (SX-3) and the second airplane with AX-3 were shipped to Brigantine for the Navy trials about the first of January 1945, only two weeks behind schedule. Fortunately, the new coding system performed well.

During the hectic month of December 1944, the project engineer, W. P. Dyke, contracted an illness which made him unavailable for the duration of the AEW program. Unfortunately, just at this time the Navy's pressure for early production deliveries increased; at a meeting in early December called by the Deputy Chief of Naval Operation (Air), the fleet's great need for AEW to combat low-flying planes and the Kamikaze attacks was officially disclosed. An overriding priority was added to the already top position of AEW in the electronics field. The Navy made available to the Laboratory crews of officers, technicians, and draftsmen as fast as they could be assimilated. A special air transport service to facilitate deliveries of parts and personnel transportation was also set up by the Navy.

Production

In July 1944, before the first of the experimental models of AEW had even been flown, the importance of solving the early warning problem had increased to such an extent that the Navy officially confirmed its earlier indication through NDRC that the Radiation Laboratory and Research Construction Co. undertake production of 40 complete systems. Production planning and a number of large production subcontracts were immediately started, thus making the program truly "crash," with research, development, and production proceeding concurrently.

In order to achieve the coordination necessary for engineering, production, and delivery by the Research Construction Company and the 30-odd major and multitudinous minor subcontractors of the Laboratory, responsibility for this phase of the activity was largely delegated to R. J. Woodrow. C. M. Kelly, previously in charge of radar research and development, was designated the production engineer for the airborne section. A. C. Byers, in addition to his duties as project engineer on other shipboard research and development, was chosen as the project engineer for the shipboard section of AEW. Weekly meetings were initiated with RCC in November 1944, and close liaison was maintained with the Laboratory subcontractors, many of whose engineers spent a substantial portion of their time in Cambridge.

In February 1945, after the formation of the Project Cadillac Committee, of which L. A. DuBridge was Chairman and M. G. White Vice-Chairman, responsibilities were reassigned. J. B. Wiesner became project engineer and the following assistant project engineers were appointed: R. E. Meagher (shipboard system); R. Rollefson (airborne system); R. M. Alexander (relay radar); J. M. Valentine (relay radar transmitter); P. A. Depaolo and D. R. Young (beacons and IFF). (See Fig. 22-3.)

It might appear impossible to schedule research and development in any detail; if the component parts and steps to be taken were known in advance, no research and development would be necessary. However, a rather unorthodox, and to a certain extent backward, process of scheduling was attempted on the five AEW experimental systems. The estimates, based on previous experience of the Division Heads and Group Leaders concerned, were combined to give target dates for the delivery of each system. Then as designs crystallized and construction began, more detailed target dates for components and subcomponents were projected backwards from the system target dates.

Bottlenecks were thus discovered, and additional effort was concentrated on them by decreasing personnel or expediting critical procurement items, by special attention in the shop, or by any other shortcuts that could be devised. Although the original target dates were missed by times ranging from two weeks for the first system up to two months for the fifth (which had many features not originally contemplated), the results were far better than most people had expected. The scheduling procedure used

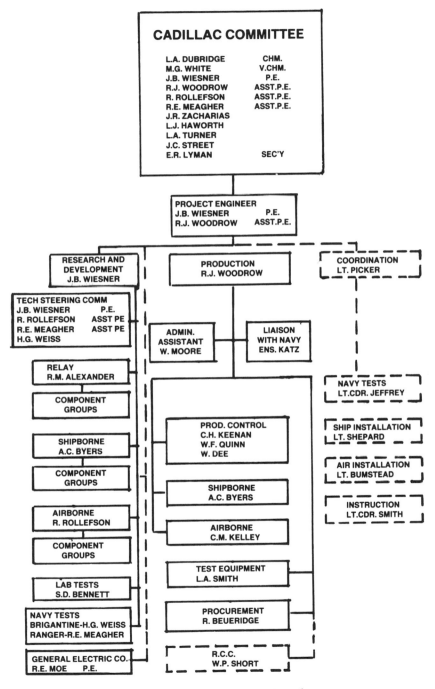

FIG. 22-3. AEW organization chart.

is believed to be the only type generally practicable for research and development; even with it, results would have been poor had the project not had a high priority.

Since research and development were proceeding in parallel with production, a very flexible method for incorporating changes was essential. A number of target dates for final freezing of design were set, each one advanced over the previous date; but actually there never was a final "freeze" date, but instead, a progressive elimination of the number of modifications to be made in designs which were in the various stages of production.

The original production schedule proposed to the Bureau of Aeronautics in June 1944, prior to the formal request for production, was to start deliveries with two complete systems in February 1945. In November 1944 a revised schedule was presented which called for the delivery of one system in March 1945, followed by four in April, and approximately eight per month thereafter. Although very great efforts were later made to advance this schedule, the final deliveries, except for items subsequently added to the system, for the most part conformed to this delivery date.

The fourth of the five experimental airborne and shipboard sections of AEW built at the Laboratory were scheduled to go to RCC as prototype units and to serve as the first complete test bench system into which early production components could be substituted and tested. It had been hoped that, prior to delivery to RCC, these units could be type tested for altitude, temperature, humidity, and vibration. However, the pressure of time and the delays in delivery of adequate test apparatus made it impossible to test more than a few of the critical components. To supplement these tests, complete type tests were run on one of two extra systems constructed. Several necessary modifications were revealed by these tests and modification kits were prepared and distributed.

Performance testing of each production component as it came off the production line was handled by complete and in some cases elaborate test procedures. Following the individual component tests, all of the critical components of the system were assembled at RCC and tested in complete bench systems, of which two and the elements of a third were eventually established. This set was a quite radical departure from previous radar production practices, but proved necessary because the space, facilities, and personnel available were inadequate for the assembly and test of all the components of each of the 40 systems at the same time. Such a procedure also made it possible to test and deliver in advance those airborne components requiring the longest time for installation in the planes.

In view of the complexity of the Cadillac system—the airborne and shipboard sections each contained approximately 200 vacuum tubes not counting those in the standardized units—a comprehensive maintenance program was planned; and at the Navy's request, the Laboratory presented in July 1944 complete recommendations for maintenance in the Fleet. These recommendations included the spare parts to be furnished, a tentative list of the test equipment believed necessary, the complement of maintenance and operational personnel which should be trained, and the importance of having a complete bench test system of the airborne section, and a standby shipboard section aboard each carrier equipped.

Between the time that these recommendations were prepared and the time of final installation and use, a year of many developments intervened. To provide greater flexibility, each equipment was provided with enough spare parts to insure approximately one year's operation. 70% of the test equipment originally proposed was either modified or replaced by other items.

It was obvious from the start of the AEW project that such complex equipment would require comprehensive instructions for maintenance and operation. The Laboratory's Publications Group (Group 35.2) and a subcontractor, the Jordanoff Aviation Co., prepared the text and illustrations for the 820-page airborne and 570-page shipboard maintenance manuals. Some writing and a substantial editing job remained, which were handled by the staffs of the airborne and shipboard production engineers and by Lieut. Robert Kellner of the Navy. Preliminary handbooks to cover the maintenance of the experimental models during Navy trials served as prototypes for the final manuals. Because of the large number of modifications ultimately made in the production systems, final shipboard instruction manuals were not available for most of the training stages and the first several installations, so that a sufficient quantity of interim handbooks were hectographed for these purposes. Handbooks of instruction had also to be prepared and published for 12 of the 18 test instruments supplied by the Laboratory and RCC.

New Trials

Navy trials of the experimental AEW systems had been contemplated early in the program; these eventually included not only operations at the CIC Group Training Center, Brigantine, New Jersey, but also installation and operation of the equipment on board the carrier USS *Ranger.*

The third airborne and shipboard experimental sections of AEW, following the flight tests at Cambridge previously described, were put into operation at Brigantine during the first part of January 1945. Many problems, some foreseen and others not, were almost immediately encountered. To accelerate their solution, a substantial reorganization of the project at the Laboratory was made about the first of February 1945. The new organization, the important aspects of which are shown in Fig. 22-3, operated essentially as a separate Laboratory division known as Project Cadillac.

Project Cadillac coordinators were appointed in the office of the Chief of Naval Operations and in the Bureau of Aeronautics and Bureau of Ships. The Bureau of Aeronautics in particular assigned a substantial staff to the project. The outlines of the organization of the Navy Liaison Office at the Radiation Laboratory are likewise indicated in Fig. 22-3.

Even before the Brigantine tests were completed, most of the improvements recommended were well under way and being tested at Bedford. Modifications were incorporated in the shipboard relay receiver and decoder to improve their performance under conditions of interference. The Naval Research Laboratory, which had previously collaborated in the solution of mutual interference problems in the airplane, worked out the design of filters for other types of shipboard electronic equipment to prevent their transmitting appreciable amounts of energy at the relay frequency. Special anti-clutter circuits were developed for the radar receiver to facilitate the distinguishing of signals through the clutter of echoes at close ranges from the surface of the sea.

While the Brigantine trials were still underway, the equipment for sea trials aboard the *Ranger* was made ready and shipped to the West Coast. When the planes joined the carrier in April 1945, the installation was essentially complete and tests started almost immediately. The *Ranger* trials lasted for two months, and appeared to establish the value of AEW beyond question. Following close upon the end of the war, at least two carriers, the *Enterprise* and the *Bunker Hill*, made trial cruises during which AEW was used.

During the trials of the AEW Cadillac I system, many data on performance were collected. It was found that single aircraft of torpedo bomber size, flying at 500 ft, could be consistently detected at ranges of 45 to 70 miles with the AEW plane flying at 2000 to 5000 ft altitudes. This was twice the range of the best shipboard radar system on similar targets. Groups of 6 to 14 planes at 500 ft were detected at ranges varying from 60 to 120 miles, or two to four times the range of shipboard sets. Surface vessels of destroyer size or larger could be detected at 200 miles with the AEW plane flying at 20,000 ft, increasing by a factor of six the previously available range. The relay equipment proved reliable out to 45 miles from the receiving ship, thus making possible a further extension of the detection range in the direction of the AEW plane.

Cadillac II

In June 1945, while the Cadillac I program was at its peak, reports from the fleet indicated the need for the type of long-range reconnaissance, warning, and control made available by AEW, but in locations unsuitable to the operation of ships having AEW shipboard equipment. Upon request from the Navy to NDRC, the Radiation Laboratory initiated the Cadillac II program. This program contemplated the development and production of the necessary equipment for an airborne Combat Information Center in a four-engine bomber. Such a system had been considered much earlier, but the Laboratory and Project Cadillac were already so heavily engaged that it required an expression of the top priority of the program before it could be undertaken.

As conceived almost from the start, Cadillac II embraced the installation of all of the previously developed Cadillac I airborne equipment plus a much increased complement of Navy-furnished communications gear. The new element in the system consisted of the CIC equipment, which was installed in the completely remodeled bomb bay of the plane. Large 12-in. off-center PPI's, equipment for ground-stabilizing the PPI presentation, and various associated apparatus were the major new contributions of the Laboratory. Plotting boards and most of the nonelectronic accessories for the CIC were developed and produced by the Special Devices Division of the Bureau of Aeronautics. Modifications of the planes and installation of the equipment were again handled by NAMU.

When Cadillac II was initiated, the construction of 11 complete systems was contemplated, for which 11 sets of the Cadillac I equipment were to be diverted. The quantity was progressively increased to 13, then 17, and finally 25; this was eventually cut back to 17 at the end of the war.

Many of the same problems of development, testing, production, etc. as were encountered in Cadillac I also appeared in Cadillac II, but to meet the very tight schedule—and since the scope of the second program was by dollar cost only a fifth of the former—substantially all the production was done at the Radiation Laboratory using Laboratory and Navy personnel. Deliveries of the 17 systems started in August and were complete by the end of October 1945.

In addition to the development and production of the 17 CIC indicator systems (AN/APA-53) plus spares and instruction manuals, Cadillac II also included the establishment at the Laboratory of a complete trainer for the system installed in a B-17 fuselage. Simulated data of the radar and IFF performance was fed into the radar indicators by a basic trainer developed earlier as a result of a Navy request on NDRC. Shortly after the end of the war the first completely equipped plane was ready to fly.

The Cadillac II program opened up a new field for operational and tactical use. The obvious possibilities of control of a fleet of aircraft from one or more airborne control centers could have revolutionized large air operations, and as ramifications of the original program, several other developments resulted which could have further extended these possibilities. An airborne moving target indication (AMTI) system had been flown and showed much promise of discriminating between moving targets and echoes from the ground or sea so that AEW operation over land had important possibilities. Designs had been completed and contracts let for complete airborne combat information center facilities for simultaneous and independent control of several combat air patrols. Means for determining the altitude of aircraft targets were likewise under study. However, the possibilities of Cadillac II, like those of Cadillac I—and like all the most advanced wartime developments, for that matter—unfortunately had no opportunity to be realized in the war.

CHAPTER 23
Trainers

As would be expected with any new weapon having a widespread use, the production of radar sets for a variety of new military applications resulted in an extensive development of radar training aids. The first demonstration of the need of radar trainers was in the Battle of Britain. Poor initial success with the AI radar on the part of RAF night-fighter operators led to the introduction late in 1940 of an AI trainer so that the night-fighter pilots could simulate an interception while still on the ground. This meant an obvious saving of time, planes, fuel, and radar sets. Moreover, students were able to make serious mistakes safely, and get classroom instruction in proper use of the radar. It was soon demonstrated that the use of the trainer rapidly increased the efficiency of the flight personnel on AI night operations.

In October 1941 work was begun by the Trainer Group (Group 64) at the Radiation Laboratory on the design of an AI trainer for the SCR-520. This marked the start of the Laboratory's contribution to the vast radar training program conducted by the Army, Navy, and Air Forces. The important part of this contribution was the development of radar trainers for microwave systems. By V-J Day the Laboratory had participated in the development of more than $9,000,000 worth of radar training equipment. Altogether radar trainers costing approximately $12,150,000 were produced by the end of the war.

The radar trainers developed by the Laboratory may be classified under the following headings: scope interpretation trainers, tracking trainers, scope adjustment or minimum discernable signal trainers, specific manipulation trainers, individual tactics trainers, small crew cooperation trainers, large crew cooperation trainers, and combination crew and tactical trainers. One design, of course, was often adaptable to a number of different purposes.

Scope interpretation trainers, as the name implies, are devices used to familiarize the student with the particular type of indicator employed in a radar system. These devices serve a particularly useful function for scope presentations where there is considerable distortion in the image. For example, a B scope shows a distorted picture since a semicircle of terrain is presented as a square or rectangle. A training device may be quite simple if it is only desired to familiarize the student with the art of scope reading. In this case it may consist of an artificial pulse generator so arranged that various ranges and azimuth angles can be set in at the discretion of the

551

instructor. A trainer of this type, for example, was designed at the Laboratory by G. R. Paine and R. L. Garman and manufactured by the Sanborn Company under the Army designation RC-225-T1.

If the device is used, however, to train the operator in the art of translating distorted signals from land masses back to true plan, a more complicated unit is required. This may consist of an optical image in which the proper distortion has been effected or it may consist of a map-reproducing device using television techniques together with a regular B scope. An ASV radar trainer, known as RC-227, which had been developed by Garman and produced by Philco in June 1942, was a trainer of this type. In this equipment a map of the region to be shown on the scope was prepared in the form of a photographic transparency with areas such as land masses which give radar return represented on the map as transparent areas, the degrees of transparency being proportional to the intensity of return expected from that particular region. The map was then illuminated by a pinpoint of light which moved across the map just as the beam from a radar antenna would sweep across the landscape. The RC-227 proved to be quite useful, and an improved model with many new features was designed by M. E. Droz and C. R. Haupt late in 1943. This equipment, AN/APS-T1, was produced by the Emerson Radio and Phonograph Corporation under Navy contract.

A photographic device which transforms a transparent photograph of a PPI or of a map into another photograph of the corresponding B-scope pattern was also developed and used during the war to produce briefing photographs. In this device the transparency of the map is placed in an enlarger and projected on a holder containing a sheet of film or enlarging paper. A slit on the holder exposes a portion of the film or enlarging paper. Manipulation of a crank causes the slit to rotate at a uniform rate about the center of the PPI projection; the paper is simultaneously made to move perpendicular to the slit. This device was designed jointly by D. B. McLaughlin and the Special Devices Division of the Navy Department.

Tracking trainers are devices which are designed to assist the student in acquiring the necessary manual dexterity and coordination to operate properly a fire-control or other radar device where manual tracking is required. These units consist of cranks or handwheels operating pointers or other devices so arranged that they may be matched with a moving object or what appears to be a moving object used to simulate the target, and provision is made for recording or indicating student performance. Garman and J. W. Stafford developed a tracking trainer device of this type that was produced by the Foxboro Company.

Specific manipulation trainers, as the name implies, are devices used to train operators to perform a given special function. Strictly speaking, tracking trainers and some others fall into this category. They may be classified separately, however, because special attention must be given to certain features of their design. A trainer for an automatic or semiautomatic fire-control radar system may be used as a typical example for devices of this sort. In such systems, typified by the AN/APG-1 radar, the operator must find the target and perform several operations in proper sequence. Often this can be accomplished in the air or on the ground using real targets, but in cases where this is not feasible a synthetic device is required. The AN/APG-T1, designed by Garman, W. K. Hodder, and G. R. Paine in 1944 and manufactured by the International Projector Corporation, was a signal generator controlled by cams, capable of presenting synthetic signals to the AN/APG-1, AN/APG-2, AN/APG-3, AN/APG-5, AN/APG-8, AN/APG-13, AN/APG-15, and AN/APG-16 radar sets.

Another device which may be classified under this heading was designed for use with ship radar equipment and designated the OCJ training device. This unit, developed by Garman, G. McClure, G. J. Maslach, and C. M. Connelly, provided six signals moved by cams in azimuth and range. It also provided jamming and other interfering signals. A contract was placed with the Raytheon Manufacturing Company during the summer of 1945 for the construction of these units for the Bureau of Ships.

Although cams are satisfactory for the generation of courses taken by synthetic targets, they do not provide the required flexibility to train students in meeting changing situations. A course computer was commonly employed to permit the course and speed (and rate of climb, if an aircraft) to be varied in a realistic manner. A trainer designated OBJ employed an electromechanical computer to generate these courses. This device, developed by Garman, J. Hicken, and R. U. Nathe in 1943 at the request of the Bureau of Ships, was manufactured by the Electrical Research Laboratories.

Another device, in which the effect of tracking influenced the course of the aircraft, was required for the AN/APQ-5 radar attachment. In this device a bombardier used a modern bombsight to track a radar target instead of a visual target as a reference point. This device, known as the AN/APQ-5T-1 and T-2, was developed by Paine and Garman, and was manufactured by the International Projector Corporation in 1943 and 1944.

Individual tactics trainers present a different problem from the types previously discussed. Devices in this category are typified by the early radar aircraft interception trainers. An aircraft interception radar operator must not only learn the art of set manipulation so thoroughly that it is an entirely automatic operation, but he must also make decisions concerning the flight and general tactics of the aircraft from data obtained with the radar system, and must be able to guide the pilot. It is difficult to obtain the needed practice in the air because targets are hard to find and to control. It is also difficult to divide the problem into several simple parts when actual airplanes are used for initial training. This is an important factor in any training program and is particularly important here since the training is so complex. A number of devices were designed and produced for various radar systems which presented these tactical problems. The RC-253-T-2 devices, designed jointly by the Link Aviation Devices Company and Garman and Droz of the Radiation Laboratory, were two examples of this type of equipment.

The H_2X bombing trainer problem was how to train radar operators to recognize ground objects for navigation and bombing. Since only one aircraft is used in air training and since the tactical problem is not as complex, for example, as for AI, considerable success can be expected by use of air training alone. Yet a synthetic trainer can pay for itself in a short time, because of the flying time it saves.

If the trainer is to be a valuable adjunct to the air training program, it must be capable of considerable realism. Although the scope interpretation devices previously described present land masses with proper shapes, they do not provide the effects of the third dimension; i.e., the shadows caused by hills and other features in the landscape. Since the pattern on a radar scope, although presented in two dimensions only, is greatly influenced by the height of the various targets, the training method should include these effects. To achieve this result the overall operation of a radar set must be simulated. This was done by use of an ultrasonic pulse-echo system, scanning a scale model of a section of the earth's surface.

Supersonic and ultrasonic simulation gives a degree of realism which is not approached by any other methods developed during the war.[1] In this type of trainer, a crystal scans a glass map under water (left polished for water areas, roughened to represent land, and built up with carborundum dust to simulate towns) in much the same way as the normal radar set scans the ground beneath it. The crystal is pulsed, generates compressional waves on ultrasonic frequencies, and receives delayed echoes which are fed to the receiver in the normal manner. A panel unit provides circuits into which course, air speed, wind speed, and direction can be arbitrarily set by the instructor to duplicate actual flight conditions; since the trainer can be "flown," the radar operator is presented with precisely the same information and problems which would appear on the set in flight.[2] Practice navigation flights and bombing runs on the trainer were an invaluable prelude to actual flights.

Several types of ultrasonic training devices were designed by the Radiation Laboratory and a large number were produced by the Research Construction Company and other manufacturers. The first of the series was designed at the Laboratory during the summer and early fall of 1943 and ten models were produced for service use by March 1944. A later model was designed at the Laboratory; 30 units were produced by the Research Construction Company, and were delivered to the services during the summer and early fall of 1944. These had the Service designation AN/APQ-13T-1. A host of "Mickey" operators (radar bombardiers) were trained on these units for the European and Japanese theaters during the fall and winter of 1944 – 45. A contract was awarded by the Army to the Bell and Howell Company for a third model known as AN/APQ-13T-2, which was in production by January 1945. The Navy followed with an order for a new model, the AN/APS-T-3; this was in production by August 1945. This model was designed to provide simulation for many Navy radar sets, such as AN/APS-2, AN/APS-3, AN/APS-4, AN/APS-10, AN/APS-15, as well as for the AN/APS-30 series. A special version of the ultrasonic trainer was developed at the Laboratory during the winter of 1944 for the AN/APQ-7 radar set and put into production by the Research Construction Company several months later. It was used to train operators of this radar for the Japanese theater and several crews so trained were in combat before V-J day.

Several other types of trainers were also developed during this period for ground and ship applications. These models included a unit for the simulation of navigation types of ship radar; a unit for the simulation of the Mark 8 and Mark 12 main-battery fire-control radar sets; and a unit for the simulation of the ground pattern produced by the GCA blind-approach radar system. The design of all of these was completed by V-J day.

Small crew cooperation trainers, as the name implies, are units in which several members of a crew must be trained together. The AI trainers were of this type, since cooperation between the radar operator and pilot is required for successful operation. The bombing trainer similarly may be used for simple crew cooperation between the radar operator and bombardier by addition of such equipment as the intercommunication system. For complete training in cooperation, however, a radar trainer should be synchronized with a visual bombing trainer. Since the accuracy required is of a high order, this is not generally possible unless the two devices are properly designed to perform this double function. A device of this type was the AN/APQ-T-1. The SCR-615 and SM training devices developed by M. E. Droz and S. B. Cohen in 1943 and 1944 are other examples of types of small crew cooperation trainers.

Large crew cooperation trainers, as the name implies, differ only in complexity from those described above. The GCA landing system, for example, employs five or six crew members, acting as a team, who give oral directions by radio to a pilot for a precision blind landing. Here split-second timing and accurate tracking information must be combined to avoid accident. It is essential, therefore, that the initial training of the crews be performed on a synthetic training device where the various operations can be broken down into simpler elements; where mistakes can be made safely and analyzed by the instructor; where records can be made and compared against the true position of the synthetic aircraft; and where, in general, various methods of training and use may be conducted with complete safety. A device of this type will, of necessity, be complex but the savings resulting from the avoidance of even one aircraft accident can more than justify the investment. Five units of a GCA trainer designed at the Laboratory by Haupt, Stafford, Garman, and C. M. Gilbert were produced by Gilfillan Brothers, and used by the Navy, the Army, and the RAF.

An example of the need for combination crew and tactical trainers is the Cadillac or AEW system. A training unit of this type was developed at the Laboratory, but was not completed before the end of the war.

The design of new radar equipment proceeded so rapidly during the war that time schedules on radar trainers were difficult to meet, particularly since the design specifications and manner of use of a particular system were usually made available only a few months before the tactical use of the radar. In 1943, therefore, an analysis was made of radar trends and radar trainer design to determine if a certain number of fundamental and highly flexible components could be designed to serve the majority of the radar trainer requirements. The analysis showed that such a scheme might be useful. Support was given by the OSRD and the Armed Services early in 1944 and a program of manufacture was begun at that time at the Research Construction Company under an OSRD contract.

The program was divided into three parts to provide components for pulse signal generation, for ultrasonic signal generation, and for computer designs. The first category included a universal trigger unit, a delay pulse generator, and a modulator-oscillator. The ultrasonic components included a modulator-pulser, an amplifier frequency converter, a universal scanning head, and a standard tank assembly. The computer group required the greatest effort but was easier to standardize since the computer design is independent of the design of the radar equipment. This group included an own-ship course unit, a target-ship course unit, an integrator amplifier, integrators, an angle computer, an antenna simulator, and a universal servo amplifier. Most of these computer elements could be used equally well for computations in spherical or rectangular coordinates.

Later experience adequately demonstrated that this program was worthwhile. 80%–90% of the later trainers were assembled from such "on the shelf" components. Some of the units were redesigned, not to achieve a greater universality, but to incorporate improved circuits and advances in the field of ultrasonics in the interests of higher accuracy and better performance.

The trainer program of the Laboratory, like the Laboratory program as a whole, was made possible by the cooperative effort of the various members of the Trainer Group, of which Garman was Leader. It included, in addition to those already mentioned in this chapter, Samuel Seely and G. F. Tape, who were Associated Group Leaders, I. H. Dearnley, G. W. Fox, S. Frankel, J. F. Johnson, R. E. Lee, W. Roth, P. Rosenberg, C. E. Teeter, R. Roberts, H. H. Wheaton, and H. S. Cutler. Numbers of Army and Navy representatives were attached to the group during much of its history.

After V-J Day a large number of radar training units in development at the Laboratory were released to the Special Devices Division of the Navy's Office of Research and Invention for use in postwar development work.

NOTES

1. To prevent confusion with regular underwater supersonic gear, the Navy introduced the term "ultrasonics" to designate ultrahigh supersonic frequencies for short-range use in either air or water. Frequencies about 100 kHz in air and frequencies above 10 MHz in liquids may be used roughly to define the ultrasonic region for radar trainers.

2. This is possible because the ratio of the velocity of supersonic waves in water to the velocity of electromagnetic energy in air is approximately 1/200,000. Hence realistic maps of enemy territory can be duplicated on small glass maps in this ratio (e.g., 30,000 square miles could be represented on a map approximately 4 to 6 ft).

Part 2

Radar Components

CHAPTER 24
Magnetrons and Modulators

Magnetrons

The purpose of the initial Radiation Laboratory program was to develop an airborne night interceptor radar using frequencies in the microwave region. The only tube available for generating microwave power of sufficient magnitude for use as a transmitter in such a radar set was the hole and slot magnetron developed by the British. Very little was known regarding magnetron operation and theory. Accordingly, in the fall of 1940, Group II, the Transmitter Group, was established at the Radiation Laboratory with I. I. Rabi as chairman. It was the responsibility of this group to study the operational and design characteristics of magnetrons, and to provide information to systems groups on magnetron operation.

A necessary task of the Magnetron Group was to develop adequate test equipment. Older techniques for lower-frequency UHF measurements were not sufficient to make the accurate measurements of wavelength, peak pulse voltage, and power input and output that were demanded in the new region of microwaves.

In addition to development of measurement techniques, the Magnetron Group in its first few months functioned as a service group for other component work. Assistance was given in measuring power output at the antenna, adjusting receivers, testing TR boxes, making local oscillator measurements, conducting research on types of antennas, etc. As the Laboratory expanded, these functions were taken over by newly organized components groups. By July 1941, G. B. Collins had assumed chairmanship of the group (Group 52) and a major proportion of the energies of the group was devoted to magnetron problems.

In 1941, the Bell Telephone Laboratory was investigating magnetrons independently of the Radiation Laboratory. Bell Telephone was responsible for the first magnetron made in this country, an exact copy of the British hole-and-slot magnetron, which at the time was to be the only available magnetron type powerful enough for use on a transmitter in a radar set. By December 2, 1940, the Transmitter Group had received five of these magnetrons, operating at 9.6 cm at a peak potential of 10 kV for a pulse duration of 1 μsec. The power output was believed to be between 10 kW and 15 kW peak. (See Fig. 24-1.)

General study of magnetron performance required the design and testing of a large number of experimental tubes at various wavelengths. For

561

FIG. 24-1. E. G. Bowen (seated), L. A. DuBridge, and I. I. Rabi
inspect a cavity magnetron similiar to the first brought by Bowen
from England in 1940.

example, there was the problem of making a magnetron which would
perform efficiently in the 10-cm band, and one for the 3-cm band. Experimental 3-cm development was a chief reason for the negotiation of a contract with the Raytheon Manufacturing Company of Waltham, Massachusetts, effective from November 1, 1940 to August 31, 1942. The
contract provided for the use of the Raytheon Company's model shop
facilities for the production of experimental magnetrons, in the absence of
similar facilities at the Laboratory. The first experimental magnetrons of
Laboratory design were under construction at Raytheon by December
1940, and by March of the following year 11 tubes had been received.
These operated at wavelengths shorter than 10 cm: four at 4.7 cm, four at
3.3 cm, two at 2.3 cm, and one at 8.8 cm. By July 1941, 63 more 3.3-cm
tubes had been received. Most of these were slot-type magnetrons, the
shift from hole and slot-type to slot-type having been made solely for
convenience in manufacturing.

The early magnetrons were invariably poor; they were gassy, operated
differently in systems and on test benches, and jumped modes of oscillation or tended to operate in two different frequency modes simultaneously. Efficiencies were low (about 15%–20%) and the cathodes of the tubes
had short lives. It was also observed that some magnetrons operated more

efficiently in one magnetic field position than in another, so that many magnetrons were labeled to show which face should be placed next to the north pole of the magnet. Often when two magnetrons with good and bad operating records were opened and examined side by side, no differences could be observed. It was apparent that a great deal of investigation would be required for a more adequate understanding of magnetron operation. Some of the variability of tubes could be traced to variations in impedance of the lead, which had a marked effect upon tube performance.

The first production tubes were the Western Electric "Dot" series, beginning with the Red Dot. These were copies of the British slot-and-hole magnetron and operated at 9.8 cm. The Western Electric Blue Dot was an improved Red Dot in which the number of cooling fins had been increased and a mounting flange had been added so that the tube could be attached to a modulator which was pressurized. The Black Dot, designed for use in aircraft operating against submarines, was a 9.1-cm magnetron obtained by widening the slots of the Blue Dot to decrease the resonant capacity and thus lower the wavelength.

Experiments with the Dot magnetrons revealed that rf power was lost by leakage on the filament leads. W. W. Salisbury of the rf group had corrected leakage at the Radiation Laboratory by putting small quarter-wavelength chokes on the filament leads of an experimental design. The Western Electric Yellow Dot was essentially a Black Dot with the addition of rf chokes on the filament leads.

By May 1941, the Raytheon E series at 3 cm was in production, in addition to the experimental tubes (mostly at 3.3 cm) which Raytheon was supplying to the Laboratory. By July 1, 1941, the E series tube was considered a standard 3-cm model; the designation was then changed to R, with the E reserved for experimental models.

In the middle of 1941, the Radiation Laboratory took the important step of breaking away from domination by British and Bell Telephone Laboratory design, largely because the Radiation Laboratory wished to make certain changes in the Bell (British design) magnetron of which BTL did not approve. In particular, White suggested changes in the fin system of the magnetron, because of the importance of shielding the modulators from rf energy. This appeared impossible to do with the British design. The Radiation Laboratory suggested that flanges be put across the magnetron where the filament leads entered, thereby shielding the magnetron from the high-voltage input. It soon became apparent that it would be desirable to have the tube constructed at right angles, a scheme which Bell Telephone opposed, but which Raytheon, already working with the Radiation Laboratory, agreed to develop and carry out. The Raytheon E204 experimental tube, tested in January 1942, was the first right-angle tube, and subsequent development of magnetrons was largely with the right-angle design. The shift from straight to right-angle design appeared to have negligible electrical effects.

On September 26, 1941, a cablegram from England reported that Sayers of the British Birmingham (Oliphant's) Laboratory had found the key to high efficiency in magnetrons. The alternate segments of the anode structure had to be plus or minus, which could be insured by a system of cross-connected wires. Strapping, as the method came to be called, resulted in a strengthening of the desired frequency mode of operation, thereby eliminating frequency jump in magnetrons. The straps essentially contributed extra inductance and capacitance to the resonator cavities—in different amounts, however, for the different possible frequency modes of operation. Thus a system of strapping designed to have a maximum effect on the desired frequency mode would separate it from other modes. The cable described how wires placed in the holes of the anode segment would produce tubes of much higher frequency than ever before (efficiency jumped from 20% to 40%).

In the Magnetron Group, pleasure at the news was tempered with shame. Group members recalled how, three or four months before, P. Spencer of Raytheon had brought to the Laboratory for testing a tube which he called his "mode-locking" tube. This tube had had its anodes connected by a recessed ring and so was the first ring-strapped tube. No one in the Laboratory was enthusiastic about it since the tube had not worked well and no one could explain why. With the British disclosure of strapping, however, it became apparent that operating conditions had not been correct for testing Spencer's strapped magnetron. The British tube had a single strap break at the point where the filament leads were to be avoided. Spencer had done a much more workman-like job and had fitted the strap into grooves so that his tube did not have a strap break. As it turned out, a tube without a strap break needed a much higher operating current and also a higher operating voltage in order to work stably. The Laboratory at the time did not have modulators capable of giving more than 15 A, and so were not equipped to test the tube. The first copy of the British strapped tube was made by the Bell Telephone Laboratories and was received at the Radiation Laboratory in November 1941. The tube operated at 13–14 kV and gave an efficiency of about 60%.

Strapping introduced a new parameter into the study of magnetron design and operation and numerous ways of strapping were explored. One of its effects was to change the frequency of the standard British magnetron from 9.8 cm to 10.7 cm. (This influenced the frequencies of future 10-cm ground sets of the Laboratory.) In the autumn of 1941, an order of the Laboratory requiring all 10-cm ground sets to operate at 10.7 cm, all 10-

cm ship sets at 9.8 cm, and all 10-cm airborne sets at 9.1 cm meant that it was desirable to be able to change the frequency of the standard 10.7 magnetron by tuning. Since the frequency of an unstrapped magnetron was changed by strapping, the possibility arose that a means of varying the strapping might permit the design of a tunable magnetron. Spencer, of Raytheon, was one of the first to try tuning in this matter. He devised a ring similar to a ring strap which could be moved near or away from the oscillator cavities, thereby changing their capacity and hence their frequency. This idea was incorporated in the 2J54 magnetron which tuned over a range of 5%. The Magnetron Group designed a tunable 10-cm magnetron in which one of the oscillator cavities was coupled to a resonant cavity mounted externally on the tube. A movable diaphragm on the side of the cavity varied its frequency, and through coupling, the frequency of the magnetron. The resonant cavity method was used in the 4J70 magnetron to give an overall tuning range of 8%. The 4J70 was also adapted to higher powers than the Raytheon tunable magnetron.

At 3 cm, the Columbia Radiation Laboratory produced the 2J51, in which plungers changed the ring of the oscillator cavities and thus their inductance. The 2J51 would tune over a range of 12%. Tunable magnetrons were highly advantageous in radar sets which were used near other radar equipment, as, for example, the SG shipboard set. In case of interference, the sets could be tuned out of the congested frequency area.

Meanwhile, the task of redesigning the then-standard 10.7 cm strapped magnetron to work well at 9.8 and 9.1 cm turned into a six-month trial of different combinations of magnetron parameters. The 9.8-cm magnetron was obtained essentially by slightly reducing the size of the resonant cavities of the 10.7-cm magnetron. The problem of a 9.1-cm strapped magnetron was not solved until early 1942, when the Group developed the technique of voltage scaling. Voltage scaling was part of the Group's overall understanding of how to build magnetrons operating at any potential; it was found out that magnetrons with large cathode size and interaction space required higher operating potentials than did those with small cathode size and interaction space. Simply reducing the size of the resonator cavities as was first done with the 9.1-cm tube resulted in a tube which, as with Spencer's first strapped magnetron, worked stably only at high potentials—for which suitable modulators did not exist at the time. The 9.8- and 9.1-cm tubes were produced by Bell Telephone Laboratories and Raytheon, and later, also by Westinghouse and Sylvania.

Voltage scaling opened the way to design of strapped magnetrons which would operate successfully at any potential. For lightweight sets, low-voltage tubes were built: a 3-cm magnetron for use with the AN/

APS-10 and a 10-cm magnetron for use in the SMTR, both operated at 5 kV input and 10 kW output.

Voltage scaling also led to the K-7 and HK-7 high-power magnetrons. A large cathode (and so a large interaction space) was required for the high current requirements. The entire tube was made wider and longer to dissipate heat. The type of strapping was revised, since the usual straps became proportionately less effective as tube size was increased. Dimensions of both input and output leads were increased to prevent voltage breakdown. In the HK-7 the anode had to be made so large that 12 resonators, instead of the usual 8, were used, and these had to be doubly strapped. The K-7 magnetron, giving a rated output power of 250 kW, was in production by the end of 1942 and was installed in many of the high-powered ground radars, such as the MEW, V-beam, and Beavertail being developed at that time. In the MEW, the K-7 was run at over twice its rated output power, with the aid of a scaling system and filament heads immersed in oil. The HK-7, giving a rated output power of 1 MW, was to replace the K-7 in the production units of the MEW and the V-beam.

Other high-power tubes to which voltage scaling contributed were the 720 (750 kW peak power output), manufactured by Western Electric and used in the SCR-615, the KT, and the SM; and the HI-5, which was an HK-7 operating at 8.9 cm.

Voltage scaling also made possible two later developments of high-power magnetrons, which never were put into actual use, however, because of the end of the war. One was the HP-10V magnetron, a gigantic tube operating at 50 kV input and 4 MW output power. The second was a 3-cm tube, the 4J50, designed by Bell Telephone in cooperation with L. R. Walker who had worked on scaling problems of the magnetron group. The 4J50, capable of 300 kW output power, was to be used in a higher-power version of the Li'l Abner height-finding set. Another 3-cm magnetron by Bell (developed not by voltage scaling but by wavelength scaling, in which all physical sizes are reduced proportionately) was the 725 magnetron which came out in 1943. This became the standard 3-cm magnetron, used on almost every Laboratory 3-cm set.

By the beginning of 1942, the practice of classifying magnetrons by means of performance charts was established. Magnetron performance had been investigated for systematic variations in magnetic field strength, peak pulse potential, and current. These quantities, together with power output and efficiency, were plotted on a single performance chart, which thus carried all information necessary to specify the performance of a magnetron.

During the first six months of 1942, a study of resonant modes of magnetrons was instituted by the Magnetron Group, under the direction of J. C. Slater of the Radiation Laboratory. The method of "cold resonance" was used, in which rf energy from an external source was fed into a nonoperating, or cold, magnetron. By means of probes, measurements were made of the amount of energy absorbed in the anode cavities; and resonant modes, or frequencies at which unusually large amounts of energy were absorbed, were determined. Furthermore, for each mode, the energy pattern inside the cavities was explored, using a tiny probe which could be moved at will. Extensive studies were made of the modes of strapped magnetrons, using tubes of different sizes and with different methods of strapping. It was concluded from these measurements that successful strapping accomplishes two fundamental things: (1) it separates the modes widely (by several percent of the wavelength of the primary mode) which is necessary for efficient operation since it becomes harder for the magnetron to slip from one mode into another; (2) it gives a good energy pattern inside the cavities for the mode in which the magnetron should operate, and a poor pattern for all other modes. Slater's study of cold resonance essentially cleared up prevailing questions of the relationship between magnetron operating frequency and operating modes. With cold resonance measurements correlated with the characteristics of operating magnetrons, it became possible to "test" experimental designs before actual construction of the tube.

There still remained the problem of the effect of load on magnetron frequency and output power. Early magnetrons were connected to double-stub tuners which were tuned to maximum power. Loading of such maximum-tuned magnetrons, however, led to frequency shift, or "pulling."[1] As early as the spring of 1942, pulling of magnetrons was a serious problem on the first ASV and AI sets, when tuning the systems before each flight led to variability in magnetron operation. Because of the marked shift in frequency that came with loading, it was important to make a systematic study of the effect of loading on magnetron operation. F. F. Rieke was made responsible for this study. The techniques which he developed subsequently became a part of the testing of all magnetrons and contributed greatly to the overall understanding of magnetron operations. A summing-up of the results of his investigation was the much-used Rieke diagram, which made it possible to determine graphically the relationship between pulling and power output. On a Rieke diagram, contours of constant frequency and power are plotted on a polar graph. The center of the graph represents a standing wave ratio of 1.0; higher standing wave ratios are represented by concentric circles of increasing radius. The phase angle, represented by the polar angle, was referred to some one point in the transmission line such as the flange on the magnetron. The Rieke diagram showed that a magnetron tuned to a system for maximum power output "pulled," i.e., operated in a region of frequency instability with respect to changes in the system. The diagram further showed, however, that high efficiency of magnetrons was associated with bad pulling figure. Thus, from inspection of the proper Rieke diagram, a region of preferred operating conditions, compromising between high efficiency and frequency instability, could be determined for any magnetron.

One consequence of Rieke's work on rf loading was "preplumbing" of magnetrons. Since, as Collins put it, a man turned loose on a Rieke diagram would tune himself into the worst part of the diagram, it was decided that system people should have less freedom in the adjustment of the magnetron. Accordingly, the magnetron was preplumbed by the addition of a slug on the central conductor. This slug was used in all pretuned magnetrons. In view of widespread Laboratory opposition to preplumbing because of the different characteristics of individual magnetrons, the Magnetron group began the practice of setting up specifications for a standard, preplumbed 10-cm magnetron.

The results obtained from performance charts, cold resonance tests, and Rieke diagrams give a fairly complete picture of magnetron performance, and also assisted in developing the theory of magnetron operation. At the end of 1942, magnetron design could proceed with a good deal more assurance of the results to be obtained than was possible in the early days.

Much of the experimental work of the group centered about the tube shop which was established at the end of 1941 to take over the duties of the Raytheon model shop. A continually important part of the tube shop program was research on cathodes, since the cathode of a magnetron was not only a chief factor governing its life, but also a critical consideration in its efficiency. Thus, too small a cathode radius results in frequency instability, while too large a radius distorts the internal fields, giving inefficient operation. E. A. Coombs and J. H. Buck began the research on cathodes in the tube shop, and their work was aided subsequently by similar investigations at the Bell Telephone Laboratories. The earliest cathodes were made of nickel tubes coated with strontium or barium oxide, since oxide coating was an efficient emitter of electrons, and so did not require heater temperature high enough to distort the internal fields of the magnetron. The mechanics of this type of construction, however, were such that the oxide coating was easily cracked off, and was too thin for long cathode life. Coombs experimented with a tube made from fine wire mesh, which was durable and which would take a thick coating of the oxide when sprayed on. This type of cathode construction increased the life of the 725 3-cm magnetron, for example, from 1 to 200 hours.

The cathode research program also included the problem of achieving high pulse current densities from cathodes. Never well understood, pulse current densities were improved, as Collins put it, by "instinct which amounted almost to witchcraft." With the nickel-tube-type cathodes, pulse current densities of 10 A/in.^{2} were about the maximum that could be obtained. By the end of 1944, Coombs and his assistants had raised this figure to 200 A/in.^{2} in the Laboratory, and values of 100 A/in.^{2} for production cathodes seemed probable. Coombs explored the entire field of cathode construction, and, in later dealings with manufacturers of magnetrons, was able to give valuable consultant advice.

With the group at the Laboratory devoting most of its time to the development of magnetrons at 10 cm and 3 cm, it appeared desirable in the spring of 1942 to have a relatively small group concentrate on the study and design of 1-cm magnetrons. Such a group was established under the direction of I. I. Rabi at Columbia University, the advantage of the New York location being accessibility to laboratories of Bell Telephone (who had designed the first 1-cm magnetron in this country), Westinghouse, Sperry, and General Electric. The new Laboratory was set up to supplement the work of the Radiation Laboratory group, not to compete with it, and close liaison was established between the two groups.

Modulators

The Modulator Group was organized at the beginning of the Laboratory under the chairmanship of K. T. Bainbridge to develop circuits which would supply narrow, square, high-voltage pulses to the radar transmitter. Modulators ("pulsers") differed from other radar pulse-forming circuits in forming their pulses at a high power and voltage level. The principles of high-voltage pulsing were developed almost completely in the field of microwave radar.

The initial problem of the group—to develop a modulator for the Laboratory's AI night interceptor radar—determined the nature of future development of pulser circuits. British work on modulators indicated the need for a pulser to produce a square wave pulse about 1 μsec long and to give about 10 kV output at 10 A, at a duty cycle of 0.1%. The Laboratory program for the AI system required development of a pulser which could be used for magnetron development, as well as an airborne modulator operable at altitudes of 20,000 feet. The first required flexibility in output voltage, current, and pulse duration. The second required efficient use of weight and power.

M. Eastham, even before the Laboratory started, arranged for a contract with Westinghouse for the development of modulators and, in October 1940, for a contract for the construction of four modulators to be used in the Laboratory. In December 1940, additional orders were placed with Westinghouse and RCA for pressurized Service airborne pulsers. By March of the following year, two or three units had been delivered on each order, but work at the Radiation Laboratory had proceeded so rapidly that no further development on the units was desirable.

Early pulser development in the Laboratory resulted in the hard-tube modulator, the first of two major types developed; the other was the line-type modulator, unknown at that time. Both types used a storage capacitor to store electrical energy between pulses and to act as a reservoir supplying all the energy to the pulse. However, in the hard-tube modulator the capacitor was large and only a small part of its energy went into the pulse, whereas in the line type, the capacitor was small, and discharged completely during the pulse. The two types also differed in the way they formed the pulse. The hard-tube modulator was so called because the switch which initiated the pulse was a vacuum (hard) tube. A square wave of the proper length applied to the grid of the vacuum-tube switch at the proper intervals of time shaped the pulse; thus the pulse length was as long as the square wave. The hard-tube switch had to fill certain requirements during the interval between pulses: it had to withstand high voltages without conduction of current; during the pulse itself the voltage across it was low, but it had to be able to conduct high currents. Vacuum tubes fundamentally fulfilled the first, but not the second, requirement. In contrast to vacuum tubes, gas (soft) tubes had the desirable characteristics of being able to pass large currents but could not withstand high voltages, and once fired, could not be turned off again by any square-wave voltage applied to the grid.

In 1941, a study was undertaken by A. E. Whitford of the Modulator Group, as well as by other companies, with a view to increasing the current which a vacuum tube could conduct, by increasing the efficiency of the cathode. Efficient cathodes were also needed to conserve power in the airborne modulator, which was at the time the important consideration of the group. By the summer of 1941 Whitford, with aid from Raytheon, had developed a tube with an oxide-coated cathode requiring only about 60 W of heater power for somewhat higher output power than could be obtained from the two Eimac 304TH tubes in parallel. (Two Eimac tubes required a total of about 500 W of heater power.) A part of Whitford's development was the use of gold-plated grids to prevent the grid emission which had proved troublesome at high voltage. A low-power modulator tube, the 3E29, produced by RCA and used in the SCR-534 and the SCR-615 sets, resulted from Whitford's work. A similar low-power cathode tube, the 715, simultaneously developed by Bell Telephone Laboratories, was used in many later sets for standardization and production reasons.

From the end of 1940 to the middle of 1941, many needed pulse-measuring techniques were developed. A vacuum-tube voltmeter which could measure the waveforms at the grid of a tube without distortion was constructed. Capacity dividers were designed for viewing output pulses. E. M. Lyman developed a synchronizer which could measure, on a scope, pulse current and pulse voltage. F. G. Dunnington and H. D. Doolittle, in the summer of 1941, completed an rf envelope viewer which could be used to examine the spectrum and amplitude of the magnetron output, and to establish the relation between magnetron current and rf output. The development of a magnetron average current meter by M. H. Kanner and M. G. White helped improve the reliability of quoted output powers.

The general solution to the problem of a driver circuit which would shape square pulses was obtained in January 1941, with J. C. Street's invention[2] of the network-controlled bootstrap driver circuit, which used tubes of low cathode power and involved no plate power consumption except during the pulse. Power requirements had been a difficulty with the Westinghouse modulators, which used plate power except during the pulse. The bootstrap part of the circuit gave a fast-rising square wave which then could be applied to the grid of the hard-tube switch. The bootstrap circuit resulted in a square wave that rose from zero to its full voltage value in a small fraction of a microsecond. The basic idea of the bootstrap circuit was used in nearly all the early Laboratory and Service modulators.

By February 1, 1941, basic circuits had been developed for Laboratory and Service modulators for outputs of the order of 125 kW peak, and the policy of using a pressurized aircraft Service pulser was agreed upon. Also, magnetron characteristics were beginning to be understood, and modulators could be designed for definite matching requirements. There was a program of expansion which included (1) life testing of all components to be used in production modulators; (2) design work leading toward higher peak powers; (3) development of high-power-level line-type modulators; (4) improvement of existing hard-tube modulator circuits.

By this date, also, the bootstrap circuit was available for immediate production use. Orders for copies of a 60-Hz-input Laboratory modulator (Mod 0) and a 400-Hz-input pressurized modulator for aircraft use (Mod 1) were placed with Raytheon in February and March 1941. These units were extensively used in Laboratory development of all types of microwave components and served in early flight tests to determine range at various repetition rates and pulsewidths. The hard-tube modulator was particularly suited to the development of the magnetron because of (1) its greater flexibility in power level, pulsewidth, and duty cycle, and (2) its more advanced state of development up to the fall of 1944.

By March 1941, two new modulators of higher peak power than the still-on-delivery Raytheon modulator had been built by the Magnetron Group. The larger of these gave peak voltages up to 20 kW, peak currents up to 13 A, and pulse widths of 1 to 3 μsec. Since the single bootstrap driver circuit was inadequate (from the power standpoint) to deal with pulses longer than about 2 μsec, a double bootstrap circuit was developed which ultimately was used in the Mod 4 modulator. Dunnington, Doolittle, Kanner, and White of the Radiation Laboratory designed this 3-MW modulator, and in the early summer of 1941, 20 of these for magnetron development had been ordered from General Electric Co. The first unit was received from General Electric in November 1942. There was one more superhigh-power modulator, the Mod 14, giving about 8 MW of peak power. Work was started on the Mod 14 about the fall of 1942 and four similar units were made by the end of July 1945. One of these was used specifically by Group 52 in developing the HP-10V magnetron.

In spring of 1941, I. A. Getting and L. N. Ridenour, working on what was to be the SCR-584, requested a modulator capable of 750 kW peak power. The unit furnished was used in the RF-1 truck during two years of test work. About this time, K. T. Bainbridge requested that ten 1-MW pulsers be built for an aircraft carrier search set. Group 51 made two sample units, and a contract was let to F. M. Link Co. for 13 additional units, which ultimately were used in the SCR-615 production set. These pulsers were also extensively used by the magnetron group for development work. The driver used in these pulsers was the double bootstrap circuit developed for and used in the Mod 4 modulator.

By March 1941, the airborne pulser program included the development of a 144-kW hard-tube modulator. The first pressurized package for this unit was completed by G. N. Glasoe in June 1941, and several of these pulsers were subsequently built by Raytheon (Mod 2). These units went into the Radiation Laboratory crash ASV sets installed in B-18's to hunt submarines. By the fall of 1941, after success with his oxide-cathode output tube, Whitford designed the Mod 3 VPP (vest pocket pulser), incorporating the similar 715 output tube, which was considerably smaller than the previous tube because of the decreased space required for heat dissipation. This modulator, originally planned for 3-cm systems, was used generally at 3 and 10 cm. An order for manufacture of this modulator was given to Stromberg-Carlson in late 1941 and delivery of the first two units was made about May 15, 1942. The vest pocket pulser was also built by Philco, and became the standard Navy aircraft modulator (used in the ASG, ASD, etc.).

The success of a network in forming pulses at the low-power pulse level of the bootstrap driver led to use, by the fall of 1941, of networks at the magnetron input power level, a use which was ultimately to overshadow the bootstrap circuit. These networks were simulated transmission lines with characteristic impedance which must be matched to the load device for maximum efficiency. The pulse was shaped by the network and all the stored energy was discharged in each pulse, the voltage falling to zero so that gaseous switching devices such as thyratrons and spark gaps could be used. Since such switches require little if any cathode power, and dissipate little plate power during the pulse and more during the interpulse interval, a line-type circuit is highly efficient.

The line-type modulator required (1) a high-voltage power source, either dc or ac; (2) a charging inductance; (3) a pulse-forming network; (4) a switching device. In the line modulator, the pulse was not shaped in the switch but in a line, or pulse-forming network, made up of lumped constants to simulate a transmission line. The storage capacitor discharged completely in driving the pulse and so had to charge completely before each pulse. The width of the pulse depends on the time required for the energy in the line to be given to the load. Thus the pulse could be given any width, and, in fact, any shape, by manipulation of line parameters. With proper design, none of the energy was left in the line. The voltage that came out of the line was, by network consideration, equal to one-half the voltage given to the network.

In the early development of networks by J. C. Street for his bootstrap driver circuit, pulse shape had not been critically dealt with, since the pulse was afterward fed to several tubes driven to saturation to steepen its leading edge. In line-type modulators, however, the network was the only shaping factor. Early improvement work on networks was performed by M. H. Kanner and J. R. Perkins, with E. A. Guillemin of MIT consulting and calculating design parameters. By the end of the year, considerable theoretical work had been done on network design, and the general outline of a satisfactory network which ultimately became known as the type E network (using equal size capacitors) was developed empirically. By the fall of 1941 it was known that network-shaped pulses which would give satifactory magnetron operation were possible. However, the pulse shape was still not as good as that produced by a hard-tube circuit. At the request of M. G. White, the Sprague Electric Co. set up a model shop for improving and building production networks of suitable sizes. Pulse shape was improved by the development of a hermetically sealed network unit in which stray capacitance to the can was much reduced. A new network was calculated which was not as sensitive to variations in manufacture, and which also could use paper instead of mica capacitors, a critical item. Paper capacitors also were lighter. At the Radiation Laboratory, Balsbaugh started development of networks using alsifilm which would stand high temperatures. The ability to withstand heat was an important requirement of airborne components because of limited space.

Much of the development work on the line-type pulsers went into the switches that initiated the pulse. Requirements for the switches were high: they had to stand, variously, a high voltage (3–60 kV) during the interval between pulses; pass a high current (up to 1000 A) during the pulse; and have long life (in excess of 500 h) at repetition rates from 100 to 2000 pulses/sec. These requirements were met by both rotary spark gaps and gas tubes. Development of both types of switches began in the summer of 1941.

The rotary spark gap switch was comprised of a succession of tungsten electrodes which rotated past a master electrode carrying the high voltage. As each electrode went past in turn, there was a spark discharge and a pulse was initiated. One disadvantage of this type of switch was that the pulse interval varied, sometimes by as much as several microseconds, which required synchronization of the indicators to the pulser. By December 1941, a rotary spark gap with a life of about 1000 h at a power of 1 MW had been incorporated in a modulator to be used on the CSCL, ultimately the SM. Rotary spark gap switches were also used in modulators for the CXBL, the SCR-632, SO, SG-3, AN/CPS-1, -4, and -6, and in Laboratory pulsers Mods 5, 6, 10, 11, and 16.

In the summer of 1941, J. R. Dillinger began work on the first of the trigatron tubes, in which the discharge between two electrodes was controlled by a third electrode acting as a trigger. The electrodes of this trigatron were sealed in glass to prevent corrosion. Although the trigatron proved to be less flexible in power handling capability than a rotary gap switch, it was more flexible and accurate in repetition rate, and was used in the AEW airborne modulator, where a rotary gap would have met with excessive pressurization and corrosion problems. Trigatron development led to the 1B41, 1B46, and 1B49 tubes produced by Sylvania and capable of handling powers from 1 to 2 MW.

In the summer of 1941, J. R. Perkins, with the help of K. J. Germeshausen, started the development of the hydrogen thyratron switch which proved to be the most important tube contribution of Group 51. On August 1, 1941, a contract had been given to Western Electric for developments of mercury thyratrons. However, the hydrogen thyratron proved better than the mercury thyratron because it was less susceptible to temperature changes, and the Western Electric contract was terminated. As early as 1942, hydrogen thyratrons had been run for 800 h at peak powers of 150 kW and repetition rates of 2000 pulses/sec. By the spring of 1944, Sylvania was producing the 4035 hydrogen thyratron tube for 300 kW peak power and, later in 1944, followed it with the 3C45 tube at 40 kW and the 5C22 tube at 1000 kW.

In June 1942, two pressurized modulators (Mods 7 and 8) were built, using hydrogen thyratron switches. The APS-30 sets used a somewhat higher power verison of the Mod 7, while the Li'l Abner height finder used a modification of the Mod 8. A hydrogen thyratron switch was used in the modulator for the AN/APS-10. In January 1945, Mods 21 and 22 pulsers, using the high-power hydrogen thyratron switch tube, the 5C22, were built as replacement pulsers for the MEW and V-beam sets, respectively. The extremely low time jitter of the hydrogen thyratron modulators permitted their use with the MTI (moving target indication) system.

There are three main methods of charging the storage capacitor of a line-type modulator: dc charging, dc resonance charging, and ac resonance charging. dc charging makes use of a battery, or dc power supply, to charge the capacitor through a resistance. Then the final voltage on the capacitor, attained during the interval between pulses, is equal to the voltage of the supply. However, twice as much energy must be taken from the supply as reaches the capacitor, since part of it is lost in the resistance. Thus dc charging, although easy, is only 50% efficient.

Dc resonance charging replaces the resistance with an inductance which has very little resistance, and efficiency goes up to 90% or 95%. The opening of the discharge switch at the beginning of the interpulse interval causes oscillations to appear across the inductance. The start of the oscillation is an upward swing; the value of the inductance and the value of the capacitance of the circuit are arranged so that the time for the complete upward swing (one-quarter wave) is equal to the interpulse interval. By having the discharge switch fire at the correct moment, the voltage which develops on the capacitor is made equal to twice the voltage of the supply. With the ac resonance charging developed by H. White in the fall of 1941, the maximum theoretical upswing in voltage was π times the original voltage. In practice, this figure was reduced to about 3. Since ac was also a more convenient supply source than dc, this method was the one chiefly used by the Modulator Group.

The success of the ac resonance charging development led to the setting up in July 1942 of a program for two Laboratory modulators known as the Mod 5 (600 kW peak power, 22 kV, 27 A) and the Mod 6 (3 MW peak power, 35 kV, 87 A). Several units of Mod 6 were assembled by the Group late in 1942, and F. M. Link Company started a production order of 20 units of each at this time. The Mod 6 was used extensively in such Laboratory sets as the MEW and the SCR-615. A total of over 100 of these modulators were made on Laboratory orders.

In the spring of 1944, a higher power version of the Mod 6 modulator was made, which gave 15 MW peak power at a repetition rate of 350 pulses/sec. This was the pulser which supplied the five magnetrons of the V-beam set. Eight of these modulators, for the eight Laboratory V beams, were completed by the middle of 1945.

In the early part of 1941, the British reported that they had shown theoretically that transformers that would pass microsecond pulses were feasible. Shortly afterward, the Modulator Group placed an order with Bell Telephone Laboratories for the production of two pulse transformers after the British designs. By the fall of 1941 the group had set up its own pulse transformer division. About this time S. Sonkin was brought from Columbia University where he had been doing work on pulse transformers to develop transformer designs at the Radiation Laboratory. Within one month, he had produced a satisfactory experimental transformer, impregnated in oil and mounted in oil-filled open beakers.

One difficulty of pulse transformers was that resonance occurred between leakage inductance and distributed capacity, producing a shock excitation on top of pulse. P. D. Crout invented a theory of "squirted inductance" which enabled the parameters of a pulse transformer to be chosen to lessen the oscillations.

When Sonkin returned to the Columbia Radiation Laboratory early in June 1942, W.H. Bostick and P. R. Gillette took over the developmental program. Model shops for development of pulse transformers were set up under NDRC contracts with General Electric at Fort Wayne (later at Pittsfield), with Raytheon, and with Westinghouse at Sharon, Pennsylvania. Bostick and Gillette made additional improvements in pulse transmitter design which included a well for mounting the magnetron on the transformer, thus shielding it from rf and pulse noise and from the atmosphere. Bostick also designed many pulse transformers for applications at low-power levels, as in indicator circuits.

Although a pulse transformer was not necessary to the working of a line-type modulator, it was often extremely useful. Thus a modulator could be designed to give required voltage and current with little attention paid to output impedance, since that could later be matched to the magnetron through a pulse transformer. The pulse transformer was also useful in installations where the modulator was located some distance from the magnetron. The distance would be covered by a low-impedance pulse cable with a pulse transformer at the end to step up the voltage to the magnetron. Fifty ohms was selected as the impedance of the standard pulse cable. This standardization resulted in a great reduction in the number of different pulse components.

The first pulse transformer was used with an experimental SF installation on the roof of the Hood Building, in June 1942. Another pulse transformer was used in the 3-cm ASD-1 airborne set, in which the modulator was located in the fuselage of the airplane, with the rf components in a wing nacelle. The bulky original lengths of waveguide were replaced by pulse transformers and cables. Pulse transformers were also used on the SCR-615, SCR-682, SM, SO, MEW, Beavertail, V beam, SG-13, etc.

NOTES

1. The "pulling" figure is defined as the maximum megahertz shift in frequency that can be produced by varying a standing wave ratio (in the coaxial or the waveguide) over all phases. Since the phase of the standing waves varies with the loading, the pulling figure may be understood to represent the degree of frequency shift with loading.

2. The idea was conceived on a Massachusetts Avenue streetcar.

CHAPTER 25
Antennas and rf Components

Antennas

The Antenna Group was organized in September 1940. Among the Group leaders who took charge at various times during the following year were W. B. Nottingham, R. Bacher, J. R. Zacharias, R. Herb, A. J. Allen, and L. C. Van Atta, who became leader of the Group in August 1941. In the succeeding years the Antenna Group grew rapidly, roughly doubling in size each year, until at VJ Day it consisted of 158 people.

The problem of the Group in 1941 was to design an antenna for the 10-cm AI radar to be used in intercepting German bombers over England. The diameter of the reflector could not be over 30 in. Since satisfactory rotary joints were not then available, the feed which illuminated the reflector consisted of a quarter-wavelength extension of the inner conductor beyond the outer conductor of the coaxial line. This was fixed in position while the reflector revolved around it. The beam pattern was poor because of unsatisfactory illumination caused by too strong illumination at the edges of the reflector and the radiating effect of currents on the outer conductor. These currents were later suppressed by a quarter-wavelength choke. The antenna was not well matched to the line; antenna dimensions were adjusted by trial and error until reflections back into the line became reasonably low. As later experience showed, inadequate matching had a marked effect on the beam pattern.

In 1941, the second Antenna Group project for a gun-laying ground radar system, later known as the SCR-584, required development of a type of scan in which the beam rotated around the axis of the reflector with a small amount of overlapping. The earliest attempt to produce this type of scan, by offsetting a dipole feed from a transmission line projecting through the vertex of the reflector, raised problems of mechanical balance. When later experiments showed that a dipole at the focus of a reflector could be adjusted to throw the beam off-axis by a small amount, it became immediately feasible to produce a conical scanning beam using a mechanically balanced feed. This method, with improvements, was used later with the SCR-584.

In 1942, another type of gun-laying antenna was introduced. The feed consisted of a circular waveguide transmission line which projected through the vertex of the parabola and a small disk which directed energy from the open end of the line back to the reflector. (Development of

578

circular waveguide was begun in 1941 by E. Purcell's Development Group.) The beam was caused to deviate off-center by a slight bend in the waveguide, which was then rotated. In application of this method to the SCR-615 ground set and the SM ship set, efficiency was found to be lower than expected. Therefore, in 1944, recommendations were made to change the antenna feed for future models of these sets to a nutating two-dipole feed. Developed originally at 3 cm, by J. S. Foster of Purcell's group, this feed consisted of two dipoles mounted in line in front of an open end of waveguide, the whole rotating around the focus of the reflector with a fixed polarization. The two dipoles acted together to reflect the energy radiated by the waveguide back to the parabolic reflector.

Before 1942 most antennas were paraboloids of revolution which radiated beams of circular cross section. An early exception was the antenna for the SG ship search set, in which the vertical beamwidth was made greater than the horizontal beamwidth in order to keep the beam on the target in spite of the roll and pitch of the ship. The vertical beamwidth was increased by reducing the vertical aperture of the parabola, making use of the optical theorem that beamwidth is proportional to aperture.

In order to stabilize design of future antennas, Van Atta in January 1942 initiated a study of the interdependence of antenna parameters. The practical goal of this investigation was maximum efficient use of rf energy by the antenna, which implied high gain, narrow beam, and low side lobes. The investigation used the ratio f/d (focal length divided by parabola diameter) as a characteristic parameter of the antenna. Variation in gain and side lobes was expressed in terms of this ratio. Thus, gain was found to be always a maximum for f/d equals 1/3. With a deep reflector (short focal length) gain was low because the edges of the reflector were too weakly illuminated. With a shallow reflector (long focal length) gain was low because energy was lost past the edges of the dish. Side lobes were observed to increase with focal length since they were produced by stronger areas of illumination at the edges of the aperture. Both the gain and side lobes showed a distinct half-wavelength effect caused by interference between radiation from the feed and dish. That is, maximum gain occurred when the feed was placed an integral number of half-wavelengths from the vertex of the reflector. Thus the most desirable location of a feed is not at the focal point but at the half-wavelength point nearest to it. Future antenna designs by the Group made use of this fact.

The relationships between primary and secondary patterns (from the feed and the reflector, respectively) were established in 1941. Most of the work on primary patterns was carried out by Foster, of Purcell's group, while the Antenna Group concentrated on secondary patterns. Correlation between the two types of patterns was at first handicapped by lack of equipment to measure the phase of primary pattern energy. Phase measuring equipment was developed by Cutler of Bell Telephone Laboratories, and used by Foster with immediate improvement in correlation between primary and secondary patterns. It was found that feeds needed to have smooth phase fronts as well as smooth amplitude patterns to produce a good secondary pattern. During this period, R. C. Spencer made a mathematical study of the relationships between primary and secondary beam patterns. Some of his results were incorporated in a handy slide rule which, for any wavelength and diameter of a paraboloid, gave approximate values of beamwidth, gain, and the position of the first side lobe. Measurement of the secondary patterns had been early provided for by the construction of a pattern measurement shack on the roof of Building 2 with measuring equipment which included a bolometer and amplifier, with a heavy wooden compass platform to hold the antenna. The 10-cm transmitter was placed on the top floor of Walker Memorial. In January 1942 a 3-cm magnetron was mounted on the roof of Building 6, thus extending the wavelength range to 3 cm.

In 1942 a second extensive program was undertaken to learn the effect on the beam of placing the feed of a paraboloid off center. This was used in conical scanning, where an offset feed was rotated; it was found possible to simplify scanning motion by moving only the feed instead of the entire antenna. In this investigation the feed was observed to produce a beam deflection of only about 80% of the tilt angle, the discrepancy increasing for deeper dishes. Gain also decreased, while beamwidth increased, and an extra lobe appeared between the main beam and the axis. There was a pronounced half-wave effect, traceable to the portion of the reflector whose perpendicular at the time pointed at the feed.

Early measurements on off-centering by R. Wright and G. Jarvis were continued in greater detail by S. Silver, who developed a formula to show the amount of energy which, in any kind of antenna, was reflected from the parabola back into the feed, causing mismatching with the transmission line. It was found that an offset feed considerably reduced this reflection. Offset feeds also cast less shadow than vertex feeds, and, in the form of horns, became standard on such sets as the SCI height finder and the AEW.

A difficulty of offset feeds was the appearance of high side lobes in the secondary beam pattern. These were caused by unevenness of illumination at the edges of the reflector. S. Mason conceived the idea of minimizing these side lobes by plotting lines of equal intensity of illumination directly on the reflector, and then cutting the reflector along one of the lines. Various lines could be chosen according to different requirements for side lobes. Continuing experiments on contour cutting, as it was called, confirmed the mathematical prediction that maximum gain would be obtained with the reflector cut along the contour line for which the amplitude of intensity was one-half the average amplitude over the entire used portion of the reflector. Contour-cut antennas were effectively used on the AEW and SO-11 antennas.

Development of shaped patterns, in which the beam had other than the normal lobe shape, was required for some applications to search radars. (Shaped patterns invariably implied csc^2 patterns.) In 1942, when the 10-cm AI set was being converted to an ASV set, it was realized that the usual pencil beam should be modified to illuminate targets more evenly at different ranges. Since this meant that a target should return equal power to the set from any angle, the antenna amplitude pattern had to be made proportional to the slant range to the target. Since slant range varied as the csc of the antenna depression angle, the power in the beam pattern varied as the csc^2 of the depression angle.[1]

H. T. Booker, of TRE, was the first to attempt a shaped beam. From theoretical considerations, he determined that the desired reflector would have a sharp peak in illumination over its central portion with a sharp change in phase. He obtained the sharp change in phase early in 1943 by displacing the two halves of a paraboloid. N. Levinson of MIT derived a formula which somewhat aided the analysis of the problem.

A rather obvious way of producing a csc^2 beam using a distributed feed was suggested by R. C. Spencer and H. J. Riblet. This was the principle that any source placed in the focal plane of an optical system would be projected out into space. Thus a row of dipoles excited according to the $csc^2 \theta$ law would result in a $csc^2 \theta$ pattern in space. The method was practical wherever beamwidth was not large. The V-beam and SCI search antennas successfully used horns arranged to give a csc^2 primary pattern. For most airborne applications where navigation and bombing specifications require a large vertical angular coverage, this method is not suitable and was replaced by the shaped reflector method proposed by L. J. Chu. This consisted in deforming a parabolic reflector at one end so as to distribute the energy as needed. Although the method had the disadvantage of producing a broader peaked beam (since the curved part of the reflector did not contribute to the peak), designs were easy to calculate and to adjust experimentally because of the one-to-one correspondence between each portion of the reflector and the corresponding direction in space. The shaped reflector principle was used by L. L. Blackmer of the MEW group to give a csc^2 pattern for high coverage.

The realization that the use of a reflector curved at one end results in a curved wave front enabled the principle of shaping to be applied to linear dipole arrays by properly phasing them. The elevation beam for GCA is obtained in this way by use of a vertical array of properly phased dipoles backed by a parabolic cylinder.

Beacon Antennas

Beacon antennas were developed by the Antenna Group to fill requirements of performance not found in other types of antennas. A homing beacon, for example, had to radiate its energy uniformly through an azimuth of 360°. This resulted in low gain which was compensated for by the fact that only one-way transmission was required.

The first Laboratory beacon was built in 1941 by G. Jarvis and employed biconical horns. In 1942 biconical horn beacons were replaced by linear array beacons with improved elevation pattern. An example was the 10-cm AN/CPN-8 (BPS) composed of 14 elements of three horizontally polarized curved dipoles each.[2] The dipoles received energy from the transmission line, as did other linear arrays such as MEW, from probes which projected into the line. Each element of three dipoles was spaced evenly around 360° to give the uniform azimuth pattern required. In a 3-cm homing beacon antenna, used in the AN/CPK-3 (BGX), the dipoles were replaced by vertical radiating slots which produced a vertically polarized pattern. A vertically polarized version of the AN/CPN-3 (BXS) employed short sections of cylinders each a half-wavelength long.

The pattern from a widely spaced dipole array contains regions of maxima where waves from succeeding dipoles differ in phase by a whole number of wavelengths. Accordingly, the spacing between elements in a beacon antenna was made a half-wavelength which, from optical diffraction theory, allows only one maximum to appear. Thus the radiation pattern of a homing antenna was constant in azimuth and limited to a narrow elevation angle, directly proportional to wavelength and inversely proportional to the number of radiating elements.

A linear array in which the elements were arranged one wavelength apart had its beam directed along the axis and was therefore known as an end-fire antenna. Such an antenna was designed for the Falcon system, which tracked targets in range only. It had 18 elements consisting of two curved dipoles each arranged to give horizontal polarization.

As part of the development of the Eagle blind bombing system and other similar systems, the Antenna Group did considerable research on rapid scanning beams (at the rate of several times per second). Rapid scanning beams were obtained either by movement of the entire antenna system (mechanical means) or by electrical means. Examples of electrical rapid scanners were moving feeds and moving paraboloids, as well as purely electrical changes. Off-centering, one method of electrical rapid scan already discussed, was limited by optical distortion effects to narrow sectors of scan. Certain other methods of rapid scan were effective over large angles. These methods invariably made use of changing the phase of the energy radiated by a series of sources, as dipoles in a linear array. The Eagle (AN/APQ-7) antenna was a typical rapid scanning array—a long waveguide whose width was variable. Rapid variation in width produced rapid variation in wavelength (phase) of the energy passing down the waveguide, which in turn caused the fan beam to wave back and forth a distance of $\pm$ 30°, three times every 2 sec. An alternative method of changing wavelength in the guide was the variable frequency generator, whose development was not sufficiently advanced at the end of the war for actual use.

Various other methods were tried for producing rapid scanning beams. G. G. Harvey and Louise Buchwalter of the Antenna Group developed the LRASV (long range ASV) antenna which was fed by a portion of coaxial transmission line having an eccentric corrugated inner conductor. The corrugations (according to their depths) slowed up the wave and, when the inner conductor was rapidly rotated, gave a varying velocity to the energy passing down the coaxial line. By the end of the war, this device had progressed to the flight testing stage, with fair success.

J. S. Foster developed a rapid scanning antenna in which the radiating element was a cone rotating within a fixed cone. Barriers, formed by intersecting teeth projecting from the adjacent surfaces of both cones, required the energy to travel different pathlengths around the cone, depending on the angle of rotation. This resulted in a scanning beam whose scan angle was, at any time, proportional to the angle of rotation. Foster's system had the advantage over previous systems of having no optical errors. Overall construction was somewhat complicated, but the advantages of a dynamically balanced mechanism and rapid scanning more than compensated.

Other types of rapid scanners, in which the feed rotated in a circle and sprayed its energy against the reflector, included such various devices as the Robinson roll scanner, developed by C. V. Robinson and used on the SCI height finder, the Lewis roll scanner developed by W. D. Lewis of Bell Telephone Laboratories, and the Lampshade scanner, developed by H. Iams of RCA. The last two had limited sector widths of scan.

In March 1942, a preliminary study was undertaken by Jarvis and Spencer to determine the effect on beam patterns of a radome (the dome-shaped housing of an antenna). Observations made with a radome surrounding a 3-cm antenna system showed negligible effect on the beam pattern, even when the radome was coated with a thin layer of ice. A layer of water, however, was found to absorb from 25%–50% of the beam energy. The study was continued about a year later by T. J. Keary and R. M. Redheffer, who observed that even small reflections from the radome into the antenna feed caused sufficient mismatch in the line to "pull" the magnetron off frequency. They found that a thin-walled radome exhibited the same thin-film interference phenomena for radar waves as a thin soap film for light waves. The problem of designing radomes of light weight but strong construction, giving minimum reflection at various angles of incidence and various wavelengths, became the effort of a large section under E. B. McMillan, who correlated the various theoretical, physical, mechanical, chemical, and manufacturing aspects of this problem.

In July 1943, a decision of the Antenna Group to build a field station for measuring antenna patterns was prompted by three factors: (1) The antennas being developed were becoming larger and more specialized in design, thus making them more of a security problem, (2) there was an increasing amount of interference between antenna pattern stations and operating systems when patterns were taken on the roof of MIT buildings, and (3) the path over which patterns were taken was too short for large dishes. Great Neck at Ipswich, Massachusetts was the site chosen for the station because it excelled others in seclusion, accessibility, radar isolation, and geometry.

rf

Prior to the organization in 1942 of the rf Group (later Group 53) under the direction of J. R. Zacharias, rf techniques had been investigated in a somewhat haphazard manner as they were needed. The importance of rf components, which critically affect the performance of a system, and so require careful design, was not always recognized in the early days of the Laboratory. The formation of the rf Group provided a center of responsibility for the design of microwave transmission line components, and brought a systematic concentration to bear on what was, at the time, a primitive art.

The first transmission line problem to be considered was construction of the line itself. The first line was the coaxial type—a circular shell surrounding a central conductor—made of soft copper tubing so as to be semiflexible and supported internally by polystyrene beads, spaced a half-wavelength apart. The addition of reflections from successive beads gave high standing waves, however, so half-wave spacing was abandoned in favor of quarter-wave spacing. The resulting line was frequency insensitive, but only for one frequency. In early 1941, J. L. Lawson, head of the Experimental Systems Group, developed the Lawson line. This used spacings of sometimes an odd and sometimes an even number of quarter-waves, according to a definite formula. The Lawson line showed a marked improvement in frequency sensitivity but, because of breakdown of the beads, was poor in power-carrying capability.

Late in 1941 there was developed at Bell Telephone Laboratories a stub-supported line, without beads, in which special quarter-wave stubs were placed along the line at frequent intervals to support it. The stub-supported line could handle comparatively high powers; the addition of a special stub-tuning device, constructed by R. V. Pound of the rf Group, made it quite frequency insensitive.

Another early development was a tuning device to fit on the coaxial transmission line, as a means of matching or balancing the line with the components it fed—in order to effect the least wasteful transmission of power. Tuners were especially needed to match the magnetron to the line, since individual early magnetrons varied greatly in their output impedances. One early tuner was the combination of a trombone (a sliding U-shaped section of coaxial line by which the physical length, and so the electrical length, of the rf line could be changed) and a T-stub with movable shorting plunger. The T-stub permitted adjustment of power matching and the trombone permitted phase-matching adjustment. Later, this combination was replaced by a double-stub tuner, with two single stubs a fraction of a wavelength apart, which had better matching power. Later (in 1941) S. Roberts designed a triple-stub tuner, with three single stubs spaced a fraction of a wavelength apart. This last was considered a universal tuner. However, its use was restricted to antenna development where were found the large standing waves with which it was capable of dealing.

As early radar components improved, many tuners were eliminated. For example, improvement in magnetron manufacture led to standardization of magnetron output characteristics and eliminated the need for an adjustable tuner to match the magnetron to the coaxial line. This was replaced with a so-called transformer (or "fixed" tuner), previously adjusted to the characteristics of an average magnetron.

Especially important in most microwave systems was the provision for duplex operation, the use of a single transmission line and antenna for both transmitting and receiving. Here the rf problem was twofold: (1) to disconnect the receiver from the line while the transmitted pulse passed through the line, and (2) to disconnect the magnetron from the line while echoes were being received.

The first need was filled by a kind of electronic switch, called a transmit–receive or, simply, a TR box. The first TR box tried in the Laboratory was a Sperry klystron microwave oscillator. This was the TR used on the *Semmes* in July 1941. The klystron had two cavities; signals returning down the transmission line were coupled into the first cavity, passed from the first into the second by means of the klystron beam, and then into the receiver. During the instant the transmitted pulse passed down the line, the klystron beam was turned off, isolating the receiver from the rest of the circuit. The system had two faults, although the receiver was protected from the transmitted pulse. The klystron absorbed about one-half the power in the transmitted pulse, and when the klystron was turned on, the electrons in the beam caused a considerable amount of noise, or static, in the receiver.

In 1941, Lawson produced a TR box consisting of a gas-filled cavity, tuned to be resonant at the operating frequency of the radar system, with a small metal spark gap at its center. The intensity of the transmitted pulse caused the gap to "break down," or spark, short-circuiting the cavity so that no energy could pass through it. After the transmitted pulse had gone, however, the gap "recovered" and the TR box became ready to pass return signals, which it did with little loss since it was tuned to the frequency of the signals. An improved gas-filled TR box, the British Sutton tube, whose method of construction had been largely copied from a British klystron, appeared in 1941. The advantage of the Sutton tube lay chiefly in its mechanical construction: the tube itself, including the spark gap and surrounding glass envelope, was separate from the cavity. Hence separate cavities could be designed to be used on a number of different wavelengths, and a bad tube could be replaced without throwing away the entire cavity. A similar development at the Bell Telephone Laboratories resulted in the 721-A tube, which eventually became the standard 10-cm TR tube in the laboratory.

TR developments during 1942 and 1943 were mainly improvements of already existing devices—making them operable with the continually increasing maximum transmitter power. In 1943, when the 10-cm band was divided into three subbands for ship, air, and ground radar systems, a tunable 10-cm TR (the 721-A modified to the 1B27) was constructed; this reduced the number of cavities needed to cover the 10-cm band from three to one.

A companion part of the work on TR boxes was mixer and crystal development by the rf Group, helped by contributions from other laboratory groups and from the British. The original mixer development in the Laboratory was done in the Receiver Group; a triode preceded by a klystron rf amplifier was used. Crystal mixers, although considered at the time for their better performance, were rejected as harder to adjust than the triode mixer. The first crystal mixer to compete with the triode mixer was brought out by the Bell Telephone Laboratories, and consisted of a coaxial line mixer with the crystal built into the assembly. Soon afterward, the Laboratory received samples of a STH crystal, previously developed by the British, which proved to be better than an American-made crystal of the same design. The British soon after produced a high-burnout crystal.

The result of the general superiority of British crystals and of ideas brought back from England by E. M. Purcell, leader of the Fundamental Developments Group, was the instituting in the fall of 1941 of a Laboratory program for improving crystals, under the general direction of C. A. Whitmer. Work along companion lines was being done under a Radiation Laboratory contract at the University of Pennsylvania (experimental measurements on noise and burnout characteristics); this institution took over the development of crystals from the Laboratory at the close of the war. In large part, the nature of the Laboratory program was an inquiry into the effects of impurities in the silicon which formed the crystal material. A major discovery was that boron and aluminum added to the silicon would give a crystal easy to manufacture, with high sensitivity and high burnout. The program led to the 10-cm LN21-B line of crystals, which had the further advantage of exhibiting the same impedance as earlier crystals, making replacement easy. Work of the group afterward concentrated on the development of 3-cm and 1-cm crystals.

The pot mixer, a modification of a British mixer, was developed to use the crystals which were being received in 1941 from the British. The type was difficult to tune but was used in early B-18 installations. Early in 1942, B. B. Cork of the Experimental Systems Group improved the BTL coaxial line-type mixer by making it an integral part of the TR box assembly and usable over a broader frequency band than earlier types. In the fall of 1942, R. V. Pound of the rf Group, with contributions from Cork, designed the Pound mixer, the first good mixer developed in the Laboratory. It had no tuning adjustment but was permanently adjusted for the average crystal. The Pound mixer was used in the Laboratory until the end of the war.

In early 1941, development work on 3-cm components began in Purcell's group. At 3 cm, radical departures from earlier techniques were demanded. Because 3-cm coaxial line was extremely small and difficult to handle, waveguide (hollow conducting pipe) was used. Waveguide was easy to manipulate and had high power-carrying capacities, but its use required developing a whole set of components for a new medium of transmission.

At the start of work at 3 cm, little was known about the properties of waveguide transmission. The previous waveguide method of carrying rf energy around corners, for example, had been to use gradual bends. S. Roberts showed that by means of a properly placed reflector rf power could be made to turn at right angles. This led to the discovery that a right-angle waveguide bend would transmit power. Roberts' method was afterward improved by the addition of a 43° bevel plate to the outer edge of the waveguide corner.

It was necessary to adapt 3-cm magnetrons, which had coaxial outputs, to waveguide. Early transition methods proved to have low power-handling capacity. One such was a probe transition, in which the inner conductor of the coaxial line projected into the waveguide, forming a small radiating antenna. This was frequency sensitive, however, and at high powers produced corona discharge and sparking. A much more successful transition, developed at GE but modified and improved by M. Clark and F. L. Niemann of the Radiation Laboratory, was the doorknob transition, so called from its appearance. The doorknob was placed inside the waveguide and fastened to the end of the inner conductor of the coaxial line. The shape of the doorknob made it suitable for a wide frequency band, and, since it had no sharp curves, it was suitable for high power requirements.

The final form of 3-cm TR tube was a scaled-down version of the 10-cm tube. The first reasonably successful 3-cm TR was devised by Roberts after Lawson's 10-cm model. The spark gap was placed at the center of a resonant cavity tuned to the signal frequency. During the transmitted pulse, when the voltage was a maximum at the center of the cavity, the gap broke down, preventing the pulse from leaking through. In a later version, invented by Preston, the gap was made adjustable so that this cavity could be tuned to a number of frequencies. This model, however, was complicated to manufacture. Bell Telephone Laboratories brought out the 724 TR tube, essentially a 3-cm version of the 10-cm 721. After a study of the problem for the Laboratory by two Westinghouse engineers, Westinghouse produced the 1B24 TR tube. This could be tuned by varying the size of the spark gap; unlike the 10-cm TR's and the 724 TR, it had an internal resonant cavity, and it included a chamber, or reservoir, of gas. The extra gas increased the life of the tube, since ordinary 3-cm tubes held so small an amount of gas that it was rapidly used up. Especially necessary in small 3-cm tubes, the internal cavity eliminated the chance of poor contact between the tube and an outside cavity. For these reasons, the 1B24 came to be the standard 3-cm TR tube of the Laboratory.

An early method of matching the magnetron to a 3-cm waveguide, a method which minimized the absorption of return signals by the magnetron, was to vary the electrical length of the line by a squeeze section. This varied the wide dimension of the waveguide, since wavelength in the guide was a function of this dimension. Three-cm magnetrons in 1941 varied greatly in their output characteristics so that no average characteristic could be assumed which would allow preplumbing. Accordingly, a solution by H. K. Farr was the anti-TR (ATR), a suitably resonant cavity placed a quarter-wavelength from the TR junction. The ATR allowed the magnetron pulse to pass uninterrupted down the waveguide to the antenna but deflected returning signals through the TR with a minimum loss which was largely independent of the magnetron.

A 3-cm coaxial line mixer was built in 1941, for use with the early Bell Laboratories crystals. A later model of the mixer was built to use BTL crystals. In 1942, Roberts designed a successful 3-cm mixer in which the crystal was mounted across the waveguide.

Major improvements in mixers during the period of 1933–44 included the design by Pound of mixers with separate compartments for several crystals. Some of these mixers used two crystals, one for signal detection and one for automatic frequency control, each coupled to a local oscillator. This arrangement gave better AFC operation than when two completely separate mixers were used. The final major Laboratory improvement in mixers was the magic-tee balanced mixer, which would reduce (or balance out) noise from the local oscillator, and which therefore could contribute to a more sensitive receiver. This noise reduction was useful at 3 cm and necessary at 1 cm, where the IF frequency was low compared with the rf frequency.

By the summer of 1942, many "fundamental discoveries" had been made, and the rf Group concentrated on improving components (often simply by reducing the number of tuning stubs) and adapting them to wavelengths other than those for which they had been originally designed. At this time, 3-cm work, which had been the responsibility of Purcell's group, was transferred to the rf Group and Purcell's group began work on 1 cm. Their work at 1 cm included the development of a TR (1B26) (the only Laboratory 1-cm TR), a scaled-down version of the 3-cm 1B24, and also the development by R. Dicke, with the help of Pound of the rf Group, of a magic-tee balanced mixer for 1 cm.

An early need of radar systems was a rotary joint (in which one length of coaxial line would rotate with respect to a second length) which could be used to carry rf energy between a rotating antenna and stationary components. The requirements for a satisfactory rotary joint were the same as those for most other transmission line components: a capacity for high powers and for passing a wide range of frequencies. The essential feature of the first and of many of the later designs was a capacity joint (one each for both the outer and inner conductors) which permitted electrical contact without physical contact. Coaxial-line rotary joints at 10 cm had been designed in 1942 by K. T. Bainbridge, A. Longacre, and H. K. Farr to handle the magnetron powers then available, but were generally too bulky and had too limited a frequency range.

It was observed that some frequency "modes" of rf energy passed through the joint more easily than other "modes" depending on the length of the joint, so that rotary joints could be used only in lengths that were favorable to desired modes. In 1942, in the new frequency region of 3 cm, Roberts began work on a rotary joint using circular waveguide. The first model was frequency narrow because of "mode impurities" and could handle only low power. W. M. Preston devised a stub which would purify the mode, but his method proved also to be frequency sensitive. Late in 1942, the British proposed a rotating joint without a stub, which appeared to give the desired characteristic of suppression at 3 cm. In early 1943, the British method was modified and improved by Farr and F. E. Ehlers to include a resonant ring filter. With the further addition of a mode absorber, the rotary joint was frequency insensitive with the added feature that it could be used in any length desired.

In early 1942, the first attempt was made in the Laboratory to use flexible waveguide. The first flexible waveguide was simply a piece of flexible cable squashed to approximately the shape of 3-cm rectangular waveguide. Although this cable was wasteful of power, it was considered promising. Bainbridge and Longacre experimented with a corrugated type of waveguide—a kind of rectangular bellows—with corrugations shallow compared to a wavelength, but deep enough to provide the required flexibility. A follow-up experiment by Purcell, using a copper corrugated waveguide formed by electroplating on a form, failed mechanically because the copper was porous and brittle. Circular bellows were also tried, with corrugations a half-wavelength deep to form a resonant structure, but these never proved satisfactory. The ultimate development in flexible waveguide was the vertebrae type (used at 3 cm and 1 cm), which consisted of successive capacity-type junctions (formed by disks mounted on a rubber jacket). Vertebrae waveguide could be bent, twisted, stretched, compressed, and sheared.

NOTES

1. Equations implying the idea of a csc^2 beam are found in the 1941 notebooks of Ray Herb, H. E. Farnsworth, and S. G. Sydoriak.

2. Radiation Laboratory Report B–64, J. J. Brady, "Antenna Catalogue," October 8, 1945. See figure on p. 70.

CHAPTER 26

Receivers, Indicators, and Precision Range Circuits

Receivers

The Receiver Group of the Laboratory (Group 61) was under the leadership of S. N. Van Voorhis. The source of material available to the Group at the beginning of its work was the entire body of receiver knowledge compiled in previous years from work on frequency modulation, television, broadcast reception, and communications, including many techniques and principles which, by selection, could be adapted to the needs of a receiver operating at microwave radar frequencies.

Two features of radar operation profoundly affect the design of a receiver. The first is the extremely wide dynamic range covered by the signals to which the receiver must respond, and the second is the very low level of the smallest signal with which it must deal properly. The first factor is due to the use by most radars of a common antenna for transmitting and receiving. Although considerable protection from the transmitted pulse is given the receiver by an electronic switch, the TR box, even the best TR boxes sometimes pass as much as a tenth of a watt of the transmitted energy. Since the weakest signals to which the receiver must respond are of the order of magnitude of 10^{-12} W, the ratio of strongest to weakest signals becomes 10^{11}:1. The problem is enhanced by the fact that the receiver must be able to respond almost instantaneously (in a few microseconds) to transient signals having this large discrepancy in strength. One section of the Receiver Group worked on the problem of improving the transient response of the receiver.

The second section of the Receiver Group was assigned the problem of reducing receiver noise, the electronic static generated inside the receiver box. In a radar system the limiting range of signals that can be received is set by the range at which the echo from the desired target becomes as weak as noise and so merges with the noise. Although range can be increased by increasing transmitted power, it is often more economical to try to decrease the noise in the receiver. In the case of receivers operating at lower than radar frequencies, the problem of receiver noise is overshadowed in importance by the problem of external noise, such as man-made static. For this reason, the study of noise reduction received little attention prior to the work at the Radiation Laboratory. With microwaves, however, externally produced noise becomes negligible, and receiver noise is the important factor. Therefore limitation of receiver noise was of utmost importance.

Besides these two primary fields of investigation, the Receiver Group investigated numerous problems involving the design of receivers to withstand military usage, to accept available power supplies, to be lightweight and small in size (for aircraft applications), and to take advantage of new developments such as miniature tubes and, later, subminiature tubes.

The first project of the Group was to build a receiver for a 10-cm airborne radar system. Since amplication of waves of so high a frequency required specialized circuit techniques, the choice was immediately made for a superheterodyne-type receiver, in which the microwave frequency could be converted into a lower, or intermediate, frequency, which then could be more readily amplified. Choice of the superheterodyne circuit entailed problems of a local oscillator tube and a mixer, as well as problems relating to the IF amplifier. The intermediate frequency to be used was arbitrarily set at 30 MHz. Although the choice was arbitrary, it afterward proved to be a fortunate compromise between higher IF frequencies which would have caused greater circuit noise and lower circuit stability, and lower IF frequencies which would have caused poor video discrimination. Before the Laboratory officially opened, an order was placed with RCA by the microwave committee for five amplifiers, based on television design practices, with an IF frequency of 30 MHz, and a bandwidth of 3 MHz. Bandwidth, the characteristic of a receiver which indicates how narrow and how square a pulse it can reproduce faithfully, was set at 3 MHz in these first units because from television experience, it was believed that this would be sufficient to reproduce the 1-μsec pulses to be used in the airborne system.

These first receivers were used on roof experimental systems, where they exposed difficulties of design and pointed the way toward further improvements. The basic fault, of course, was that they were built to television ideas, with small realization of the requirements of microwave frequencies. The sets had poor transient response, and the signals were rounded. Shielding was insufficient, and an aluminum chassis did not provide a good common ground. Circuits were complicated and difficult to adjust. The local oscillator was a negative-grid type, difficult to tune and with poor frequency stability. Its operating frequency range so nearly reached a limit at 10 cm that it was barely possible to tune it to work with the early magnetrons, operating at 9.5 cm.

At the beginning of the Laboratory, possibilities for local oscillator tubes included the klystron, a velocity-bunching type of oscillator which required such a high-voltage power supply that it was unsuitable for the narrow confines of receiver work. A smaller unit, developed soon after, still required higher voltages than were suitable. Finally, in the beginning of 1941, two newer types of velocity-bunching tubes were developed by the Bell Telephone Laboratories. One of these, called the Samuel tube, was similiar to the klystron. The second, called the McNally tube, was of the reflex velocity modulation type, and used a reflector which produced electron bunching by turning the electron beam back upon itself. Both tubes had external cavities, with the resultant advantages that the same glass envelope could be used with different cavities. The Samuel tube was installed on the *Semmes* in July 1941. Since it was frequency stable and had no provision for electronic tuning, it was favored at the time by the Group. However, the McNally tube later proved to be the most universally applicable for all radar sets. It needed relatively low voltages to operate, and its ability to change frequency by change of an applied voltage proved later (when a 3-cm system was being developed) to be a key to one method of automatic frequency control.

A whole series of tubes built to this fundamental operating principle later appeared. The first was the 707A built by Bell Telephone Laboratories, later followed by the 707B which was temperature compensated (the 707A sometimes drifted out of its tuning range, because of temperature effects), and still later by the 2K28 which was mechanically smaller. The final series included the 726 A, B, and C (developed by Bell Telephone), which were analogous to the original klystron in that they had the entire tuned circuit within the vacuum. A considerable part of the Receiver Group's efforts went into measurements on local oscillator types. These measurements included power output, tuning range, stability, load effects, and the like, and, as was the case for crystals, included setting up specifications suited to quantity manufacture of the tubes.

The mixer that was used in the early receivers was a crystal with an adjustable "cat's whisker." The drawback to this mixer became apparent, however, when it was realized that such an assembly could not be used in any field application where adjustment by relatively unskilled personnel would be required. Accordingly, a triode vacuum-tube mixer was next tried. Essentially, this was an altered negative-grid oscillator tube. It had the faults of having a much more complicated rf circuit than a crystal and of having a lower signal-to-noise ratio; its filaments, of peculiar design, were susceptible to vibration so that the tube could not be used for ship or airplane applications.

Late in 1941, various Laboratory members returning from England reported that the British had developed cartridge crystal units, in which the crystal and cat's whisker were permanently installed inside a ceramic case, and required no adjustment. In test, these units duplicated the satisfactory performance of the adjustable-type crystal and, as a result, soon became standard for practically all 10-cm systems. A series of measurements were made by the Receiver Group, evaluating the effects of different construction and composition of the crystal material, and establishing suitable specifications for the manufacture of the units in quantity. This work was begun in the Receiver Group but was taken over in the fall of 1942 by the rf Group.

To reduce noise and to gain signal amplification, much effort was concentrated on the intermediate-frequency amplifier circuits. In the summer of 1941, measurements by E. J. Schremp and his co-workers showed that most receiver noise was produced in the IF amplifier stages. Earlier investigation of the same subject by the Bell Telephone Laboratories had resulted in a new amplifier circuit which reduced amplifier noise by a factor of about 20%. However, the new circuit was hard to set up and adjust so that it never saw much service. In the first part of 1944, the Radiation Laboratory developed a circuit which employed two amplifiers in combination, consisting of a grounded-grid triode and a grounded-cathode triode. This combination turned out to have the satisfying property of reducing receiver noise level to a point about three times as close to the theoretical level as before. At the close of the Laboratory, the circuit was being built into many systems, either on replacement unit basis or in the original design.

Design of proper circuits to couple successive IF amplifier tubes was another important problem. The first IF amplifier "strip" built by the Receiver Group was designed to have the maximum possible overall gain. This resulted in circuits which were difficult to adjust originally, and so sensitive as to require realignment whenever tubes were changed. Accordingly, a thorough investigation of the best type of coupling circuit (taking into account needs for gain, bandwidth, and ease of adjustment) was begun under the direction of H. Wallman. He showed that a separate single-tuned circuit (the simplest possible), when used as a coupling circuit, gave a bandwidth as nearly ideal as could be expected for the faithful reproduction of radar signals. He showed, however, that for a given amount of gain several such coupling circuits would combine to limit the overall bandwidth of the receiver. Where the receiver had to pass narrow, square pulses requiring wide bandwidth, this would be a decided disadvantage. Wallman therefore developed a method of stagger tuning of successive coupling elements, i.e., tuning them to frequencies on opposite sides of the center frequency, the pattern of tuning depending upon the desired overall bandwidth. By stagger tuning, it became possible to increase the maximum attainable bandwidth by a factor of about 3 or more.

The first IF amplifiers built by the Group used full-size tubes. By 1942, miniature tubes (the 6AK5 was the first widely used, replacing the

6AC7), about one-half as large as the others, became available. It was at once apparent that miniature tubes would make possible substantial savings both in space occupied by the IF amplifier and in power consumed by it. Electrical characteristics of these tubes were approximately the same as before so that no circuit changes were required. However, the former method of mounting resistors and condensers directly on the tube sockets to reduce stray capacitance (a factor limiting gain and bandwidth) had to be changed, because of smaller tube sockets, in favor of mounting parts away from the sockets but with the same reduced stray capacitance. Resistors and capacitors also were developed to compare in size with the miniature tubes. Many amplifiers using these tubes were put into production.

A further development in the same direction was subminiature tubes, an outgrowth of the very large-scale development of the proximity fuze, although the tubes themselves were not at all similar to those used in the proximity unit. At the close of the Laboratory, work was still in a preliminary stage as far as IF amplifier use was concerned, but the potentialities of the tubes were such that whole amplifiers built as a unit and impregnated against moisture and fungi, etc., were foreseen.

A major problem, as far as the Receiver Group was concerned, was the development of a receiver to work well at 3 cm. The problem was stabilization of the local oscillator; at this high frequency, the local oscillator could not be relied upon by itself to give the percentage stability required. The solution was AFC, automatic frequency control. The first successful AFC method used for its local oscillator a kind of McNally tube called the 723A, in which a change in reflector voltage produced a change in output frequency. The circuit designed transformed any undesirable change in output frequency into a compensating voltage, which then could be added to the reflector voltage of the McNally tube. When newer types of local oscillators appeared in 1944, including thermally tuned tubes, extensions were made on the original AFC techniques. G. H. Nibbe and E. Durand, W. M. P. Strandberg, and A. E. Whitford worked on different AFC methods, using thermally tuned tubes. Whereas the ordinary reflex velocity-modulated tube could be tuned electronically over a small fraction of its mechanical tuning range, thermally tuned tubes could be adjusted electronically over their entire range. This meant that they could correct large errors in frequency. Because their range was so wide, circuits had to be developed to keep them on the proper side of the magnetron frequency. Ways were also found to stabilize the frequency of a local oscillator by using a cavity as the controlling element. This method was called absolute frequency stabilization, as compared to the relative stabilization with respect to the transmitter employed in normal AFC systems. Absolute frequency stabilization found important use in receivers for beacon signals, where the transmitter was often as much as 100 miles away, and so was not available to furnish tuning information for the receiver oscillator.

Special problems were undertaken by the Receiver Group at various times. Protection against jamming was important: it was necessary to design circuits to minimize the danger of large spurious signals overloading, and momentarily paralyzing, the receiver. Such circuits were also useful in blanking out ground clutter which appeared on the indicators, and in the case of experiments conducted on the roof of the Laboratory, helped in protecting one system against the concentrated jamming of other systems which invariably operated at the same time and on the same frequency. Some fundamental work on protection against jamming was done by Group 44, under the direction of J. L. Lawson. He also cooperated on another Receiver Group project—that of shielding the rf amplifier from strong local-oscillator output power.

Radar beacon receivers were another project of the Group which required attention to special requirements. From 1942 until the end of the Laboratory, beacon receivers were designed and built to respond to signals within as wide a frequency band as 70 MHz (for 10 cm) and 100 MHz (for 3 cm). Although IF bandwidths in these receivers had to be made wide for interrogation purposes, video bandwidths could be small since beacon pulses were usually long and so contained only a small range of frequencies. Furthermore, beacon receivers had to be made unable to respond to strong, narrow pulses, as from a powerful search radar in the vicinity. Special circuits to prevent beacon receivers from "pulse stretching" were designed by M. I. Crouch, H. J. Lipkin, and J. H. Tinlot.

Indicators

The responsibilities of the Indicator Group of the Laboratory were the design of indicators and their adaptation to systems, and the design of computers whose use involved indicators. The indicator section was not concerned with such elements as fire-control directors where transmission of information was completely automatic and the radar served only to furnish range and direction. The Indicator Group was part of Division 6, which was concerned with the receiving components of the radar system.

There were two groups concerned with indicator problems, the Indicator Group, which concentrated on developing cathode-ray tubes and accessories and their displays, and the Precision Group which developed precision ranging equipment, and such indicator-computers as bombsights. There was much overlapping between these groups; for example, the Indicator Group displays often required range measuring devices and the Precision Group equipment often required special displays.

Although when the Laboratory began the British had developed a cathode-ray tube with long-persistence screen, and had applied it to both the A scope and the PPI, satisfactory tubes and circuits for producing displays did not exist in this country. Hence, almost immediately upon formation of the Laboratory, L. A. DuBridge opened negotiations with the General Electric Company and RCA for the development and production of cathode-ray tubes and indicators, with stress placed on the development of a long-persistence screen. One of his first steps was to call in W. B. Nottingham of the MIT Physics Department, to serve as consultant on screen development, principally to make intensity measurements on screens produced by the two companies and to suggest improvements. The contract for cathode-ray tube development remained in force with RCA throughout the war; with GE, until 1944. Late in 1940, all indicator work was merged in the Indicator Group, with W. M. Hall as Group leader. R. F. Bacher, then Associate Group leader, replaced Hall when the latter left the Laboratory in 1941.

With the early Laboratory interest in the 10-cm AI radar, the Indicator Group concentrated on building suitable indicating displays. The first display was an A scope, which showed target range information only as a vertical deflection in a horizontal line (amplitude modulation); this used cathode-ray tubes up to 12 in. in diameter. At the same time, work began on intensity-modulated displays (in which the signal appeared as a spot) so that angle information might also be produced. The first such display was the type C, which presented elevation angle vertically against azimuth horizontally. Since this display did not show range, there could be no "spreading out" of noise, with the result that noise "piled up" and appeared as spurious target signals. The signal-to-noise ratio thus was very low. The first improvement was the type D display, identical to the type C except that the trace on the scope face also included a short upward sweep in range. The next development was the type B display in which the elevation coordinate was eliminated; range was shown vertically and azimuth horizontally. So satisfactory was the B display both for search and for high signal-to-noise discrimination that it was widely used by systems throughout the existence of the Laboratory.

In application to the AI equipment, the B scope worked in conjunction with a C scope through a control on the B scope panel. This control made the C scope inoperative except within a short range interval, the start of which was determined by setting the control to see the range of the target as it appeared on the B scope. Thus the C scope showed only the desired target. A similar combination of these two types of indicators served throughout the war in SCR-520 and SCR-720 sets.

In early 1941, E. C. Pollard began the design of a rotating-coil PPI, copied somewhat after that of the British. He was assisted by R. E. Meagher who soon assumed entire responsibility for the development in the Laboratory of this type of PPI. During the late spring of 1941, Bacher initiated the investigation of electronic PPI's, in which the rotation of the scope trace as it followed the radar antenna required no moving parts but was produced by electronic means. When preliminary experiments by J. Koehler showed that an electronic PPI was feasible, investigation of three methods began. The general idea was to treat the sweep in terms of the sine and cosine components of a circle, later combining these components in the deflection coils of the cathode-ray tube. Sinusoidal potentiometers were used to generate the sweep initially in terms of its sine and cosine components (M. A. Starr), and also used to reduce an already generated sweep to these components (G. E. Kron). Koehler and V. C. Wilson worked on a method of generating components of a sweep by means of selsyns.

By the end of 1941 the electronic PPI had progressed to the point where a successful indicator had been produced by both of the potentiometer methods. One of these indicators was installed in the Dumbo I. The selsyn methods had been less successful since loss of dc voltage in passing through the selsyn caused the trace to jump around the center of the scope face. When RCA suggested to the Laboratory that clamping circuits, such as were used in certain applications of the television trade, might help, C. W. Sherwin was assigned to adapt their use to the selsyn PPI. By the end of 1941, development of the selsyn PPI, using clamping, had advanced to the point where this method was more successful than either kind of potentiometer PPI. A model with 60-Hz power supply was adapted early in 1942 for use with the SCR-584, and PPI's of this type were included in all later versions of that set.

A second selsyn-type PPI designed by Sherwin to incorporate a 400-Hz power supply and lightweight components for use in airborne equipment was later rated by L. J. Haworth, as head of Division 6 and leader of the Indicator Group, as the most important airborne indicator designed by the Radiation Laboratory. In the spring of 1942, it was incorporated in the ASG, the DMS-1000, and several experimental Laboratory sets. The indicator incorporated in the ASG was subsequently used in the H_2X. It was the first really successful electronic PPI and the methods used in it proved, as time went on, to be basically the most effective ever developed.

A consequence of the work on potentiometer PPI's was the forming of the potentiometer division of the Indicator Group, under P. Rosenberg, which developed a variety of high-precision, long-wearing potentiometers for PPI and other purposes.

During the summer of 1941, other indicators besides electronic PPI's were being developed. An order for several engineered combination B- and C-scope units was placed with the Raytheon Company for Laboratory use. Ultimately, these indicators were modified (by W. A. Higinbotham and others) and incorporated in the B-18 ASV-10 systems. Indicator Group members aided in the installation and spent time in the field with the Sea Search Attack Group.

In November 1941 the Indicator Group was called upon for indicators for the AIA, an interception radar to be placed in single-seat fighters. The problem was to design an indicator which would give all necessary information to the pilot, who also had to fly the airplane. Haworth, later helped by R. M. Walker, developed the type H indicator, a B scope on which the signal appeared not as a dot but as a line, called a semaphore, the orientation of which gave a rough measure of the elevation angle of the target. This was later modified to the "double-dot" presentation, in which dots appeared in the position of the ends of the original semaphore, the interpretation being the same as before. The type H scope, with double-dot indication, was used in all single-seat aircraft interception sets throughout the war. For blind firing, the indicator was supplemented at close ranges by a spot-error indicator with wings for indicating range.

With the first few months of 1942 came a rush for equipment for specific military purposes which was to continue at an accelerated pace and absorb most of the time of the Group for the rest of the war. Indicators were built for ASD (by Haworth and A. W. Rawcliffe); for AGL-1 (by H. W. Babcock); for the experimental MEW (Sherwin and L. D. Ellsworth), etc. In June, the Indicator Group assumed the task of building the indicators for the Eagle precision radar bomb sight. Because of the scanning peculiarities of the equipment which included the intersection of the radar beam with the earth in a hyperbola, the indicator problems were complex. They were studied by Higinbotham, and continued as a major indicator development for nearly two years.

The GCA blind-landing equipment represented another major problem. In this connection, Sherwin developed two expanded-sector indicators, one to show height of the target above the runway, the other to show its position along the runway. On the faces of these scopes a desired glide path could be superposed by means of partially silvered mirrors. For the high-speed scanning requirements of GCA, the new electronic technique was developed of having a variable capacitor furnish the rotation of the scope trace by making use of variation in the capacity of the capacitor as one set of plates rotated with respect to a second set. Two each of these indicators with different ranges were ultimately incorporated in the GCA. The vertical type of display, when later modified, became the range-height indicator, or RHI, used in all height finders having a vertical scan, including the AN/TPS-10, AN/CPS-6, AN/CPS-4, and SX sets.

Another new development, begun late in 1942, was the indicator for the SCR-598 coastal gunnery control set. Since this set involved high-speed scanning, approximately 16 scans/sec, the variable capacitor technique for furnishing azimuth rotation of the scope trace was used.

The first of 1943 brought an increase in the number of requests for indicators for specific systems. The P-31, developed jointly by the Indicator and Precision Groups, was a combination PPI and expanded B scope contained in one console and useful for accurate tracking and target designation. The VE was an accurate rotating coil-PPI, engineered by the Indicator Group, designed to serve as the main PPI in the SR equipment and also as a remote PPI for general shipboard use. The indicators for the MEW systems consisted of two PPI'S and four B scopes, one of the latter highly expanded.

While work on the development and engineering of specific indicators was progressing, the cathode-ray tube section, under J. T. Soller, continued to advise tube manufacturers in the development of long-persistence screens and in the general construction of tubes, particularly of the magnetic deflection type. By late summer of 1941, screens and electron guns had been settled upon for the 5, 7, 9, and 12 in. types of magnetic-deflection tubes. The screens were all of the cascade type of persistent screen, development of which had been started by the British; it ultimately became known as the P-7. In the autumn of 1941, the Indicator Group placed orders with the General Electric Company and the Radio Corporation of America for some 500 tubes, principally of the two smaller sizes. This educational order had a profound effect upon the future availability of all cathode-ray tubes used for intensity modulated types of displays.

In the spring of 1942 the British reported that they had developed a new type of cathode-ray tube screen, called a skiatron, which presented signals of a dark magenta color against a white background. Its value in projecting a large image of the tube face onto a projection screen was immediately realized for situations such as a central control room, where it was desirable for several people to observe radar data simultaneously. Arrangements were quickly made by the Laboratory with both RCA and GE to build experimental skiatron tubes, while Soller assumed the responsibility for testing them. Meanwhile, equipment which W. B. Nottingham had designed to measure characteristics of the standard, long-persistence screens (P-7 screens) was being built.

In the new skiatron field, reasonably successful tubes were quickly made, and early in 1943, the Indicator Group succeeded in interesting the Bureau of Ships in the advantages of using a skiatron in combat information centers. Starr and H. O. Marcy designed the optical and electrical parts of the equipment, while Soller designed the cabinet and the general layout of parts. Development continued until an engineered model was obtained which could be turned over to GE for manufacture. A number of the Laboratory designed units were also produced by RCA for Laboratory and Service use. Ultimately, those units were installed on several American warships.

In the rush to build equipment the Group had tended to neglect the important duty of development of components. Hence, in the summer of 1944, specific developmental assignments were made. One important investigation, conducted by Rawcliffe, was to find ways of improving the focus and deflection characteristics of 12 in. cathode-ray tubes. Both the importance and later success of his work were demonstrated in one instance by the 12 in. tubes in use at the MEW Florida field station, whose focus, at first poor, was later greatly improved. The important work of designing indicator test equipment and of testing and developing circuit parts was given to M. D. Fagen.

The most important circuit technique developed during 1944 was the MS-PPI, in which the deflecting coils which controlled the movement of the cathode-ray tube beam were driven from a selsyn, without the intermediate use of amplifier tubes. This method economized in both number of tubes and power required and so effected savings in weight of both the indicator and its power supply. The drawback to the method was the difficulty of clamping the scope trace, that is, of starting the trace at the center of the circle. Methods of clamping as previously developed by Sherwin were not applicable to the new method of "direct drive." As a result, the technique was developed of creating a trace which would "wait" at the exact center of the scope face until the transmittal pulse had gone out, after which the trace would sweep in range. This method was applied during the fall of 1944 to the AN/APS-30 search set. A little later British workers came out with a more elegant method of clamping, which outmoded the Laboratory technique. The British method was not developed in time for use with any equipment designed for production by the Radiation Laboratory.

An important job of the Indicator Group from the middle of 1944 until the end of the war was the provision of improvements and attachments to existing equipment. An example of this was the application of the off-center PPI principle to the MEW indicators in the field. Work with MEW's in active theaters had shown that the combination of ordinary PPI and B scope was insufficient for the highly expanded, highly accurate sector displays required. As a result, J. E. Hexem, who had worked on the off-center PPI development in 1943 and who was stationed at BBRL, developed modification kits to effect the off-center change, as well as to change the indicators from 7 in. to 12 in. cathode-ray tubes. Hexem's plans for the kit were used for application to the AN/CPS-6, the V-beam radar, and also by General Electric for the production version of the MEW and V-beam sets.

The off-center PPI principle was adapted to the AEW set. In this application, the method served to cancel out the relative motion of the plane. In a later version, adapted to the AEW II, the off-centering principle was controlled from a ground position indicator.

A three-tone PPI, designed to afford contrast between land and villages, as well as between land and water, was applied to the AN/APS-15 and AN/APS-15A. This was finished too late, however, to have appreciable effect on the war.

Precision Group

The Precision Group was originally part of Group B, but was set up as a separate group (Group 63) under Britton Chance, after the reorganization of the Laboratory in 1942. The Group's first job was the development of precision ranging circuits for ground and airborne fire-control systems. The first attempt at automatic tracking was the circular sweep ranging system, which proved not very effective. However, the time modulation circuits which were a part of this system found important use in the SF ship search system, and in airborne fire-control systems (AGL and ARO). A more successful method of automatic tracking was developed in 1942. This used completely electrical tracking systems and medium-precision ranging circuits, and was widely used in airborne fire-control systems. The automatic range tracking units were adapted as target selectors in the radar homing bombs Pelican and Bat.

Ranging circuits, mainly for fire-control systems, but also for other systems (the HPG system, and HR), continued to occupy a large part of the Group's efforts. In the fall of 1942 there was a good deal of interest in developing more accurate ranging circuits for fire control, and work was begun on circuits for the SCR-598 (AN/MPG-1) and on an all-electric automatic range tracking circuit for the SCR-584, which was later used as the range unit for the SCR-784. These techniques were also applied to the SP and P^3I (precision indicator) range circuits.

In the spring of 1944 the techniques of automatic range tracking were applied to precision data transmission in the relay radar project. This was used in the AEW system, and much work was done on the problem by the Group in 1944 and 1945. Several methods were developed and used not only in the AEW but also in the ground relay application of MEW. Pulse remote control circuits were also designed for steering a guided missile from a ground radar system (AN/APW-3).

During 1944 and 1945 a great increase in the reliability of precision ranging circuits was obtained through studies of the stability of timing oscillators and delay circuits. There were three important applications of these studies: first, to the extremely high-precision circuit of the Mark 35 radar system (part of the Mark 56 gun director); second, to the intricate problem of muzzle velocity determination in the T-5 chronograph; and third, to a lightweight airborne direct-reading Loran indicator.

In 1942 there was increased interest in methods of obtaining precision data from scanning radar systems. The first approach was to employ automatic range and azimuth tracking. This was used in the Dolphin project for blind torpedo firing with the SO set, and in H_3X. A second approach was the use of the expanded B scope with a continuously resettable spot (P^3I). This latter project continued actively during 1943 in collaboration with Group 62 (Indicators) and Division 10.

A considerable amount of work was done on manual and automatic angle tracking systems for conical scanning systems. This work was originally part of the AGL program, and led to the development of lightweight, simple servos which were used in the AN/APG-3. Since similar servo mechanisms were needed for other automatic range tracking systems, and for aided tracking systems with mechanical output, work on servos was continued during 1943 and 1944. This work finally resulted in the development of the servos used in the Mark 35 system.

Bombing systems, as well as fire-control systems, occupied a considerable amount of Group 63's attention. The first work done by the Group on bombing circuits was in 1942, when Group 63 collaborated with Division 9 and Group 62 on the LAB system. This work soon led to the development of precision range circuits for Eagle.

During the spring of 1943, when there was a lull in the bombing program as a result of the conference at which it had been decided that only the precision Eagle would meet military requirements, the Group concentrated on developments in medium-precision timing circuits. The whole level of circuit technique was improved by a series of visits between the Radiation Laboratory and TRE. The British phantastron circuit was studied carefully and a reasonably good design obtained which had an overall accuracy of 3%–10%. In addition, the construction and use of small pulse transformers and of pulse generators and frequency dividers using blocking oscillators were studied.

As a result of these studies the Precision Group was able to contribute a compact, simple range unit and bombing computer to the H_2X crash program. This range circuit was also used in the production of AN/APS-

15 and AN/APQ-13, and was adapted to H bombing as part of the Micro-H and H_3X systems.

The H_2X program resulted in the development by the Group of a number of precision circuits for bombing computers. Several methods of bombing were studied; the outcome was the GPI system, first developed in 1943. This finally led to the navigation-only GPI (AN/APA-48), which was designed as a peacetime navigational instrument. From these studies there were evolved various methods of triangle resolving from rectangular to polar coordinates of integration and differentiation. The GPI technique was used in an attachment to the MEW for precision bombing with a slowly scanning system, and was also used to stabilize the PPI displays of Cadillac II.

One use of simple computation techniques was the application of differentiation to the solution of the time-to-collision of two aircraft. This led to the experimental use of automatic range tracking and range rate determination with division to determine time-of-collision for toss bombing. Computation techniques were also used in the navigational problems in Loran and H systems, where without some form of computer a continuous plot of the present position of the aircraft would be extremely difficult. Several types of Loran plotting boards were built and tested.

In conjunction with the bombing program many studies were made upon the possibility of employing Doppler effect for the determination of aircraft ground speed. When, however, Ruby Sherr discovered that drift angle could readily be determined by Doppler effect, this technique was combined with a simple aided tracking device as a bombing computer for H_2X (Nosmo).

The Doppler studies also led to consideration of the use of supersonic delay lines, in the summer of 1944. The techniques developed served as a basis for the work of Group 65 on MTI. Work was undertaken on the use of storage tubes as a substitute for the supersonic delay line, and a storage tube MTI was successfully demonstrated.

The increased interest of the Laboratory in field service, which began with the H_2X program, made itself felt in the components groups by the spring of 1944, and less elaborate and more readily reproducible circuits were the result. For example, precision ranging was adapted to manual tracking to make the Falcon system, where the emphasis was on simplicity and compactness. In the ballistic range converter AN/APA-30, a simple ranging circuit was adapted for aided tracking of a ship for rocket firing.

Test equipment for precision circuits was needed and several units were built. The most important were the TS-100 (A/J scope) and the A/R scope (Dumont 256B). The latter had a wide application to various problems, particularly in nuclear physics. Another technical contribution of some lasting importance was the utilization of the subminiature tubes of the proximity fuse to more complicated circuits, such as the AN/APA-58 and a lightweight Loran indicator.

CHAPTER 27

Power Equipment and the
Test Laboratory

Power Equipment

In the middle of 1941, it was realized that the growing list of Laboratory projects called for a service group, with personnel qualified to furnish design and development guidance in mechanical problems. The Engineering Group was set up under F. S. Dellenbaugh, becoming Group 56, Power Equipment, when the Laboratory as a whole was reorganized in the spring of 1942. When Dellenbaugh left to accept a commission in the Army, the group was put under the leadership of M. M. Hubbard.

An important activity of the group in the time of Dellenbaugh was the study of devices for furnishing fundamental power to airborne radar systems. Motor alternators were originally used, capable of converting the 24-V dc aircraft supply to 400-Hz energy at the level of a few hundred watts. Four hundred hertz was chosen since it was already an Army standard aircraft frequency used for automatic pilot equipment and for some instrumentation. The higher the frequency, the lighter in weight the generator, but the harder the generator is to make; 400 Hz was a compromise. The choice of 400 Hz enabled the group to design transformers with a saving in weight over commercial 60-Hz units of from $\frac{1}{2}$ to $\frac{1}{4}$. Later design enabled further reductions in weight of transformers to be made, although emphasis leaned more toward the production of more stable and reliable units.

In 1942, the group felt the need of augmenting its own design facilities, as well as of being able to obtain sample production, and model-shop contracts were let to the General Electric Company, the Westinghouse Electric Company, and the Raytheon Manufacturing Company. In 1943, a model-shop contract was let to the Utah Radio Products Company. An example of the need then and later for quick-production facilities was the development of the Laboratory's shop during the course of the war into an organization which was capable of producing 200 to 300 transformers a week, and which could undertake crash production of components where quantities of fewer than 50 units were involved. The Laboratory shop also had the advantage of two to three months time saved over the delivery time of outside companies.

Later problems of airborne power units were quantity procurement problems. The first alternators (dc and ac) were obtained from the Fort Wayne works of General Electric, in sizes from 250 VA to 1000 VA output power. When these proved too small because of rapid advances in radar art at the Laboratory, a 1500-VA size was developed which proved adequate for a large number of applications during the war. Supply of alternators from Fort Wayne alone was inadequate, however, and Leland Electric Company was called in. Finally in 1942, the Signal Corps took over the responsibility of developing and standardizing this type of alternator, and subsequent improvements in the type came chiefly from Wright Field.

At powers higher than a few hundred watts, motor alternators drained too heavily from an aircraft's power system, especially since these alternators were only 50% efficient. In April 1942, therefore, the Joint Aircraft Radio Board recommended replacing airborne motor alternators with alternators direct-driven from the aircraft engines. The drawback that output frequency would thus depend on the speed of the engine mattered less to the radar set than to certain smaller components of the system, such as selsyns, which required a fixed frequency. Accordingly, with many aircraft installations of direct-driven alternators, a small motor alternator was installed to provide a constant frequency. The development of direct-driven alternators was at first slow because of lack of interest on the part of airborne radar groups, but jumped when application to the high-power airborne search set, AEW, came in sight in early 1944. NDRC contracts were let to both Leland Electric and General Electric Company (for a later improved model) for direct-driven generators suitable to B-17 and B-24 installations of AEW. Generators obtained from GE were rated at about 10 kVA and installed in TBM planes.

There was developed an all-ac system of 400-Hz constant frequency at a 40-kVA power level for anticipated use in large, high-flying aircraft. This used a direct-driven alternator connected to the engine through a hydraulic servo constant-speed system. A constant-speed servo system had not been applied to previous smaller alternator installations since its relative weight was excessive. This development was chiefly by General Electric working with Wright Field, with the Radiation Laboratory performing a variety of consultant services, especially on the performance of radar equipment at high altitudes. One member of the group, L. E. Ross, handling the effect of high altitudes on generator brush operation, was subsequently appointed an NDRC representative, coordinating the investigation program at Wright Field. Reports covering findings of the Engineering Group's consultant work were issued to the services.

Working with General Electric and the Modulator Group (51), the Engineering Group developed generators for ac resonance charging modulators, having repetition rates of 350, 400, and 600 Hz, for ground and ship systems such as the MEW, SP, and V beam. The particular requirements for such generators were good speed regulation and a stable wave form, the latter implying low output impedance. This development work led later to the production in quantity of these generators for the AN/CPS-1, the AN/CPS-4, and the SP. In connection with its work, the group at this time issued a report covering the characteristics of all generators ordinarily used for radar purposes.

Although many systems divisions in the Laboratory had their own engineering groups, considerable assistance in mechanical design—much of it on antenna mounts and reflectors—was furnished by members of Group 56. Sometimes, engineers were loaned directly to a system group during a period of important mechanical development, although more usually the group considered itself to form a body of expert engineering advice on the problems of the Laboratory. Thus, for pattern measurement work, an SCR-547 mount was modified and installed at the Ipswich Field testing station of the antenna group. SCR-584 antennas were modified for use as Oboe ground stations. Work was done on "pill-boxes" for the H_2K scanner. Scanners and reflectors for the AEW (Cadillac) project were designed, making use of aircraft construction techniques, a departure from earlier group practice. The AEW reflector size, 3×8 ft, represented a new upper limit on the size of airborne antennas at the time of its design. The production model, produced by the Columbia Machine Company, used magnesium castings. At the close of the war, an even larger AEW antenna, 4×14 ft, making use of similar methods of construction, was being designed, for installation in B-17's. A scanner for ASV (3-cm) antisubmarine operation was designed by Group 56, with Philco, the manufacturer, taking over the Laboratory's design. Another high-speed scanning antenna, rotating at 180 rpm and carrying four 4-ft reflectors, was constructed for a prototype system to be used for mortar-fire detection by Division 8 of the Laboratory.

Following the development in 1943 of selsyn transmission of data for PPI's and in view of the procurement problems that resulted from the increasing use of selsyns in other parts of the Laboratory, S. Noodleman, and later E. R. Perkins, constructed typical selsyn combinations and measured their performance. By the compiling of results, overall accuracies of selsyn systems could be predicted.

Among the few developments of Group 56 which may be described, apart from systems application, was a pressure unit requested by the Navy, and which, with only small leakage, allowed atmospheric pressure to be maintained in rf lines at high altitudes. Since no satisfactory commercial unit was then available, all parts had to be developed by the group. The first design was produced by Zenith Associates at the rate of 1500 per month. Shortly afterward, a new design was made, with improved efficiency, more reliable performance, and ten times the pressurizing capacity. Although the new design was accompanied by a weight increase of 40%, its better performance was more than enough compensation. The new model was applied to the AEW radar. An order for 3000 pressurizing units, placed by Wright Field, was met by the Romec Pump Company of Cleveland, Ohio.

Also in 1943 came a request from the Services for automatic PPI cameras for use by the Radar Photo Reconnaissance Group. These were to take indexed photographs of the scope picture at controlled rates. The cameras, to be used for briefing and training, were designed by Group 56 and manufactured by Research Construction Company, pending development of a more satisfactory design by Fairchild.

In cooperation with the potentiometer shop set up in 1943 by the Indicator Group to produce potentiometers for a variety of indicator needs, Group 56 designed and assembled a card-winding machine which made it possible to produce in quantity a sinusoidal potentiometer having high precision. The group also designed a potentiometer construction which permitted linear operation of 0.05% or better. This construction was so satisfactory that sample potentiometers retained their order of linearity after 2 or 3 million operations. Arrangements for quantity production were made with the Muter Company and Fairchild.

For voltage regulation of airborne motor alternators and airborne ac generators, the group designed a voltage stabilizer, operable at 60 Hz or higher frequency, which would reduce an input voltage variation of plus or minus 10% to a variation of plus or minus 1%. The stabilizer was independent of frequency change within a range of 20%.

The Test Laboratory

In January 1944, the Engineering Group acquired the Test Laboratory, which up to that time had been part of Division 10. With this acquisition came the test laboratory's facilities for certain mechanical engineering studies—problems of heat rise in enclosed pressurized containers, vibration effects on components, humidity and altitude tests reactions, effect of temperature changes on operation, conditions of voltage breakdown—all necessary to meet requirements for service operation of equipment laid down by the Navy, the Army, and the Joint Army-Navy and American War Standards Specifications.

The test laboratory was originally part of Group 108, which was organized in July 1942 for the purpose of making operational analyses of radar equipment in the field; formulating engineering specifications of standards which both components and total assemblies had to meet; and operating a test laboratory whose purpose would be to study in the experimental stage the durability of equipment later in the field.

The original equipment of the test laboratory comprised a high-altitude chamber, capable of a pressure range from atmospheric pressure to 50,000 ft above sea level, a temperature range of -40 to $+75°C$, and a relative humidity range between 45% and 95%; a vibration table capable of shaking chassis weighing less than 25 lbs. between 600 and 2400 times a minute at amplitudes between $\frac{1}{64}$ and $\frac{1}{16}$ inch; and a high-voltage power supply capable of voltages up to 40 kV.

Much early testing of circuit components such as resistors for extremes of temperature, humidity, and electrical overload was requested from time to time by the purchasing and standards sections of the Laboratory. Small-size high-voltage components were operated at simulated high altitudes to test for corona or other electrical leakage. Vacuum tubes were vibrated under normal electrical loads. Few large pieces were dealt with, partly because the group often did not have equipment big enough to handle them and partly because the Services had not yet written specifications for testing microwave equipment.

In December 1942, small hot and cold chambers were installed for temperature tests of resistors and capacitors. Power supplies up to 100 kV ac and dc were acquired, and used to test the dielectric strength of new insulating materials. In January 1943, the group acquired rf test equipment, including a signal oscillator, and two impedance-measuring bridges. Early in 1943, the test group began to exchange information with the Naval Research Laboratory and the Army Signal Corps regarding the integration of test procedures.

The first large-scale development program of the test group was a mechanical rubber shock absorber for protection of delicate equipment installed on ships and aircraft. The development, undertaken late in 1942 by E. Pietz, head of the test group, continued for about a year. In its early stages, it benefited from a combined impact-shock and vibration machine built by D. Drake of the test group, after a similar model in possession of the Naval Research Laboratory. In December 1942, the shock program received additional vibration equipment able to give accelerations up to ten times gravity, at frequencies of vibration between 600 and 3000 times a minute, and amplitudes up to $\frac{1}{2}$ inch. The result of the program, which included first-hand observations by Pietz of field requirements,[1] was the successful development of a shock absorber not resonant within the range of ship vibration frequencies. As a result, BuShips requested the test group to aid in a comprehensive testing program, to lead to later review and possible revision of Navy shock requirements. In September 1943, new shock and vibration specifications were announced by the Navy, and Pietz's shock mounts were recommended for use in shipboard equipment.

When Division 10 changed from a service division for the entire Laboratory to a ship research division, services of the test group (now Group 102) were limited largely to activities within the new division. This limitation slowed the test group's growth, since large expenditures of money could not be justified for services for a single division. In January 1944, the test group was transferred to Group 56. New equipment included:

(1) a high-altitude chamber big enough to test entire radar systems under any conditions of vibration, low temperature, and altitude;

(2) a $14 \times 10 \times 8$-ft low-temperature chamber;

(3) a vibration table capable of satisfying requirements prescribed in Joint Army-Navy specifications for vibration and shock tests;

(4) a new humidity chamber;

(5) a battery of test panels (recommended by the Bureau of Ships), for transmitting any desired power anywhere in the test group from a centrally located power room; and

(6) a large impact-shock and vibration machine, purchased on recommendations by BuShips that this machine be bought from the General Electric Company, Schenectady, for use in testing equipment being produced by the Radiation Laboratory, Submarine Signal Company, and Raytheon Manufacturing Company.

The test group was now divided into sections, D. G. Bagley responsible for operations and personnel, Pietz in charge of shock and vibration tests. Testing of electric motors also shifted from the Engineering Group to the new test section. Many vibration tests were performed on whole components. In fact, the test group's chief value to the Radiation Laboratory after its transfer to Group 56 and its subsequent enlargement consisted in its tests of temperature, humidity, altitude, vibration, and shock characteristics of complete radar systems. This was especially true of the AEW radar, which was almost wholly tested in the test laboratory. The war ended, however, before the results of this type-testing could be recorded in the field.

NOTES

1. One trip is memorable because of its results. On 28 December 1942, Pietz and T. A. Farrell set out in a PT boat from Melville, R. I., to measure amplitudes and accelerations of shock and vibration aboard the boat. They were accompanied by Walter Keyes of the U. S. Rubber Company, who was to operate his company's measuring equipment. Measurements were made at the height of a storm outside the bay under severe sea and weather conditions. From the standpoint of information obtained, the trip was a success. Between bouts of seasickness, it was learned that vibration aboard a PT boat is almost negligible, while shock is a serious problem. Unfortunately, personal tragedy attended the trip. As the result of internal injuries received from being buffeted around the boat's tank room in heavy seas, Walter Keyes died three weeks after the trip.

Part 3

Fundamental Research

CHAPTER 28

Experimental Systems Groups

MTI (Moving Target Indication)

Moving Target Indication is the presentation of moving targets on a PPI scope while simultaneously eliminating permanent echoes due to ground clutter, buildings, hills, or other fixed objects. The Laboratory had found that the usefulness of many radar sets was seriously affected by the masking effects of ground clutter, and had set up a committee, composed of E. C. Pollard, Chairman, A. G. Emslie, R. A. McConnell, W. B. Sheriff, J. A. Smith, and R. F. Thomson, to study the problem in June 1943.[1] The Army had been thinking along the same lines and had two specific needs: (1) a warning set against low-flying strafing planes for the use of ground troops, and (2) a long-range air-warning set to supplant pulse radar in mountainous territory, such as China and Burma. Discussions were held by E. C. Pollard with representatives of the Office of the Chief Signal Officer in Washington and at Camp Evans Signal Laboratory during August and September 1943, and on August 30, 1943 the Laboratory accepted a project entitled "Elimination of Ground Clutter".[2] The Laboratory's project was directed toward supplementing the microwave program, but the Army was anxious to have circuits developed which would be applicable also to the long-wave sets SCR-271 and SCR-588.

A survey of existing developments and available techniques was made, and a preliminary report was written for General Roger B. Colton. It was eventually decided that the Laboratory should work on pulse Doppler methods and secondly on coherent pulse technique. It was realized that the second method involved long-term research and, as it turned out, the manpower shortage and the consequent low priority of the project in the Laboratory resulted in delaying the completion of a successful MTI system until May 1944. Research was carried on in Divisions 12, 10, and 6 of the Radiation Laboratory. It was not until December 1944 that a special group (Group 65) was set up for MTI, headed by R. A. McConnell.

To carry out preliminary tests on the application of pulse Doppler to a radar system, the Laboratory borrowed a CXCA (400-MHz) set from RCA and set it up on the roof of Building 6 under the supervision of B. R. Curtis. The problems were fairly simple, since the principle involved is that transmitted signals from fixed objects will be reflected without change of frequency of the transmitted rf pulse, whereas signals from moving targets will be reflected with a change of frequency proportional to their radial velocity. By December 1943, the conclusion had been reached that, while it would be feasible to add a pulse Doppler kit to the CXCA or to a similar system, this would constitute only a stopgap. Tests showed that beat frequencies from moving targets would be present only if a fixed target were fairly close to moving aircraft.[3] The Laboratory decided to terminate its activities along these lines and to concentrate on coherent-pulse Doppler techniques. In the second part of the work, the Laboratory could rely to some extent on British experience. The British had begun their research in 1942 and proceeded at very low priority, but they had achieved some success.

Coherent-pulse technique uses the phase information inherent in every radar echo pulse and depends on the storage and comparison of echo patterns. It was decided to use the SCR-615 radar set for experimental purposes since it could be easily modified, rather than to design a new set with MTI as its primary function. Because freedom from jitter is a prime requisite, systems using spark-gap modulators or lighthouse tubes were not suitable, so the choice was limited.

In the MTI system, as finally achieved in May 1944, signals are locked in phase with transmitted pulses by mixing with a stable local oscillator, called the STALO, which is beaten against a coherent oscillator, called the COHO, at the IF stage. If the target is fixed, the signal will be in constant phase relation with the coherent oscillator, but if it is moving, the phase difference will change from pulse to pulse. It is not possible to detect a change in phase difference from the beginning to the end of one pulse because of the shortness of the pulse; therefore pulses must be compared in order to detect a change. For this purpose, a supersonic delay line with a delay time equal to the transmitter repetition time is used. Successive pulses due to fixed targets are cancelled because their constant phase gives constant rf amplitude. Moving targets, on the other hand, change phase between pulses; the rf amplitude also changes, and cancellation is not complete.

The British had developed a water delay line but it had been found to be affected by attenuation and temperature changes. A mercury supersonic delay line had been developed before the war at the University of Pennsylvania in connection with acoustical research. The Radiation Laboratory, following this lead, developed a mercury delay line adaptable to the MTI system, which was extremely successful. McConnell and Emslie succeeded in establishing IF coherence as early as October 1943, but it was not until later that P. R. Bell achieved self-synchronization of a complete MTI system using a delay line constructed by G. D. Forbes and H. Shapiro.[4]

An attempt to use a method of storage other than the delay line ("kinetic storage method") came from the suggestion of McConnell that a storage tube might be developed by using a television mosaic as a moving target selector by scanning it with a deflection-modulated, high-velocity electron beam ("static storage method").[5] This method would have the advantage of being applicable to radar systems using the spark-gap modulator, which was the principal limitation on the use of the kinetic storage method. Experiments were not successful, however, and all effort was put on the development of the delay line.

News of some spectacular demonstrations of the MTI came to General H. M. McClelland, and in August 1944 he urged that the project be given higher priority and suggested its possible use with a ground radar system to follow the flight of guided missiles.[6] The Army Ground Forces and the Laboratory decided that the set which stood the best chance of being modified by the addition of MTI and put into active service before the end of the war was the SCR-584. A crash program was set up with J. A. Smith as project engineer and the Army eventually requested 20 kits for testing and use. A contract was placed with Federal Radio and Telephone Company, but they were not able to make deliveries before V-J Day. The Laboratory, however, built and delivered two "crash" kits by June 1, 1945.

In January 1945, the Navy requested that a project be set up at the Radiation Laboratory to adapt MTI to the SP radar set with the object of providing a partial solution to Kamikaze attacks. Because the MTI group was heavily burdened with the SCR-584-MTI program, no great amount of effort could be put on the SP. Some experiments were carried out at Spraycliff under the direction of R. M. Emberson.

Since the SP uses a spark-gap modulator, another modulator had to be procured, and this was not available to the Laboratory until August 1, 1945. In terminating the SP-MTI project, the Laboratory recommended that SP was not particularly suitable for MTI application and that development should be started on a new high-power search set.

Before the end of the war, projects were under way to adapt MTI to MEW, GCA, and Project Cadillac, but these programs were barely started before the end of the war. Although MTI did not see combat service, it was one of the most significant contributions of the Laboratory's fundamental research and it provides one of the most fruitful fields for post-war research.

Project Sambo

When J. L. Lawson and L. B. Linford were making observations on the photography of successive pulse reflections from moving targets in June 1942, they obtained some traces showing what they found to be caused by aircraft propeller modulation of the radar echoes.[7] Further theoretical study and observation led to the belief that if propellers could be treated with a special nonreflecting material, subharmonics of the normal propeller modulation could be produced which would serve as a means of identification of friendly aircraft.

The Bureau of Aeronautics became interested, and in April 1943 the Radiation Laboratory formally accepted a project, surrounded with great secrecy, for an aircraft identification system which later became known as "Sambo". The Bureau of Aeronautics agreed to make aircraft available to the Laboratory through Project Cast, and at the Naval Research Laboratory a technician experienced on problems of paint chemistry was assigned to maintain liaison with the DuPont Experimental Station, at Wilmington, Delaware, where the Navy already had a contract for experimental work, with a view to establishing pilot-plant production of the Sambo material at the proper time.

Four lines of endeavor were carried on at the Laboratory: general research on propeller modulation in the Experimental Systems Group, headed by Otto Halpern; development of a special indicator by R. M. Walker in the Indicator Group; and the application of such equipment to ground and ship systems and the evaluation of its tactical usefulness in the Ground and Ship Division, coordinated by E. C. Pollard.[8]

By May 1944, a nonreflecting paint had been developed and tested, and plans had been made for further research and development at the DuPont Company. The important problem of indication had been solved by the development of an indicator utilizing resonant reeds similar to those used in power-line frequency meters and tachometers. Tests of these indicators at the Spraycliff Field Station had shown detectable modulation from single aircraft at ranges out to 30 miles.[9] Tests using propellers coated with the nonreflecting paint had not been entirely successful since some nontreated propellers had shown effects identical to those treated with the Sambo material. This "Pseudo-Sambo" effect was one of the most disturbing problems, and Lawson had found out that as small an asymmetry as 1/8 in. in a rotating propeller could cause this effect.

A demonstration was held at the Radiation Laboratory on June 6, 1944, for Army Air Forces as well as Navy representatives, since by this time the Army had been brought into the project.[10] Sambo-treated and nontreated aircraft were flown simultaneously and Sambo indicators were used for identification. An untreated F6F aircraft showed strong Sambo signals which shook the confidence of the spectators in the system. In the discussions which followed, it was apparent that in the opinion of Navy representatives the system could not be accepted as it was, but Commander L. V. Berkner insisted that the fate of the Sambo material should be separated from that of detection of propeller modulation which, as he expressed it, had "opened up a new dimension in radar work".

The Laboratory decided to discontinue all other work on the project except the development of systems for modulation detection. Some further studies were carried out using the SM system.

After the German V-1 attacks started, the British Branch of the Radiation Laboratory teletyped a request for Sambo indicators to be sent to England for use with SCR-615 or British Type-24 equipment. It was hoped that Sambo might be used to differentiate between aircraft and a propellerless Buzz-Bomb.

One set of indicators was shipped to BBRL on June 21 and four others were shipped later. The first set arrived but the four others were missent to a manufacturer in Newcastle who looked them over, thought they had been sent to him by a crank, and turned them over to Scotland Yard, who finally got them to BBRL.

A Sambo installation was made on the Type 13 radar set at Beachy Head, and a large number of scientists and military personnel witnessed tests. Results were unsatisfactory since only 25% of the aircraft showed reliable modulation. Most of the aircraft used were five-bladed, Type 14 Spitfires, and it was thought that their modulation was above the limit of the indicators. It was conclusively shown, however, that Sambo would have no value in V-1 identification.[11]

In September 1944, the Army Air Forces asked the Laboratory to develop an airborne identification set based on the propeller modulation detection principle. Since it was a fairly simple problem, limited to fire control systems and to be applied only to the identification of heavy bombers against fighters, the Laboratory accepted the project for which B. L. Birchard and W. M. Breazeale served as project engineers.

Research in the Laboratory resulted in the design of a satisfactory unit called Ella, later designated AN/APX-15, to work in conjunction with AN/APG-15B equipment installed in the tail stations of B-29 aircraft. Operation was based on the fact that B-29 aircraft propeller modulation lies in the band between 40 and 63 Hz, while propeller modulation from all Japanese fighters was known to lie above 70 Hz.[12]

Tests of this equipment were satisfactory and the set was in production at Bell Sound Systems, Inc., Columbus, Ohio at the end of the war. It had been planned to supply AN/APX-15 sets to each B-29 equipped with AN/APG-15B.

The Experimental Systems Group

The Experimental Systems Group (Group 44, also known as the Advanced Development Group) was organized in the fall of 1941 under J. L. Lawson, who remained in charge during the entire life of the group. It was essentially a research service group for the benefit of the development groups, which tried to foresee and undertake studies which would lead to the improvement of the radar art. The group maintained several carefully calibrated research radar systems, it cooperated closely with various other Laboratory groups in the development of testing of new components, and it undertook several independent studies, particularly the study of the anti-jamming problem. For a short time, a small subgroup, later separated as Group 45, worked within the group on modulation studies which led to Project Sambo.

Research Radar Systems

When the Experimental Systems Group was organized, it inherited from the Roof Group (Division 10) the so-called S-1 system, designed by L. B. Linford and A. Longacre. This was a 10-cm system with a 30-in. parabolic reflector, soon replaced by a 48-in. reflector. The original bead-supported transmission line was replaced in January 1943 by 2-7/8-in. stub supported line. This system was carefully calibrated and very reliable, with the best available components, and was widely used in demonstrations, in testing new components, and in research studies.

Early in 1942, a similar 3-cm system, the X-1, was set up by M. W. P. Strandberg. The S-1 and X-1 systems were very useful in making comparative tests of 10-cm and 3-cm operation.

Coincident with the X-1 there was also being built the S-2, 10-cm "back-of-dish" system designed by R. Rollefson, in which the transmitter tube, the rf section, and the receiver were all mounted on the back of the reflector to minimize rf loss and do away with the need for long rf transmission lines. The solution of the many problems involved in the construction of this system was a necessary preliminary to the development of many high-power systems, particularly the MEW.

Besides the problem of rf design for high-power operation, many other similar problems were worked out with the aid of these research systems. The EEI (expanded elevation indicator), very like the PPI, with the angular position of the sweep determined by the angle of elevation of the scanner, proved to be extremely accurate for height finding. During the winter of 1941–1942, there was undertaken an elaborate series of experiments on the best method of determining the maximum range at which a target is observed on any particular radar system. In the course of many observations on airplanes, the standard cross sections of various aircraft were measured; these measurements became standards throughout the Laboratory. Both the S-1 and X-1 systems were used and it was found that cross sections for the 3-cm band were the same as those for the 10-cm band. Since photographic records of experimental work were highly valuable, methods of photographing various types of indicators were worked out within the group, with the assistance of H. E. Edgerton of MIT.

Cooperation with Other Groups

Some of the most valuable work of the Experimental Systems Group was done in connection with the work of other groups, particularly the rf, Propagation, and Indicator Groups. This work is described in more detail in the histories of these groups. Some of the more important problems on which Group 44 worked were the development of improved methods of tuning between the magnetron and the TR junction, the improvement of the resonant-cavity TR switch, the development of the "pot" mixer and the "pen" mixer, and the testing of indicator tubes, particularly the persistent-screen cathode-ray tube and the dark-trace tube. In addition, the Experimental Systems Group made many observations and recorded a considerable amount of data for the Propagation Group. Group 44 was always ready to furnish information and perform specific tests for groups within the Laboratory and visitors from other organizations. The work of the group on modulator studies has already been recounted in connection with Project Sambo.

Anti-Jamming Studies

The experimental Systems Group became interested in the problem of anti-jamming early in 1943. Before the development of "window", the group had tested the effect of metal "nails" of various lengths, dropped from an airplane, on the S-1 and X-1 systems. These nails had very much the same jamming effect as Window. Thereafter the Group devoted considerable effort to the study of anti-jamming circuits which could "see" through window.

In the summer of 1943, work began on the use of shorter pulse lengths. Theory suggested that short pulse lengths should improve the visibility of a discrete target in the midst of a diffuse target (window, cloud, or sea return), as well as give increased range discrimination. Short-pulse modulators, wideband receivers, and special video and IF amplifiers were designed by members of the group for both the S-1 and X-1 systems. Fairly extensive tests with various pulse lengths supported the theory, and led to wider use of shorter pulse lengths in production systems. In addition, it was found that these new components gave startling improvements in target mapping.

One of the best anti-jamming techniques is the use of tunable transmitters and receivers, since if the enemy jams on one frequency the system can then be operated on a different frequency. In the first system, known as the Octopus, two transmitter tubes and two TR switches were connected to a five-way coaxial junction, whose fifth terminal was connected to the antenna. By the proper selection of frequencies and appropriate pretuning it was possible to match all the elements. Later, an attempt was made to develop a continuously tunable system. The first method of solving this problem was to gang the tuning of the TR switch, the coarse adjustment of the McNally-tube local oscillator, and the mechanically tuned magnetron. Various refinements to this system were discovered during succeeding months, and by the end of the war tunable magnetrons were available which, with the components then developed, permitted systems that could be continuously tuned over a range of 40 MHz.

The reduction of pulse interference is important since, in addition to being a method of jamming, this kind of interference is often unwittingly caused by other nearby radar sets. Most early methods of reducing pulse interference used pulse-amplitude discriminators which removed all signals above some usually adjustable limit. This inevitably causes the loss of some desirable but strong radar echoes. Group 44 developed a pulse-interference suppressor, which used a separate omnidirectional, beacon-type antenna and a separate receiver channel, and was a practical and effective device. Pulses larger than a certain adjustable limit blanked the video output of the radar receiver and no longer obscured the presentation.

A necessary part of the anti-jamming program was the measurement of the visibility of signals in noise in order to determine the relationship between the power of the noise and the power of the minimum signal that could be discerned through the noise. This involved a series of statistical correlation experiments in order to establish a standard for a minimum discernible signal. Studies were made to determine how the minimum discernible signal varies with various characteristics of the system such as pulse length, bandwidth of the IF and video amplifiers, pulse repetition frequency, the presentation, etc.

The visibility of signals in various types of jamming were also extensively studied. The data obtained in this study were correlated with the data obtained from the study of the visibility of signals in noise.

The study of the theory of rejection filters, which would eliminate narrow-band jamming, led to the design of such a filter. It was only moderately successful, being too difficult to keep tuned for use in the field.

A considerable amount of work was done on receivers which would be less sensitive than the ordinary receiver to overloading as a result of jamming. Various methods, such as a receiver with a very small IF section, and back-bias circuits, were tried with scant success. More successful was the detector balanced bias circuit, used in the AEW receiver. This circuit applied a delayed bias to the second detector, using a cathode-follower amplifier with controlled positive feedback. The amplification was a function of frequency, so that the low video frequencies associated with cw jamming were not much amplified, whereas the high frequencies associated with clutter and Window jamming were greatly amplified.

The determination of the effect of video bandwidth on off-frequency cw jamming proved complex. Theoretical and experimental analysis showed the value of the usual practice of having the video passband at least as wide as the full IF bandwidth.

Building of Special Equipment

As a part of their studies and tests, the Experimental Systems Group built a number of components and pieces of test equipment. Most of these were developed in connection with the anti-jamming studies. In particular, there was the development of very stable wideband receivers, the development of a reliable attenuator which would not leak and which would have a minimum insertion loss, and the development of a very dependable signal generator. In connection with the rf work there was designed a magnetron spectrometer which determined the quality of the rf output of magnetrons under varying conditions.

NOTES

1. K. T. Bainbridge to I. A. Getting, Secretary Steering Committee, 10 July 1943, "Appointment of Committee."

2. E. C. Pollard, "Report on Discussion with Lt. Col. McRae on Elimination of Ground Clutter (MTI), 3 August 1943, at Pentagon," 4 August 1943.

3. V. Olson, B. R. Curtis, Lt. F. Cunningham, and R. Miller, "Pulse Doppler," Radiation Laboratory Report 103-12/16/43.

4. Radiation Laboratory Notebook No. 2827, issued to R. A. McConnell, pp. 23-28, 7 May 1945, "History of the MTI Self-Triggering Delay Line Feature."

5. R. A. McConnell, A. G. Emslie and F. Cunningham, "A Moving Target Selector Using Deflection Modulation on a Storage Mosaic," Radiation Laboratory Report 562, 6 June 1944.

6. Brig. Gen. H. M. McClelland to Dr. L. A. DuBridge, 28 August 1944.

7. J. L. Lawson, "Photography of Successive Pulse Reflections from a Moving Target," Radiation Laboratory Report 64-5, 12 June 1942.

8. E. C. Pollard to L. A. DuBridge, 3 February 1944, "Sambo."

9. J. M. Sturtevant, "Summary of Work on Propeller Modulation at the Radiation Laboratory," Radiation Laboratory Report 103-3/21/44.

10. H. E. Guerlac, "Memorandum of Conference on the Sambo Project Held at the Radiation Laboratory, 6 June 1944," 12 June 1944.

11. E. C. Pollard and L. A. Hartman, "Preliminary Report on Attempted Operational Use of Sambo," BBRL Ref. No. 45, 4 August 1944.

12. J. V. Holdam, "Final Report AN/APX-15", October 1945.

CHAPTER 29

History of the Theoretical Group at the Radiation Laboratory[1]

By Albert E. Heins

The Theory Group of the Radiation Laboratory was organized early in 1942, first under the joint leadership of J. C. Slater and S. A. Goudsmit and later (July 1943) under the leadership of G. E. Uhlenbeck. Its aim was to provide theoretical information which would assist in the design and development of certain types of radar components. In the course of this work calculations were often carried out which contributed to the understanding of various phenomena encountered by experimentalists in the laboratory. In addition to this fundamental work, there was also some work carried out on systems and operational analysis.

In order to get a proper picture of the physics which one encounters in the theory of waveguides, resonant cavities, and similar rf equipment, it is necessary to find solutions of the Maxwell equations for those various structures. For structures which are bounded by simple coordinate surfaces, solutions are quite straightforward. Unfortunately, many of the components do not have such simple geometries. A simple example would be the presence of a diaphragm in a guide. While it is an elementary problem to find the transmission properties of a guide without the diaphragm, its presence complicates matters considerably. One of the major projects undertaken by the group dealt with just such questions. Many of the fundamental techniques employed were suggested and developed by J. S. Schwinger, H. A. Bethe, H. Hurwitz, Jr., and N. Marcuvitz and were presented by Schwinger in a series of lectures which began in the spring of 1944. These lectures have since been published.[2] This work was carried out in detail by the following men: A. Baños, Jr., H. A. Bethe, W. A. Bowers, J. F. Carlson, L. J. Chu, P. D. Crout, N. H. Frank, H. Goldstein, W. W. Hansen, A. E. Heins, H. Hurwitz, Jr., H. M. James, H. Levine, P. M. Marcus, R. E. Marshak, N. Painter, D. S. Saxon, J. S. Schwinger, J. C. Slater, and R. M. Whitmer. This group provided solutions for many practical problems which arose in connection with research and development

625

being done in the fundamental research and the rf groups of the Radiation Laboratory.

An important aspect of all this work was the concept of an equivalent circuit. It is possible without the detailed solution of these problems to provide an equivalent circuit, that is, the circuit which represents the effect of the obstacle in the guide. This equivalent-circuit technique is quite important because it relates the theory of guided waves to classical transmission-line theory. Thus, if we have a complicated geometric structure, it may be possible to decompose it into a series of elementary structures under certain appropriate electrical and geometrical restrictions. If now the equivalent circuit of each of these elementary structures is known, it is possible to find the composite effect by classical transmission-line theory. Many cavities were treated on this basis, for example, the TR box, the rhumbatron, etc. In this treatment the cavity is viewed as a waveguide structure with short-circuiting plates in appropriate positions.

In order to find the parameters of the equivalent circuits already described, it is necessary to formulate and solve a boundary-value problem. One can express the magnetic or electric field in the region in terms of the electric and magnetic fields on the boundary of the region and hence by imposing appropriate boundary conditions on these field quantities, one obtains an integral equation. In the early work of the group, some of the integral equations which arose could be approximated by other integral equations which had arisen in connection with problems in hydrodynamics and aerodynamics many years earlier. Thus, one could solve these asymptotic integral equations and provide a first-order result to the problem. In due time, it turned out that it was possible to give corrections to these asymptotic solutions so that one was relatively certain of getting within a few percent of what the Maxwell equations predicted.

Unfortunately there were many integral equations which could not be solved in any sense of the word. In particular, one should mention those problems which arose in the one-dimensional change of cross section of a rectangular guide. The technique of solving such problems was the so-called equivalent static technique first used by Lamb and Rayleigh in connection with acoustical and hydrodynamic problems. Here one made use of the methods of function theory, and this technique was applied also to the study of thin and thick windows in rectangular guides as well as in the theory of right-angle bends and T-sections.

Starting with the problem of diffraction of electromagnetic radiation by a small circular aperture in a perfectly conducting plane screen, a theory of the performance of a small aperture of simple shape for arbitrary exciting fields was developed. The aperture was described by electric and magnetic polarizabilities, quantities proportional to the dimensions of the aperture and independent of the wavelength. With knowledge of the primary (exciting) field in a particular location of the aperture and the polarizabilities, the electric and magnetic dipole moments of the aperture are obtained, and these determine the secondary (scattered) field. The theory was applied to the coupling of cavities and waveguides by a window in a common wall, the effect of a window in the side wall of a waveguide, the direct coupling of two waveguides through a TR-box, and the energy radiated through a window into free space or a waveguide.

It is always desirable to be able to estimate the errors encountered in theories which have approximations contained within them. Either the integral equation formulation or the equivalent static formulation can be expressed as a variational principle. This was an important advantage, for once the first-order solutions were provided, they could be inserted in appropriate variational expressions, and it was possible to smooth out the results and improve the final answer. Another advantage of these variational principles became evident when it was found that the integral equation technique provides two formulations. One of these appears in terms of a tangential electric field in some aperture, the other in terms of a surface current density on some obstacle. If one applied variational principles to both of these classes of integral equations, one discovered that the desired quantity (a susceptance or impedance) could be formulated as the lower or upper bound of some quadratic functional. With these upper and lower bounds it was possible to estimate the errors incurred as a result of the approximations in the theory.

Another big advance came with a second means of formulating these boundary-value problems. Rather than use a finite matching surface in the formulation of the integral equation as had been done in the earlier work of the group, an infinite matching surface was used. For those problems in which this was possible, one got an integral equation which was a generalization of the Wiener–Hopf type, first considered by these men in 1932. In many cases this equation appeared directly and the problem could be solved rigorously. This theory was applied to the following geometric structures: bifurcation of a rectangular guide (E and H plane); the coupling of a circular guide to an open-end coaxial line; the radiation of semi-infinite parallel plates into free space; the radiation of a semi-infinite circular guide into free space; etc. Thus was found a mathematical technique, not 15 yrs old, which provided for the first time a rigorous solution to many problems that arose in connection with rf components. On the basis of this theory it was possible to give solutions to many of the problems which were considered in the early years of the group. The forbidding factor was the lack of time on the part of the analytical assistants.

In the fall of 1942 a first edition of what was termed the *Wave Guide Handbook* appeared. Here a number of elementary properties of waveguides were brought together and the equivalent circuits due to the effect of various types of obstacles in guides were given, as well as the graphs of the circuit elements. A revised edition appeared in February 1944 and was eventually published.[3] This publication presented in useful engineering form the equivalent circuits as well as graphs of the various circuit elements. In connection with the practical results put forth in the *Wave Guide Handbook*, the work of the analytical assistants must be mentioned. The work in this group was carried on by Helen Arens, Ruth Dickan, Norma Gould, Mida Karakashian, Ruth Krock, Dorothy Perkins, Barbara Siegle, Sylvia Stillman, Helen Spence, Helen Stearns, Anna Walter, and Elizabeth Oelschlegel.

If one assumes that methods are available to calculate the effects of discontinuities (irises, obstacles, etc.) in cavities, an equivalent-circuit picture can give much valuable information. Thus, using perturbation techniques and equivalent circuit methods, one can analyze the coupling between certain degenerate modes of an echo box and also calculate its ringing time. Similarly, these circuit methods can be used to study the problem of magnetron modes, the effect of strapping on mode separation, coupling of magnetrons to lines, effect of anode length, lid height, etc. Many of the circuit parameters described in the *Wave Guide Handbook* are useful in these studies.

A flux-plotting method was introduced by P. D. Crout for the calculation of the resonant frequencies of a magnetron. This method supplies useful information when analytical methods become forbidding and it is a generalization of one found in potential theory.

The scattering of a plane electromagnetic wave by objects of various shapes was treated by several mathematical methods. This work was carried out by J. F. Carlson, L. J. Chu, P. D. Crout, H. Levine, P. M. Marcus, J. S. Schwinger, and A. J. F. Siegert. It was the aim of this group to find the nature of the scattered field and especially the radar cross section. Those objects having sufficiently simple geometries were done rigorously. The more complicated forms of scatterers were treated by approximation methods, such as variational principles, the Huyghens–Kirchhoff principle, and geometrical optics. Some of these results were applied to such scatterers as "Schnorkel," the breather device on German submarines.

Electronic problems were considered by W. P. Allis, A. Baños, Jr., E. H. B. Bartelink, W. W. Hansen, J. K. Knipp, D. S. Saxon, A. J. F. Siegert, J. C. Slater, G. E. Uhlenbeck, and G. Vineyard. These problems were concerned with the interaction between electrons and the fields through which they move. The general aim of the work was to understand and improve the operation of klystron and lighthouse-tube amplifiers and oscillators and magnetron oscillators.

Klystron and lighthouse tubes are essentially one-dimensional devices. Transit-angle effects leading to beam loading and frequency modulation were studied. Effects of load were investigated. Local-oscillator noise was calculated. A general theory was developed for the small-signal behavior of a narrow uniform gap having a beam, the electrons of which are distributed in velocity and some of which are reflected. In this theory, space-charge effects lead to a Fredholm integral equation for the rf field.

The frequency modulation of a magnetron by an electron beam injected along the axis of a cylindrical cavity, operating in any one of its H-modes and tightly coupled to the magnetron, was developed. The frequency change is proportional to the beam current. In special cases the cavity can be merely one of the slots of the magnetron or an external cavity in which the field has only a single rectangular component, this component being transverse to the beam.

The electronics of magnetrons is complicated by the geometry and high fields. The theory of electron orbits was investigated for time-independent fields and fields having various assumed time variations. The consistency of the resulting space charge and the field was sought after and analytical and numerical methods of solution were outlined. Starting currents, efficiency, frequency of oscillation, and effects of load were investigated as functions of the tube parameters. Much work was done to understand starting conditions and mode jumping in magnetrons. A model was developed for the latter based on an equivalent circuit having several modes of oscillation.

A number of statistical problems were undertaken by H. Goldstein, M. Kac, A. J. F. Siegert, G. E. Uhlenbeck, and Miss M. Wang. These people worked on the theory of receiver noise and outside interference (clutter). Receiver noise is a name for random voltages created by the thermal agitation of the electrons and by statistical fluctuations (shot effects) in the number of electrons in vacuum tubes. Clutter is a fluctuating return signal received from rain, "window," or the surface of the sea (sea echo).

The investigations on receiver noise were directed toward predicting the smallest received power of a signal necessary for its detection against a background of noise (the minimum detectable signal). It was shown theoretically how this quantity depends on the outside parameters such as the bandwidth of the IF amplifier and of the video amplifier, the pulse repetition frequency, and the range scale of the Type "A" scope. For special devices such as Sambo the minimum detectable modulation of the incoming signal was investigated. The influence of clipping circuits and the effect of mixing at various stages of the receiver were calculated for other special apparatus. The calculations were mainly based on the intuitive criterion that the change in the value of the observed quantity caused by the signal must be equal to a constant of order unity times the fluctuation of the observed quantity in the absence of the signal, in order that the signal be detectable. The value of the constant has to be determined experimentally. In certain limiting cases it has been possible to derive the criterion deductively, assuming an "ideal observer," that is, an observer who is capable of performing any calculations with the observed data and thus use the information contained in the data to its full extent. A good deal of this material has since been published.[4]

The theory of random functions was also applied to clutter echoes. The echo from rain, "window" dipoles, or scatterers on the surface of the sea (sea return) was treated as a two-dimensional random walk problem, in which the return from individual scatterers, represented as vectors in the phase diagram, are considered as the steps of the random walk. The fluctuations of the returned signal power, in large measure, are caused by changes in the relative phases of the scatterers due to their motion. In such cases the correlation function and power spectrum of the signal can be expressed in terms of the velocity distribution of the scatterers. Ground clutter consists of two components: a steady echo from stationary scatterers (rocks, tree trunks, etc.), and the contribution from scatterers which may move in the wind (branches or leaves). This latter contribution is also treated as a random walk problem.

The picture of clutter echoes thus arrived at has been verified by extensive experiments conducted by the Propagation Group. These theoretical studies also have a bearing on the limitations of MTI and pulse Doppler devices, and it is hoped that information on meteorological turbulence can be obtained from the study of the fluctuations of echoes from rain.

Some of this material has appeared in the Radiation Laboratory Series.[5]

W. H. Furry worked closely with the Propagation Group at the Radiation Laboratory and was concerned mainly with the calculation of field strengths in the so-called diffraction zone. All of the work was based on the convenient concepts of a flat earth and a modified index of refraction. This work fell into two fairly distinct categories, one of which was the theory of characteristic functions which arose in the calculation of anomalous propagation into the diffraction zone; the other was the work on a particular model for the modified index-of-refraction curve. The work on the theory of characteristic functions was concerned with the establishment of the fundamental use of these functions in the theory of propagation and also with a critique of the phase-integral method of determining these functions and of the characteristic values which were devised by

Eckersley. The special model which was studied most extensively in this laboratory was that commonly referred to as the bilinear model, in which the modified index curve was composed of portions of two straight lines. The tabulation of a special class of Hankel functions was required for the work on the bilinear model. These tables were prepared by the Harvard Computation Laboratory on the authorization of the U. S. Navy and have been published by the Harvard University Press. Some of this work has been included in the Radiation Laboratory Series.[6]

H. M. James was concerned with several problems, one of which was the stabilization of radar shipborne systems. Analyses were made of the stabilization problems involved in two- and three-axis stabilization. The importance of computers in correcting data was studied and calculations on the rates of accelerations required of the driving servos were carried through. The use of computing devices representing the geometrical situation, instead of the analytical computers used in the stabilization of gun directors, was suggested. This analysis of stabilization problems led to the choice of the three-axis mounts for the SM system. Later studies had to do with similar problems connected with the design of the Mark 56 Director and other shipborne systems. The distortions on the indicator pattern due to lack of stabilization were predicted and comparative studies were made of the advantages of line-of-sight and roll stabilization. The behavior of airborne stable verticals was studied, and recommendations were made for the elimination of troubles encountered in systems in which the stable vertical was carried on the scanner. The use of computers in correcting the indication of unstabilized systems was studied.

In addition to the work of stabilization of radar systems, H. M. James did some theoretical work on the AEW system. Coverage patterns were worked out and the observed intermittent sighting of targets was explained. The use of the diversity antennas at the shorter wavelengths was found to be necessary to make such systems practicable. The principles of the diversity antenna design were worked out and specifications were laid down for antennas which were built. The theories of harmonic distortions, frequency modulation, and relay transmissions were worked out. The relative advantages of frequency modulation and amplitude modulation relay lengths were analyzed and the effects of various types of interference were predicted.

E. H. B. Bartelink, F. B. Hildebrand, H. M. James, and A. J. F. Siegert carried through the analysis of bombing devices under design to predict their accuracy (H_2K, GPI, and LAB Mark II). An example of this work was the study of bombing errors to be expected in the use of LAB Mark II in high-altitude bombing. A complete analysis of the sources of errors led to the estimation of the overall accuracy and to the suggestion of means of minimizing errors. F. B. Hildebrand and A. J. F. Siegert developed computation forms to compute target-to-beacon distance in the field for bombing and parachuting problems. This work was carried out so that the least amount of computation labor was necessary to obtain results compatible to the accuracy desired.

The work described above is the main contribution of the group. There were numerous other problems which do not fit into any particular category used above and they will now be described briefly. H. A. Bethe worked on problems of crystal rectification. The influence of the shape of the contact between the "cat's whisker" and crystal was discussed and the advantages of a knife-edge contact were pointed out. A modification of this suggestion in which the "cat's whisker" and the crystal form knife edges at right angles to each other was employed in practice. H. M. James studied decay of phosphorescence in sulfide phosphor. An explanation of the decay laws was developed on the assumption that trapping of electrons occurs on surfaces of discontinuity of crystallites in the material.

F. E. Bothwell, P. D. Crout, and N. H. Painter worked on the design procedures and methods of calculating the behavior of pulse transformers. They devised a method of resonant charging which reduced the size and weight of the modulator power supply. At the request of the Bureau of Ships they worked out methods for handling various mechanical problems caused by gun blast and vibration. This led to the design by E. Pietz of shock mounting equipment which was used for the protection of one of the main radar batteries of the USS *South Dakota*.

At the request of many experimentalists in Divisions 4, 5, and 6, numerous problems in standard transmission line theory and circuitry were set up and worked out in sufficient detail to be useful.

NOTES

1. The space allotted to any given topic discussed here is by no means indicative of its importance although it does reflect roughly the man-hours allocated to those various problems. In some cases the information was readily available while in others it could only be pieced together. The author is indebted to his colleagues for their valuable assistance without which he would not have been able to give so accurate an account.

2. They form the basis of much of N. Marcuvitz, *Waveguide Handbook*, Radiation Laboratory Series No. 10 (McGraw-Hill, New York, 1951).

3. Marcuvitz, *op. cit.* Note 2.

4. J. L. Lawson and G. E. Uhlenbeck, *Threshold Signals*, Radiation Laboratory Series No. 24 (McGraw-Hill, New York, 1950).

5. D. E. Kerr, *Propagation of Short Radio Waves*, Radiation Laboratory Series No. 13 (McGraw-Hill, New York, 1951); Lawson and Uhlenbeck, *op. cit.* Note 4.

6. Kerr, *op. cit.* Note 5.

CHAPTER 30
Research on the Propagation of Microwaves

It was well known that the transmission of energy at microwave frequencies through the atmosphere differs from transmission at lower frequencies, in that there is neither a ground (or surface) wave at appreciable distances nor any reflection from the Kennelly–Heaviside layer. However, no real investigation of the character of microwave transmission through the atmosphere was made during the first year of the Radiation Laboratory's existence. There was more concern with getting radar systems into operation than with undertaking fundamental research in areas like that of microwave propagation. External factors affecting the performance of radar systems seemed of secondary importance.

It was not until the fall of 1941 that an active research program in microwave propagation began at the Laboratory at the insistence of Ralph Bown and Melville Eastham of the Microwave Committee, and of J. A. Stratton, that such studies were necessary for the proper development of microwave equipment.

Actual organization of a group did not take place until January 1942 when Donald E. Kerr, who had been working under Eastham on Project III, organized the nucleus of what came to be known as the Propagation Group. In February the Group was placed in Rabi's Division and, after the Laboratory's reorganization in April 1942, was designated Group 42 with Kerr as Group Leader.

Actual research began in February 1942. The principal emphasis was to be placed on the wavelengths then in use or contemplated by the Laboratory and the work was to be done in close cooperation with J. A. Stratton and his associates in Project III who were preparing transmission curves for standard atmospheric conditions in connection with Loran. Stratton's group was later absorbed by Group 42.

For predictions about general radar coverage to be possible, it was necessary to observe the influence of the electrical and geometrical properties of the surface over which microwaves are transmitted and to study the effects of refraction, scattering, and absorption in the atmosphere through which the waves travel. In other words, the two major problems were to determine (1) radar coverage under conditions of standard re-

633

fraction and (2) the effects of changing atmospheric conditions upon this coverage.

Coverage under Conditions of Standard Refraction

The earliest studies of radar coverage appeared as a series of reports[1] by J. A. Stratton and his associates. Simplification of the early calculation methods was reported by W. T. Fishback late in 1943.[2]

By the end of 1944 Group 42 had evolved several methods for calculating coverage diagrams under standard conditions. Charts were developed for certain idealized conditions which show for different wavelengths and transmitter heights the regions in space in which satisfactory radio reception or radar detection can be expected (cf. Fig. 30-1), and general methods were developed for use in complicated problems.

Kerr and his associates were frequently asked by the systems groups at the Radiation Laboratory and by Army and Navy groups to predict radar coverage for new radar systems or to explore coverage difficulties of exist-

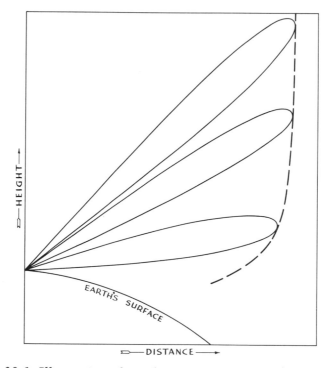

FIG. 30-1. Illustration of a radar coverage pattern. (Via D. Kerr.)

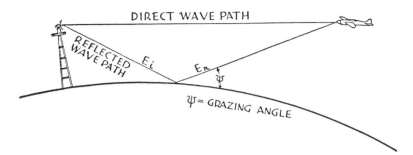

FIG. 30-2. Reflection coefficient equals ratio of field strength reflected from the Earth (E_r) to that striking the Earth (E_i) (always 1 or less for a flat surface). (Via D. Kerr.)

ing systems. It was essential for such predictions to have information on the ability of the earth to reflect microwave radiation, as expressed by the reflection coefficient, i.e., the ratio of the field strength reflected from the reflecting surface to the field strength striking the reflecting surface. (See Fig. 30-2.)

An actual attack on this problem was made early in 1943 when Group 42 obtained equipment that permitted them to measure the reflection coefficients. This equipment consisted of stable, well-calibrated 10- and 3-cm receivers and recorders, installed in an aircraft, and a portable stabilized cw transmitter with a monitor.

Beginning in May 1943 the group made numerous flights out to distances of about 60 miles to obtain data on the reflection coefficient of fairly smooth sea water at grazing angles to about 5°. Ten-centimeter reflection coefficients were also measured over land—Florida, Long Island, and Massachusetts—using the same technique as was used over water, that is, flying a recording receiver at constant altitude toward a cw transmitter.[3]

It was discovered that over fairly smooth water, regular interference patterns exist, and that the values of the reflection coefficient are somewhat less than predicted by theory for a smooth sea, the extent of the departure depending upon polarization of the radiation and upon the degree of surface roughness. Over land the reflection coefficient is usually quite small. The information in these reports was of great importance because reflection from the earth is one of the major factors affecting the coverage diagram.

The study of the fluctuations of radar echoes was a problem of interest because fluctuations did not seem to be due to changes in atmosphere refraction, and because it was important for several radar developments, such as fire control radar and moving-target detectors.

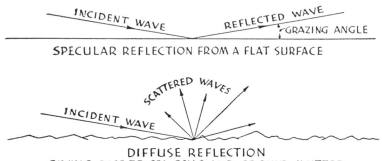

FIG. 30-3. *Specular and diffuse reflection.*

In the summer of 1943 a program for the study of these fluctuations was begun, using the two experimental truckborne systems, XT-3 on 10 cm, and XT-2 on 3 cm, as nucleus of the project. Attention was first concentrated on overwater targets, which were observed continuously over periods of several days. By the fall of 1943 it had become clear that the fluctuations of such echoes are connected with changes in the reflection from the water surface. (Cf. Fig. 30-3.) There were also indications that other types of echoes fluctuate not because of changes in the atmospheric path, but because of the relative motions of unrelated parts of the targets resulting in interference between the component echoes. It was felt that a study of "clutter" echoes—"chaff", sea echo, ground clutter, etc.—which contained large numbers of targets should show the phenomena most clearly. Preliminary to such a study, during the winter 1943–1944, the fluctuations inherent in the systems were carefully reduced, and techniques were devised for the high-speed photographic recording of echo intensities.[4] When preliminary results corroborated the hypothesis,[5] extensive measurements of clutter fluctuations were made, first of chaff echo in the spring of 1944 and later of sea echo and ground clutter at Bar Harbor, Maine, in July of that year.

Simultaneously, the mathematical description of the echo from targets made up in whole or in part of randomly moving scatters was being formulated in terms of the theory of random functions as developed by the Theory Group. The experimental results were analyzed in the summer of 1944, in cooperation with the Sperry Research Laboratories, and good agreement with theory was obtained on almost all points. Typical echo probability distributions and frequency spectra characterizing the various types of clutter signals were computed from the data. By the fall of 1944 it was felt that the general outlines of clutter fluctuations were well understood and a paper on the subject was presented at the third Propagation Conference, November 1944.[6] Using this picture a theoretical analysis was made of the effects of such fluctuations on MTI performance.

The efforts of this section of the Group were now turned to a large extent to the vexing problem of the nature of the scatterers responsible for the sea echo which seriously handicapped many types of radar sets. In preparation of this investigation further improvements in the existing truck systems were made. In addition, a third truck system on 1 cm, T-14, and a "photographic central" were built in the winter and spring of 1945. Data on sea echo were obtained in the late spring and early summer of 1945. Unfortunately no clear solution to the problem was reached. Although the measurements on the frequency dependence of sea echo appeared to exclude small spray drops as the scatterers, the polarization dependence seems explicable only on the basis of some drop mechanism.[7]

During this last period measurements were also obtained on the intensity of storm echoes, which further confirmed the precipitation origin of these echoes. More refined and extensive data on clutter fluctuations were also gathered, without, however, affecting the general conclusions reached previously.

Submarine periscopes were given special attention. Early in 1942 Kerr's group had conducted tests of periscope detection at Deer Island using cylinders of varied size as targets for the XT-2 (3-cm) truck system and SCR-582 (10-cm) harbor defense system. Curves were prepared to show how the strength of signal echo depends upon the angle of tilt, the ratio of height to range, and the length and the diameter of cylinders. Calculations were also made in 1942 of the reflecting properties of spheres and cylinders.

Effects of the Atmosphere on Standard Propagation

No prediction of general radar coverage is possible without taking into account the influence of atmospheric refraction which varies with changes in pressure, temperature, and humidity. The study of this problem led to investigations of the effects of moisture and temperature upon the distribution of the atmospheric index of refraction and of the effect of a given distribution of index upon coverage.

In November 1943 J. E. Freehafer published a Laboratory report dealing with (1) the application of geometrical optics to the investigation of the effect of atmospheric refraction on the propagation of short radio waves and (2) the use of wave optics to treat certain simple cases of refraction.

In the case of (1) Freehafer presented criteria for determining when the results of ray tracing can be relied upon and when they may be questionable. He evolved a method of ray tracing which permits a rapid estimation of certain refractive properties of the atmosphere, and used this method to show that, under certain refraction conditions, energy can be trapped in horizontal layers, giving rise to effects of possible strategic importance.

In the case of (2), where wave optics is used to treat certain simple cases of refraction, Freehafer showed that a linear dependence of the refractive index upon the height may be allowed for by use of a fictitious earth's radius and a homogeneous atmosphere.[8]

In the course of developing the ray-tracing formulas, Freehafer introduced a new quantity which greatly simplifies the mathematical problems and serves as the logical boundary between the meteorological and electromagnetic sides of his problem. This quantity, now called the *modified index* (following the terminology of British scientists who independently discovered the importance of this quantity at about the same time), reduces the complicated spherical earth problem to the much simpler one for a flat earth.

The mathematical problems arising from atmospheric refraction led to considerable complications, and required the borrowing of techniques from quantum theory. Aid from the Theory Group was obtained from W. H. Furry, who explored the advanced phases of the problem, while the theoretical section of Group 42 attempted to reduce the theory to a workable form and apply it to specific cases in connection with the experimental program.[9]

Meanwhile, the investigation of actual conditions in the atmosphere and their effects on microwave propagation had been continuing since the spring of 1942. Beginning in the summer of 1942 radiosonde information and other data from meteorological sources were analyzed in an attempt to determine the cause of specific instances of supernormal ranges (often referred to as "anomalous propagation") which were reported from operational theaters as the use of radar became widespread. The Group was obtaining its meteorological data from aircraft and blimp flights, from the Weather Bureau, and from its own instruments. Steep temperature and water-vapor gradients above the Earth caused bending of waves around the earth's surface, which at first was explained in terms of an apparent increase in the Earth's radius.[10] Calculations of this effect from the weather data then available failed to explain the supernormal ranges. Other channels of information were being opened up, notably the Navy and the Army, whose reports aided in a general way in the study of transmission problems and weather analysis. Some simple conclusions on the effects of pressure, temperature, and relative humidity on ray curvature were drawn at this time:

(1) When the temperature is sufficiently high, moderate gradients of temperature and relative humidity can affect ray curvature very easily, but in the region of freezing temperature these quantities have less effect. This explained why supernormal ranges are so much more common in the summer than in the winter in temperature climates.

(2) With a decrease of refractive index with height, rays tend to follow the curvature of the Earth and the Earth tends to seem flattened. In warm weather the gradient may fluctuate widely and may even decrease the effective radius of the Earth.

These first tentative conclusions were greatly modified in the succeeding years and were put on a sound basis as both the meteorological and electromagnetic phases of the problem were studied quantitatively.

As part of this program for obtaining meteorological data, Kerr and his co-workers inaugurated a program of airplane soundings of the atmosphere in the fall of 1942. Soundings were made in the vicinity of Boston and over Massachusetts Bay. Unfortunately, the temperature and humidity instruments which were immediately available were unsatisfactory. The airplane meteorographs then in use employed hair elements for humidity measurements. Attempts to make detailed vertical soundings with these meteorographs were unsuccessful because the hairs and the temperature-measuring elements were too slow in their response, and because they were limited in accuracy. Available radiosondes were not much better since they sampled the atmosphere at too widely spaced intervals and also used a bank of hairs to measure humidity.

Neither the meteorograph nor the radiosonde measured temperature differences with the accuracy desired. The meteorograph used a bimetallic strip whose scale was not sufficiently expanded to measure small temperature differences, while the radiosonde employed an electrolytic thermometer in the circuit which did not permit reading temperatures to the accuracy desired.

To meet the needs of microwave propagation studies, early in 1943 Isadore Katz, Robert Burgoyne, and Lewis Neelands, all of the Propagation Group, developed an electronic instrument whose essential elements were a pair of temperature sensitive resistors which served as resistance thermometers for the determination of wet- and dry-bulb temperatures.[ii] Their accuracy and speed of response were much superior to that of other available instruments. This instrument, known as the electronic psychrograph, was mounted on an airplane or carried aloft by a captive balloon. A record was kept by the observer of the altimeter indications, in the case of aircraft measurements, or of the length of cord paid out and the inclination angle in the case of balloon measurements. The measurements of dry- and wet-bulb temperature as a function of height constituted a vertical sounding. Balloons were used as a vertical for the psychrograph up to about 500 ft, while airplanes were used from 20 ft to several thousand feet when necessary.

The development of the psychrograph enabled the Group to study the "fine structure" of the atmosphere in a manner never possible before, and, beginning in the Spring of 1943, the instrument was used to produce the first detailed low-level overwater soundings. In the latter part of 1943 some soundings were also made with the psychrograph from a boat in cooperation with the Woods Hole Oceanographic Institution and the Signal Corps.

A major difficulty encountered in applying the results of the atmospheric measurements was the amount of computational work necessary to obtain the modified index from the original record produced by the instrument. However, by February 1944, R. H. Burgoyne had prepared nomograms which simplified the calculation to the point where only a straightedge ruler was necessary for graphical analyses. As a consequence, the modified index profile could be plotted almost simultaneously with the measurement of temperature and humidity. Several hundred soundings were made in New England[12] and Florida by the several techniques in use. These studies and many more constructed from crude data from other parts of the world, when correlated with meteorological analysis, made possible a tentative classification of the basic types of modified index profiles encountered, and of the meteorological processes causing them.[13]

In the spring of 1944 the Group undertook a large-scale experiment to study the effect of refraction resulting from continental air blowing over the waters of the ocean. Transmission circuits were set up from Provincetown to Eastern Point, Massachusetts, as well as the one used earlier from Deer Island to Eastern Point, and one-way measurements were made continuously during the summer and fall on wavelengths of 256, 10, 3, and, for a short time, 1.25 cm. Soundings were made simultaneously. The psychrograph was carried by aircraft and by balloons launched from on shore and from a boat. All together, about 1000 soundings and 2500 hr of transmission were obtained. Also, radar observations were made during part of the time on 10, 3, and 1.25 cm. This experiment resulted in a great extension of previous knowledge of the meteorological processes responsible for producing nonstandard refraction, and provided opportunity for testing the theory of nonstandard propagation. Although chiefly useful because it clarified the mechanisms involved, it also gave much specific and immediately useful information. The U.S. Weather Bureau lent the part-time services of two of its forecasting staff at the East Boston Airport, and the Army and Coast Guard generously furnished personnel and equipment in large amounts to help in this project. The work could not have been done without this aid.

In January 1945, the study of the preceding summer's meteorological data was far enough advanced to permit drawing some preliminary conclusions regarding vertical distribution of modified index close to the ocean surface.[14]

At the end of May, the group began a program of meteorological measurements to distances of 50–200 miles offshore. This new program, in contrast to most of the previous sounding projects which had furnished information about the low-level structure of the atmosphere over the sea close to shore or over the open ocean, was undertaken because soundings were desired in air that had traveled greater distances over water. Theory indicated that important effects might be expected. This sounding program was made possible by the active support of the Navy. Quonset Naval Air Station supplied a PBY aircraft and a 100-ft boat together with the services of their crews. Between June 1 and June 22, 32 soundings were made over the ocean in the region between Nantucket and New Jersey. Weather forecasts and weather data were made available by the Aerological Office, and were used to construct hourly weather maps of the local region as an aid in the analysis of the soundings. The space data from this program were analyzed in detail at the Woods Hole Oceanographic Institution.

Radar Storm Detection

Out of the main problem of identifying spurious targets and linking them to previous meteorological phenomena grew the art of radar storm detection, which during the war became a valuable tool in planning tactical operations of the air forces in all parts of the world. As early as June 1942[15] the Propagation Group was reporting programs in identification of meteorological disturbances seen by radar.

These echoes had often been described as "clouds" because of their appearance on PPI's and B scopes. In a report covering observations made in New England from March 1942 through January 1943, A. E. Bent described the characteristics of such echoes, illustrating his descriptions sufficiently to enable radar operators to identify similar observations on the scopes of systems being operated by them.[16]

By August 1942 the application of radar techniques to weather analysis and forecasting appeared to be a distinct possibility, and the military services and the Weather Bureau expressed strong interest. In September a thunderstorm was seen on a powerful radar system at the then record distance of 164 miles.

In December 1942 representatives of the Weather Directorate, Headquarters Army Air Forces, appeared at the Laboratory to participate in experiments intended to evaluate the usefulness of radar in military meteorology. This was the beginning of an important collaboration with the Army Air Forces on the problem of storm detection and wind measurements.

Late in the fall of 1942 Bent demonstrated the capabilities of microwave radar as a storm detector to the officers of the Sixth Weather Region and the Aircraft Warning Service in Panama. He helped the Air Forces set up a system for reporting storm echoes from all radar in the Panama region to the Weather Section of the Army's Fighter Command, thereby greatly increasing the efficiency of the Army's Fighter Command in routing planes around storms and saving life and property.

When Bent left Panama his weather reporting program was in operation, precipitation areas being reported hourly from 1100Z to 0400Z by the SCR-615 radar at Taboga Island. The reports were being sent to the Weather Section of the Fighter Command for the information of the Operations Officers. Coded reports were then sent by teletype to the Bomber Command Control Room and the Albreek Field Weather Station as well as the weather stations at air bases in the Republic of Panama.[17]

The Army Air Forces in February 1944 issued a manual for the use of their meteorological personnel called *Radar Storm Detection*. This manual was based on work begun and fostered in Group 42 of the Radiation Laboratory, and carried on with the aid of the Army Air Forces Weather Service.

The tactical value of being able to locate storms and plot their course with accuracy need hardly be stressed here. As the result of the development of this aspect of radar, the air arms of the services were able to save their squadrons much time and trouble by locating storms and guiding the planes around them, either from the ground or from the air.

In addition to reports from ground radar, information was available to pilots in planes equipped with airborne microwave radar, who were able to note storm signals on the face of their radar scopes and navigate around storm areas.

NOTES

1. Radiation Laboratory Reports C-1–C-11.

2. W. T. Fishback, "Simplified Methods of Field Intensity Calculations in the Interference Region," Radiation Laboratory Report 461, December 8, 1943.

3. This work was summarized in Pearl J. Rubenstein and W. T. Fishback, "Preliminary Measurements of 10-cm Reflection Coefficients of Land and Sea at Small Grazing Angles," Radiation Laboratory Report 478, December 11, 1943. They included data on specular reflection coefficients of seawater and land, which had been measured for the first time on X and S bands for grazing angles up to about 5°.

4. H. Goldstein and P. D. Balen, "High Speed Photography of the Cathode-Ray Tubes," Rev. Sci. Instrum. **17**, 89 (1946).

5. H. Goldstein, "Preliminary Fluctuations of Radar Signals," Radiation Laboratory Report 569, May 16, 1944.

6. H. Goldstein, "Fluctuation of Radar Signals," Report of Third Conference on Propagation (Wave Propagation Group, Division of War Research, Columbia University), November 1944, pp. 155-166.

7. Phys. Rev. **69**, 695 (abstract) (1946).

8. J. E. Freehafer, "The Effect of Atmospheric Refraction on Short Radio Waves," Radiation Laboratory Report 447, November 29, 1943. This result was obtained years earlier, but was based on ray methods and was valid only for the interference region.

9. Radiation Laboratory Reports 680 and 795.

10. Minutes of the Coordination Committee, Vol. 1, pp. 19, 20, June 24, 1942.

11. I. Katz, "Instruments and Methods for Measuring Temperature and Humidity in the Lower Atmosphere," Radiation Laboratory Report 487, April 12, 1944.

12. *Ibid.*

13. I. Katz and J. M. Austin, "Qualitative Survey of Meteorological Factors Affecting Microwave Propagation," Radiation Laboratory Report 488, June 1, 1944.

14. R. B. Montgomery and R. H. Burgoyne, "Modified Index Distribution Close of the Ocean Surface," Radiation Laboratory Report 651, February 16, 1945.

15. Minutes of the Coordination Committee, *loc. cit.* Note 10.

16. A. E. Bent, "Climate in Relation to Microwave Radar Propagation in Panama," Radiation Laboratory Report 476, February 25, 1944.

17. *Ibid.*